Carbon Dioxide Sensing

Carbon Dioxide Sensing

Fundamentals, Principles, and Applications

Edited by
Gerald Gerlach
Ulrich Guth
Wolfram Oelßner

Editors

Prof. Dr. Gerald Gerlach
Technische Universität Dresden
Faculty of Electrical and Computer Engineering
Institute of Solid-State Electronics
01062 Dresden
Germany

Prof. Dr. Ulrich Guth
Technische Universität Dresden
Faculty of Chemistry and Food Chemistry
01062 Dresden
Germany

Priv.-Doz. Dr. Wolfram Oelßner
Kurt-Schwabe-Institut für Mess- und Sensortechnik e.V. Meinsberg
Kurt-Schwabe-Straße 4
04736 Waldheim
Germany

Cover
Background image: Creativ Collection

Library of Congress Card No.:
applied for

British Library Cataloguing-in-Publication Data
A catalogue record for this book is available from the British Library.

Bibliographic information published by the Deutsche Nationalbibliothek
The Deutsche Nationalbibliothek lists this publication in the Deutsche Nationalbibliografie; detailed bibliographic data are available on the Internet at <http://dnb.d-nb.de>.

Print ISBN: 978-3-527-41182-5
ePDF ISBN: 978-3-527-68829-6
ePub ISBN: 978-3-527-68827-2
oBook ISBN: 978-3-527-68830-2

Cover Design SCHULZ Grafik-Design, Fußgönheim, Germany
Typesetting SPi Global, Chennai, India
Printing and Binding CPI books GmbH, Germany

Printed on acid-free paper

10 9 8 7 6 5 4 3 2 1

Contents

Scientific Biographies of the Authors

Manfred Decker studied organic and analytical chemistry at the University of Münster, Germany. He gained industrial experience in the field of medical and environmental sensors before he started his employment at Kurt-Schwabe-Institut in 2007. His research topics cover amperometric, coulometric, and potentiometric sensors and their applications in environmental, biotechnological, and medical needs. This involved R&D of Clark-type biosensors for the determination of glucose and lactate in blood and of probes for the H_2O_2 analysis in exhaled breath. Further research covered the development of pH measuring devices for agricultural needs and the application of Severinghaus-type CO_2 sensors for bioprocess control and medical purposes.

Gerald Gerlach is a professor and the head of the Solid-State Electronics Laboratory at the Technische Universität Dresden (TUD), Germany. After obtaining his MSc and PhD degrees in electrical engineering from TUD in 1983 and 1987, respectively, he spent almost ten years in sensor industry before taking up his present appointment at TUD. Professor Gerlach has authored over 400 scientific publications and was granted more than 50 patents. He is author or co-author of nine monographs and textbooks. From 2007 to 2010, Prof. Gerlach was President of the German Society for Measurement and Automatic Control (GMA). From 2007 to 2008, he served as Vice President and President of EUREL (The Convention of National Associations of Electrical Engineers of Europe), respectively. He is Associate Editor-in-Chief for the *IEEE Sensors Journal* and Founding Chief Editor for the *Journal of Sensors and Sensor Systems* (JSSS).

Ulrich Guth received his PhD in Physical Chemistry from the University of Greifswald, Germany, in 1975. From 1989 to 1993, he worked in the industry and at Battelle Institute, Frankfurt am Main. In 1993, he became a professor of solid-state chemistry at the University of Greifswald. From 1999 until his retirement in 2010 he was Director of the Kurt-Schwabe-Institut für Mess- und Sensortechnik e.V. Meinsberg and Professor for Physical Chemistry, especially Sensor and Measuring Technology at Technische Universität Dresden. He is now Professor emeritus. His principal research interests include solid electrolyte sensors, fuel cells, and new materials for high temperature sensors and high temperature fuel cells. He authored more than 230 scientific publications including 5 book contributions and more than 40 patents.

Frank Kühnemann received his MSc (Diplom-Physiker) and PhD degrees in physics from Humboldt University in Berlin in 1987 and 1991, respectively, and his habilitation in experimental physics from Bonn University in 2000.

After several research and teaching positions at universities in Germany and Egypt, he has been working at the Fraunhofer Institute for Physical Measurement Technologies IPM (since 2011) and teaches at the Institute of Physics, Freiburg University (since 2017).

His research activities include high-sensitivity laser spectroscopic methods for both trace gas analysis and residual absorption in optical materials and nonlinear optical frequency conversion for tunable light sources and infrared detection. He is (co-) author of more than 30 papers and co-owner of five patents in these areas.

Armin Lambrecht has received his PhD in Physics from the University of Karlsruhe in 1985. Since 1986 he has been working at the Fraunhofer Institute for Physical Measurement Techniques IPM in Freiburg, Germany. Starting with R&D projects on molecular beam epitaxy for mid-infrared lasers and thermoelectric devices, he later focussed on infrared sensors systems for gas and liquid analysis. His main research interests are directed towards laser spectroscopy applications for process analytics. He has 15 years of experience as department head at Fraunhofer IPM and has acquired and managed several industrial and public funded R&D projects. He is author of more than 100 publications and several patents. He is a member of DPG, VDI, and GDCh focussing on process analytics (www.arbeitskreis-prozessanalytik.de).

Detlev Möller was born in 1947 in Berlin. He studied chemistry at Humboldt University in Berlin (HUB) from 1965 to 1970. He gained his PhD from HUB on electrochemical kinetics in 1972. Since 1974 he has been working on air chemistry in different institutes of the former Academy of Sciences of GDR (until 1991); his habilitation on the atmospheric chemistry he obtained in 1982. In 1992 he became the head of a branch of the Fraunhofer Society in Berlin in atmospheric environmental research. In 1994 he became a full professor for atmospheric chemistry and air pollution control at Brandenburg University of Technology Cottbus-Senftenberg. In 2012 he retired but still works as guest professor. He is/was member of many editorial boards, national and international committees, and three academies of sciences; he authored more than 200 scientific publications including 25 book contributions and 3 monographs. His principal research interest were all aspects of chemical climatology such as rain and cloud chemistry, ozone formation, and sulphur and nitrogen cycling by means of modelling, monitoring, and field experiments.

Wolfram Oelßner studied nuclear physics at the Technische Universität Dresden from 1956 to 1962, where he received his PhD in Electrochemical Measuring Technology in 1968. Since 1963 he has been working at the Meinsberg Kurt-Schwabe Research Institute, currently still as a scientific consultant. Furthermore, he is appointed at the Technische Universität Dresden as a private lecturer for electrochemical measuring technique as well as for environmental measuring technology and corrosion research. His diverse research and development activities have been focussed on pH measurement, encapsulation and application of ISFET pH sensors, corrosion measurement technique, and electrochemical carbon dioxide sensors and measuring devices.

Gerald Urban was born in Vienna, Austria, and studied physics at the Technical University (TU) Vienna. Afterwards he was employed at the neurosurgical department of University Hospital Vienna. In 1985 he received his PhD in electrical engineering at the TU Vienna. He was co-founder of the venture company OSC in Cleveland. In 1994 he received the Venia Legendi for Sensor Technology and in 1995 he became scientific director of the Ludwig Boltzmann Institute for Biomedical Microengineering in Vienna, Austria. In 1997 he became a full professor of sensors at the faculty of engineering at the University of Freiburg/Germany. He was dean of the faculty and speaker of the academic senate from 2009 to 2011. From the year 2002 till now he is member of the directorate of the Freiburg Material Research Center. He is co-founder of the company Jobst Technologies which was sold to the Endress group in the year 2015.

His research interests are the development of miniaturized and integrated sensor systems and complete miniaturized analysis systems for biomedical applications. Additionally, he is interested in the development of microbiosensor arrays for measuring RNA and protein markers by highly sensitive microcapillary immunoassays in body fluids.

He published more than 100 journal papers and received four awards; he currently holds 70 patents, and is a series editor of the Springer series on chemical sensors and biosensors and is corresponding member of the Austrian Academy of Sciences.

Jens Zosel received his diploma in physics from the University of Greifswald in 1990 and his PhD from the Freiberg University of Mining and Technology in 1997. Since 1992 he has been working at the Meinsberg Kurt-Schwabe Research Institute. His basic research interests are directed towards the behaviour of electrochemical sensors in liquid and gaseous flows and the development of electrochemical sensors, based on solid and liquid electrolytes for different applications. Especially the development of materials, designs and measuring methods for solid electrolyte devices like gas sensors, fuel cells, and electrolyzers as well as the exploitation of new applications for those are the focus of his research activities.

Scientific Biographies of the Co-Authors to Chapter 16

Josef Guttmann studied Applied Precision Engineering at the FH Furtwangen in 1971–1975 and Biology/Biophysics at the Universität Freiburg in 1975–1981 where he received in 1986 his PhD on Biophysics. In 1994 he habilitated in Biomedical Engineering at the University of Basel. In 2001 he was appointed as an APL Professor of Biomedical Engineering at the University of Freiburg. In 1994–2014 he was the head of the working group "Clinical Respiratory Physiology" at the Department of Anesthesiology and Critical Care, Medical Centre – University of Freiburg, with the following main research areas: biomechanics of the lung, control of ventilators, and respiratory monitoring; in 2008–2012 he was Director of the course "MasterOnline Technical Medicine (TM)".

Prof. Josef Guttmann contributed to Section 16.6.

Jochen Kieninger received his diploma in microsystems engineering from the University of Freiburg in 2003. Afterwards, he worked in the Laboratory for Sensors at the Department of Microsystems Engineering (IMTEK) and in the School of Soft Matter Research at the Freiburg Institute for Advanced Studies (FRIAS). In 2011, he completed his PhD on "Electrochemical microsensor system for cell culture monitoring". In 2012, he was nominated as a lecturer and has since been working as a senior scientist in the Laboratory for Sensors. His research interests are electrochemical sensors, biosensors, microsensors for neurotechnology, cell culture monitoring, electrochemical methods, and microfabrication.

Dr.-Ing. Jochen Kieninger contributed to Sections 16.1–16.4.

Andreas Weltin received his diploma and doctoral degrees in microsystems engineering from the University of Freiburg. Since 2015, he has been a group leader at the Laboratory for Sensors. In 2016, he received the 2nd Klee prize from DGBMT for his dissertation on microfabricated, electrochemical in vivo sensors. Among his research interests are (bio-)analytical microsystems, bio- and chemo-sensors, electrochemistry, microfluidics, and biomedical applications. Current activities include the development of novel electrochemical sensor principles and platforms, organs-on-chip in cancer research, and sensors at neural interfaces.

Dr.-Ing Andreas Weltin contributed to Sections 16.2–16.4.

Jürgen Wöllenstein received his degree in electrical engineering from the University of Kassel in 1994. In 1994 he joined the chemical sensors group at the Fraunhofer Institute for Physical Measurement Techniques in Freiburg. He is head of a department at Fraunhofer-IPM. In 2009, he became a full professor at the Department of Microsystems Engineering of the University of Freiburg. He is author and co-author of more than 50 publications and holds several patents.

Prof. Jürgen Wöllenstein contributed to Section 16.5.

Preface

Carbon dioxide (CO_2) is one of the key components of life. Without any doubt, it is the most important chemical substance in the global climate system. For instance, it is considered as one of the crucial sources of the dangerous greenhouse effect and the corresponding global warming. On the other side, it governs photosynthesis and, hence, is – beside oxygen and carbon – the basis for the existence of life on Earth.

Sensing and monitoring of carbon dioxide is fundamental to get knowledge on CO_2-affected mechanisms and to control them. Observation of CO_2 in the atmosphere and in the oceans yields important data for long-term predictions of the world's climate. Monitoring of CO_2 in industrial processes is a decisive tool to control their efficiency. Carbon dioxide in higher concentration can be lethal for human beings, so warning detectors for dangerous CO_2 concentrations are needed. This short listing gives proof for the general importance to sense CO_2 both in gases and in liquids over a wide range of concentrations from the ppm level up to the pure carbon dioxide.

The manifold application fields as well as the huge range of CO_2 concentration to be measured make CO_2 sensing a challenging task. This was the reason that a part of the author team of this book decided in 2011 to write a review paper *The measurement of dissolved and gaseous carbon dioxide concentration* [1]. The intention was to give an overview of the state of the art and the new developments to measure CO_2 and of the different fields where CO_2 monitoring and detection is of interest and which methods are used there. Shortly after being published we were heavily surprised by the large download numbers of this paper. At this point we came to the conclusion that it might make a lot of sense to give researchers and scientists, confronted with the detection of CO_2 in the most diverse application fields, a guide to support them in their decision which method should be used for which application-related demands and requirements. We are convinced that this goal could only be achieved by a much broader approach, i.e. by a monography instead of a review article.

The present book is the result of our efforts with respect to these objectives. It is organized in three parts. After a general consideration in Part I on the properties of CO_2 and the CO_2 cycle, Part II gives an overview of the different chemical and physical measuring methods and sensors for the determination of carbon dioxide in liquids and gases (Chapters 3–11). Afterwards, Part III describes the most important application fields of CO_2 sensing – from environmental monitoring and safety control via biotechnology and industrial processes to the measurement in biology and medicine (Chapters 12–16). In view of the great diversity of CO_2 measurement tasks, it is not intended and not possible at all to present a solution for each individual measurement problem. Rather, it is about showing by means of a great number of typical and also somewhat exceptional examples where everywhere and how CO_2 is measured.

However, the heart of the book is Chapter 11 'Survey and Comparison of Methods'. It gives a concise overview of the characteristics of all these CO_2 measurement methods including their advantages and disadvantages and provides a decision support for choosing the most suitable analysis method for a certain measurement task.

The editors of this book hope that the contents depict both the state of the art and the most recent progress in sensing and monitoring CO_2 in the many fields of application. We are convinced that this book can fill the gap between scientific research in measurement technology and its application in practice. Let the book be an inspiration to all the colleagues involved in this area!

Most of the authors of this book are scientists from the Meinsberg Kurt-Schwabe-Institut and the Technische Universität Dresden having been involved in the 'sensor business' for many years. However, to achieve our goal, we had to form a team far beyond these two institutions to become capable to deal with all aspects of such a complex matter. We found this expertise in our colleagues from the Universities of Freiburg and of Cottbus-Senftenberg as well as of the Fraunhofer Institute for Physical Measurement Techniques IPM in Freiburg. We are deeply indebted to them and would like to thank them for contributing their comprehensive knowledge and particular competence to this book. We would also like to thank those companies and institutions that allowed us to use figures and material and which are named in the captures of the individual figures. Furthermore, we would very much like to thank VCH-Wiley and in particular Mrs. Nina Stadthaus and Mrs. Abisheka Santhoshini for the cordial cooperation, but also for the patience when faced with repetitive delays due to the authors' workload. We are deeply grateful to the VCH-Wiley staff for their support during the entire process from the first idea all the way through to the final book.

Dresden, July 2018

Gerald Gerlach
Ulrich Guth
Wolfram Oelßner

Reference

1 Zosel, J., Oelßner, W., Decker, M. et al. (2011). The measurement of dissolved and gaseous carbon dioxide concentration. *Meas. Sci. Technol.* 22: 072001. https://doi.org/10.1088/0957-0233/22/7/072001.

Part I

General

1

Introduction

Wolfram Oelßner

Kurt-Schwabe-Institut für Mess- und Sensortechnik e.V. Meinsberg, Kurt-Schwabe-Straße 4, 04736 Waldheim, Germany

Carbon dioxide (CO_2) is one of the key components in our life and without any doubt the most important chemical substance in the global climate system. It presents the feedstock for plant assimilation and for the growth of plants and phyto cells in the biological carbon cycle as well as buffer system in the blood of humans and animals, and hence, it is important for total life on Earth. On the other hand, CO_2 is the "waste" produced by the metabolism of most of the living creatures on our Earth and by combustion of fossil fuels using carbon stocks or biomass, and it is widely understood to be one of the crucial sources of the dangerous greenhouse effect. Observation of CO_2 in the atmosphere and in the oceans yields important signals for long-term predictions of the world's climate. Furthermore, in process technology, CO_2 is an important reagent for the manufacture of a variety of products. Monitoring of CO_2 in chemical as well as in biotechnological processes is a valuable tool to control the efficiency of the production processes. Since carbon dioxide in higher concentration can be lethal for human beings, CO_2 warning devices are needed. This short listing indicates the general importance of CO_2 and the need to determine it in gases as well as in liquids over a wide range of concentrations from ppm level up to 100% [1].

In publications and regulations the CO_2 concentration is indicated in different units. The general formula for converting the units vol ppm to mg m^{-3} and vice versa is:

$$c[\text{mg m}^{-3}] = c[\text{vol ppm}] \cdot M/V_\text{m} \quad (1.1)$$

with M being the molar mass in g mol^{-1} (e.g. for CO_2 $M = 44\,\text{g mol}^{-1}$) and V_m being the molar volume in l mol^{-1} (e.g. for ideal gases at 25 °C $V_\text{m} = 24.5\,\text{l mol}^{-1}$).

Table 1.1 is intended to simplify the conversion.

In medicine, the carbon dioxide partial pressure pCO_2 is usually indicated in the unit mm Hg instead of in the SI unit Pa. The conversion is done according to Table 1.2.

Carbon Dioxide Sensing: Fundamentals, Principles, and Applications,
First Edition. Edited by Gerald Gerlach, Ulrich Guth, and Wolfram Oelßner.

Table 1.1 Conversion factors for CO_2 concentrations.

To convert from the units on the left to the units on top, multiply by		To		
		vol%	vol ppm	mg m^{-3}
From	vol%	1	10^4	1.8×10^4
	vol ppm	10^{-4}	1	1.8
	mg m^{-3}	5.56×10^{-5}	0.56	1

Table 1.2 Conversion factors for CO_2 partial pressures.

To convert from the units on the left to the units on top, multiply by		To		
		mm Hg	Pa	bar
From	mm Hg	1	133.32	1.33×10^{-3}
	Pa	7.50×10^{-3}	1	10^{-5}
	bar	750.06	10^5	1

This means that typical CO_2 partial pressures in the range pCO_2 = 35–45 mm Hg correspond to 4.6–6.0 kPa.

Depending on the medium in which CO_2 needs to be measured and the requirements for measuring range, accuracy, long-term stability, selectivity, and maintenance, different methods can be applied [1]:

a) Standard test methods for the analytical determination of total and dissolved carbon dioxide in water require the titration of test samples.
b) CO_2 sensors that are based on various chemical or physical measuring methods are more user-friendly and therefore preferably applied:

- Because of their simple set-up and the resulting low costs, membrane-covered electrochemical CO_2 sensors according to the Severinghaus principle have been manufactured and widely applied already for a long time. Unlike other types of CO_2 sensors, Severinghaus sensors can be applied not only in gases but also for direct measurements in liquid media.
- In comparison with these sensors with aqueous electrolytes, solid electrolyte CO_2 sensors operating at high temperatures have the advantages of a short response time and maintenance-free operation without calibration. They are used successfully in all cases of long-term measurements in air, in breath gas analysis, and in the process monitoring especially at higher temperatures.
- As an economic alternative to the electrochemical CO_2 sensors, detector tubes have been used in a broad range of applications, in particular for control of the concentration at the workplace.
- Nowadays IR, NDIR, opto-chemical, and acoustic CO_2 sensors, which use physical measuring methods, are being used increasingly.
- In several fields of application, a variety of other CO_2 sensor principles, based on conductometric, thermal conductivity, hydrogel expansion, and mass spectrometric measurements, have been tested and partly commercially applied.

Compared to spectrometric (FTIR, UV-VIS), mass spectrometric (MS) and chromatographic techniques (GC, HPLC), electrochemical sensors (Severinghaus and solid electrolyte sensors) are simple in their set-up as well as in the electronic equipment necessary for operation and for data acquisition. The effort for maintenance and calibration is low. Since sensor signals are obtained directly (*in situ*), real-time information for process control is delivered. Therefore, they are preferred tools for screenings in field application. On the other hand, electrochemical sensors cannot completely replace the standard methods in laboratories in terms of precision, detection limit, etc.

The current development activities in CO_2 sensor technology and application are focused on [1]:

- Miniaturization of electrochemical sensors based on the Severinghaus principle, e.g. for measurement in liquid biological systems, cell cultures, cell tissues, and living organisms;
- Development of sterilizable and even CIP (cleaning in process)-resistant sensors for measurement of dissolved CO_2 in biotechnological processes and foodstuff production;
- Extension of the measuring ranges to higher or lower concentrations, as required;
- Extension of the sensor service lifetime and the calibration intervals;
- Application of thin-film and thick-film manufacturing technologies for the mass production of low-cost sensors;
- Development of solid electrolyte CO_2 sensors with short response time for *in situ* breath analysis;
- Miniaturization and improvement of selectivity and sensitivity of IR sensors; and
- Utilization of ultrasonic sensors for breath gas analysis in medical and sportive applications.

Depending on the special field of application and the goals of the investigation, the measuring conditions and technical requirements on the sensors can be very different. Each application has its own scientific background without which the results of measurement cannot be interpreted. A detailed knowledge of the basic detection principles and the frames for their applications is necessary to find an appropriate decision on the technology to be applied for measuring dissolved CO_2. Especially the pH value and the composition of the analyte matrix may exert important influence on the results of the measurements, and sampling of liquids in which CO_2 is dissolved is often a source of errors. Sensors for safety control should be mechanically robust and long-term stable and have low maintenance, whereas for measurements in boreholes or in the deep sea, challenging demands on pressure resistance and compensation of rapid temperature changes have to be fulfilled. In biology and medicine often small dimensions and short response times are required. In biotechnology precise, real-time data on CO_2 concentration fosters the understanding of critical fermentation and cell culture processes and can help in gaining insight into cell metabolism, cell culture productivity, and other processes within bioreactors. But in this case the sensor must be sterilizable. When being applied online in food industry, it is required that the sensor is non-breakable and even survives the rigorous CIP cleaning procedures [1].

After a general consideration on the CO_2 cycle, the book gives an overview of the different chemical and physical measuring methods and sensors for the determination of CO_2 in liquids and gases and their manifold applications in environmental control, biotechnology, biology, food industry, and medicine to a certain extent without claiming to cover completely the whole phenomenon. The wide variety of applications is illustrated by some typical and also somewhat original examples ranging from measurements in the higher atmosphere to the depth of the ocean. The advantages and drawbacks of the different sensor principles will be outlined with the main focus directed on electrochemical sensors, which means on devices that can be applied directly (*in situ*) without sampling. There is no CO_2 sensor available to date that meets these partly contrary requirements all at the same time. For this reason, the book should not only be a source of information about CO_2 measurement, but it is also intended to be an invitation to the reader to accept the challenge to continue developing and improving CO_2 sensors and to be a motivation to open up new areas for their application.

Reference

1 Zosel, J., Oelßner, W., Decker, M. et al. (2011). The measurement of dissolved and gaseous carbon dioxide concentration. *Meas. Sci. Technol.* 22: 072001. https://doi.org/10.1088/0957-0233/22/7/072001.

2

Carbon Dioxide in General

Detlev Möller[1], Manfred Decker[2], Jens Zosel[2], and Wolfram Oelßner[2]

[1] *Brandenburgische Technische Universität Cottbus und Senftenberg, Fakultät für Umwelt, Verfahrenstechnik, Biotechnologie und Chemie, Institut für Umweltwissenschaften, Platz der Deutschen Einheit 1, 03046 Cottbus, Germany*
[2] *Kurt-Schwabe-Institut für Mess- und Sensortechnik e.V. Meinsberg, Kurt-Schwabe-Straße 4, 04736 Waldheim, Germany*

2.1 Chemical and Physical Properties of Carbon Dioxide

2.1.1 Chemical Properties of Carbon Dioxide

2.1.1.1 Chemical Properties

Carbon dioxide (CO_2) is a linear molecule with the oxygen atoms placed opposite to the carbon centre. The carbon atom itself is sp-hybridized with each of the sp orbitals forming a σ-bond to an oxygen atom. The p_y orbital of the carbon is establishing a π-bond with the available p orbital of one O atom, whereas the p_z orbital completes the second double bond with the other oxygen partner. This results in the high stability of the gaseous CO_2 molecule with a standard formation enthalpy $\Delta H_r°$ of $-393.522\ \text{kJ mol}^{-1}$ at 298.15 K [1]. This high value causes remarkable dissociation reactions of pure carbon dioxide to C, CO, and O_2 only at temperatures far above 1000 K [2, 3].

The two π-bonds of CO_2 show a length of 116.2 pm [4]. The two p orbitals connecting the oxygen atoms are oriented perpendicular to each other. Although each C=O bond is polarized caused by the differences of electronegativity of carbon and oxygen, the linear symmetry of the CO_2 molecule results in a net dipole moment of zero. However, the polarization of the carbon–oxygen bond is essential for the explanation of the CO_2 reactivity. The pull of π-electrons to the oxygen atoms causes an electron deficiency at the carbon atom and enables the attack of nucleophiles and electron donors to the carbon centre of the molecule. This step is essential for the formation of carbonic acid and hydrogen carbonate in water [5]. This nucleophilic addition changes the sp hybridization of the C atom to a sp^2 state. Otherwise the partial negative polarization of the oxygen can be targeted for the coordination of electron deficient metals. An intensive overview of the reaction paths for catalysed chemical transformations of CO_2 is comprehended in detailed publications [6–8]. The use of carbon dioxide in industrial syntheses is mainly concentrated on applications where CO_2 is produced as waste

Carbon Dioxide Sensing: Fundamentals, Principles, and Applications,
First Edition. Edited by Gerald Gerlach, Ulrich Guth, and Wolfram Oelßner.

product or by-product of other chemical transformations, mainly by combustion processes for energy generation.

2.1.1.2 Industrial Use of Carbon Dioxide

The industrial consumption of carbon dioxide can roughly be estimated as up to 100 Mt a^{-1}. Table 2.1 presents the dominating products as has been listed in the IPCC Special Report on Carbon Dioxide Capture and Storage (2005) [9].

The main product based on carbon dioxide by far is still urea, which is mostly needed as fertilizer in agricultural markets. The first step of the two-stage process of the reaction of ammonia with CO_2 comprises the exothermic formation of ammonium carbamate. The following endothermic process step leads to the dehydration of the intermediate resulting in an overall exothermic manufacturing of urea:

$$CO_2 + 2NH_3 \rightleftarrows H_2N{-}CO{-}ONH_4 \quad (\Delta H_r^\circ = -117\ \mathrm{kJ\ mol^{-1}}) \tag{2.1}$$

$$H_2N{-}CO{-}ONH_4 \rightleftarrows H_2N{-}CO{-}NH_2 + H_2O \quad (\Delta H_r^\circ = +16\ \mathrm{kJ\ mol^{-1}}) \tag{2.2}$$

Although the formation of urea is a carbon dioxide-consuming process, the complete synthesis causes an overall negative carbon dioxide balance because of the need of ammonia. This is due to the fact that the hydrogen required for the reaction with molecular nitrogen in the Haber–Bosch process has to be produced by the highly endothermic steam reforming process. The energy is generated by the combustion of fossil fuels and exceeds the introduced amount of CO_2 more than twice [10]. In general, ammonia and urea production are placed together to connect the formation and the consumption of carbon dioxide.

A remarkable amount of carbon dioxide is introduced in the large-scale synthesis of methanol. The industrial formation of CH_3OH is mainly based on the catalysed reaction of carbon monoxide with hydrogen both produced by the synthesis gas process. These steps are accompanied by an additional advantageous reduction of remarkable amounts of CO_2 to methanol. However, the direct reduction of carbon dioxide with electrochemically produced hydrogen to methanol with metal catalysts is still intensively investigated.

Table 2.1 Industrial use of carbon dioxide ([9]; the author points out the large uncertainty of the data).

Product class or utilization	Yearly demand (Mt yr^{-1})	Applied carbon dioxide (Mt yr^{-1})
Urea	90	65
Methanol (addition to CO)	24	<8
Inorganic carbonates	8	3
Organic carbonates	2.6	0.2
Polyurethanes	10	<10
Technological uses	10	10
Food applications	8	8

Furthermore, carbon dioxide is used for the industrial production of inorganic chemicals such as carbonates and bicarbonates based on sodium, potassium, ammonium, and other salts.

Although most conversion processes of the highly oxidized and thermodynamically stable carbon dioxide need catalysts and high temperatures for the reaction, some organic synthesis steps for important intermediates and products have still been established in industry [5]. The reaction of sodium phenolate with CO_2 leads to the production of salicylic acid needed in pharmaceutical industry. And by the addition of reactive synthons like epoxides or the use of catalysts, carbon dioxide is a valuable feedstock for the preparation of polycarbonates and polyurethanes – a perspective way to avoid the use of the highly toxic phosgene in chemical plants.

Besides the introduction of carbon dioxide as reactant in chemical synthesis, the compound is involved in a lot of further applications [11], such as:

- In food processing, e.g. freezing and refrigeration, decontamination of foodstuff, extraction of caffeine from coffee, packaging in an inert atmosphere, carbonation of drinks, supercritical fluid extraction, and cooling of transport of goods (dry ice);
- In metal industry as shielding gas during welding or cleaning reagent (supercritical fluid) for precision parts in electronic industry;
- For agricultural purposes, such as fumigation of grain silos, greenhouse atmosphere, and cultivation of algae;
- In water treatment for pH control as well as for recarbonation of sweet waters;
- For enhancement of oil recovery in mostly depleted oilfields;
- For cylinder fillings, e.g. for fire extinguishers, cartridges for carbonation of drinks; and
- In medical applications.

The main sources for introduced carbon dioxide are situated in the industry – especially in the ammonia production – where it is a by-product of other processes. Additionally, CO_2 can be isolated during fermentation processes, e.g. in the beer industry, or it is sampled from natural wells. Depending on the source, the quality of the gas differs remarkably. Especially the demands concerning the purity of CO_2 for beverage, food, and medical purposes require a sophisticated removal of other ingredients before application.

2.1.2 Physical Properties of Carbon Dioxide

2.1.2.1 Mechanical Properties

From the various mechanical properties of pure carbon dioxide in its three states of aggregation, only the densities, the viscosities, the surface tensions, and the speed of sound are described here with their impacts on applications and environmental processes.

The density of solid carbon dioxide depends on temperature and pressure. Till now two forms of solid carbon dioxide are known, which are referred to as 'dry ice' [12] and 'amorphous carbonia', the latter being also called 'a-carbonia' or 'a-CO_2' [13]. The density of dry ice amounts to 1562 kg m^{-3} at a pressure of

101.325 kPa and a temperature of −78.5 °C. In contrast to water, it is higher than the density of the saturated liquid with 1101 kg m^{-3} at −37.0 °C [14] and that of the gas amounting to 1.977 kg m^{-3} at standard conditions [15]. Since this value is much higher than that of air, carbon dioxide is accumulated always at the bottom when introduced into closed air-filled volumes, which is important, for instance, for safety issues in underground salt mining as described in Chapter 13.

The viscosity of carbon dioxide depends also on temperature and pressure. As it is described in detail in [16], this dependency can be calculated by an equation containing the viscosity at the zero-density limit, the excess viscosity that represents the increase at elevated density, and a critical enhancement accounting for the increase on viscosity in the immediate vicinity of the critical point. From this complex behaviour, which is especially important for gas separation processes and supercritical CO_2-driven extraction processes, only the value of the dynamic viscosity of $\eta = 13.7$ μPa s at standard conditions is provided here.

The surface tension of liquid CO_2 is an important parameter for CO_2 utilizing extraction processes, for instance, the decaffeination of coffee. Similar to the parameters described above, it depends also on temperature and amounts to 4.57 mN m^{-1} at 273 K [17].

The speed of sound in gaseous CO_2 amounts to 267 m s^{-1} at standard conditions and is therefore significantly lower than the speed of sound of 331.54 m s^{-1} in dry, CO_2-free air [18]. Due to that difference the speed of sound in real air decreases with increasing CO_2 content. This phenomenon is used for rapid measurement of CO_2 content, i.e. in breathing air with ultrasonic sensors [19].

2.1.2.2 Thermally Related Properties

The temperature and pressure ranges of the three states of aggregation of pure CO_2 are shown in the phase diagram of Figure 2.1. The important points in this parameter field are:

- The triple point at $p_t = 5.185$ bar and $T_t = 216.58$ K [20] and
- The critical point at $p_c = 73.825$ bar and $T_c = 304.200$ K [21].

According to [22] the sublimation line between 154.26 and 195.89 K can be described by the following Antoine equation:

$$\log_{10}(pCO_2/\text{bar}) = A - (B/(T + C)) \tag{2.3}$$

with pCO_2 the vapour pressure of CO_2, $A = 6.812\,28$, $B = 1301.679$ K, and $C = -3.494$ K.

A very broad range of phase change data of pure CO_2 can be calculated Internet-based with sufficient accuracy for most applications [23].

The thermal conductivity λ of CO_2 is a parameter with sensory relevance. It increases with temperature at standard pressure and amounts to 14.7 mW m^{-1} K^{-1} [24]. Since this value is significantly lower than that of air with 24.3 mW m^{-1} K^{-1}, it is also possible to measure higher CO_2 concentrations in air by thermal conductivity sensors.

The values for the specific heat and the phase change enthalpy are provided in Table 2.2. They are important for many technical processes that utilize CO_2.

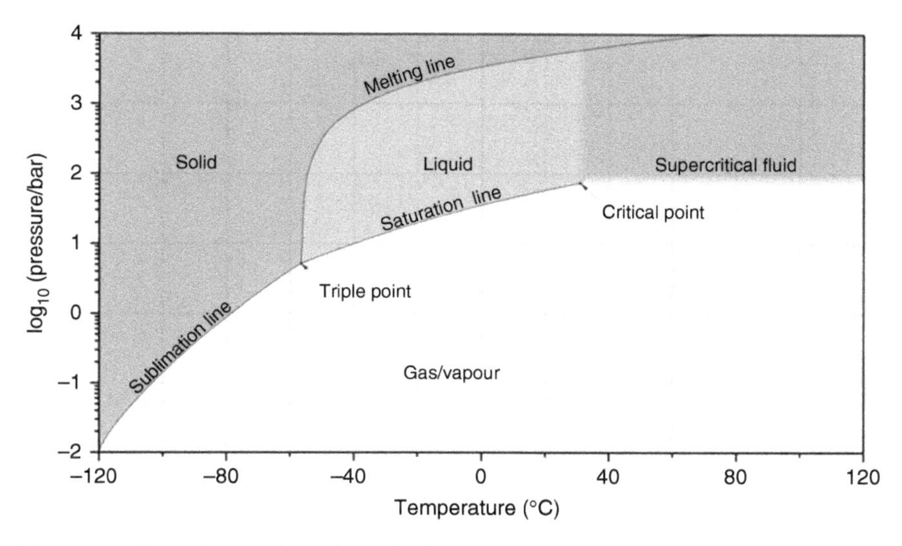

Figure 2.1 Phase diagram for carbon dioxide.

Table 2.2 Specific heats and phase change enthalpies of carbon dioxide at specified temperatures and pressures.

Parameter	Gas-phase heat capacity	Heat capacity at triple point	Enthalpy of vaporization	Enthalpy of sublimation
Value	$37.12\,J\,mol^{-1}\,K^{-1}$	$59.13\,J\,mol^{-1}\,K^{-1}$	$16.7\,kJ\,mol^{-1}$	$26.1\,kJ\,mol^{-1}$
Temperature (range) (K)	298		273–304	198–216
References	[25]	[26]	[27]	

The physical solubility in different solvents, especially in water, and its diffusion in gases and liquids are two important parameters with respect to physiological effects of CO_2. Both parameters change with temperature and pressure. The solubility in water with different salinities, which is of particular interest for environmental processes, is given by the following equation [28]:

$$\ln\left(\frac{K_0}{\mathrm{mol/l\cdot atm}}\right) = A_1 + A_2\frac{100}{T/\mathrm{K}} + A_3\ln\left(\frac{T/\mathrm{K}}{100}\right) + S‰\left[B_1 + B_2\left(\frac{T/\mathrm{K}}{100}\right) + B_3\left(\frac{T/\mathrm{K}}{100}\right)^2\right] \tag{2.4}$$

where K_0 is the solubility in mol/l atm, $S‰$ the salinity in parts per thousand, T the absolute temperature, $A_1 = -58.0931$, $A_2 = 90.5069$, $A_3 = 22.294$, $B_1 = 0.027766$, $B_2 = -0.025\,888$, and $B_3 = 0.005\,057\,8$. Equation (2.4) covers the temperature range 0– 40 °C and the salinity range 0–40‰.

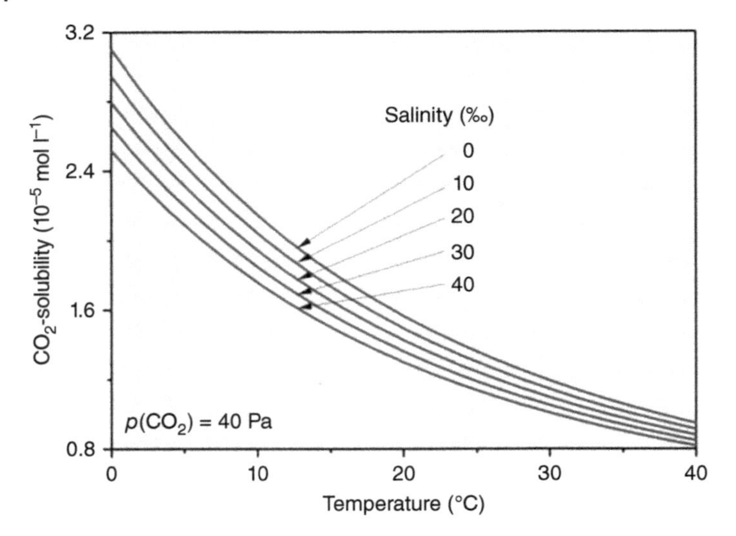

Figure 2.2 Solubility of carbon dioxide in water at different temperatures and salinities at equilibrium with air. CO_2 concentration 400 vol ppm.

As depicted in Figure 2.2, airborne CO_2 with a concentration of $c = 400$ vol ppm at 25 °C and standard pressure leads to a concentration of dissolved CO_2 in distilled water and seawater amounting to

$$c_{CO_2} = \begin{cases} 1.36 \times 10^{-5}\ \text{mol l}^{-1} \text{ in distilled water } (S = 0) \\ 1.16 \times 10^{-5}\ \text{mol l}^{-1} \text{ in seawater } (S = 35‰) \end{cases} \quad (2.5)$$

The diffusion coefficient of CO_2 in gaseous and liquid environments can differ by a factor of c. 10^4 and depends strongly on temperature and – in the case of gas diffusion – also on pressure [29]. At 25 °C and standard pressure, it amounts to

$$D_{CO_2} = \begin{cases} 1.92 \times 10^{-5}\ \text{cm}^2\ \text{s}^{-1} \text{ in water} \\ 0.161\ \text{cm}^2\ \text{s}^{-1} \text{ in air} \end{cases} \quad (2.6)$$

2.1.2.3 Electrical Properties

The most important electrical parameter of carbon dioxide with respect to the sensor behaviour is the dielectric constant ε_r. According to [30], it increases with pressure. For 0.019 °C and 3.29 MPa, it amounts for gaseous CO_2 to 1.046. Liquid and solid CO_2 are electrical insulators.

2.1.2.4 Optical Properties

If light travels through gaseous atmospheres, then the electromagnetic wave interacts with the gas molecules by an energy exchange. Very often, the intensity of light is decreasing caused by energy absorption by the gas molecules and increasing their thermal or chemical energy. This phenomenon is responsible for warming the Earth's atmosphere at increasing carbon dioxide and methane concentrations (see Section 3.1) and for the synthesis of ozone in the upper atmosphere.

As described in Chapter 7, optical absorption is also used for spectroscopically high selective and sensitive measurements of gas concentrations. Since the energy transfer occurs at selected vibrations of the molecules or shell electrons, every gas has characteristic absorption lines. The two most important analytically utilized absorption peaks of carbon dioxide range in the mid-infrared region at the wave numbers ν of 2349 cm^{-1} (asymmetric bond stretch) and 667 cm^{-1} (bond bending) [31].

2.2 The Carbon Cycle

2.2.1 Sources of Carbon on Earth

Our galaxy is probably 13.7 ± 0.2 billion years (Ga) old and was formed by the hot Big Bang, assuming that the whole mass of the galaxy was concentrated in a primordial core. Hydrogen and helium produced in the Big Bang served as the 'feedstock' from which all heavier elements were later created in stars. The fusion of protons to form helium is the major source of energy in the solar system. This proceeds at a very slow and uniform rate, with the lifetime of the proton before it is fused to deuterium of about 10 Ga (note that the proton lifetime concerning its decay is $>10^{30}$ a). Carbon and oxygen were formed during the fusion process from helium within the first steps after the Big Bang:

$$^{4}He + {}^{4}He \rightarrow {}^{8}Be$$
$$^{8}Be + {}^{4}He \rightarrow {}^{12}C$$
$$^{12}C + {}^{4}He \rightarrow {}^{16}O$$

It is likely that most of the oxygen (and other volatile elements) in the particulate matter of the solar nebula was chemically bonded with metals forming oxides [32]. The most abundant elements up to iron (Fe) are multiples of ^{4}He (^{12}C, ^{16}O, ^{24}Mg, ^{28}Si, ^{32}S, etc.). Hence, carbon (beryllium in between is extremely rare) is the first common element in stars' fusion process having (together with oxygen) the highest abundance in cosmos (neglecting H and He). The formation of molecules is impossible in stars because of the high temperature, but in the interstellar medium with temperatures between 10 and 20 K, chemical reactions are possible, which have the potential to create molecules. At present, nearly 200 molecular species are listed from which only about 20 are inorganic species (including carbon species CO, CO_2, CH_4, HCO, HCN, and OCS) and all other are organic carbon molecules; even life-supporting organics such as amino acids have been detected. It is logical that based on the molar ratios of H, O, and C, hydrocarbons (H_xC_y) and water (H_2O) are the most abundant molecules in space.

The Earth, like the other solid planetary bodies, was formed by the accretion of large solid objects in a short time between 10 and 100 million years. During the formation of the Earth by the accumulation of cold solids, very little gaseous material was incorporated. There is evidence that most of the Earth's volatiles may have been supplied by a 'late heavy bombardment' (LHB) of comets and carbonaceous meteorites, scattered into the inner solar system following the formation of the giant planets. At least 80 organic compounds are known to occur in carbonaceous meteorites [33]. How much in the way of intact organic molecules of potential prebiotic interest survived delivery to the Earth has become an increasingly debated topic over the last several years. The principal source for such intact organics was probably the accretion of interplanetary dust particles of cometary origin. Outer-belt asteroids are the probable parent bodies of the carbonaceous chondrites, which may contain as much as 5% organic material, and they contain it mainly in unoxidized form, a substantial fraction in the form of solid, heavy hydrocarbons [34, 35].

The dominant fraction of carbon on Earth (see Table 2.3) is termed kerogen, a mixture of organic chemical compounds that make up a portion of the organic matter in sedimentary rocks. When heated to the right temperatures in the Earth's crust, some types of kerogen release crude oil or natural gas, collectively known as hydrocarbons (fossil fuels). When such kerogens are present in high concentration in rocks such as shale and have not been heated to a sufficient temperature to release their hydrocarbons, they may form oil-shale deposits. Kerogens and coals in evolved stages are very similar. The main difference is that coals are found in the form of bulk rocks and kerogens in dispersed form (sandlike). The idea that kerogens have been transported to Earth by carbonaceous chondrites is supported by the finding of kerogen-like material, mainly as solvent-unextractable macromolecular matter, analogous to terrestrial

Table 2.3 Reservoir distribution (in 10^{19} g element, resp. water); after [36], if not other noted.

Reservoir	C	C[a)]	C[b)]	O	H_2O
Atmosphere	0.075	0.0766	—	119	1.3
Ocean	3.8/0.07[c)]	3.8–4.0/—[c)]	—/0.06[c)]	12 500[d)]	135 000[i)]
Land plants	0.06	0.054–0.061	0.095	Neglectable	Neglectable
Soils, organic	0.15	0.15–0.16	0.16	? (neglectable)	? (neglectable)
Fossil fuels	0.7	0.4		Neglectable	Neglectable
Sediments	~5 000/1 500[e)]	6 600–10 000[f)]	6 000/1 500[e)]	4 745[g)]	[m)]
Clathrates[h)]	1.1			—	—
Rocks	3200–9 300[j)]			1200[k)]	>600 000[l)]

a) Pidwirny [37].
b) Pédro [38].
c) Carbonate/dissolved organic carbon (DOC).
d) In water molecules.
e) Carbonate/buried organic (kerogen).
f) Not specified into carbonate and organic C.
g) Held in Fe_2O_3 and evaporitic $CaSO_4$.
h) Methane hydrates; after [39].
i) Additional in ice: 3 300.
j) Estimated by using mean element abundance [40] and assuming a mass of crust being 4.9×10^{25} g; sediments likely are included (the chemical form is not specified).
k) Held in silicates.
l) Murakami et al. [41].
m) Groundwater holds 15 300 and soils 12.2.

kerogen or poorly crystalline graphite in different meteorites [42]. Some of those hydrous carbonaceous chondrites show evidence for heating during ageing, and the kerogen-like amorphous carbonaceous materials lose their labile fractions and become more and more graphitized.

As discussed above, carbon is among the interstellar gases, in form of hydrocarbon, deposited on carbonaceous chondrites, but it is also found in the form of carbides in meteorites and on Earth. Elemental carbon in the form of graphite can react with many elements (especially alkali and alkaline earth metals) to form graphitic mixtures and carbides. Carbides, produced at high temperatures (>1000 °C) from carbon and the metal or its oxide, and have been also found in nature (they are produced synthetically for industrial purpose). In molten iron (above 1100 °C), carbon can be dissolved up to 4.3%, about double the carbon content in cohenite (2.2%). The carbon solubility increases with temperature, and when the solution slowly cools down, carbon in excess of 4.3% separates as graphite. There is no information on carbon in the Earth's core, but it can be speculated on it and assumed that some was also mixed within other upper layers. A carbon content of 1.4–2.3% has been found in native iron [40].

It also can be speculated that unstable carbides existed on the early Earth and converted according to methane (CH_4). Today, the presence of carbides in the Earth's crust can safely be discounted (although there is no evidence for the absence of carbides at an early stage). Hence, there must have been carbon in other compounds in the crust that can be oxidized to CO_2 or reduced to CH_4 depending on the reaction conditions. The small mean Clarke number of carbon in rocks (0.02%) denotes the geochemical abundance of elements in sense of a mixing ratio. It probably increases with depth and could be very inhomogeneously distributed and is the only source of carbon gases. With the assumption that all rocky carbon exists as hydrocarbons, it is easily to explain that under pressure, heat, and hydrogen, CH_4 is produced. It has been shown that the coexistence of water vapour, hydrogen, and both oxides of carbon is possible depending upon varying conditions of temperature and concentration [40].

Oxygenated hydrocarbons also produce oxygen under high temperature and pressure:

$$C_nH_mN_xO_y \xrightarrow{T,p} CH_4, C, H_2, CO, CO_2, N_2, H_2O, O_2 \tag{2.7}$$

and when water is added:

$$C_nH_mN_xO_y \xrightarrow{H_2O,T,p} CH_4 + O_2 + H_2 + N_2 \tag{2.8}$$

This O_2 is used subsequently (assuming overall reducing conditions), for example, to oxidize hydrocarbon with increasing yield of CO and CO_2. Another initial production of free oxygen in the Earth's mantle can be also explained by the thermal decomposition of metal oxides, transported to hotter regions, for example, FeO, giving oxygen and metallic iron, the heavy iron moving towards the Earth's core, leaving the oxygen to escape.

The free oxygen, however, could have oxidized the reduced carbon existing in heavy hydrocarbons into carbon dioxide and water:

$$C_6H_2O + 6O_2 \rightarrow 6CO_2 + H_2O \tag{2.9}$$

or generally

$$(CH_2O)_n + nO_2 \rightarrow nCO_2 + nH_2O \tag{2.10}$$

The destruction of hydrocarbons under pressure and higher temperatures produces CH_4 as well as elemental C (in oxygen-poor conditions) and CO_2 as well as H_2O (in oxygen-rich conditions) as a continuous process over geological epochs. Under oxygen-free conditions the product from thermal dissociation according to Eq. (2.7) is $C + CO_2 + H_2$. Hydrogen can also be produced via Reactions (2.7) and (2.8) and transforms deep carbon into CH_4 and H_2O as seen in Eq. (2.7). Reaction (2.8) can invert under the conditions deep within the Earth:

$$CO_2(+H_2O) \underset{O_2}{\overset{H_2}{\rightleftharpoons}} CO(+H_2O) \underset{O_2}{\overset{H_2}{\rightleftharpoons}} C_nH_m\,(+H_2O) \tag{2.11}$$

In other words, the process shown in Eq. (2.11) represents an inorganic formation of hydrocarbons ('fossil fuels'). Although the biogenic theory for petroleum was first proposed by Georg Agricola in the sixteenth century, various abiogenic hypotheses were proposed in the nineteenth century, most notably by von Humboldt, Mendeleev [43], and Berthelot, and renewed in the 1950s (e.g. [44] and later by Gold [45]; see below).

Under pressure and high temperature (>900 °C), equilibriums are established between CO, CO_2, H_2O, H_2, and CH_4:

$$2CO + 2H_2O \rightleftarrows 2CO_2 + 2H_2 \tag{2.12}$$

$$4CO + 2H_2 \rightleftarrows 2H_2O + CO_2 + 3C \tag{2.13}$$

$$2CO + 2H_2 \rightleftarrows CO_2 + CH_4 \tag{2.14}$$

$$CO_2 + H_2 \rightleftarrows CO + H_2O \tag{2.15}$$

The degassing of CO_2 from primordial carbonates is not improbable. However, carbonates are very rare in meteorites. Hence, this was probably not a dominant source of CO_2 in the early Earth and may have become dominant in the volcanic source due to carbonate subduction. The rocks give off on average 5–10 times their own volume of gases (excluding H_2O vapour). Before heating, the rocks were dried in the experiments to remove hygroscopic moisture. The produced steam (H_2O) however was dominant and exceeds by a factor of 4–5 all other gases from the rocks. The evolved gases from granite and basalt are given in Table 2.4.

Table 2.4 Degassing of gases from heated rocks ([46] and citations therein).

Gas	Degassing (%)
H_2	10–90
CO_2	15–75
CO	2–20
CH_4	1–10
N_2	1–5

Table 2.5 Evolution of the Earth's atmosphere.

Atmosphere	Time ago (Ga)	Composition	Origin	Fate
Primordial[a)]	4.5–4.6	H_2, He	Solar nebula	Erosion to space
Primitive (first)	~4.5	NH_3 (?), CH_4 (?), CO_2, H_2O	Degassing	Photolysis
Secondary	4.5–4.0	N_2, CO_2, H_2O	Degassing	Washout
Intermediate (third)	4.0–2.3	N_2, CO_2	Secondary phase	Remaining
Present (fourth)	2.3–0.5	N_2, O_2	Photosynthesis	Biosphere–atmosphere equilibrium

a) Speculative.

In the following the chemical fate of the degassed compounds, such as CH_4, CO_2, and H_2O in the first atmosphere (Table 2.5), will be considered. Methane photolysis occurs at the α-Lyman wavelength (121.6 nm) into H, H_2, CH, CH_2, and CH_3, and in the atmosphere of the Titan, it is thought to promote the propagation of hydrocarbon chemistry [47]. A small presence of O_2 and OH radicals is a result of H_2O photolysis. Assuming CH_4 and H_2O to represent at first the degassing products, the formation of CO_2 is an irreversible subsequent step:

$$CH_4 + 2O_2 \xrightarrow{h\nu} CO_2 + 2H_2O \text{ (via } CH_x) \tag{2.16}$$

CO_2 cannot be reduced under lower atmospheric conditions; it will scavenge while forming carbonic acid.

The lifetime of CH_4 with respect to oxidation by OH radicals is around 10 a in today's atmosphere. In the early atmosphere OH levels would be so low that only the photolysis remains as a sink, resulting in a residence time of CH_4 in the order of 10^5 a [48]. It is assumed that methane remained until formation of the oxygenic environment in air (2.2–2.7 Ga ago) at relatively high concentration (ppm level) to maintain a warming potential (greenhouse effect). Therefore, CH_4 must be produced from the crust at rates compensating its atmospheric oxidation. It is unlikely that H_2 was available at high concentrations for hydrogenations, e.g.

$$CO_2 + 4H_2 \rightarrow CH_4 + 2H_2O \tag{2.17}$$

CO_2 photolysis occurring in the upper atmosphere produces O_2, which can fix hydrogen back into water. The presence of oxygen in even small concentrations (10^{-8}), however, would inhibit the formation of more complex organic molecules due to CO formation (and subsequent CO_2) as proposed for the origin of life (the Miller–Urey hypothesis).

The photolysis of carbon dioxide (besides that of water) also provides small amounts of oxygen (and related radicals). Hence, the presence of CO_2 and H_2O as well as UV radiation can also produce simple hydrocarbons. As mentioned, strong UV radiation with energy $h\nu$ limited the synthesis of more complex molecules, and radical reactions led back to the radiative relatively stable molecules $CO_2 + H_2O$:

$$CO_2 + H_2O \xrightarrow{h\nu} H + O_x + H_xO_y + CO + (C_xH_y\ (?)) \rightarrow CO_2 + H_2O \tag{2.18}$$

Table 2.6 Main composition of the atmospheres of the inner terrestrial planets (%).

Substance	Venus	Earth	Mars
Carbon dioxide	96	0.03	95
Nitrogen	3.5	77	2.7
Oxygen	<0.001	21	0.13

Source: Adapted from Brimblecombe 1996 [49].

with the following main elementary reactions behind:

$$CO_2 \xrightarrow{h\nu} CO + O \tag{2.19a}$$

$$H_2O \xrightarrow{h\nu} OH + H \tag{2.19b}$$

$$OH + O \rightarrow O_2 + H \tag{2.19c}$$

$$O + O_2 \rightarrow O_3 \tag{2.19d}$$

$$O_3 \xrightarrow{h\nu} O + O_2 \tag{2.19e}$$

Only with the occurrence of photosynthesis about 2.7 Ga ago O_2 was available in a stepwise excess (compared to the low photolytic production in air), but first it was consumed by oxidizing Fe^{2+} and other reduced compounds. Only after reaching a redox equilibrium the seawater became saturated with O_2 and oxygen may have escaped to the atmosphere. This certainly had quite an impact on further evolution. Small amounts of oxygen abiotically produced in the atmosphere had been tolerated for the first 2.5 Ga. Besides free oxygen in the lower atmosphere, oxygen was deposited due to oxidation of reduced materials on the crustal surface of the continents. The further fate of biological evolution separated the earlier similar atmospheric composition of the Earth-like planets (Table 2.6).

2.2.2 Carbon Pools and Global Cycling

The carbon on the surface or in the sediment of the Earth (note that this excludes possible carbon stocks in rocks and deep in the Earth) is estimated to be around 80% in the form of carbonate rocks and 20% in unoxidized form, frequently referred to as organic carbon (possible carbon stocks deep in the Earth are not counted) (cf. Table 2.3). If the vast amounts of present carbonate sediments (5000×10^{19} g C) had originally been exhaled in reduced forms (CH_4 or CO), it would have required oxygen from the atmosphere for oxidation. This huge amount of required oxygen, however, is not balanced with the O_2 produced via photosynthesis, but it is 2–3 times greater. Thus, the Earth's present carbonate reservoir was probably initially exhaled from the Earth as CO_2, which was oxidized from reduced carbon derived from chondrite-type debris. There is no argument against the hypothesis that this is a more or less continuous process over the entire history of the Earth, explaining the deposition of carbonate sediments. In a process of continuous recycling, the proportion of ^{13}C would continuously increase in the atmosphere, and because the younger carbonates

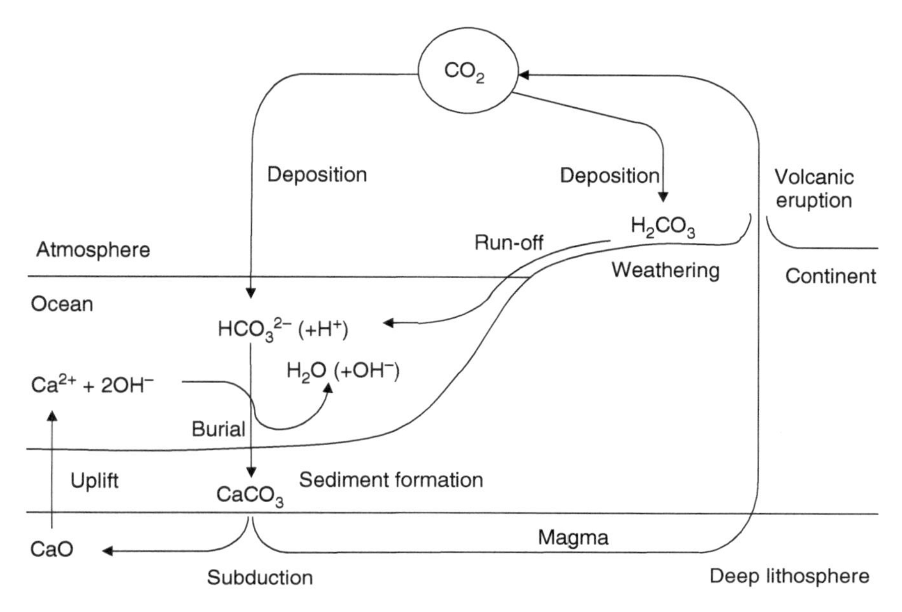

Figure 2.3 Scheme of the inorganic carbon cycle (inorganic carbon burial).

should be isotopically heavier than the old ones, this is not the case. Marine carbonates of all ages back to the Archean show the same narrow range of the carbon isotopic ratio [50]. What it shows is a reasonably continuous process of laying down carbonate rocks according to

$$CaO + CO_2 \rightarrow CaCO_3 \qquad (2.20)$$

with no epoch having enormously more per unit time, nor enormously less (Figure 2.3). Most, but not all, of this carbonate has been an oceanic deposit, deriving the necessary CO_2 from the atmospheric–oceanic CO_2 store. The amount that is at present in this store is, however, only a very small fraction of the amount required to lay down the carbonates present in the geological record. The atmospheric–oceanic reservoir at present holds only about 0.05% of the Earth's carbon (cf. Table 2.3).

The carbonate sediments also support the hypothesis that CO_2 was a product of the Earth's inner processes (not only atmospheric CH_4 oxidation), thermal dissociation of primordial carbonates – which is not very likely due to its spare finding in meteorites – or other carbon species, already formed within the solar nebula. Transforming back the CO_2 stored in inorganic and organic sediments to the atmosphere, one would have 60–80 bar CO_2 in the first atmosphere. Another 140 bar is provided by the water from the ocean and the sediments transformed back into the atmosphere.

The amount of carbon estimated in the rocks of the lithosphere (Table 2.3) corresponds to about 3% carbon content. This is very close to the assumption [45] of carbonaceous chondrite material, which may comprise 20% of the material in the depth range between 100 and 300 km. In this material, carbon amounts to 5%. Hence, this layer would provide some 3×10^{24} g C from which only 2% have been released to provide the surface carbon. A supply of hydrocarbons at depth may thus provide CO_2 in three different ways:

- Through volcanic pathways and oxidation with oxygen supplied by the magma,
- By the ascent of hydrocarbons through solid rocks and oxidation at shallow levels, most likely by bacterial action, with subsequent escape of CO_2 to the atmosphere, and
- By the escape of methane and other hydrocarbons into the atmosphere, where, in the presence of atmospheric O_2, they would oxidize to CO_2.

Without life on Earth, probably most of the geochemical redox potentials would become in equilibrium due to tectonic mixing of all redox couples over the entire Earth history. Generally, the process of photosynthesis is written as

$$CO_2 + H_2O \xrightarrow{h\nu} CH_2O + O_2 \tag{2.21}$$

where CH_2O is a synonym for organic matter (a building block of sugar $C_6H_{12}O_6$). The creation of a photosynthetic apparatus capable of splitting water into O_2, protons, and electrons was the pivotal innovation in the evolution of life on Earth. The cycle is closed by respiration, the process of liberation of chemical energy in the oxidation of organic compounds:

$$CH_2O + O_2 \rightarrow CO_2 + H_2O \tag{2.22}$$

It is remarkable that in this way a stoichiometric ratio of 1 : 1 between fixed carbon and released oxygen is established. Therefore, a net oxygen production is only possible when the rate of Reaction (2.22) is smaller than that of Reaction (2.21), or in other words, the organic matter produced must be buried and protected against oxidation. This was the first closed biogeochemical cycle.

As can be assumed in the modern world that photosynthesis is balanced with respiration, one can also write the formation and degradation of 'biomass' in the overall reaction:

$$CO_2 + H_2O + NH_3 + H_2S \underset{\text{oxidation}}{\overset{\text{reduction}}{\rightleftharpoons}} C_xH_yN_nS_mO_o + O_2 \tag{2.23}$$

The biospheric carbon turnover is so vast that small (never measurable) annual variations will be buffered and result in a well-balanced redox and acidity state of the climate system over geological time. It seems that the climate system itself has established a buffer system to stabilize it even in the case of (not too large) catastrophic effects. Consequently, disturbing the redox state of the environment, for example, by human activities, can lead to changing reservoir distributions.

Modelling the long-term carbon cycle shows strong relationships between the key parameters, atmospheric oxygen and carbon dioxide (inverse correlated) as well as carbon burial, which correlates with oxygen [51]. The (calculated) ranges of the amounts are:

- Atmospheric O_2 concentration: 13–30%,
- Atmospheric CO_2 concentration (factor to pre-industrial level): 1–25%, and
- C burial: $(2.3–6.2) \times 10^{18}$ mol Ma^{-1} (or 28–74 Tg a^{-1}).

From the dates given by Berner [51], only within the period of the last 410 Ma the global organic carbon burial has amounted to be ~2200×10^{19} g carbon. Using the low burial rate in the period back to first oxygenic photosynthesis (2.7 Ga), it would give a value of more than 6000×10^{19} g carbon burial, together with a multiple of the estimated organic carbon sediments (~1500×10^{19} g). This suggests that organic carbon is also cycling through subduction and volcanic release to the atmosphere, at an average rate of 15 Tg C a^{-1}.

In the carbon cycle we have to consider long-term cycling, including rock weathering and volcanism. Over much longer timescales, atmospheric CO_2 concentrations have varied tremendously due to changes in the balance between the supply of CO_2 from volcanism and the consumption of CO_2 by rock weathering. Over geological timescales, large (but very gradual) changes in atmospheric CO_2 result from changes in this balance between rock weathering and volcanism. CO_2 in the atmosphere is consumed in the weathering of rocks:

$$CO_2 \xrightarrow{H_2O} H_2CO_3[H^+ + HCO_3^-] \xrightarrow{CaCO_3} Ca(HCO_3)_2[Ca^{2+} + 2\,HCO_3^-] \tag{2.24}$$

This comes about by the first global reaction, first deduced by Ebelmen [52]:

$$CO_2 + (Ca, Mg)SiO_3 \rightarrow (Ca, Mg)CO_3 + SiO_2 \tag{2.25}$$

Carbonic acid is strong enough to dissolve silicate rocks – in small quantities, of course, and over long timescales. For example, an orthosilicate is dissolved into orthosilicic acid (where SiO_2 is the anhydride) and bicarbonate:

$$CaH_2SiO_4 + (2H^+ + 2HCO_3^-) \rightarrow H_4SiO_4\,(SiO_2 + 2H_2O) + Ca(HCO_3)_2 \tag{2.26}$$

SiO_2 is moderately soluble in water (5–75 mg l^{-1} in river water and 4–14 mg l^{-1} in seawater). The products are then transported in river water to the oceans. There organisms such as foraminifera use calcium carbonate to make shells. Other organisms such as diatoms make their shells from silica. When these organisms die, they fall into the deepest oceans. Most of the shells redissolve, but a fraction of them are buried in sediments of the seafloor. The overlaying sediments are carried down to the depths by subduction. Temperature and pressure transform the shells back to silicate minerals, in a process releasing CO_2 back to the surface of the Earth through volcanoes and into the atmosphere to begin the cycle again, over a geological timescale. This inorganic (no photosynthesis) but biotic (mineral production) carbon cycle is not linked with the oxygen cycle but with water (H_2O) and acidity (H^+). Simply said, insoluble rock carbonate is transformed into more soluble bicarbonate where atmospheric CO_2 is fixed as dissolved bicarbonate. The volcanic carbon dioxide released is roughly equal to the amount removed by silicate weathering. The two processes, which are the chemical reverse of each other, sum to roughly zero and do not affect the level of atmospheric carbon dioxide on timescales of less than about 10^6 a. As a planet's surface becomes colder, however, atmospheric CO_2 levels should tend to rise. The reason is that removal of CO_2 by silicate weathering followed by carbonate deposition should slow down as the climate cools and would cease almost entirely if the planet were to glaciate globally. On planets like Earth that have abundant carbon (in carbonate rocks) and some mechanism, like plate tectonics, for recycling this carbon, volcanism should provide a more or less continuous input of CO_2 into the atmosphere.

Additionally, one has to include in the budget still permanent CO_2 degassing (unknown value) from the crust (accepting the deep carbon hypothesis). There is also no doubt that degassed CH_4 is partly a product of deep rock chemistry (cf. Eqs. (2.7) and (2.8)).

Four revolutionary changes have occurred in the chemical evolution of carbon:

(a) The first change (and what separates the Earth's atmosphere from those of the other terrestrial planets) was the absorption of atmospheric CO_2 by rain in the early Earth during the formation of the oceans with subsequent seawater carbonate equilibriums.
(b) With the modern photosynthesis by cyanobacteria (~2.3 Ga ago), a huge consumption (equivalent to the atmospheric oxygen increase by a factor of 300) of CO_2 into biomass occurs within a few million years.
(c) A third drop in atmospheric CO_2 was associated with the O_2 increase in the late Silurian (420 Ma ago) due to the evolution of land plants.
(d) Today atmospheric CO_2 is increasing at a rate that has probably never occurred over the entire history of the Earth due to the release by human activity of carbon in 'fossil fuels' buried over millions of years in a period of barely hundreds of years.

The amount of CO_2 removed from the atmosphere each year by oxygenic photosynthetic organisms is massive (Figure 2.4). It is estimated that photosynthetic organisms remove about 120×10^{15} g C a^{-1}. This is equivalent to 4×10^{18} kJ of free energy stored in reduced carbon, which is roughly 0.1% of the visible radiant energy incident on the Earth per annum. Each year the photosynthetically reduced carbon is oxidized, either by living organisms for their survival or by combustion. The burial rate is small, approximately between 60 and 200 Tg C a^{-1} [51] (see Figure 2.4). Only the present atmospheric oxygen requires a carbon equivalent via photosynthesis of 44×10^{19} g C, 2 orders of magnitude more than present estimates of economically extractable fossil fuels (0.7×10^{19} g C in the twentieth century, of which coal represents more than 90%). However, by burning all the fossil fuels that are still available in a short time, the oxygen content of the atmosphere would be reduced by only 1%. According to an estimate [57], the atmospheric CO_2 concentration would rise to about 800 ppm, twice the present level and about three times more than the pre-industrial level. This is large compared with the CO_2 variation over the last few 100 000 a but small concerning the timescale over epochs. The amount of combustible fossil fuels is only 40 times larger than the yearly biological turnover of carbon. The problem, however, consists in the human timescale of a few 100 a and the vulnerable infrastructural systems of mankind.

Burning of fossil fuels amounts now to ~8×10^{15} g C a^{-1}, which is a mere 10% of the terrestrial carbon uptake by photosynthesis. However, it interrupts the carbon cycle due to the large residence time of CO_2 in the atmosphere. The oceans mitigate this increase by acting as a sink for atmospheric CO_2. It is

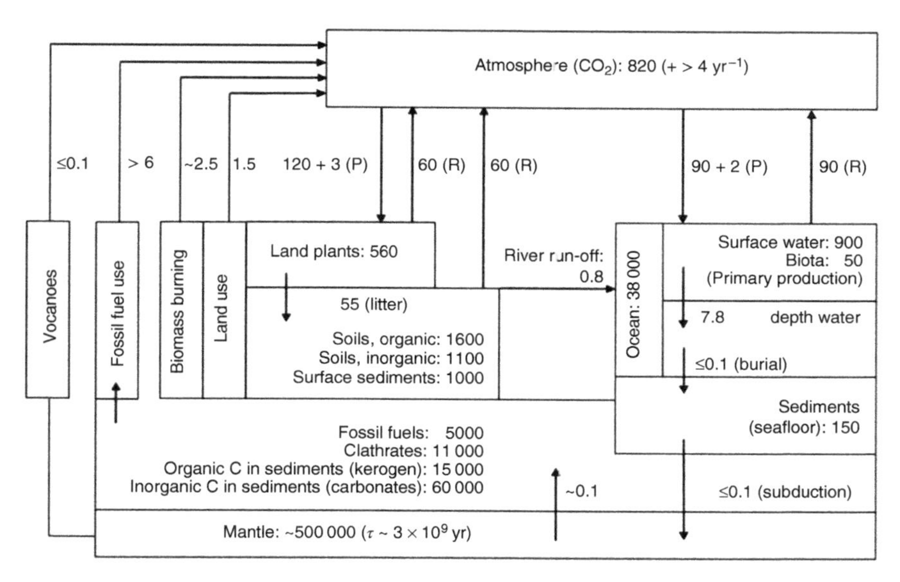

Figure 2.4 Scheme of the carbon cycle and reservoirs; fluxes in 10^{15} g a^{-1} and pools in 10^{15} g. Source: Data adapted from [36, 38, 53–55]. R, respiration; P, photosynthesis. River run-off (0.8 Tg C a^{-1}) consists of (all in Tg C a^{-1}) 0.2 rock weathering, 0.4 soil weathering, and 0.2 hydrogen carbonate precipitation [54]. The pre-industrial ocean–atmosphere exchange amounts to 70 Tg C a^{-1} [54] (20 anthropogenic additional), and the oceanic surface water contains 112 ± 17 Pg accumulated anthropogenic C [56].

estimated that the oceans remove about 2×10^{15} g C a^{-1} from the atmosphere. This carbon is eventually stored on the ocean floor. Although these estimates of sources and sinks are uncertain, the net global CO_2 concentration is increasing. Direct measurements show that currently each year the atmospheric carbon content is increasing by about 3×10^{15} g. Over the past 200 years, CO_2 in the atmosphere has increased from about 280 ppm to its current level of nearly 400 ppm. Based on predicted fossil fuel use and land management, it is estimated that the amount of CO_2 in the atmosphere will reach 700 ppm by the end of this century. The consequences of this rapid change in our atmosphere are unknown. Since CO_2 acts as a greenhouse gas, some climate models predict that the temperature of the Earth's atmosphere may increase by 2–8 °C. Such a large temperature increase would lead to significant changes in rainfall patterns. This could enhance weathering and subsequent CO_2 fixing as described above; it seems that the key question lies in the different timescales of the interactive processes contributing to liberation and fixing of CO_2. Little is known about the impact of such drastic atmospheric and climatic changes on plant communities and crops. Current research is directed at understanding the interaction between global climate change and photosynthetic organisms.

2.2.3 Carbon Budget

According to Eq. (2.19), solar energy is transformed into chemical bonding energy within biomass $(CH_2O)_n$ via photosynthesis by autotrophs (plants, blue-green algae, autotrophic bacteria); approximately 475 kJ of free energy is stored in plant biomass for every mole of CO_2 fixed during photosynthesis (compare it with the bonding energy of C–H to be 416 kJ mol^{-1}). Although photosynthesis is fundamental to the conversion of solar radiation into stored biomass energy, its theoretically achievable efficiency is limited by both the limited wavelength range applicable to photosynthesis and the quantum requirements of the photosynthetic process. Only light within the wavelength range of 400–700 nm (photosynthetically active radiation, PAR) can be utilized by plants, effectively allowing only 45% of total solar energy to be utilized for photosynthesis. Furthermore, fixation of one CO_2 molecule during photosynthesis necessitates a quantum requirement of 10 (or more), which results in a maximum utilization of only 25% of the PAR absorbed by the photosynthetic system. Based on these limitations, the theoretical maximum efficiency of solar energy conversion is approximately 11%. In practice, however, the magnitude of photosynthetic efficiency observed in the field is further decreased by factors such as poor absorption of sunlight due to its reflection, respiration requirements of photosynthesis, and the need for optimal solar radiation levels. The net result is an overall photosynthetic efficiency of 3–6% of total solar radiation.

About 120 Gt of carbon is fixed from atmospheric and dissolved aquatic CO_2 per annum by terrestrial plants (Figure 2.4). This is the gross primary production (GPP), the rate at which an ecosystem's producers capture and store a given amount of chemical energy as biomass in a given length of time. Some fraction of this fixed energy is used by primary producers for cellular respiration and maintenance of existing tissues. Whereas not all cells contain chloroplasts for carrying the photosynthesis, all cells contain mitochondria for oxidizing organics, i.e. the yield of free energy (respiration). The remaining fixed energy is referred to as net primary production (NPP):

$$\text{NPP} = \text{GPP} - \text{plant respiration} \tag{2.27}$$

NPP is the primary driver of the coupled carbon and nutrient cycles and is the primary controller of the size of carbon and organic nitrogen stores in landscapes.

Above-ground NPP ranges from 35 to 2320 g m^{-2} a^{-1} (dry matter) and total NPP from 182 to 3538 g m^{-2} a^{-1} [58]. However, quantifying primary production at a global scale is difficult because of the range of habitats on Earth and because of the impact of weather events (availability of sunlight, water) on its variability. Direct observations of NPP are not available globally, but computer models based on remote sensing and derived from local observations have been developed to represent global terrestrial NPP showing a range from 40 to 80 Gt C a^{-1} [59]. For example, a model [60] estimates a global terrestrial NPP of 48 Gt C a^{-1} with a maximum light use efficiency of 0.39 g C MJ^{-1} PAR. Over 70% of the terrestrial net production takes place between 30 °N and 30 °S latitude. It is estimated that the total (photoautotrophic) NPP of the Earth was 104.9 Gt C. Of this, 56.4 Gt C a^{-1} (53.8%) was the product of terrestrial organisms, while the remaining 48.5 Gt C a^{-1} was accounted for by oceanic production [61]. For the 1997–2004 period, it was found [62] that on average approximately 58 Gr C a^{-1} was fixed by plants. Consistent data on terrestrial net primary productivity NPP is urgently needed to constrain model estimates of carbon fluxes and hence to refine our understanding of ecosystem responses to climate change. Recent

Table 2.7 Global estimates of *NPP* and biomass (standing crop).

Ecosystem	NPP in 10^{15} g a^{-1}		Biomass in 10^{15} g		Area in 10^6 km^2
References	**[64]**	**[55]**	**[64][a)]**	**[55][b)]**	**[55]**
Tropical forests	49.4	40 (26–60)	1025	206	18.1
Extratropical forests	24.5	20 (6–40)	625	216	24.2
Grassland, bushland, savanna	25	15 (4–40)	124	92.6	39.7
Cropland (cultivated)	9.1	6.5 (1–40)	14	21.5	15.9
Others	7	—	48	—	—
Wetlands	—	30 (10–60)	—	7.8	2.9
Inland waters	—	4 (1–15)	—	—	2.0
Subtotal terrestrial	115	115	1836	544	147.2[c)]
Open ocean	41.5	—	1	—	—
Other marine ecosystems	14	—	3	—	—
Total global	160	—	1840	—	—

a) As biomass (dry mass).
b) As carbon.
c) Not the sum of above ecosystems.

climatic changes have enhanced plant growth in northern mid-latitudes and high latitudes. Research findings [63] indicate that global changes in climate have eased several critical climatic constraints to plant growth, such that NPP increased by 6% (3.4 Gt C over 18 years, 1982–1999) globally. The largest increase was in tropical ecosystems.

A summary of both NPP and biomass is given in Table 2.7 [64]. It seems that a value of ~60 Pg C a^{-1} is consistent with estimates by several authors over the last 20 years. In contrast, the oceanic NPP there shows high uncertainties. But it is agreed on that ocean phytoplankton is responsible for approximately half the global NPP. Oceans (especially at high latitudes) typically represent a net sink of atmospheric carbon. These regions are dominated by diatoms that typically grow and sink faster than other phytoplankton groups and, thus, can represent an important carbon transfer mechanism to the deep sea. The low latitudes, conversely, represent a source of CO_2 to the atmosphere. Satellite-based ocean chlorophyll records indicate that the global ocean annual NPP has been declining more than 6% since the early 1980s [65]. The reduction in primary production may represent a reduced sink of carbon here via the photosynthetic pathway.

The fate of NPP is heterotrophic respiration (R_h) by herbivores (animals) who consume 10–20% of NPP and respiration of decomposers (microfauna, bacteria, fungi) in soils. The largest fraction of NPP is delivered to the soil as dead organic matter (litter), which is decomposed by microorganisms under release of CO_2, H_2O, nutrients, and a final resistant organic product, humus. Hence, the net ecosystem production NEP is defined by

$$\mathrm{NEP} = \mathrm{NPP} - R_h \tag{2.28}$$

Table 2.8 Loss of *NPP*.

Sink type	Sink flux in 10^{15} g C a^{-1}	References
Biomass burning	~4.0	[66]
Human use	18.7	[67]
VOC emission	1.2	[68]

where R_h is the consumers' respiration. Another part of NPP is lost by fires (biomass burning; see Section 2.1.1), by emission of volatile organic substances (VOC), and through human use (food, fuel, and shelter) (Table 2.8).

NEP finally represents the burial carbon, a large flux at the beginning of the biospheric evolution but nowadays limited at about zero. Most of the NPP goes in litter and is finally mineralized back to CO_2. For the 1997–2004 period, it has been estimated [62] that approximately 95% of the NPP was returned back to the atmosphere via R_h; another 4% (or 2.5 Gt Y a^{-1}) was emitted by biomass burning (see Section 2.1.1) – the remainder consisted of losses from fuel wood collection and subsequent burning.

Historically, global estimates of litter production have ranged from 75 to 135 Gt C a^{-1} (often cited as dry mass or matter, dm, not identical with carbon). In [62] a dry matter carbon content of 45% is used. Steady-state pools of standing litter represent global storage of around 174 Pg C (94 and 80 Pg C in non-woody and woody pools, respectively), whereas the pool of soil C in the top 0.3 m, which turns over on decadal timescales, comprises 300 Pg C. Several estimates [69] suggest values in the middle of this range, from 90 to 100 Gt C a^{-1}, accounting for both above-ground and below-ground litter. Above-ground litter production may be 5–10 Gt C a^{-1} including mainly forest, woodland, and grassland. The global litter pool estimated from the measurement compilation amounts to 136 Gt C [69]. Inclusion of the remaining ecosystems may add c. 25 Gt C, raising the total to c. 160 Gt C. Some additional 150 Gt C is estimated for the coarse woody detritus pool. Global mean steady-state turnover times of litter estimated from the pool and production data range from 1.4 to 3.4 a; mean turnover time from the partial forest/woodland measurement compilation is ~5 a, and turnover time for coarse woody detritus is c. 13 a.

Considering only the terrestrial NPP (~60 Gt C a^{-1}), the biomass energy amounts to c. 30×10^{20} J stored per year (simply calculated from C bonding), which corresponds to 75 TW.

When colonization of the continents by plants did occur, with still reduced oxygen, respiration was not yet significant and large quantities of GPP could be buried. As a result, coal, oil, and natural gas have been formed. Nonetheless, it seems that there are strong reasons to believe that besides a natural process of carbonization of buried former biomass, there were abiogenic processes in the formation of oily and gaseous hydrocarbons, as already remarked in discussing gases occluded and produced from rocks. Now, with little doubt, large carbon inputs to the early Earth from heavy bombardment have to be considered. It is only a question of understanding the geochemical processes that occurred under high temperature and pressure, and over a long timescale, to see how the carbon compounds delivered to Earth have been turned into petroleum (liquid crude oil and long-chain hydrocarbon compounds). The possible formation of methane (CH_4) needs no further comment.

It is however remarkable that various abiogenic hypotheses were proposed long before the presence of evident organic matter from space was proved (e.g. von Humboldt, Mendeleev, Berthelot, Kudryavtsev and his colleagues, as well as Gold, as described in detail in [70], Chapter 3). Currently, there is little direct research on abiogenic petroleum or experimental studies into the synthesis of abiogenic methane. However, several research areas, mostly related to astrobiology and the deep microbial biosphere and serpentinite reactions, continue to provide insight into the contribution of abiogenic hydrocarbons into petroleum accumulations. Similarly, such research is advancing as part of the attempt to investigate the concept of panspermia and astrobiology, specifically using deep microbial life as an analogue for life on Mars.

It is worth noting here that both abiotic (chemical) and biological evolutionary processes are occurring simultaneously at different times and sites. Therefore, arguments such as that biomarkers are indicative of the biological origin of petroleum dissipate when it is accepted that the biosphere is integrated with the geosphere and, hence, chemical and biological processes are overlapping.

Human use of biomass has become a major component of the global biogeochemical cycles (Table 2.9). The use of land for biomass production (e.g. crops) is among the most important pressures on biodiversity. At the same time, biomass is indispensable for humans as food, animal feed, raw material, and energy source. It seems that biomass – with the exception of a source of organic substances and food – will not play any role in the future as an energy source. The global annual food supply is equivalent to c. 3×10^{19} J and will probably increase by a factor of 2–3 in the next 50–100 a. Then it would amount a percentage of 10% to the global NPP ($\sim 5 \times 10^{20}$ J a^{-1}).

Table 2.9 Global biomass use as of the year 2000.

Usable biomass	Amount (Pg C a^{-1})
Global usable biomass	18.7
Destroyed using harvesting	6.6
Used biomass	12.1
Feed	7.0 (58%)
Food	0.8 (7%)
Raw materials	2.4 (20%)
Biofuels	1.2 (10%)
Others	0.7 (5%)

Percentage related to totally used biomass.
Source: Adapted from Krausmann et al. 2008 [67].

2.2.4 Subsurface CO_2 Monitoring

Soils are both the largest terrestrial sinks and the largest terrestrial sources of CO_2. The total global emission of CO_2 from soils is estimated as one of the largest fluxes in the global carbon cycle and has a significant influence on the atmospheric CO_2 levels [71–76]. In soils CO_2 is produced by root respiration, microbial respiration, and oxidation of organic matter. Its concentration typically ranges between 1 and 6 vol%, spatially and temporally strongly varying according to temperature, moisture, soil type, cultivation, vegetation, and land use. Industrial processes, deforestation, and land-use changes have led to the destruction of the natural terrestrial carbon sinks, such as forests, and consequently to higher net carbon fluxes into the atmosphere, resulting in increased greenhouse gas effect. The emission of CO_2 from natural sources, geological repositories, or a leaking gas pipeline represents serious risks in industrial and urban areas [77]. Furthermore, CO_2 is also a vital indicator for the biological activity, soil fertility, and many physiochemical properties of soils [73]. Therefore, reliable detection of CO_2 within the shallow subsurface is required to observe critical gas accumulations before degassing into the atmosphere. A detailed investigation of the underlying processes and dynamics in soils through long-term field studies and continuous monitoring could be helpful for better understanding of the exchanges of the CO_2 stored in and emitted out from terrestrial ecosystems as well as between them and the atmosphere [78].

A wide range of techniques have already been developed and field-tested for soil CO_2 measurements at different temporal and spatial scales, for instance, manual periodic sampling of soil gas from different depths, fixed and dynamic chamber-based shallow subsurface techniques to directly measure CO_2 concentrations in soils periodically or continuously [79], and temporally highly resolved multipoint measurements using small solid-state sensors on field scale [80]. In order to detect or observe degassing structures in large regions, the surveying of the subsurface by geophysical methods and space-integrating gas measurements in the lower and upper atmosphere has been suggested and applied [81]. This type of atmospheric monitoring is adaptable in horizontal and uniform terrain with the aid of techniques such as open path infrared sensors in conjunction with eddy-covariance-based micrometeorological techniques allowing for a non-invasive measurement of CO_2 exchange between atmosphere and terrestrial ecosystems within the atmospheric boundary layer [82, 83]. Whereas manual sampling has tremendous labour and cost repercussions, chamber techniques are subject to uncertainties related to so-called chamber effects as well as suffer from limited spatial resolution [84]. The measurement through solid-state gas sensors is much inadequate to correctly represent spatial variation of CO_2 concentrations in soils. Techniques employed in lower atmosphere (up to 100 m above surface) like eddy-covariance or atmospheric sampling methods may solve limitations regarding spatial constraints of measurement but suffer major inaccuracies when there are unstable atmospheric conditions (like wind, temperature, humidity, etc.) or in the presence of vegetation or complex landscape between measurement heights [80]. Moreover, the eddy-covariance method is mathematically complex, requires a lot of care for setting up and processing data, and must be calibrated using ground-truth measurements in terms of actual concentrations. Numerical simulation shows that monitoring of CO_2 in the subsurface has greater potential to detect and more accurately quantify gas dynamics in heterogeneous ground than above-ground techniques [85].

Considering the large heterogeneity of soil systems and some of the limitations of airborne measurement techniques, there is a need of robust monitoring systems that work efficiently within the subsurface and can gather information on a relevantly large scale and respond rapidly. To suit the advantages of measuring directly and continuously in the subsurface, membrane-based monitoring techniques have been gaining increasing importance [78]. Such a technique, as described in Section 10.4, uses tubular gas-selective membranes as sensors with which it is possible to measure an average gas concentration value over a certain line and to gain scale-dependent insights into the spatial variability of gas behaviour with negligible impact of the sensor on its environment. The measuring tubes can replace a large number of individual sensors, thus reducing the cost for representative measurements. They can also be used advantageously for safety monitoring of CO_2 sequestration sites, abandoned gas wells, gas pipelines, or sewers [86]. Lazik et al. applied membrane-based sensor networks favourably for *in situ* gas monitoring in the subsurface [77, 81, 86, 87]. In [81] they reported a field test with linear membrane-based gas sensors. Fourteen 40 m long sensors were installed in two horizontal nets one above the other within a homogenized soil and scanned from a control station. Before line-sensor installation, a pointlike gas injection port was inserted in a cavern of 25 cm below the lower sensor net. This set-up enabled the comparison of the response of different line sensors and the observation of the lateral and vertical spread of gas. The paper [87] focuses on the early detection of CO_2 leakages in the shallow subsurface with increased lead times to safety warnings. In a field experiment ten 40 m line sensors were arranged in a meander-like fashion horizontally at depths between 18 and 89 cm below ground level with nearly equal vertical distances above each other in a 40 cm wide, 10 m long trench. For leakage simulation, a small gas injection port was installed at a depth of 59 cm below ground level between two line sensors. The experiment demonstrated the applicability of the characteristic length approach in the subsurface as well as for rapid leak detection. Besides a high ratio between the lateral spread and the rise of CO_2 from a leakage, a vertical displacement of the CO_2 accumulation downwards to greater depths, probably driven by density differences, was demonstrated during the test by the whole sensor network.

2.3 Anthropogenic CO_2

Humans are a part of nature and now drive the chemical evolution in the order of geological forces but in much shorter time, i.e. with higher rates. The chemical composition of air is now contributed by both natural and man-made sources. The human problem lies in the interruption of biogeochemical cycles. For that reason, a sustainable development must reorganize global cycles such as the carbon dioxide economy proposed [88].

2.3.1 Biomass Burning

Biomass burning is the oldest known man-made source of carbon into the atmosphere and was responsible for significant local and regional air pollution even before the present era of fossil fuel combustion [89, 90]. Several source categories are included with biomass burning (percentages in parenthesis according to [91]): savanna (35%), forests (21%; tropical 14% and extratropical 7%), agricultural residues (13%), charcoal making and burning (2%), and biofuel burning (29%), totalling 9200 Tg dry mass a^{-1} (based on the late 1990s) (Table 2.10).

The term 'biofuel' used in combination with biomass burning considers not only primary biomass, such as wood, but also peat, savanna grass, and animal excrements (to avoid misunderstanding it should better be renamed as biomass fuel). Biofuel in a technical sense is defined as solid, liquid, or gaseous fuel obtained from living organisms or from metabolic by-products (organic or food waste products). Global biofuel production consists of ethanol (90%) and diesel (10%); the production tripled from 1.7×10^{10} l in 2000 to about 5.6×10^{10} l in 2007 (roughly 50 Tg C a^{-1}), but still accounts for less than 3% of the global transportation fuel supply [94]. This amount is negligible compared with wood burning.

In contrast to open (wild) fires, domestic fires (mainly burning wood from farmland trees, indigenous forests, woodlands, and timber offcuts from plantations; also termed firewood) for cooking and (less importantly) heating are an important source of air pollutants. Little is known about the contribution of domestic fires to global biomass burning emissions. The only estimate is given by Andreae [91], which assessed this source to be ⅓ of the total burning. Three-fourth of the world's population use wood as their main source of energy. Of the global wood fuel production (1996: 1.9×10^{15} g), 50% is used in Asia, 27% in Africa, 10% in southern America and 13% in the rest of the world [95, 96].

Table 2.10 Global estimates of burned biomass and released carbon into the atmosphere.

Source	Burned biomass (Tg dry mass a^{-1})		Total carbon release (Tg C a^{-1})		
	[92]	[66]	[92]	[66]	[93]
Tropical forest	1230	1230–2430	550–1090	570	1260
Savanna	3470	1190–3690	540–1660	1660	1650
Boreal forest	520	280–1620	130–230	130	205
Temperate forest	—	—	—	—	185
Wood fuel	1880	620–1880	280–850	640	—
Charcoal	—	21	30	30	—
Agricultural waste	1360	280–2020	300	910	—
Tundra	4	—	2	—	—
Total[a)]	7660	3625–11665	1800–4740	3940	3300[b)]

a) 9200 Tg dm after [91].
b) 1700–4100 Tg C a^{-1} [54], corresponding to 3–8% of total terrestrial NPP.

The FAO (the *Food and Agriculture Organization* of the United Nations) estimates that wood fuel consumption rose by nearly 80% between 1961 and 1998, slightly trailing world population growth of 92% over the period. There are indications that as much as 2/3 of wood fuel worldwide probably comes from non-forest sources (woodlands, roadside verges, residues, wood recovered, and waste packaging). However, this amount also turns finally in the cycle of harvested forest wood.

Biomass burning is known to be a significant source of many air pollutants, as estimated from a large number of measurements [90, 97, 98] (Table 2.11). Biomass burning caused by fires is an important source of soot, particulate matter, and gases. Generally, there is a distinction between different types of biomass: wild fires (savanna, forest), agricultural burning, wood, and charcoal burning (for cooking and heating), and waste (agricultural residues) burning. Almost all wild biomass burning (90%) is caused by humans (whether careless or intentional), of which 80% happens in the inner-tropical zone and 45% is due to savanna fires [99].

Biomass burning shows large interannual variations: between 1997 and 2004 the variation was estimated from 2.0 $Pg\,C\,a^{-1}$ (minimum) and 3.2 $Pg\,C\,a^{-1}$ (maximum in 1998) with an average of (2460 ± 366) $Pg\,C\,a^{-1}$ [62] (see Table 2.10). The factors influencing the emissions from biomass burning are combustion completeness, fuel loads, moisture, and burned area, integrated over time and space scale of interest. Burned area is usually considered to be the most uncertain parameter in estimates of these emissions (it has become available on a global scale only recently using satellite data). Total carbon emission tracks burning in forested areas (including deforestation fires in the tropics; see Section 2.2.2), whereas burned area is largely controlled by savanna fires that responded to different environmental and human factors.

It is important to note that over a long timescale, biomass burning is neutral in the atmospheric CO_2 budget, but in the last 200 years a steadily increasing amount of biomass burning has been occurring [92]. It is likely that only

Table 2.11 Biomass burning emissions in $Tg\,C\,a^{-1}$ (related to the late 1990s).

Substance	Forest and wood		Savanna burning
	Total[a]	Forest fire	
CO_2	2000	850	1400
CO	260	85	85
CH_4	25	9	5
NMVOC	60	22	20
Particulate C	40	14	12
Total C	2386	980	1522

a) Includes charcoal making and burning (amounts to c. 2% of total burned carbon).

Source: Adapted from Andreae 2004 [91].

savanna biomass burning lies within a timely closed carbon cycle because of rapid bush and grass growth in the next growing season. Therefore, forest burning contributes to the increase of atmospheric CO_2. The amount of burned forest wood, however, is in the order of global coal consumption (2–4 Tg a^{-1}; see Figure 2.5). From 1850 to 1980, 90–120 Gt of CO_2 was released into the atmosphere from tropical forest fires (Figure 2.6). Comparatively, during the same time period, an estimated 165 Gt of CO_2 was added to the atmosphere by industrial nations through the burning of coal, oil, and gas [102]. There have been major changes in the regional distribution of emissions from fires in the last century, as a consequence of (i) increased burning in tropical savannas and

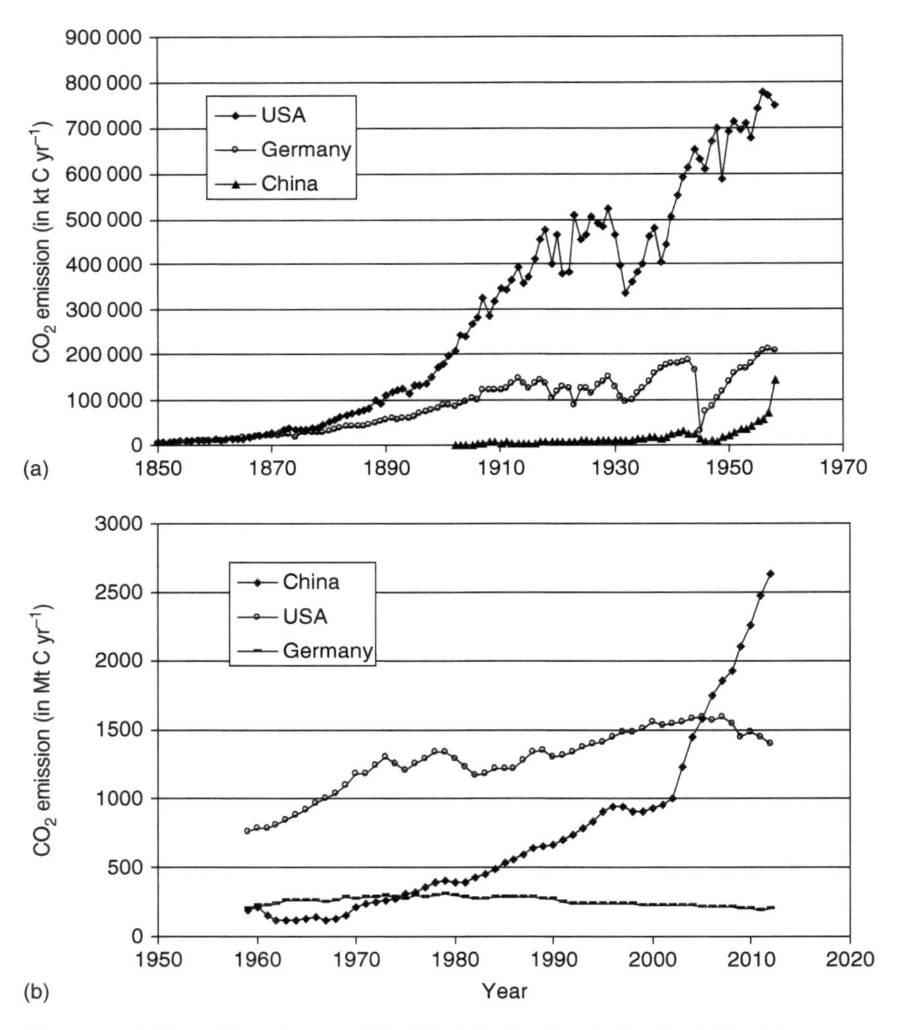

Figure 2.5 (a, b) World production of fossil fuels in Mt oil equivalent (toe). The IEA (International Energy Agency) defines 1 toe = 41.868 GJ or 11.630 kWh. Source: Data from [100].

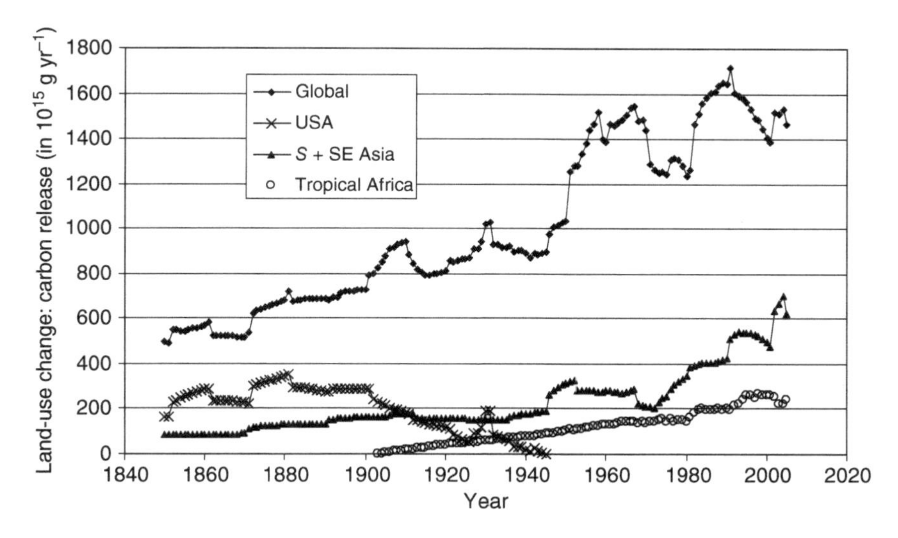

Figure 2.6 Carbon release due to land-use change. Source: After [101].

(ii) a switch of emissions from temperate and boreal forests towards the tropics. At the end of the twentieth century, the annual volume of biomass burned was about 5000 Tg C a^{-1}, compared with about 2000 Tg C a^{-1} at the beginning of the century. Therefore, biomass burning contributes significantly to the greenhouse effect, estimated to be 25% [89]. Today, an estimated 3 Gt of carbon is released into the atmosphere each year due to coal burning (about 7 Gt including all fossil fuels). Burning of forests and wood contributes another 2 Gt of carbon per year.

2.3.2 Land-Use Change and Deforestation

Land use is the human modification of natural environment or wilderness into built environment such as fields, pastures, and settlements. Land-use change represents the anthropogenic replacement of one land-use type by another, e.g. forest to cultivated land (or the reverse), as well as subtle changes of management practices within a given land-use type, e.g. intensification of agricultural practices, both of which are affecting 40% of the terrestrial surface [103]. Agriculture has been the greatest force of land transformation on Earth [104]. A global increase of cropland area from 265×10^{10} m^2 in 1700 to 1471×10^{10} m^2 in 1990 and a more than six times increase of pasture area from 524×10^{10} m^2 to 3451×10^{10} m^2 have been estimated [105]. Another estimate [106] is similar; for the year 1750, the global area of cropland and pasture has been assessed to be $(7.9–9.2) \times 10^{12}$ m^2 (corresponding to 6–7% of the global land) and for the year 1990 to be $(45.7–51.3) \times 10^{12}$ m^2 (35–39 % of global land: 131×10^{12} m^2 excluding Antarctica). This land-use change was partly due to deforestation; the forest area decreased in this time period only by 11×10^6 km^2 (9% of global land) [107]. A recent estimate by D'Annunzio et al. [108] shows that the total forest area in 2010 was $3.89 \cdot 10^{13}$ m^2 ($4.00 \cdot 10^{13}$ m^2 and $3.95 \cdot 10^{13}$ m^2 in 1990 and 2000, resp.), which is around 30% of the global land area. Between 1990 and

2010, there was a net reduction in the global forest area of around $5.3 \cdot 10^{10}$ m^2 yr^{-1}. Worldwide, the gross reduction in forest land use caused by deforestation and natural disasters over the 20 year time period ($15.5 \cdot 10^{10}$ m^2 yr^{-1} from which 84% is attributed to tropical domains) was partially offset by gains in forest area through afforestation and natural forest expansion ($10.2 \cdot 10^{10}$ m^2 yr^{-1}). Interestingly, the ratio 6.7 of the world population increase from 1750 to 1990 is only somewhat higher than the increase of the agricultural area of 5.8 (5.0–6.5) from the data cited above, suggesting that the crop yield per area only slightly increased globally.

The major effect of land use since 1750 has been deforestation of temperate regions. Of particular concern is deforestation, where logging or burning is followed by the conversion of the land to agriculture or other land uses. Even if some forests are left standing, the resulting fragmented landscape typically fails to support many species that previously existed there. It is estimated that roughly 50% of the original forest (existing c. 8000 years ago) has been lost [94]. In addition to the intrusion of humans and their activities throughout the Earth's land area, degradation of soil has emerged as a critical agricultural problem [109]. Erosion on agricultural land is estimated to be 75 times greater than that occurring in natural forest [110]. About 50% of the global land is devoted to agriculture. Of this, about one-third (15×10^{12} m^2) is planted with crops and two-thirds is pasture land. As a result of erosion, during the last 40 years, about 30% of world's arable land has become unproductive (roughly 200×10^{10} m^2). Throughout the world, current erosion rates are greater than ever before. Further, people in developing countries have been forced to burn crop residues for cooking and heating, and this exposure of the soil to wind and rainfall energy intensifies soil erosion as much as 10 times. Most of the additional cropland needed yearly to replace lost land now comes from the world's forest areas (roughly 4000×10^{10} m^2 in 2000). Forests cover about 30% of the land surface and hold almost half of the world's terrestrial carbon; if only vegetation is considered (ignoring soils), forests hold about 75% of the living carbon [101]. Per unit area, forests hold 20–50 times more carbon in their vegetation than the ecosystem that generally replaces them. The above-ground living biomass has been estimated to be about 200 t C ha^{-1} for tropical and temperate forest and 60 t C ha^{-1} in boreal forests [111]. The global estimate results in 60 t C ha^{-1} [112]. According to different scenarios [113], forest area in industrialized regions will increase between 2000 and 2050 by about 60–230 million ha. At the same time, the forest area in the developing regions will decrease by about (200–490) 10^6 ha. According to FAO [96], a forest is defined as an ecosystem that is dominated by trees (defined as perennial woody plants taller than 5 m at maturity), where the tree crown (or equivalent stocking level) exceeds 10% and the area is larger than 0.5 ha. The term includes all types of forests but excludes stands of trees established primarily for agricultural production and trees planted on agroforestry systems.

Figure 2.6 shows that deforestation is a significant CO_2 source to the atmosphere. It is important to note that carbon release is due to two principal processes: (i) by the vegetation replacement (mainly by fires – biomass burning) and (ii) by release from the soil's carbon pool after land-use change. About the same amount of carbon is emitted from soil into the atmosphere, a considerable amount additionally to other anthropogenic sources. The historic loss of soil carbon (to be emitted as CO_2 to the atmosphere) from cultivated cropland soils of the world ranges from 41 to 55 Pg C [63]. The global organic carbon amount in soil ranges from 1500 to 2000 Pg C (to 1 m depth) and is in various forms, from recent plant litter to charcoal and very old humified compounds. Inorganic carbon (carbonate) amounts to 800–1000 Pg C.

The estimated net flux ($\sim$1 Gt C a^{-1}) suggests that the global forest is a sink for atmospheric CO_2. However, this value, based on forest inventories, is only a fraction of the sinks inferred from atmospheric data and models, estimated to be (2.1 $\pm$ 0.8) Pg C a^{-1}, referred to the northern mid-latitudes and suggesting that more than half of the northern terrestrial carbon sink is in non-forest ecosystems [53]. However, the uncertainties are large.

2.3.3 Fossil Fuel Burning

Fossil fuels, coal, crude oil (petroleum), and natural gas are non-renewable sources of energy and materials. Other fuels are under investigation, such as bituminous sands and oil shale. The difficulty is that they need expensive processing before being used.

Still more than 85% of the total primary energy supply is provided by fossil fuels. It seems that in the next 20 years all growth in energy carriers will be linear: oil (1.25% a^{-1}), coal (3.25% a^{-1}), gas (2.25% a^{-1}), renewable (6.75% a^{-1}), and nuclear (1.75% a^{-1}). From the world electricity production of 18.9 GWh in 2006, 67% is due to fossil fuels, but it corresponds to a total fossil fuel energy supply of about 41.1 GWh. This means that the global conversion efficiency (not to be confused with the efficiency of power plants) is only 30%. From the electricity production of 18.9 GWh, only 16.4 GWh was the world net electricity consumption, i.e. 13% of electricity produced was lost before use.

It is obvious that increasing efficiency leads to reduced fuel consumption and specific emissions. All fossil fuels consumed in 2006 provided about 9.1 Gt of carbon; a mere 9% were converted into materials, such as chemicals, plastics, elastomers, etc., which delays its return as CO_2 into the atmosphere after its final combustion. Therefore, more than 90% of fossil fuel carbon is released as CO_2 (8.2 Gt in 2006) into the air. Around 30% of fossil fuel carbon is used as engine fuels (transport) and the same amount (30%) for heat and industrial processing (10%, or about 1 Gt C for steel and cement production). The different uses of fossil fuel carbon also show that in future carbon cannot be fully replaced by renewable energy sources. About 1.5–2 Gt C is needed for non-energetic (materials) and process-specific use [100].

About 95% of all anthropogenic industrial CO_2 emission is caused by fossil fuel use; 4% is from cement production (limestone burning and CO_2 release from past carbonate sediments). China is the world's largest hydraulic cement producer. In 2006 China produced over 1.2×10^9 t of hydraulic cement, or roughly 47% of the world's production. Emissions from cement production account for 9.8% of China's total industrial CO_2 emissions in 2006.

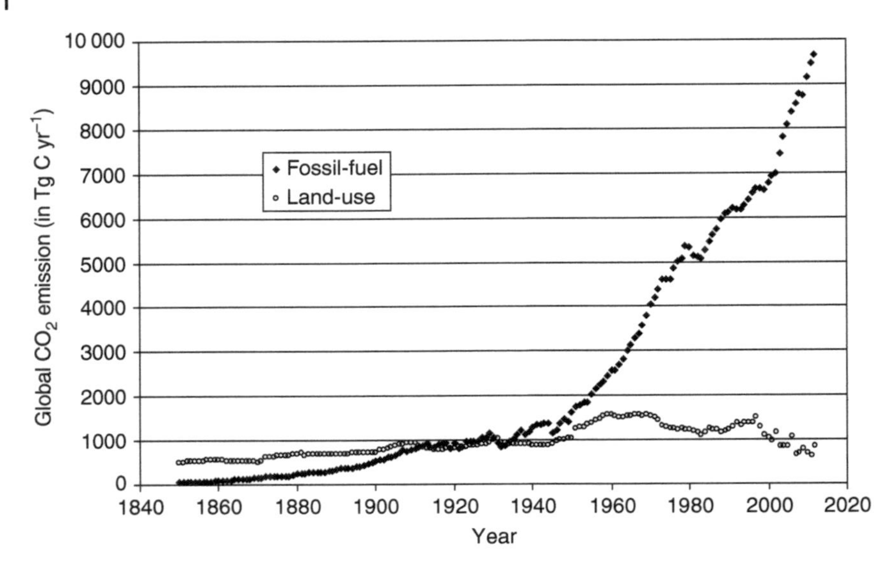

Figure 2.7 Global CO_2 emission trend; fossil fuel data from [114], land-use change data from [94].

It lies in the nature of the combustion processes of fossil and biofuels (gaseous, liquid, and solid) that all carbon will be oxidized into CO_2 to obtain a maximum heat of reaction. Naturally, depending on the completeness of the oxidation process, unburned carbon (BC – black carbon), volatile organics, and CO are emitted too. Besides wood, coal was the dominant carbon carrier for centuries. The rise of coal-released CO_2 after the Industrial Revolution around 1850 is expressive (Figure 2.7). Only after 1900 liquid fuels (dominant for transport) became significant and after 1950 natural gas was substituted for coal in stationary combustion processes (also due to reduced SO_2 emissions). In 2005 the percentage of the global CO_2 emission was about 39% for liquids, 38% for solids, and 18% for gas, not very much changed since 1980 [115]. According to reported energy statistics, coal production and use in China has increased since the early 1960s by a factor of 10. As a result, Chinese fossil fuel CO_2 emissions have grown remarkable by 79.2% since 2000 alone. At 1.66 billion metric tons of carbon in 2006, China has surpassed the United States as the world's largest emitter of CO_2 due to fossil fuel use and cement production.

Per capita CO_2 emission is particularly interesting (the regional differences range over more than three orders of magnitude; Figure 2.8): from 1950 to 1980 it increased in a linear way from 0.64 to 1.20 t per capita and year, whereas in 1980–2005 this value remained approximately constant at (1.15 ± 0.05) t per capita and year. Despite the fact that this figure represents only socio-economic and political relationships, it can be used as an abatement target only for larger regions, e.g. continents, when producer and consumer are balanced.

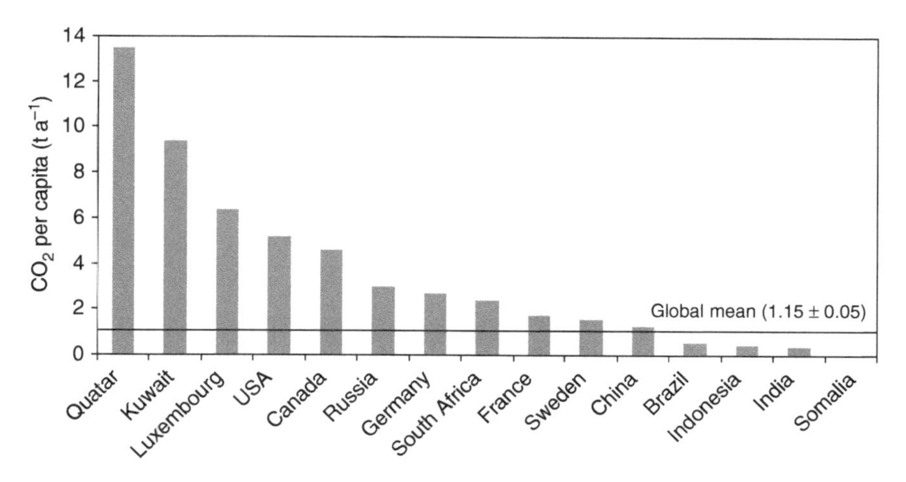

Figure 2.8 Per capita CO_2 emission for selected countries; data from [115].

References

1 Chase, M.W. Jr. (1998). NIST-JANAF thermochemical tables. 4th edition. *J. Phys. Chem. Ref. Data, Monogr.* 9: 643.

2 Yau, A.W. and Pritchard, H.O. (1979). Unimolecular reactions of N_2O and CO_2 at high pressure. *Can. J. Chem.* 57 (13): 1731–1742.

3 Lietzke, M.H. and Mullins, C. (1981). The thermal decomposition of carbon dioxide. *J. Inorg. Nucl. Chem.* 43 (8): 1769–1771.

4 Glockler, G. (1958). Carbon-oxygen bond energies and bond distances. *J. Phys. Chem.* 62 (9): 1049–1054.

5 Pocker, Y., Davison, B.L., and Deits, T.L. (1978). Decarboxylation of mono-substituted derivates of carbonic acid. Comparative studies of water- and acid-catalyzed decarboxylation of sodium alkyl carbonates in H_2O and D_2O. *J. Am. Chem. Soc.* 100 (11): 3564–3567.

6 Mikkelsen, M., Jorgensen, M., and Krebs, F.C. (2010). The teraton challenge. A review of fixation and transformation of carbon dioxide. *Energy Environ. Sci.* 3 (1): 43–81.

7 Palmer, D.A. and van Eldik, R. (1983). The chemistry of metal carbonato and carbon dioxide complexes. *Chem. Rev.* 83 (6): 651–731.

8 Omae, I. (2012). Recent developments in carbon dioxide utilization for the production of organic chemicals. *Coord. Chem. Rev.* 256 (13–14): 1384–1405.

9 Metz, B., Davidson, O., de Coninck, H.C. et al. (eds.) (2005). *IPCC Special Report on Carbon Dioxide Capture and Storage.* Prepared by Working Group III of the Intergovernmental Panel on Climate Change, 442. Cambridge, New York: Cambridge University Press.

10 Wilcox, J. (2012). *Carbon Capture.* Dordrecht, Heidelberg, London: Springer Science + Business Media New York.

11 Aresta, M. (ed.) (2003). *Carbon Dioxide Recovery and Utilization*. Boston, MA, London: Kluwer Academic Publishers Dordrecht.

12 Roller, D.H.D. and Thilorier, M. (1952). Thilorier and the first solidification of a "permanent" gas (1835). *Isis* 43 (2): 109–113. https://doi.org/10.1086/349402.

13 McMillan, P.F. (2006). Solid-state chemistry: a glass of carbon dioxide. *Nature* 441 (7095): 823.

14 Vargaftik, N.B., Vinogradov, Y.K., and Yargin, V.S. (eds.) (1996). *Handbook of Physical Properties of Liquids and Gases*, 3e. New York: Begell House Publishers.

15 Hodgeman, C.D., Weast, R.C., and Selby, S.M. (eds.) (1963). *CRC Handbook of Chemistry and Physics*, 44e, 2560–2561. Boston, MA: Chemical Rubber Publishing Company.

16 Fenghour, A., Wakeham, W.A., and Vesovic, V. (1998). The viscosity of carbon dioxide. *J. Phys. Chem. Ref. Data* 27 (1): 31–44.

17 Muratov, G.N. and Skripov, V.P. (1982). Oberflächenspannung von Kohlenstoffdioxyd (surface tension of carbon dioxide). *Teplofiz. Vys. Temp.* 20: 596–598 (in Russian).

18 Zuckerwar, A.J. (2002). *Handbook of the Speed of Sound in Real Gases*, vol. III, 35. San Diego, CA, London: Academic Press.

19 Folke, M. and Hök, B. (2008). A new capnograph based on an electro acoustic sensor. *Med. Biol. Eng. Comput.* 46 (1): 55–59.

20 Angus, S., Armstrong, B., and de Reuck, K.M. (1976). *International Thermodynamic Tables of the Fluid State – 3 Carbon Dioxide*. New York: Pergamon Press.

21 Morrison, G. (1981). Effect of water on the critical points of carbon dioxide and ethane. *J. Phys. Chem.* 85 (7): 759–761.

22 Giauque, W.F. and Egan, C.J. (1937). Carbon dioxide. The heat capacity and vapor pressure of the solid. The heat of sublimation. Thermodynamic and spectroscopic values of the entropy. *J. Chem. Phys.* 5 (1): 45–54. https://doi.org/10.1063/1.1749929.

23 MegaWatSoft Inc. (2009–2016). http://www.carbon-dioxide-properties.com (retrieved 28 March 2017).

24 Vesovic, V., Wakeham, W.A., Olchowy, G.A. et al. (1990). The transport properties of carbon dioxide. *J. Phys. Chem. Ref. Data* 19 (3): 763–808.

25 Chase, M.W. Jr. (1998). NIST-JANAF thermochemical tables, 4th edition. *J. Phys. Chem. Ref. Data Monogr.* 9: 1–1951.

26 Jäger, A. and Span, R. (2012). Equation of state for solid carbon dioxide based on the Gibbs free energy. *J. Chem. Eng. Data* 57 (2): 590–597.

27 Stephenson, R. and Malanowski, S. (1987). *Handbook of the Thermodynamics of Organic Compounds*. New York: Elsevier Science Publishing. doi: https://doi.org/10.1007/978-94-009-3173-2.

28 Weiss, R.F. (1974). Carbon dioxide in water and seawater: the solubility of a non-ideal gas. *Mar. Chem.* 2 (3): 203–215.

29 Cussler, E.L. (1997). *Diffusion Mass Transfer in Fluid Systems*, 102, p. 112. Cambridge: Cambridge University Press.

30 Schmidt, J.W. and Moldover, M.R. (2003). Dielectric permittivity of eight gases measured with cross capacitors. *Int. J. Thermophys.* 24 (2): 375–403.
31 Martin, P.E. and Barker, E.F. (1932). The infrared absorption spectrum of carbon dioxide. *Physiol. Rev.* 41 (3): 291–303.
32 Farquhar, J. and Johnston, D.T. (2008). The oxygen cycle of the terrestrial planets: insights into the processing and history of oxygen in surface environments. *Rev. Mineral. Geochem.* 68 (1): 463–492.
33 Zinner, E., Amari, S., Wopenka, B., and Lewis, R.S. (1995). Interstellar graphites in meteorites: isotopic compositions and structural properties of single graphite grains from Murchison. *Meteorit. Planet. Sci.* 30 (2): 209–226.
34 Briggs, M.H. and Mamikunian, G. (1963). Organic constituents of the carbonaceous chondrites. *Space Sci. Rev.* 1 (4): 647–682.
35 Botta, O. and Bada, J.L. (2002). Extraterrestrial organic compounds in meteorites. *Surv. Geophys.* 23 (5): 411–467.
36 Schlesinger, W.H. (1997). *Biogeochemistry – An Analysis of Global Change.* San Diego, CA: Academic Press.
37 Pidwirny, M. (2008). The carbon cycle. In: *Encyclopedia of Earth* (ed. C.J. Cleveland). Washington, DC: Environmental Information Coalition, National Council for Science and the Environment (NCSE). http://www.eoearth.org/article/Carbon_cycle (retrieved 28 March 2017).
38 Pédro, G. (2007). *Cycles biogéochimiques et écosystèmes continentaux* (biogeochemical cycles and continental ecosystems). Paris: Académie des sciences France (in French).
39 Kvenvolden, K.A. and Lorenson, T.D. (2001). The global occurrence of natural gas hydrate. In: *Natural Gas Hydrates: Occurrence, Distribution, and Dynamics*, AGU Geophysical Monograph Series, vol. 124 (ed. C.K. Paull and W.P. Dillon), 1–23. Washington, DC: American Geophysical Union.
40 Clarke, F.W. (1920). *The Data of Geochemistry.* Washington, DC: Government Printing Office.
41 Murakami, M., Hirose, K., Yurimoto, H. et al. (2002). Water in earth's lower mantle. *Science* 295: 1885–1887.
42 Nakamura, T. (2005). Post-hydration thermal metamorphism of carbonaceous chondrites. *J. Miner. Petrol. Sci.* 100 (5): 260–272.
43 Mendeleev, D. (1877). L'origine du petrole. (origin of petrol) Revue Scientifique, 2e Ser. VIII, 409–416 (in French).
44 Kudryavtsev, N.A. (1959). Geological proof of the deep origin of petroleum. *Trudy Vsesoyuz. Neftyan. Nauch. Issledovatel Geologoraz Vedoch. Inst.* 132: 242–262 (in Russian).
45 Gold, T. (1999). *The Deep Hot Biosphere.* Berlin: Springer-Verlag.
46 Möller, D. (2014). *Chemistry of the Climate System*, 2e. Berlin, New York: De Gruyter.
47 Wilson, E.H. and Atreya, S.K. (2000). Sensitivity studies of methane photolysis and its impact on hydrocarbon chemistry in the atmosphere of titan. *J. Geophys. Res.* 105 (E8): 20263–20273.
48 Kasting, J.F. and Siefert, J.L. (2002). Life and the evolution of earth's atmosphere. *Science* 296 (5570): 1066–1068.

49 Brimblecombe, P. (1996). *Air Composition and Chemistry*. Cambridge: Cambridge University Press.
50 Schidlowski, M., Eichmann, R., and Junge, C.E. (1975). Precambrian sedimentary carbonates: carbon and oxygen isotope geochemistry and implications for the terrestrial oxygen budget. *Precambrian Res.* 2 (1): 1–69.
51 Berner, R.A. (2003). Overview. The long-term carbon cycle, fossil fuels and atmospheric composition. *Nature* 426 (6964): 323–326.
52 Ebelmen, J.J. (1845). Sur les produits de la décomposition d'espèces minérales de la familie des silicates (the decomposition products of silicate species). *Ann. Mines* 7: 3–66 (in French).
53 Houghton, R.A. (2005). Aboveground forest biomass and the global carbon balance. *Global Change Biol.* 11 (6): 945–958.
54 Denman, K.L., Brasseur, G., Chidthaisong, A. et al. (2007). Couplings between changes in the climate system and biogeochemistry. In: *Climate Change 2007: The Physical Science Basis. Contribution of Working Group I to the Fourth Assessment Report of the Intergovernmental Panel on Climate Change* (ed. S. Solomon, D. Qin, M. Manning, et al.), 499–588. Cambridge: Cambridge University Press.
55 Nieder, R. and Benbi, D.K. (2008). *Carbon and Nitrogen in the Terrestrial Environment*. Berlin: Springer-Verlag.
56 Sabine, C.L., Feely, R.A., Gruber, N. et al. (2004). The oceanic sink for anthropogenic CO_2. *Science* 305 (5682): 367–371.
57 Warneck, P. (2000). *Chemistry of the Natural Atmosphere*. New York: Academic Press.
58 Scurlock, J.M. and Olson, R.J. (2002). Terrestrial net primary productivity – a brief history and a new worldwide database. *Environ. Rev.* 10 (2): 91–109.
59 Cramer, W. and Field, C.B. (eds.) (1999). The potsdam NPP model intercomparison. *Global Change Biol.* 5 (Suppl. 1): 1–76.
60 Potter, C.S., Randerson, J.T., Field, C.B. et al. (1993). Terrestrial ecosystem production: a process model based on global satellite and surface data. *Global Biogeochem. Cycles* 7 (4): 811–841.
61 Staley, J.T. and Orians, G.H. (2002). Evolution and the biosphere. In: *Earth System Science*, International Geophysics Science, vol. 72 (ed. M.K.J. Jacobson, R.J. Charlson, H. Rodhe, et al.), 29–61. San Diego, CA: Academic Press.
62 van der Werf, G.R., Randerson, J.T., Giglio, L. et al. (2006). Interannual variability of global biomass burning emissions from 1997 to 2004. *Atmos. Chem. Phys.* 6 (11): 3423–3441.
63 Ramakrishna, R.N., Keeling, C.D., Hashimoto, H. et al. (2003). Climate-driven increases in global terrestrial net primary production from 1982 to 1999. *Science* 300 (5625): 1560–1563.
64 Townsend, C.R., Begon, M., and Harper, J.L. (2008). *Essentials of Ecology*, 3e. Malden, MA: Blackwell Publishers.
65 Gregg, W.W., Conkright, M.E., Ginoux, P. et al. (2003). Oceanic primary production and climate: global decadal change. *Geophys. Res. Lett.* 30 (15): 1809. 4 pp. doi: https://doi.org/10.1029/2003GL016889.

66 Malingreau, J.-P. and Zhuang, Y.H. (1998). Biomass burning: an ecosystem process of global significance. In: *Asian Change in the Context of Global Climate Change* (ed. J. Galloway and J. Melillo), 101–127. Cambridge: Cambridge University Press.

67 Krausmann, F., Erb, K.-H., Gingrich, S. et al. (2008). Global patterns of socioeconomic biomass flows in the year 2000: a comprehensive assessment of supply, consumption and constraints. *Ecol. Econ.* 65 (3): 471–487.

68 Guenther, A., Hewitt, C.N., Erickson, D. et al. (1995). A global model of natural volatile organic compound emissions. *J. Geophys. Res.* 100 (D5): 8873–8892.

69 Matthews, E. (1997). Global litter production, pools, and turnover times: estimates from measurement data and regression models. *J. Geophys. Res.* 102 (D15): 18771–18800.

70 Glasby, G.P. (2006). Abiogenic origin of hydrocarbons: an historical overview. *Resour. Geol.* 56 (1): 83–96.

71 Raich, J.W. and Schlesinger, W.H. (1992). The global carbon dioxide flux in soil respiration and its relationship to vegetation and climate. *Tellus B* 44 (2): 81–99.

72 Eswaran, H., Den Berg, V., and Reich, P. (1993). Organic carbon in soils of the world. *Soil Sci. Soc. Am. J.* 57 (1): 192–194.

73 Raich, J.W. and Potter, C.S. (1995). Global patterns of carbon dioxide emissions from soils. *Global Biogeochem. Cycles* 9 (1): 23–36.

74 Mosier, R.A. (1998). Soil processes and global change. *Biol. Fertil. Soils* 27 (3): 221–229.

75 Houghton, R.A. (1999). The annual net flux of carbon to the atmosphere from changes in land use 1850–1990. *Tellus B* 51 (2): 298–313.

76 Schlesinger, W.H. and Andrews, J.A. (2000). Soil respiration and the global carbon cycle. *Biogeochemistry* 48 (1): 7–20.

77 Lazik, D. and Ebert, S. (2012). Improved membrane-based sensor network for reliable gas monitoring in the subsurface. *Sensors* 12: 17058–17073.

78 Sood, P. (2016) Concept-based improvements of membrane based linear gas sensors for the measurement of CO_2 concentrations in wet soils. Master thesis. Karlsruhe, Germany: Karlsruhe University of Applied Sciences.

79 Norman, J.M., Garcia, R., and Verma, S.B. (1992). Soil surface CO_2 fluxes and the carbon budget of grassland. *J. Geophys. Res. Atmos.* 97 (D-17): 18845–18853.

80 Tang, J., Baldocchi, D.D., Qi, Y., and Xu, L. (2003). Assessing soil CO_2 efflux using continuous measurements of CO_2 profiles in soils with small solid-state sensors. *Agric. For. Meteorol.* 118 (3–4): 207–220.

81 Lazik, D. and Ebert, S. (2013). First field test of linear gas sensor net for planar detection of CO_2 leakages in the unsaturated zone. *Int. J. Greenhouse Gas Control* 17: 161–169.

82 Janssens, I.A., Kowalski, A.S., and Ceulemans, R. (2001). Forest floor CO_2 fluxes estimated by eddy covariance and chamber-based model. *Agric. For. Meteorol.* 106 (1): 61–69.

83 Madsen, R., Xu, L., Claassen, B., and McDermitt, D. (2009). Surface monitoring method for carbon capture and storage projects. *Energy Procedia* 1: 2161–2168.

84 Mosier, A.R. (1990). Gas flux measurement techniques with special reference to techniques suitable for measurements over large ecologically uniform areas. In: *Soils and the Greenhouse Effect* (ed. A.F. Bouwman), 289–301. Chichester, West Sussex: Wiley.

85 Lewicki, J.L. and Oldenburg, C.M. (2005). Near-surface CO_2 monitoring and analysis to detect hidden geothermal systems. In: *Proceedings 13th Workshop on Geothermal Reservoir Engineering*. Stanford, CA: Stanford University.

86 Lazik, D., Ebert, S., Leuthold, M. et al. (2009). Membrane based measurement technology for in situ monitoring of gases in soil. *Sensors* 9 (2): 756–767.

87 Lazik, D., Ebert, S., Bartholmai, M., and Neumann, P.P. (2016). Characteristic length measurement of a subsurface gas anomaly - a monitoring approach for heterogeneous flow path distributions. *Int. J. Greenhouse Gas Control* 47: 330–341.

88 Möller, D. (2012). SONNE: solar-based man-made carbon cycle, and the carbon dioxide economy. *Ambio* 41 (4): 413–419.

89 Andreae, M.O. (1991). Biomass burning: its history, use, and distribution and its impact on environmental quality and global climate. In: *Global Biomass Burning: Atmospheric, Climatic, and Biospheric Implications* (ed. J.S. Levine), 3–27. Cambridge: MIT Press.

90 Andreae, M.O. and Merlot, P. (2001). Emission of trace gases and aerosols from biomass burning. *Global Biogeochem. Cycle* 15 (4): 955–966.

91 Andreae, M.O. (2004). Assessment of global emissions from vegetation fires. *Int. For. Fire News* 31 (1): 112–121.

92 Dignon, J. (1995). Impact of biomass burning on the atmosphere. In: *Ice Core Studies of Global Biogeochemical Cycles*, NATO ASI Series, vol. 30 (ed. R.J. Delmas), 299–346. Berlin: Springer-Verlag.

93 Mouillot, F., Narasimha, A., Balkanski, Y. et al. (2006). Global carbon emissions from biomass burning in the 20th century. *Geophysical Research Letters* 33: L01801. https://doi.org/10.1029/2005GL024707.

94 Worldwatch Institute (2006). *Biofuels for Transportation. Global Potential and Implications for Sustainable Agriculture and Energy in the 21st Century*, 38 pp. Washington, DC: Worldwatch Institute.

95 FAO (2001). *Global Forest Resources Assessment 2000*. Rome: UN Food and Agriculture Organization.

96 FAO (2003). *State of the World's Forests 2003*, 151 pp. Rome: UN Food and Agriculture Organization. http://www.fao.org/DOCREP/005/Y7581E/Y7581E00.HTM (retrieved 28 March 2017.

97 Simoneit, B.R.T. (2002). Biomass burning – a review of organic tracers for smoke from incomplete combustion. *Appl. Geochem.* 17 (3): 129–162.

98 Koppmann, R., von Czapiewski, K., and Reid, J.S. (2005). A review of biomass burning emissions, Part I: Gaseous emissions of carbon monoxide, methane, volatile organic compounds, and nitrogen containing compounds.

Atmos. Chem. Phys. Discuss. 5: 10455–10516. https://doi.org/10.5194/acpd-5-10455-2005.

99 Cachier, H. (1998). Carbonaceous combustion aerosol. In: *Atmospheric Particles* (ed. R.M. Harrison and R. van Grieken), 295–348. Chichester: Wiley.

100 EIA (2009). *International Energy Outlook 2009*. United States Energy Information Administration. http://www.worldenergyoutlook.org/media/weowebsite/2009/WEO2009.pdf (retrieved March 28, 2017).

101 Houghton, R.A. (2005). Tropical deforestation as a source of greenhouse gas emission. In: *Tropical Deforestation and Climate Change* (ed. P. Moutinho and S. Schwartzman), 13–21. Belém, BR: Amazon Institute for Environmental Research.

102 Parkinson, C.L. (1997). *Earth from Above: Using Color-Coded Satellite Images to Examine the Global Environment*, 192 pp. Sausalito, CA: University Science Books.

103 Fischlin, A., Midgley, G.F., Price, J. et al. (2007). Ecosystems, their properties, goods, and services. In: *Climate Change: Impacts, Adaption and Vulnerability. Contribution of Working Group II to the Fourth Assessment Report of the Intergovernmental Panel of Climate Change* (ed. M.L. Parry, O.F. Canziani, J.P. Palutikof, et al.), 211–272. Cambridge: Cambridge University Press.

104 Lambin, E.F. and Geist, H.J. (2006). *Land-Use and Land-Cover Change: Local Processes and Global Impacts (Global Change – The IGBP)*. Berlin: Springer-Verlag.

105 Goldewijk, K.K. (2001). Estimating global land use change over the past 300 years: the HYDE database. *Global Biogeochem. Cycle* 15 (2): 417–433.

106 Forster, P., Ramaswamy, V., Artaxo, P. et al. (2007). Changes in atmospheric constituents and in radiative forcing. In: *Climate Change 2007: The Physical Science Basis. Contribution of Working Group I to the Fourth Assessment Report of the Intergovernmental Panel on Climate Change* (ed. S. Solomon, D. Qin, M. Manning, et al.), 129–234. Cambridge and New York: Cambridge University Press.

107 IPCC (2007). *Climate Change 2007: The Physical Science Basis. Contribution of Working Group I to the Fourth Assessment Report of the Intergovernmental Panel on Climate Change* (ed. S. Solomon, D. Qin, M. Manning, et al.), 24–26. Cambridge and New York: Cambridge University Press.

108 D'Annunzio, R., Lindquist, E.J., MacDicken, K.G. (2017). Global forest land-use change from 1990 to 2010: an update to a global remote sensing survey of forests. Forest Resources Assessment Working Paper 187. FAO, Rome, 14 pp.

109 Pimentel, D. and Kounang, N. (1998). Ecology of soil erosion in ecosystems. *Ecosystems* 1 (5): 416–426.

110 Pimentel, D. and Pimentel, M.H. (2007). *Food, Energy, and Society*, 3e. Boca Raton, FL: CRC Press.

111 Keith, H., Mackey, B.G., and Lindenmayer, D.B. (2009). Re-evaluation of forest biomass carbon stocks and lessons from the world's most carbon-dense forest. *Proc. Natl. Acad. Sci. U.S.A.* 106 (28): 11635–11640.

112 Kindermann, G.E., McCallum, I., Fritz, S., and Obersteiner, M. (2008). A global forest growing stock, biomass and carbon map based on FAO statistics. *Silva Fenn.* 42 (3): 387–396.

113 Nabuurs, G.J., Masera, O., Andrasko, K. et al. (2007). Forestry. In: *Climate Change 2007: Mitigation. Contribution of Working Group III to the Fourth Assessment Report of the Intergovernmental Panel on Climate Change* (ed. B. Metz, O.R. Davidson, P.R. Bosch, et al.), 541–584. Cambridge and New York: Cambridge University Press.

114 Boden, T.A., Marland, G., and Andres, R.J. (2013). *Global, Regional, and National Fossil-Fuel CO_2 Emissions, Carbon Dioxide Information Analysis Center*. Oak Ridge, TN: Oak Ridge National Laboratory, U.S. Department of Energy. https://doi.org/10.3334/CDIAC/00001_V2013.

115 Boden, T.A., Marland, G., and Andres, R.J. (2009). *Global, Regional, and National Fossil-Fuel CO_2 Emissions, Carbon Dioxide Information Analysis Center*. Oak Ridge, TN: Oak Ridge National Laboratory, U.S. Department of Energy. https://doi.org/10.3334/CDIAC/00001_V2009.

Part II

Principles of Carbon Dioxide Sensors and Measuring Methods

3

Analytical Methods for the Detection of Gaseous CO_2

Gerald Gerlach[1], Armin Lambrecht[2], and Wolfram Oelßner[3]

[1] *Technische Universität Dresden, Faculty of Electrical and Computer Engineering, Institute of Solid-State Electronics, 01062 Dresden, Germany*
[2] *Fraunhofer Institute for Physical Measurement Techniques IPM, Heidenhofstraße 8, 79110 Freiburg, Germany*
[3] *Kurt-Schwabe-Institut für Mess- und Sensortechnik e.V. Meinsberg, Kurt-Schwabe-Straße 4, 04736 Waldheim, Germany*

Methods of analytical chemistry [1–5] are used to separate, identify, and quantify chemical components (solid, liquid, or gas) with respect to chemical structure and composition. Often the first step of the analytical methods is the preparation of the samples to be analysed and the separation of its components. This can be achieved by classical methods, such as precipitation, extraction, and distillation, or by more advanced instrumental methods such as chromatography or electrophoresis. In the second step, the quantities of the analyte components will be determined using such physical properties as evaporating temperature, light absorption, fluorescence, or conductivity.

The two most important techniques to measure CO_2 analytically using physical methods, explained in Sections 3.1 and 3.2, are:

- Spectroscopy and
- Gas chromatography.

In Section 3.3 the analytical determination of dissolved carbon dioxide in water with chemical methods is described.

3.1 Spectroscopy

Spectroscopy is a term used to refer to the measurement of radiation intensity as a function of wavelength. Gas spectroscopy – as it is used also for the detection of CO_2 concentrations in gases – exploits the behaviour of gas molecules to act as mechanical resonators. The resonator is formed by the mass of the involved atoms and the chemical bonds holding the atoms together in the molecule. The force, which is needed to move an atom away from its equilibrium distance, constitutes a spring with spring constant k. The energy to excite the vibration of the atoms within the molecule originates from optical radiation. At ambient temperature, the molecules are vibrating around their rest position mostly in their fundamental

Carbon Dioxide Sensing: Fundamentals, Principles, and Applications,
First Edition. Edited by Gerald Gerlach, Ulrich Guth, and Wolfram Oelßner.

vibrational energy levels. The amplitudes of these vibrations amount to a few nm and will increase if energy is transferred to the molecule.

3.1.1 Molecular Vibrations of Molecules, in Particular CO_2

Table 3.1 shows fundamental oscillations of CO_2 and H_2O as simple molecules. If a molecule has N atoms, the position of each atom is determined by coordinates

Table 3.1 Fundamental oscillations of CO_2 and H_2O.

Molecule type	Molecule		Wave number $\tilde{\nu}$ (cm^{-1})	IR active
Linear molecule DOF = $3N-5$	Carbon dioxide (CO_2)  O–C–O	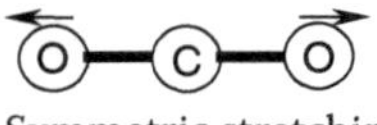 Symmetric stretching	1334	No
	$N=3$ DOF = 4	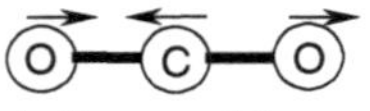 Antisymmetric stretching	2283	Yes
		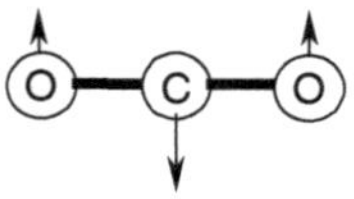 Bending (scissoring)	649	Yes
		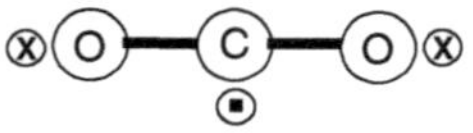 Bending (scissoring) perpendicular to the drawing plane	649	Yes
Non-linear molecules DOF = $3N-6$	Water (H_2O) 	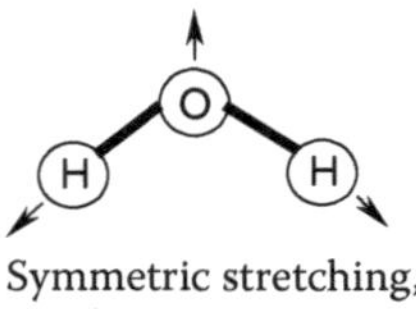 Symmetric stretching, parallel	3657	Yes
	$N=3$ DOF = 3	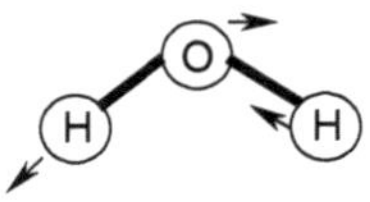 Antisymmetric stretching, perpendicular	3756	Yes
		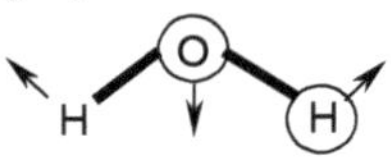 Symmetric bending (scissoring)	1595	Yes

Source: According to [9] from [6]; values from [7, 8].

x, y, and z. The molecule has $3N$ degrees of freedom (*DOF*). The $3N$ DOF determine the bonding distances and angles between the atoms of the molecule. The molecule itself can move in space in three directions and can turn around three axes (for linear molecules around two axes, as the longitudinal axis is the symmetry axis). This means that a total of $3N - 5$ (for linear) or $3N - 6$ (for non-linear molecules) DOF remain. Each *DOF* corresponds to a possible fundamental oscillation.

Spectroscopy often considers both vibrational oscillations and rotations. Transitions involving changes in both vibrational and rotational states can be abbreviated as ro-vibrational (or rovibrational) transitions.

Fundamental oscillations of linear molecules that do not generate a dipole moment, e.g. symmetric stretching oscillations of CO_2, are infrared (IR) inactive. This means that this oscillation mode does not absorb energy from incident radiation. The others absorb energy at the respective resonance frequency or resonance wave number energy and hence are called IR active.

Practically, not only one single oscillation occurs but an immense number of different oscillations. This regards in particular the harmonics, i.e. multiples of the basic resonance frequencies listed in Table 3.1, with intensity decreasing with increasing frequency. In addition, the oscillations affect each other, causing combination oscillations.

Due to $3N - 6$ or $3N - 5$ DOF, the IR spectrum of larger molecules shows numerous normal oscillations that can be divided into two classes:

- *Skeletal oscillations*: Such oscillations comprise many atoms of the molecule. For organic molecules, they usually lie in the wave number range of 1400 to 700 $\mathrm{cm^{-1}}$.
- *Group oscillations*: They concern only a small part of the molecule, mainly certain molecular groups. The rest of the molecule remains in a more or less quiescent state. It is common in spring–mass systems that the larger the resonance frequency or wave number ν becomes, the lighter the oscillating atoms in the terminal groups are. Due to their small oscillating mass, group oscillations are different from skeletal oscillations regarding their wavelength range.

3.1.2 Characteristic Wave Numbers and Wavelengths of Gases

Regarding to Newton's mechanics, a CO_2 molecule as a diatomic molecule can be considered as a mass–spring oscillator where the carbon atom (mass m_c) in the centre is connected with the two oxygen atoms (mass m_o each) on both sides by springs with a given spring constant k (Figure 3.1). For single bonds k amounts to c. $5 \times 10^2\ \mathrm{N\,m^{-1}}$ [9]. For double and triple bonds k values double or triple approximately.

The acceleration a of mass m in the direction x follows Newton's law:

$$F = ma = m\frac{\mathrm{d}^2x}{\mathrm{d}t^2} \tag{3.1}$$

where F is the exciting force. The restoring force is proportional to the spring constant k:

$$F = m\frac{\mathrm{d}^2x}{\mathrm{d}t^2} = -kx \tag{3.2}$$

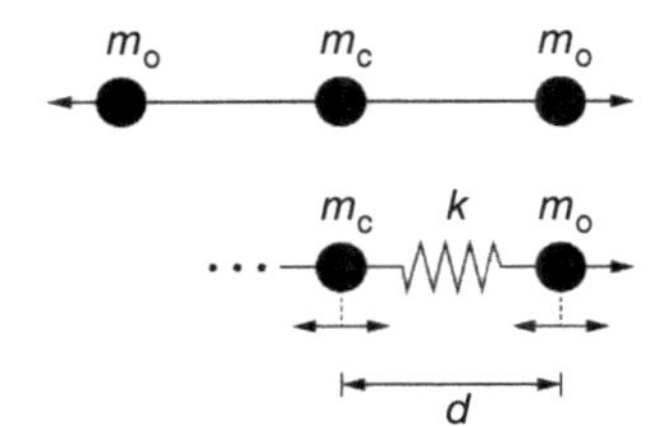

Figure 3.1 CO_2 molecule as simple mechanical spring–mass resonator. The bonds between the carbon atom and the oxygen atoms (masses m_c and m_o) are considered as springs with spring constant k. A restoring force is required to deflect the atoms from its resting position.

The instantaneous deflection of the mass m has to be a harmonic function:

$$x = A\cos(2\pi \nu t) \tag{3.3}$$

with A the maximum amplitude of the mass and ν the natural oscillation frequency. Inserting Eq. (3.3) and its second derivative into Eq. (3.2) yields the natural oscillation frequency:

$$\nu = \frac{1}{2\pi}\sqrt{\frac{k}{m}} \tag{3.4}$$

where ν depends on m and on k, but does not depend on the excitation energy. Changes of the energy result only in a change of the amplitude A.

One has to consider that the oscillation regards both masses m_c and m_o, which are connected by the spring. The resulting reduced mass becomes

$$m = \frac{m_c m_o}{m_c + m_o} \tag{3.5}$$

With $m_c = 2.0 \times 10^{-26}$ kg and $m_o = 2.7 \times 10^{-26}$ kg, one gets an effective mass of $m = 1.1 \times 10^{-26}$ kg and hence a natural frequency of $\nu = 3.3 \times 10^{13}$ Hz.

In spectroscopy usually the wavenumber

$$\tilde{\nu} = \nu / c \tag{3.6}$$

and the wavelength

$$\lambda = 1/\tilde{\nu} = c/\nu \tag{3.7}$$

are used. Here, $c = 2.997\,924\,58 \times 10^8\ \text{m s}^{-1}$ is the speed of light in vacuum. Inverse centimetres (cm^{-1}) are used as unit for $\tilde{\nu}$ so often so that $\tilde{\nu}$ may be stated "in wave numbers."

Using the values of $k = 5 \times 10^2\ \text{N m}^{-1}$ and $m = 1.1 \times 10^{-26}$ kg yields $\tilde{\nu} = 1100\ \text{cm}^{-1}$, which is quite close to the actual wave number ($\tilde{\nu} = 1334\ \text{cm}^{-1}$; see Table 3.1) for the symmetric stretching vibration of the CO_2 molecule. The calculated wavelength amounts to $\lambda = 9.1$ μm, which corresponds to radiation in the mid-infrared (MIR) range.

The corresponding energy E needed to cause a vibration in the bond is according to Eq. (3.4)

$$E = h\nu = \frac{h}{2\pi}\sqrt{\frac{k}{m}} \tag{3.8}$$

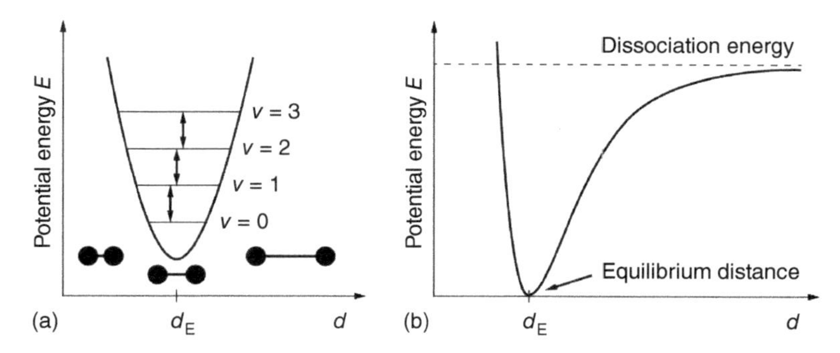

Figure 3.2 Potential energy $E(v)$ of a diatomic molecule with (a) linear and (b) non-linear characteristics. d_E, equilibrium distance.

where h is Plank's constant. However, in reality, a molecule can only exhibit discrete energy levels E_v:

$$E_v = h\nu_v = \left(v + \frac{1}{2}\right) \frac{h}{2\pi} \sqrt{\frac{k}{m}} \tag{3.9}$$

where v is a quantum number that can take values of 0, 1, 2, ...

(Unfortunately, the widely accepted symbol v for the quantum number of the quantum states is quite similar to the Greek ν (nu) symbolizing frequency. It has to be distinguished strictly between both variables in Eq. (3.8) and following equations to not erroneously mistake them.)

Transitions between the energy levels can only occur between adjacent levels, i.e. $\Delta v = \pm 1$, at a time. The energy of the electromagnetic radiation, i.e. the photon energy, that is absorbed in order to promote the molecule to an excited level has to match the difference between two adjacent energetic levels (Figure 3.2a):

$$\Delta E = E_{v2} - E_{v1} = \Delta v h \nu_v \tag{3.10}$$

It has to be noted that – even in the ground state at $v = 0$ and at absolute zero temperature, i.e. at the so-called zero-point vibration – molecules are not completely at rest. In real systems, due to the non-linear character of the oscillators, the harmonic potential becomes anharmonic (Figure 3.2b). A purely empirical function to describe this anharmonic energy is the so-called Morse potential:

$$E(d) = E_d(1 - \mathrm{e}^{-a(d-d_E)})^2 \tag{3.11}$$

where d is the interatomic distance, d_E the equilibrium bond distance, E_d the energy where the molecule dissociates, and a a parameter describing the "width" of the function. Using a Taylor expansion of $E(d)$ around $d = d_E$ to the second derivative of the potential energy function, the force constant k of the bond according to Eq. (3.2) can be calculated as

$$k = 2E_d a^2 \tag{3.12}$$

As can be seen in Figure 3.2b, energy spacings are equal only for the lowest levels where the simple mass–spring model for the potential energy holds. The level spacing decreases as the energy approaches the dissociation energy. The energy E_0 really required for dissociation is lower than the dissociation energy E_D due to the zero-point energy of the lowest vibrational level for $v = 0$.

3.1.3 Absorption of Radiation in Molecules

Atoms and molecules show a large variety of interactions with electromagnetic radiation. This refers to all frequency and wavelength ranges (Table 3.2).

The usage of IR radiation is of particular importance because the interaction is based on the spontaneous polarization in an electric field and it does not necessarily need permanent dipoles.

This makes it advantageous for analysing gases or liquids. Due to the large DOF of oscillations of atoms or groups of atoms within a molecule, the IR spectrum of large molecules shows numerous oscillations. Fundamental oscillations of linear molecules that do not generate a dipole moment, e.g. symmetric oscillations of CO_2 (see Table 3.1), are IR inactive; the others absorb energy at the respective resonance frequency or resonance wave number and are IR active. The different oscillations occur simultaneously.

3.1.4 Molecular Absorption in the Infrared Range

Consider a plane wave of electromagnetic radiation with an intensity I_0 that travels in vacuum in z-direction. At $z = 0$ it hits an optical medium, e.g. the gas to be measured. Due to energy absorption in the medium, the intensity of the wave decreases by $\mathrm{d}I(z)$ between z and $z + \Delta z$:

$$\mathrm{d}I(\lambda, z) = -\varepsilon(\lambda, z)I(\lambda, z)\mathrm{d}z \tag{3.13}$$

Table 3.2 Interaction of atoms and molecules with electromagnetic radiation.

Wavelength range	Effect
Radio frequency range 100 m–1 cm	Electron spin causes a very small magnetic dipole; inverting the spin causes a change of the dipole's spatial direction that interacts with the magnetic part of the radiation and leads to absorption or emission
Microwave range 1 cm–10 μm	Molecules with permanent polar dipole moment align according to the electric component of the radiation (rotation); this leads to a wavelength-dependent absorption or emission
Infrared (IR) range 100–0.78 μm	Oscillations of the atoms in the molecules that cause a change of the dipole moment interact with the electric component of the radiation (IR active). When the dipole moment in the molecule remains constant for specific oscillation modes, it is IR inactive
Visible (Vis)/ ultraviolet (UV) range 0.78 μm–10 nm	Alternating electric field of the radiation induces a periodic deflection of the electrons in the molecules and thus a change in the dipole moment; it causes the absorption of the radiation at resonance frequency of the deflection

where the proportionality constant $\varepsilon(\lambda, z)$ denotes the (monochromatic) extinction or absorption coefficient. Separation of variables and integration then yields

$$I(z) = I_0(\lambda)e^{-\int_0^z \varepsilon(\lambda,z)dz} \tag{3.14}$$

The dimensionless quantity $\int_0^z \varepsilon(\lambda, z)dz$ in the exponent of Eq. (3.14) is known as the optical thickness or opacity of the path traversed by the beam [10]. An optically thick medium will strongly attenuate radiation, whereas an optically thin medium will attenuate radiant flux only slightly. If the extinction coefficient $\varepsilon(\lambda)$ does not depend on position, then Eq. (3.14) simplifies to

$$I(z) = I_0 e^{-\varepsilon z} \tag{3.15}$$

Equation (3.15) is called the Bouguer–Lambert law. It was discovered in 1729 by Pierre Bouguer [11]. Johann Heinrich Lambert, who cited Bouguer's work, stated in 1760 that the absorbance of a material sample is directly proportional to its thickness (Lambert's law) [12].

In 1852 August Beer discovered another attenuation relation, namely, that absorbance is proportional to the concentrations of the attenuating species in the material (Beer's law) [13]. Today, both findings – the absorbance due to both the thickness d of the material sample and the concentrations of the attenuating species – are combined in the Lambert–Beer law:

$$I_1 = I_0 e^{-\varepsilon c d} \tag{3.16}$$

Actually, the Lambert–Beer law is valid just for monochromatic radiation. Since the relationship for high gas concentrations is non-linear, Eq. (3.16) can be applied also only for dilute solutions and gases. It has to be noted that the Lambert–Beer law of Eqs. (3.15) and (3.16), respectively, is not compatible to Maxwell's equations. Strictly speaking it describes the propagation of light within a medium instead of the transmission [14].

The Lambert–Beer law can also be applied in the case of remote reconnaissance for the absorption and scattering in the atmosphere:

$$I = I_0 e^{-m(\tau_a+\tau_g+\tau_w+\tau_{NO_2}+\tau_{O_3}+\tau_r)} \tag{3.17}$$

Here, m is the atmospheric mass and τ are the optical thicknesses for the different species in the atmosphere. The indices stand for aerosols (a; both absorbing and scattering), absorption of homogeneous gases (g; CO_2 and O_2), water vapour (w; absorbing), NO_2 (absorbing), ozone (O_3; absorbing), and Rayleigh scattering of molecular O_2 and N_2 (r).

3.1.5 Line Shapes for Molecular Absorption in the Infrared Range

As considered in Section 3.1.2, molecules interact with radiation at certain frequencies or wavelengths. However, because this interaction is based on statistical processes, one can find that the interaction causes a statistical distribution of values centred about the hypothetical absorption line. For that reason, one can also express the Lambert–Beer law as

$$I(\nu) = I_0 e^{-\alpha(\nu)c} \tag{3.18}$$

which uses a frequency-dependent gas-specific absorption coefficient

$$\alpha(\nu) = S(T, \nu_0)g(p, \nu - \nu_0) \tag{3.19}$$

and describes the line shape by the temperature-dependent line strength $S(T, \nu_0)$ at the centre frequency ν_0 and the pressure-dependent line shape function $g(p, \nu - \nu_0)$.

3.1.5.1 Line Broadening

Absorption lines are broadened by several effects depending on temperature and pressure [10, 15]:

- Collision broadening or, more commonly, pressure broadening. The energy states of atoms or molecules are perturbed when they collide. The larger this effect is, the higher the pressure is. Pressure broadening is the dominant source of line broadening in most application cases in the IR radiation range.
- Doppler broadening due to the translational velocity of the molecules. It results from frequency shifts caused by the motion of molecules under the influence of thermal agitation. It is important at very high temperatures and dominates the line shape at pressures below c. 1 kPa (10 mbar).

There are other effects also leading to a broadening but can be neglected for practical purposes. For instance, natural broadening takes place as consequence of the uncertainty principle. The uncertainty $\Delta\nu$ of the radiation frequency amounts to c. 10^8 Hz, which is 6–8 orders of magnitude smaller than the frequencies at which the transitions usually take place (10^{14}–10^{16} Hz) [9].

3.1.5.2 High-Resolution Transmission Molecular Absorption Database

For small molecules, the line strengths $S(T, \nu_0)$ are tabulated and collected in spectral databases. The most common database is HITRAN [15–18]. HITRAN is an acronym for *high-resolution transmission molecular absorption database* and is a compilation of spectroscopic parameters that a variety of computer codes use to predict and simulate the transmission of light in the atmosphere or gas mixtures. Furthermore, it allows one to estimate qualitatively the expected absorption spectrum for a given application and to determine gas concentrations without cumbersome gas calibration procedures. With the additional knowledge of a given measurement set-up, the best measurement strategies can be derived for detection with high sensitivity and selectivity.

3.1.5.3 Lorentzian Line Shape

At pressures above c. 1 kPa (10 mbar), where pressure-dependent collisional broadening in gases is dominating, the line shape can be described by a Lorentz function:

$$\alpha(\nu) = \frac{S}{\pi}\frac{\beta}{(\nu - \nu_0)^2 + \beta^2} = \frac{S}{\pi\beta}\frac{1}{1 + \left(\frac{\nu - \nu_0}{\beta}\right)^2} \tag{3.20}$$

where

$$\beta = \frac{2D^2 p}{c\sqrt{\pi M k T}} \tag{3.21}$$

is the Lorentz half-width [10]. Here, D is the molecular diameter, p the partial pressure, c the velocity of light, M the mass of the molecule, k the Boltzmann constant, and T the temperature. Equation (3.21) can also written as

$$\beta = \beta_0 \frac{p}{p_0} \sqrt{\frac{T_0}{T}} \tag{3.22}$$

where the index 0 denotes a reference pressure and temperature, respectively.

For CO_2 in air, pressure broadening amounts to linewidths of 0.05–0.15 cm^{-1}.

3.1.5.4 Gaussian Line Shape

When Doppler broadening dominates, i.e. in gases for pressure values below 1 kPa = 10 mbar, the line shape becomes Gaussian:

$$\alpha(\nu) = \frac{S}{\pi\beta} e^{-(\ln 2)\left(\frac{\nu-\nu_0}{\beta}\right)^2} \tag{3.23}$$

The variable $x = (\nu - \nu_0)/\beta$ determines here the characteristic function $e^{-(\ln 2)x^2}$ in the same way as for the Lorentzian shape $1/(1+x^2)$. Both the Gaussian and the Lorentzian have a maximum value of 1 at $x = 0$ and a value of 1/2 at $x = \pm 1$.

Compared with the case of CO_2 in atmosphere, pressure reduction down to about 0.1 atm will result in a Doppler-dominated Gaussian line shape with linewidths of c. 100 MHz or 0.003 cm^{-1}.

3.1.5.5 Applicability of Line Shapes

As mentioned above, line shapes depend on the ruling line-broadening effect and the corresponding probability function. In general, it applies for practical applications [19]:

- The Lorentzian profile works best for gases but can also fit liquids in many cases.
- The Gaussian profile works well for solid samples, powders, gels, or resins.
- The best function for liquids is the Voigt profile (convolution of the Lorentzian and Gaussian profile) or the combined Gaussian–Lorentzian profile where the fractions A_{gauss} and $A_{Lorentz}$ (with $A_{gauss} + A_{Lorentz} = 1$) of the Gaussian and the Lorentzian character are summarized: $\alpha(\nu) = A_{gauss} \cdot \alpha_{gauss} + (1 - A_{gauss}) \cdot \alpha_{Lorentz}$.

3.1.5.6 Spectral Resolution of Individual Ro-vibrational Absorption Lines

To resolve individual ro-vibrational absorption lines, the spectral resolution of an IR spectrometer has to be better than 0.1 cm^{-1}. Laser spectroscopy has an ultimate high spectral resolution limited by the spectral linewidth of the employed laser, which easily can be stabilized to the megahertz range or less. Thus, a spectral resolution of 10^{-3} cm^{-1} can be achieved, allowing one to record the actual shape of individual lines.

3.1.6 CO_2 Absorption in the Infrared Range

As a linear symmetric molecule, CO_2 has two IR-active fundamental vibrations (Table 3.1; both bending oscillations behave similar and hence are counted only

Table 3.3 Isotopologues of CO_2.

Isotopologue	AFGL code	Abundance	HITRAN 2012 spectral coverage (cm^{-1})	HITRAN 2012 number of transitions
$^{12}C^{16}O_2$	626	0.984 204	345–12 785	128 170
$^{13}C^{16}O_2$	636	0.011 057	406–12 463	49 777
$^{16}O^{12}C^{18}O$	628	0.003 947	0–9 558	79 958
$^{16}O^{12}C^{17}O$	627	$7.339\,890 \times 10^{-4}$	0–9 600	19 264
$^{16}O^{13}C^{18}O$	638	$4.434\,460 \times 10^{-5}$	489–6 745	26 737
$^{16}O^{13}C^{17}O$	637	$8.246\,239 \times 10^{-6}$	583–6 769	2 953
$^{12}C^{18}O_2$	828	$3.957\,340 \times 10^{-6}$	491–8 161	7 118
$^{17}O^{12}C^{18}O$	827	$1.471\,800 \times 10^{-6}$	626–5 047	821
$^{12}C^{17}O_2$	7 276	$1.368\,470 \times 10^{-7}$	536–6 933	5 187
$^{13}C^{18}O_2$	838	$4.446\,000 \times 10^{-8}$	4599–4 888	121

Source: De Bievre et al. 1984 [20]. Reproduced with permission of AIP Publishing.

as single oscillation). The corresponding ro-vibrational absorption bands are centred around 15.0 μm (ν_2 band) and 4.24 μm (ν_3 band).

To determine the CO_2 concentration, a particular wavelength has to be chosen:

- That corresponds to the characteristic absorption bands of the gas components and
- That is sufficiently far away from the characteristic wavelengths (wave numbers) of the other gases or the other components in aqueous solutions.

Spectra of CO_2 differ for their different isotopologues (i.e. isotopic compositions with at least one atom having a different number of neutrons). Table 3.3 shows the 10 known isotopologues of CO_2, their abundance, and their spectral coverage in HITRAN.

Due to its abundance of 98.42%, $^{12}C^{16}O_2$ (AFGL code 626) is the most decisive isotopologue of CO_2. Figure 3.3 shows the distribution of absorption lines of this isotopomer in the wavelength range between 1 and 20 μm. Additional bands are visible in the IR spectrum, which are due to combinational transitions (2.7 μm, 2.0 μm) or difference bands (10.6 μm). However, these bands are much weaker than the fundamental bands by at least one order of magnitude. Therefore, for a given concentration, they will result in a much weaker absorption signal. The strong variation of the line strengths of the different IR bands of the CO_2 molecule in Figure 3.3 enables one to select an optimum spectral range for the sensitivity depending on a given application.

As can be seen in Figure 3.3, the spectrum is clearly dominated by the ν_3 band of $^{12}C^{16}O_2$. For that reason, CO_2 is usefully detected at a wavelength of c. 4.24 μm corresponding to a wave number of 2357 cm^{-1}. Nevertheless, the bands originating from the other isotopologues must not be ignored (Figure 3.4). In turn, they offer the possibility for isotope-resolved measurements.

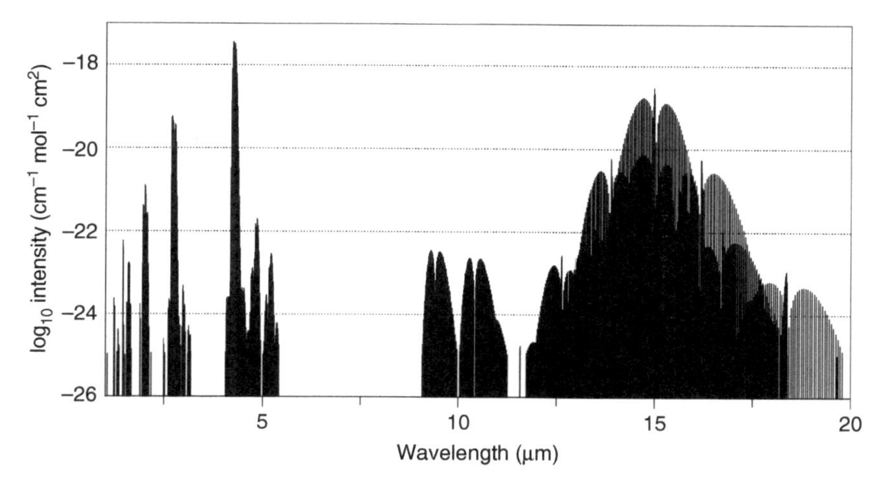

Figure 3.3 Line strengths *S* for ro-vibrational transitions of the most abundant CO_2 isotopologue $^{12}C^{16}O_2$ in the spectral range 1–20 μm, calculated with [21]. Source: Adapted from SpectralCalc.

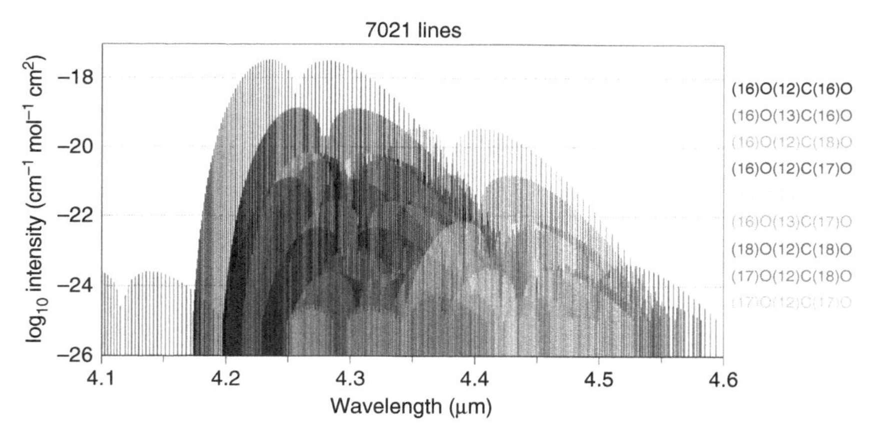

Figure 3.4 Line strengths of CO_2 transitions for different CO_2 isotopomers around the strongest absorption band of $^{12}C^{16}O_2$ at 4.24 μm at room temperature, calculated with [21]. Source: Adapted from SpectralCalc.

Figure 3.5 shows the absorption bands of both CO_2 and H_2O in the mostly used wavelength range 4.243–4.244 μm at higher resolution. It can be seen that the line strengths of the H_2O absorption bands are several orders of magnitude smaller than that of CO_2. The largest H_2O line occurs at 4.243 58 μm. However, in the case of trace gas detection, CO_2 might occur only with a very small concentration compared with H_2O. At ambient pressure the absorption lines are strongly overlapping, and the weak absorption lines of the target gas may be hidden due to pressure-broadened H_2O lines.

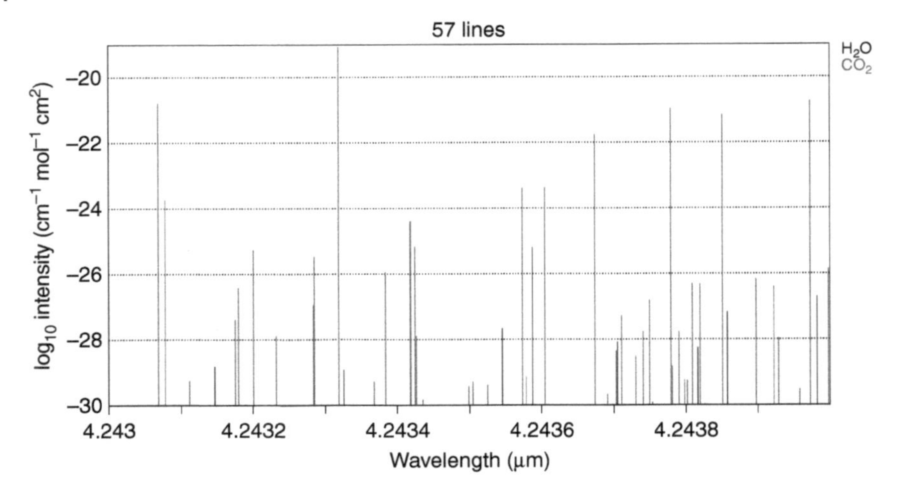

Figure 3.5 Line strengths of CO_2 and H_2O at higher resolution between 4.243 and 4.244 μm, corresponding to 2356 and 2357 cm^{-1}, calculated with [21]. Source: Adapted from SpectralCalc.

A typical way to reduce the cross-interference is to diminish the pressure in the sample cell and hence, according to Eq. (3.22), to decrease the linewidths of the gas absorption lines. This leads to a reduced spectral overlap of lines and suppresses cross-interference of the different constituents of the mixture to be evaluated. As an example, Figure 3.6 shows the transmission of CO_2 with a concentration of 100 ppb against the background of H_2O (concentration 10%). At atmospheric pressure, $p = 1013\,\text{hPa} = 1013\,\text{mbar}$, the CO_2 lines are hardly visible against the H_2O background. After reducing the pressure to 50 mbar, the CO_2 line at 2356.65 cm^{-1} becomes clearly separated from the strong water background at 2356.87 cm^{-1}. This avoids the interference of water vapour on the CO_2 measurement result.

This approach of pressure reduction is advantageous for many applications, in particular when the sample gas to be analysed is a mixture of many different gases. For example, in automotive exhaust gases, trace amounts of CO and NO_x have to be detected in the presence of high H_2O concentrations.

The strong variation of the line strengths of the CO_2 molecule at different wavelengths in Figures 3.4 and 3.5 enables to select an optimum spectral range for a given application. This is illustrated in Figure 3.7 where the absorption spectra of CO_2 are shown for three different concentrations. The wavelength range around 4.3 μm (MIR) is most suitable for trace measurements of low ppm concentrations. For measurements at concentrations above 1000 ppm (e.g. combustion or greenhouse gas applications), a wavelength of 2 μm may be the better choice. This is of practical relevance because the costs for optical components, light sources, and detectors drastically increase with the wavelength.

3.1.7 Laser Spectroscopy

Laser absorption spectroscopy (LAS) is the suitable method to exploit the narrow absorption lines of the ro-vibrational transitions of molecular gases

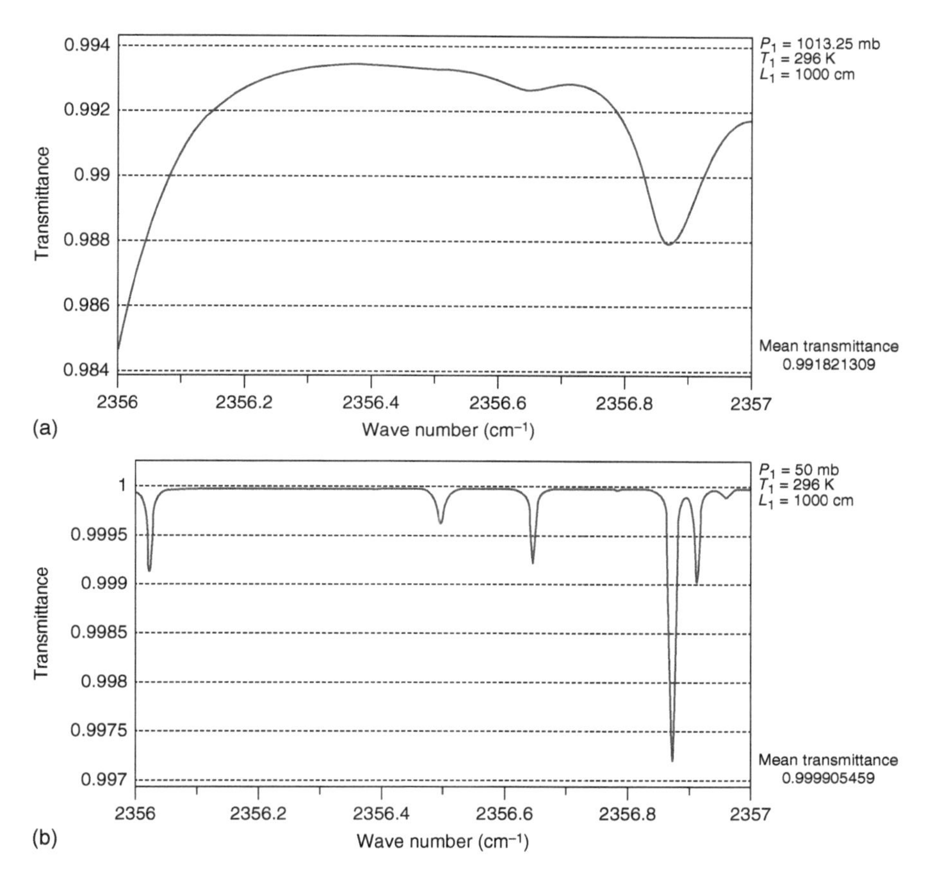

Figure 3.6 CO_2 linewidth reduction by pressure reduction in a sample cell of a mixture of 100 ppb CO_2 and 10% water vapour in N_2 at a pressure of (a) 1013 mbar and (b) 50 mbar. Absorption length 10 m, temperature 296 K, calculated with [21]. Source: Adapted from SpectralCalc.

(see Figure 3.5). This is due to the fact that a laser linewidth of less than 30 MHz (or 10^{-3} cm^{-1}), which is needed to resolve a Doppler-limited CO_2 absorption line at $\lambda = 4.24\,\mu m$ (thus corresponding to a wavelength resolution of less than 2 pm), can be easily obtained with state-of-the-art laser devices. Additionally, the wavelength of most of these lasers can be continuously tuned over a sufficient spectral range to scan across a single absorption line. Thus tunable laser absorption spectroscopy (TLAS) is the predominantly exploited technique.

Most frequently commercially available semiconductor lasers such as compact distributed-feedback (DFB) laser or vertical-cavity surface-emitting laser (VCSEL) are used for this purpose. Due to their internal structure, single-mode operation in specified operating current and temperature regimes is ensured. Using more complex laser modules, e.g. in an external resonator configuration, broader tuning ranges spanning more than $100\,cm^{-1}$ are also commercially available [22, 23]. However, for compact and rugged industrial sensors, less costly DFB lasers and VCSELs with rather narrow tuning ranges are preferred.

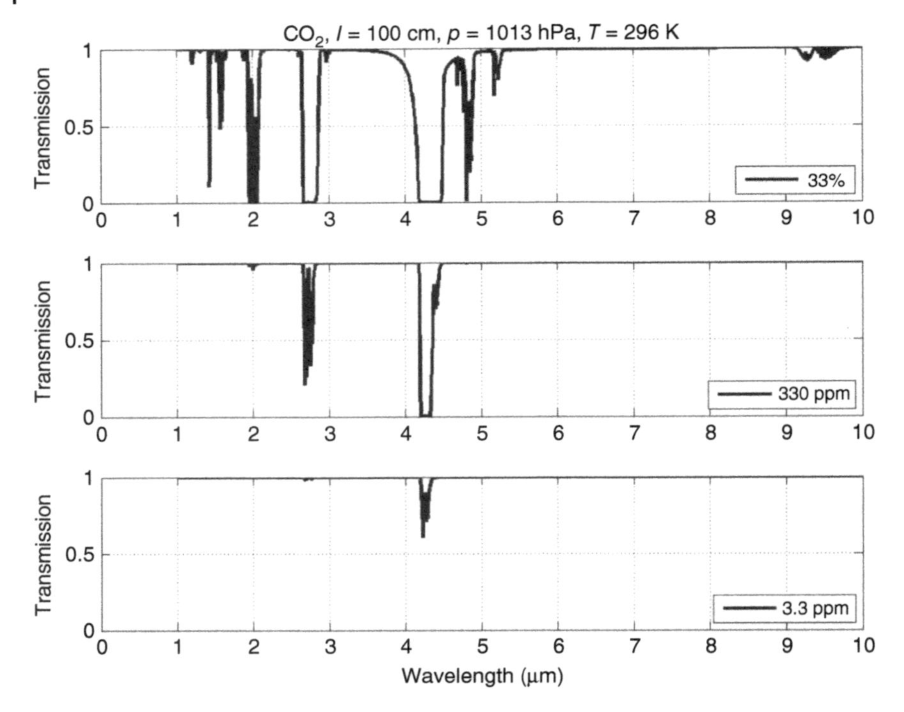

Figure 3.7 Transmission of CO_2 with different concentrations. Absorption length 1 m, atmospheric pressure 1013 hPa, room temperature, calculated with [21]. Source: Adapted from SpectralCalc.

To determine a CO_2 concentration in a gas mixture, a suitable absorption line has to be chosen according to Section 3.1.6. The line has to be sufficiently far away from the absorption lines of the other gases in the mixture. To reduce further the cross-interference, pressure reduction has to be considered (see Figure 3.6).

Recently, frequency comb spectroscopy (FCS) has been established as a new concept of laser-based high-resolution spectroscopy, since the corresponding *Nobel Prize* has been awarded to T. W. Hänsch in 2005. A frequency comb consists of an array of (e.g. more than 100.000) evenly spaced sharp laser emission lines spanning more than an octave. A frequency comb can be locked to an atomic clock, yielding an extremely precise frequency ruler. Meanwhile, FCS is also extended to the MIR range, and several measurement techniques and applications are investigated for narrow and broadband molecular spectroscopy [24, 25]. CO_2 is one of the target molecules [26]. Currently, exclusively used in R&D labs, it is foreseen that soon FCS will also be exploited for special industrial applications.

3.1.7.1 Basic Measurement Concepts

Figure 3.8 illustrates the basic laser spectroscopy set-up and the observed detector signal. The basic feature of TLAS is that a single laser emission line is spectrally tuned across a characteristic absorption feature of the gas. In the case of

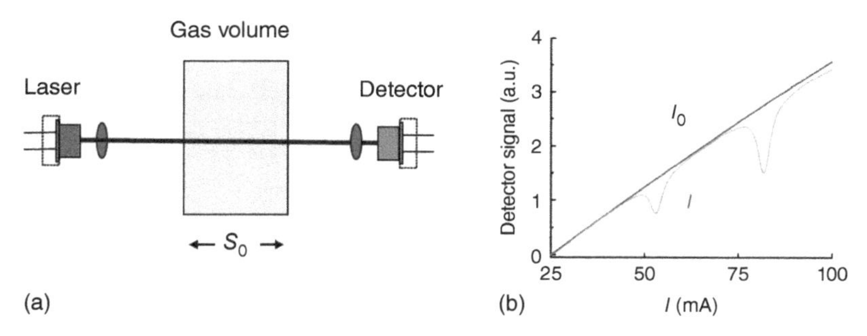

Figure 3.8 (a) Basic laser spectroscopy set-up. (b) Observed detector signal when tuning the laser emission wavelength without (dark grey) and with gas absorption (light grey).

semiconductor lasers, tuning is achieved by changing the laser chip temperature or the operating current in a defined way. This tuning behaviour of the laser, once determined for a given set of parameters, is assumed to be stable. However, it has to be checked regularly to ensure proper operation of an instrument especially for long-term application conditions.

Spectra are usually recorded as a function of time and later converted to a wave number or wavelength scale by the data evaluation software using the established laser tuning parameters.

Tuning can be achieved in a continuous way by a sawtooth modulation of the laser temperature via modulation of the operating current. Alternatively, the laser temperature is increased when a stepwise current pulse is applied to a semiconductor laser. In addition to this basic tuning method, modulation schemes are frequently employed for optimum low-noise signal acquisition as will be described in Section 3.1.7.3.

As mentioned before, it is important that the tuning behaviour of the laser is well known and stable during the gas measurements.

3.1.7.2 Properties of Lasers Suitable for TLAS

For 'spectroscopic' lasers, several requirements have to be fulfilled, which are listed in Table 3.4. The corresponding complex laser qualification and selection represents a considerable effort for the manufactures and strongly affects device prices.

For gas spectroscopy of CO_2, the semiconductor lasers may be used depending on the application wavelength according to Figure 3.7. Table 3.5 shows that for all CO_2 absorption bands, semiconductor lasers operating at room temperature are available. Only for the 14–16 μm band pulsed operation of DFB quantum cascade lasers (QCLs) is required. In all other spectral regions, continuous wave (cw) operation can be used. In the NIR range, compact VCSELs with lower power consumption compared with edge emitters are available. The rapid development of QCLs and interband cascade lasers (ICLs) in the last two decades has opened up a new era of laser spectroscopy. The MIR range is now easily accessible and industrial measurement systems are starting to enter the market.

Table 3.4 Typical requirements for lasers suitable for gas spectroscopy.

Requirements for spectroscopic lasers	Typical value
Gas-specific single-mode emission wave number within operational parameter range (temperature, current)	$\nu_{gas} \pm 2\,cm^{-1}$
Effective linewidth (depending on measurement method)	$0.001\,cm^{-1}$ (30 MHz)
Operational temperature range	$260\,K < T < 350\,K$
Side-mode suppression ratio (SMSR)	>30 dB
Continuous mode-hop-free mode-tuning range	$>1.0\,cm^{-1}$
Tuning rate	$>1\,cm^{-1}$ Hz
Maximum modulation frequency (depending on measurement method)	≈1 MHz
Average power	>1 mW
Far-field polarization	TM_{00}/TE_{00}
Lifetime	>20 000 h

Table 3.5 Commercially available single-mode semiconductor lasers for the different CO_2 absorption bands.

CO_2 wavelength (wave number) range, μm (cm^{-1})	Semiconductor laser type	cw operation at 300 K
1.21 (8264)	DFB diode [27]	x
1.45 (6896)	DFB diode [27, 28]	x
1.59–1.61 (6289–6211)	VCSEL [15]; DFB diode [27, 29]	x
1.98–2.12 (5050–4717)	VCSEL [15]; DFB diode [27, 29]	x
2.6–2.8 (3846–3571)	DFB diode [27]	x
4.23–4.35 (2364–2300)	DFB-ICL [27]; DFB-QCL [30–32]	x
9.2–9.6 (1087–1042)	DFB-QCL [30, 32]	x
10.2–10.7 (980–934)	DFB-QCL [30, 32]	x
14–16 (714–625)	DFB-QCL [30]	

References are given to typical supplier webpages.

3.1.7.3 Laser Absorption Spectroscopy Measurement Schemes

Intrapulse Measurement Scheme A straightforward way to perform tunable laser spectroscopy using semiconductor lasers is to drive the laser with a rectangular current pulse. During the pulse the active layer of the laser is heating up, and subsequently the effective refractive index of the whole internal waveguide structure is changing according to employed semiconductor material properties. This is the main reason for the observed frequency wavelength tuning. Heating also changes the relatively broad laser gain profile. For large temperature sweeps, this may result in mode hopping.

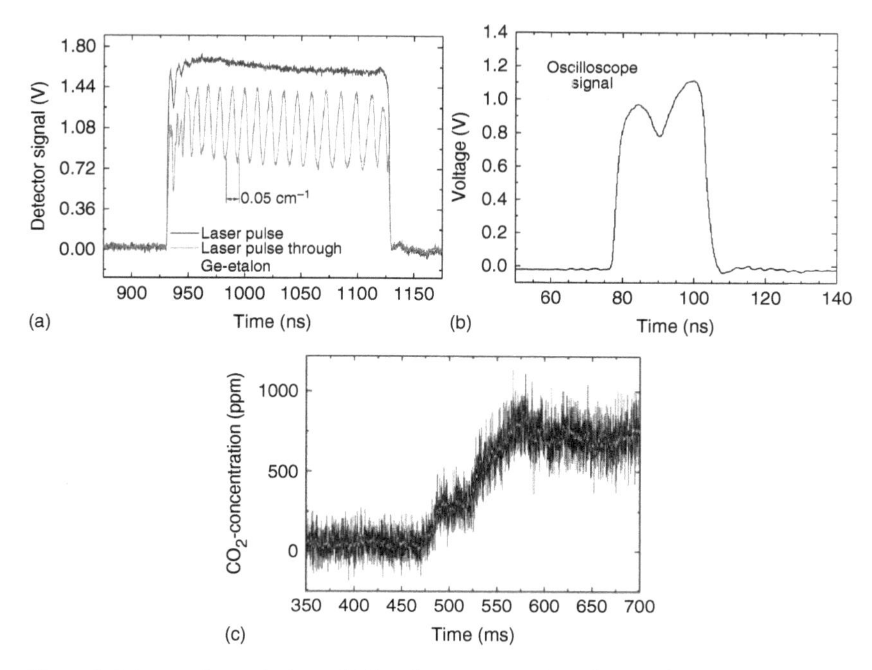

Figure 3.9 Intrapulse technique. (a) Fast frequency tuning of a QCL during a 200 ns current pulse. A 1″ Ge etalon inserted in the beam shows continuous tuning of $0.75\,cm^{-1}$. (b) Absorption line of CO_2 at $2335\,cm^{-1}$ scanned during a 35 ns pulse. (c) Resulting fast change of the CO_2 concentration by exhaling into the beam path between laser and detector (single pulse data are dark grey, a moving average is light grey).

A positive wavelength tuning coefficient is standard. For example, a typical DFB laser for measuring CO_2 at 2004 nm has a tuning coefficient of $0.041\,nm\,mA^{-1}$ [27].

Figure 3.9a) shows the wavelength tuning of an MIR QCL operated with a 200 ns current pulse, and Figure 3.9b) shows an absorption line scan of CO_2 at $2325\,cm^{-1}$ in a 10 cm gas cell during a 35 ns QCL pulse. By integration of the area between background and absorption line, the CO_2 concentration can be determined and calibrated with reference gas mixtures. The pulse repetition rate was 10 kHz. Using that technique very fast concentration measurements can be performed, which can be applied to study combustion processes (see Figure 3.9c). Every single pulse contains the full spectral information. If the laser beam is blocked in a high dust load application for a number of pulses, these pulses can be discarded.

Further details on the intrapulse technique can be found in [33–35]. However, it requires the use of fast detectors, signal electronics, and software. By a fast frequency chirp of the laser, the effective linewidth is increased to values of more than $0.01\,cm^{-1}$. With cw methods higher sensitivities and better spectral resolution can be achieved.

Direct Absorption Spectroscopy By using a cw sawtooth current ramp to operate the laser emission, intensity and frequency are tuned across a CO_2 absorption

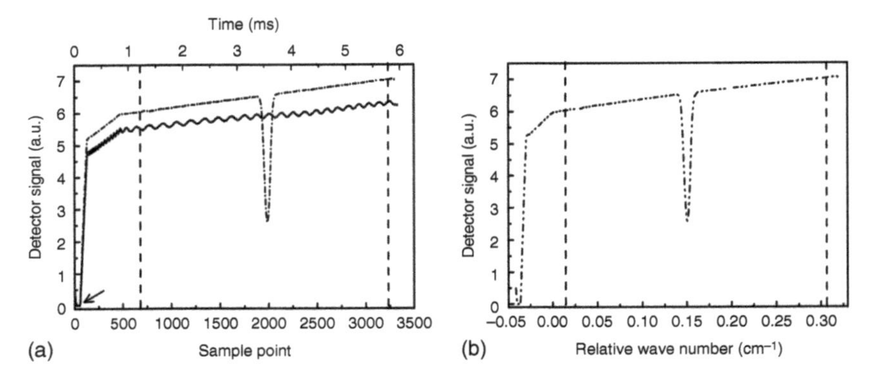

Figure 3.10 Example of a laser scan across the CO_2 P36e absorption line (dash-dotted line) and signal recorded by light passing through a Fabry–Pérot etalon (solid line). (a) Detector signal as a function of sample point or time and (b) detector signal as a function of wave number. The arrow indicates the laser-off period of the modulation cycle that was used for corrections for background radiation and electric offset of the detector. Vertical dashed lines indicate the spectral window used for fitting. Source: Pogany et al. 2013 [36]. Reproduced with permission of Elsevier.

line as can be seen in Figure 3.10. A detailed description of this technique is given in [36].

The fringes obtained by the etalon in Figure 3.10a are used to obtain a precise expression for the frequency tuning curve $\nu(t)$, which is used to transform the measured $I(t)$ to $I(\nu)$. A background curve $I_b(\nu)$ then is estimated from the measured $I(\nu)$ either by a zero gas measurement (taken at a different time) or by a fitting procedure using only left and right data points of $I(\nu)$ with a "safe" offset from the line centre to be not influenced by the absorption. This background line fitting is a critical step because the background may be affected by the changing measurement conditions (e.g. temperature, pressure).

Finally, the absorbance A is calculated from

$$A(\nu) = -\ln \frac{I(\nu)}{I_b(\nu)} \tag{3.24}$$

Using the absorption length L, the gas concentration c is obtained with

$$c = \frac{A(\nu)}{\alpha_f(\nu) \cdot L} \tag{3.25}$$

by fitting with an theoretical absorption coefficient $\alpha_f(\nu)$, e.g. based on HITRAN [15–17] parameters according to Eq. (3.19).

The line fitting procedure based on a physical model yields a powerful noise suppression and low detection limits. If high-quality line parameters are available, calibration-free measurement systems solely relying on a spectral database are feasible by direct absorption spectroscopy (DAS).

Wavelength Modulation Spectroscopy To improve the sensitivity of LAS, modulation techniques have been developed. Most frequently the $2f$ wavelength modulation spectroscopy (WMS) technique is applied. Today, it is used in most

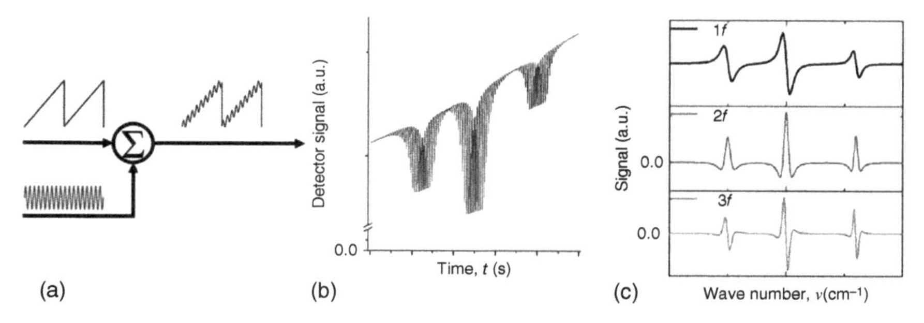

Figure 3.11 Principle of 2f WMS. (a) Modulation with frequency f is added to the current ramp of a semiconductor laser. (b) Detector signal for three absorption lines. (c) Lock-in signals of first three harmonics with time axis converted to a wave number scale. The second harmonic is used for further data analysis, i.e. concentration determination.

industrial process monitoring applications. Generally, lower detection limits can be achieved with 2f WMS than with DAS.

In addition to the slow current ramp, an additional modulation at higher frequency is superimposed on the injection current. The interaction between the modulated wavelength and a non-linear absorption feature gives rise to higher harmonic components in the detector signal, which can be isolated with lock-in amplifiers (Figure 3.11).

To achieve good separation of harmonics for more effective filtering, the modulation frequency is usually over two orders of magnitudes greater than the scanning rate. The detected modulated absorption signal is processed through a hardware lock-in amplifier or a corresponding software routine to obtain the harmonics (2f, 3f, etc.). A theoretical investigation shows that an optimum of sensitivity is achieved when the modulation amplitude is 2.2 times the half width at half maximum (*HWHM*) of the absorption line to be measured [37]. More detailed information on this technique and recent developments can be found in [37–41].

Despite its higher sensitivity, WMS generally requires a calibration of the measurement system with defined gas mixtures due to the more complex signal processing compared with DAS.

Cavity-Enhanced Techniques The LAS measurement techniques can be applied in an in-line or *in situ* configuration or in an extractive way. In the first case the gas mixture is probed in an open path, which can be several kilometres long, e.g. for measurements of the atmospheric CO_2 concentration [42], or the laser radiation is passing through an industrial pipe, reactor, or furnace [43]. Pressure and temperature cannot be controlled in this case, and concentrations of the analyte may strongly fluctuate in time or position along the optical path established by the instrument. The advantage of this method however is a rapid response without any delay caused by sampling procedures. On the other hand, calibration could be a challenge, and sensitivity and selectivity are lower compared with sampling methods.

Extractive techniques as the second case have the advantage that the sample gas mixture can be conditioned to optimum pressure and temperature ranges, interfering components may be removed (e.g. particles, humidity, etc.), and calibration

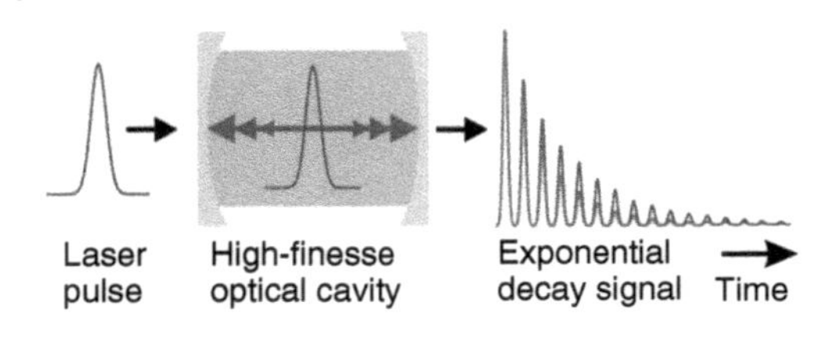

Figure 3.12 Principle of cavity ring-down spectroscopy (CRDS).

can be performed by zero and calibration gas mixtures. Thus, by extractive techniques, high sensitivity and selectivity can be achieved. However, probing and sample preparation may cause long delays and artefacts.

By proper design of the measurement cell, sensitivity can be greatly enhanced. An established method is the application of multi-reflection cells. White and Herriott cells are most frequently used [44, 45].

In an extractive set-up a huge sensitivity improvement can be achieved by the use of cavity methods. The basic idea is to harness the long optical path length of high-finesse optical resonators. Cavity ring-down spectroscopy (CRDS) is the most familiar technique.

In the original CRDS approach [46], radiation of a pulsed laser is injected into a high-finesse cavity with mirror reflectivities $R > 99.9\%$ (Figure 3.12). While the pulse is bouncing back and forth in the cavity, an exponential decay of the intensity is observed at the output:

$$I = I_0 \exp\left(-\frac{t}{\tau}\right) \tag{3.26}$$

The decay time constant τ_0 of an empty cavity (i.e. when the intensity falls to $1/e$ of the initial value) is compared with the value τ_g for the cavity filled with a gas. Absorption causes a reduction of $\tau_g < \tau_0$. When tuning the laser wavelength, absorption spectra can be obtained.

The main advantage of the method is the huge increase of the absorption path, for instance, a 25 cm cavity can effectively yield a 20 km optical path length. Another advantage of the method is that it is immune against drift and fluctuations of the laser intensity.

Today, cw lasers are mainly used for CRDS. In this case the cavity is first filled with laser radiation and intensity build-up is monitored by a photodetector. When a certain level is reached, the laser is rapidly switched off – e.g. by an acousto-optical modulator – and the decay of the intensity is observed.

CRDS systems are commercially available, e.g. by Picarro, Inc. [47]. The method is especially adopted for isotopically resolved measurements of CO_2 in environmental measurements [48].

Meanwhile many different cavity methods have been developed. A review on CRDS can be found in [49]. Off-axis cavity-enhanced absorption spectroscopy (OF-CEAS) [50] and off-axis integrated cavity output spectroscopy (OF-ICOS) [51] have found a broad range of applications, the latter being also used in industrial process control [52].

Thus, by overcoming the initial bias due to their sensitivity to mirror degradation and contamination, cavity methods are gaining increasing acceptance in trace gas analysis.

3.2 Gas Chromatography

Gas chromatography (GC) is a technique that is used in analytical chemistry for separating and analysing compounds that can be vaporized without decomposition [53–57]. Generally, samples can be liquids, gases, or solids. They can also be part of a solvent matrix that also has to be vaporized. Main applications of GC comprise:

- Determination of the purity of a particular substance,
- Separation of the different components of a mixture including determination of the relative amounts of that components, and
- Identification of compounds.

Column chromatography was invented already in 1903 by the Russian botanist Mikhail Semyonovich Tsvet [1, 58]. He used this technique for the separation of different plant pigments, chlorophylls, and xanthophylls by running a solution of these compounds through a glass column filled with finely dispersed calcium carbonate. Since the separated substances appeared as coloured bands in the column, he coined the name chromatography (in Greek *chroma* or χϱϖμα means colour and *graphein* or γϱαφειν means write) [59].

3.2.1 Functional Principle

Gas chromatograph uses a flow-through element (e.g. a column), through which a sample with a certain composition passes in a gas stream of carrier gas, which is called the *mobile phase* (Figure 3.13). The mobile phase consists of the sample being separated and analysed and the solvent that carries the sample. It moves through the chromatography column (the stationary phase) where the sample interacts with the stationary phase and becomes separated. For that, substances can be used that vaporize below c. 300 °C and hence are stable up

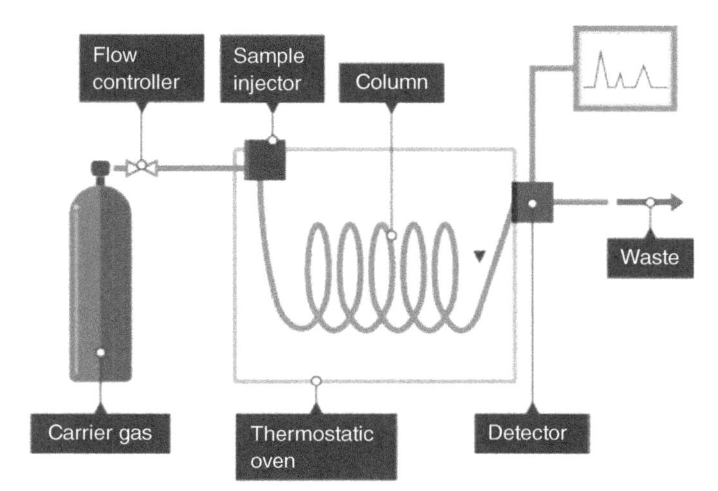

Figure 3.13 Functional principle of gas chromatography.

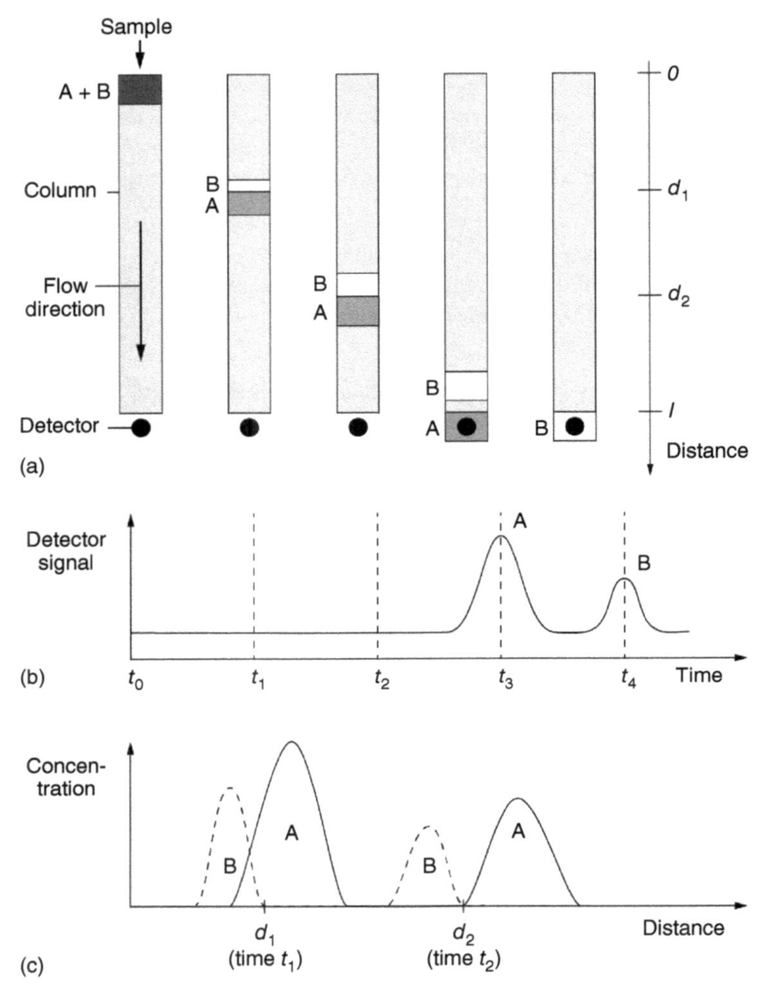

Figure 3.14 (a) Separation of a composition containing the components A and B in the GC column. (b) Corresponding time-dependent detector signal for the elution. (c) Concentration profiles for the components at two different times (locations) during the flow through the column.

to that temperature. A known volume of gaseous or liquid analyte is injected into the inlet of the column. During the flow through the column, the chemical components react or interact with a specific column filling, called the *stationary phase*. These reactions take place at different rates depending on their various chemical and physical properties. Since reactions and interactions between the mobile and the stationary phase are different for each component, they show a different rate of progression in the column and become separated into discrete bands (Figure 3.14). Finally, each individual component exits the column outlet at a different time, the so-called retention time. When emerging (eluting) from the column, the components are identified and analysed by means of appropriate detectors.

In general, substances are identified qualitatively by the order in which they elute the column and by the retention time of the analyte in the column. Quantitative evaluation is possible by analysing the integral of the detector signal from a certain component that is proportional to the amount of analyte.

The retention time is the most decisive characteristic to separate the components of the sample to be analysed. Parameters that can be used to alter the order or the retention time are the carrier gas flow rate, the column length, and temperature. Changing the temperature programme within the column could further improve the differentiation between substances that behave similarly during the GC process.

Figure 3.14 shows exemplarily that the movement of the component B is slower than that of component A. Therefore, with increasing time, both bands for A and B continue to separate. At the same time, a broadening of both bands takes place, which decreases the separating capacity and hence the efficiency of the column. If the right conditions for the separation process will be chosen, then the broadening develops slower than the separation of the bands. This allows usually the complete separation of both bands, given that the column is sufficiently long. Measures to decrease the band broadening are [1]:

- Narrower columns and smaller particle size of the medium filling the column.
- *Gaseous mobile phases*: Reduction of the longitudinal diffusion within the column by decreasing temperature and hence the corresponding diffusion coefficient.
- *Liquid stationary phases*: Minimization of the thickness of the adsorbing liquid film in the column.

3.2.2 Classification of Chromatographic Methods

Chromatographic methods can be classified with respect to (i) the type of contact between the stationary phase and the mobile phase and (ii) the nature of the stationary and mobile phase including the equilibrium for the transition of the dissolved substances between both phases.

For the former, one can distinguish between column chromatography and planar chromatography. For the column chromatography, the stationary phase is located in a narrow column where the mobile phase is pumped through by pressure or by gravity. The planar chromatography uses a flat plate or the interspaces between the pores of paper where the stationary phase is bound. Here, the mobile phase moves through the stationary phase by means of capillarity or by gravity. Generally, the physical basics for both methods are identical.

A more general classification concerns the mobile phase that can be a gas, a fluid, or a supercritical fluid (Table 3.6). GC in principle is limited to column techniques.

3.2.3 Gas Chromatography Instrumental Components

GC – in particular gas–liquid chromatography (GLC) – is based on a vaporized sample that is injected into the chromatographic column. The sample is transported through the column by the flow of a gaseous mobile phase, which has to

Table 3.6 Classification of gas and fluid chromatographic methods.

General classification	Method	Stationary phase	Corresponding equilibrium reaction
Gas chromatography (GC)	Gas–liquid chromatography (GLC) (GC at liquid phases)	Liquid adsorbed at solid	Distribution between gas and liquid
Mobile phase: gas	GC at chemically bound phases	Organic species chemically bonded to solid surface	Distribution between gas and chemically bound phase
	Gas–solid chromatography (GSC) (GC at solid phases)	Solid	Adsorption
Liquid chromatography (LC)	Liquid–liquid or partition chromatography	Fluid adsorbed at solid	Distribution between two non-mixable fluids
Mobile phase: liquid	LC at chemically bound phases	Organic species chemically bonded to solid surface	Distribution between liquid and chemically bound phase
	Liquid–solid or adsorption chromatography	Solid	Adsorption
	Ion exchange chromatography	Ion exchange resin	Ion exchange
	Gel permeation or exclusion chromatography	Liquid in the interspaces within a polymer solid	Molecular sieve effect

Source: After [1].

be chemically inert to avoid any chemical reaction. The column itself contains a liquid stationary phase adsorbed at the surface of an inert solid. After separation of the sample's components, they exit the column outlet and are detected component by component. In the following, the main parts of the GLC set-up will be considered in detail with respect to gas detection and, in particular, CO_2 detection.

3.2.3.1 Autosampler

The function of the autosampler is to introduce the sample to be analysed automatically into the inlets. Autosamplers differ with respect to sample number and capacity, to the degree of automation (for instance, robotic technologies, e.g. rotating robots), and to the type of analysis (e.g. syringe technology, solid-phase microextraction [60]).

3.2.3.2 Sample Injection Port

The column inlet (or injector), which is attached to the column head, introduces a sample into a continuous flow of carrier gas. To avoid band broadening and hence loss of resolution, the sample should be injected fast and like a plug of vapour.

The choice of the injection technique depends on the type of the sample:

- Dissolved samples can be injected directly onto the column.
- If a solvent matrix has to be vaporized, then split/splitless (S/SL) injectors are used. A microsyringe injects the sample with sample matrix though a rubber septum into a flash vaporizer. The temperature of the sample port is usually c. 50 K higher than the boiling temperature of the least volatile component of the sample matrix. The carrier gas then sweeps either the entire sample (splitless mode) or a portion of the sample (split injection) into the column. In split mode, a part of the sample/carrier gas mixture in the injection chamber is exhausted through a split vent. Split injection is preferred when working with samples with high analyte concentrations (>0.1%), whereas splitless injection is best suited for trace analysis with low amounts of analytes (<0.01%).
- Gaseous samples are mostly injected via gas switching valves.
- For adsorbed samples, e.g. on adsorbent tubes, purge-and-trap systems are used where an inert gas is bubbled through an aqueous solution. Insoluble volatile chemicals are trapped on an adsorbent column at room temperature. The volatiles are brought into the carrier gas stream by heating.

3.2.3.3 Column

The column serves as the mean – in conjunction with the stationary phase – for separating the components of the sample [61, 62]. In principle, two types are mainly used. Both have an adsorbent, and their surfaces are covered with a liquid film as stationary phase:

- Packed columns are tubes completely filled with the adsorbent, a finely divided and inert solid material, and are coated with the liquid stationary phase. They have a length of 1.5–10 m and an inner diameter of 2–4 mm.
- Capillary columns are composed of a narrow tube and a stationary phase coated on its interior surface. Mostly, fused silica and stainless steel are used as material for the tube. The column length can be up to 100 m, and the inner diameter around 0.05–0.53 mm. Such columns are flexible and can be wound into coils. The outer surface has often a polyimide coating, which enables the winding by protecting the column from breaking. Capillary columns will give much better separation for complex samples as normal packed columns.

Thick-film columns with a film thickness of 1–5 μm are especially suited for analytes with low boiling temperatures (e.g. volatile organics and gases). Thin-film columns with a film thickness of 0.10–0.25 μm are better for analytes with high boiling temperatures and low volatility. The analytes elute at lower temperature and in better time.

Different compounds have different retention times. Retention time depends on:

- *The boiling temperature of the compound*: For column temperatures below the boiling temperature, compounds spend almost all of its time condensed as a liquid at the beginning of the column. The higher the boiling temperature, the longer the retention time.

- *The solubility in the liquid phase*: The more soluble a compound is in the liquid phase, the less time it will be gaseous and is carried along the column by the gas. High solubility in the liquid phase means a high retention time.
- *The temperature of the column*: If the temperature increases, then the probability of molecules for being in the gas phase raises. A high column temperature shortens retention times for everything in the column.

As mentioned above, separation improves with lower column temperature, but retention time increases decisively. On the other hand, using a high column temperature, separation efficiency suffers. As a compromise between high separation efficiency and relatively short retention time, one can start with the column relatively cool and then gradually and very continuously increase the temperature.

3.2.3.4 Carrier Gas

The carrier gas serves as the mobile phase and hence is decisive for the separation process. Carrier gases should have a purity of typically at least 99.999% or higher (10 ppm or 5.0 grade).

The mostly used carrier gases are helium, nitrogen, argon, hydrogen, and air. The choice of gas that should be used is usually determined both by the detector and by the sample or sample matrix (Table 3.7). For example, thermal conductivity detectors require helium or nitrogen as the carrier gas because of their relatively high thermal conductivity keeping the sensor's filament cool. On the other hand, the carrier is often selected with respect to the sample's matrix. For example, when analysing a mixture, argon as carrier gas is preferred, because argon in the sample will not be seen in the chromatogram. Other aspects are safety and availability. For example, hydrogen is flammable, and certain high-purity gases (e.g. helium) are difficult to obtain in some areas of the world. Here, hydrogen is often being substituted for helium. However, helium provides mostly the best separation. Helium is non-flammable and can be applied to most of the detectors. Therefore, helium is the mostly used carrier gas. Hydrogen is necessary if a flame ionization detector (FID) is used.

3.2.3.5 Stationary Phase

Separation occurs due to interactions of the mobile phase with the stationary phase. These interactions are basically of three types:

- *Polar interactions*: Differences in polarity between the sample components and the bonding entities on the stationary phase result in preferential retention.
- *Ionic interactions*: Separation is based on the charge properties of the sample molecules that have a distinctive affinity for oppositely charged ionic centres on the stationary phase.
- *Molecular size*: Separation takes place due to entrapment of small molecules in the pores of the stationary phase. Therefore, larger molecules pass through first, followed by the elution of the smaller trapped molecules.

Choosing the best stationary phase is the most important decision when selecting a capillary column. Unfortunately, it is also the most difficult and ambiguous decision. The most reliable method is to consult the large collection of example

Table 3.7 Detectors for gas chromatography.

Detector	Measured quantity	Selective to	Support gases	Detectability	Linear range	Temperature (°C)
Thermal conductivity detector (TCD)	Concentration	All compounds except carrier gas	Make-up: carrier gas	5–20 ng	10^5–10^6	150–250
Flame ionization detector (FID)	Mass flow	Compounds with C—H bonds	Combustion: H, air Make-up: He, N	0.1–10 ng	10^5–10^7	250–450
Electron capture detector (ECD)	Concentration	Halogens, nitrates, conjugated carbonyls	N_2 or Ar/CH_4	0.1–10 pg (halogenated compounds); 1–100 pg (nitrates); 0.1–1 ng (carbonyls)	10^3–10^4	300–400
Flame photometric detector (FPD)	Mass flow	S- or P-containing compounds (only one at a time)	Combustion: H, air Make-up: N	10–100 pg (S); 1–10 pg (P)	Non-linear (S); 10^3–10^5 (P)	250–300
Photoionization detector (PID)	Concentration	Aromatics and olefins (10 eV lamp energy)	Make-up: carrier gas	25–50 pg (aromatics); 50–200 pg (olefins)	10^5–10^6	200
Electrolytic conductivity detector (ELCD)	Concentration	Halogens, S- or N-containing compounds (only one at a time)	H (halogens, N); air (S)	5–10 pg (halogens); 10–20 pg (S); 10–20 pg (N)	10^5–10^6 (halogens); 10^4–10^5 (N); $10^{3.5}$–10^4 (S)	800–1000 (halogens); 850–925 (N); 750–825 (S)
Nitrogen–phosphorus detector (NPD)	Concentration	N- and P-containing compounds	Combustion: H and air Make-up: He	1–10 pg	10^4–10^{-6}	250–300
Mass spectrometer (MS)	Mass flow	Any compound producing fragments within the selected mass range	None	1–10 ng (full scan); 1–10 pg (for selected ions)	10^5–10^6	250–300 (transfer line), 150–250 °C (source)

(Continued)

Table 3.7 (Continued)

Detector	Functional principle
Thermal conductivity detector (TCD)	The detector cell contains a heated filament where a current is applied. As the carrier gas containing solutes passes through the cell, the filament current changes, which is compared against the current in a reference cell
Flame ionization detector (FID)	Compounds are burned in a hydrogen/air flame. Carbon-containing compounds produce ions that are attracted to the collector. The number of ions hitting the collector is measured
Electron capture detector (ECD)	Electrons are supplied from a ^{63}Ni foil lining the detector cell and generating a current in the cell. Electronegative compounds capture electrons, resulting in a reduction in the current
Flame photometric detector (FPD)	Compounds are burned in a hydrogen/air flame. S- and P-containing compounds produce light-emitting species (S at 394 nm and P at 526 nm). A monochromatic filter allows only one of the wavelengths to pass. A photomultiplier tube is used to measure the amount of light. Each detection mode requires a different filter
Photoionization detector (PID)	Compounds eluting into a cell are bombarded with high-energy photons emitted from a lamp. Compounds with ionization potentials below the photon energy are ionized. The resulting ions are attracted to an electrode and are measured
Electrolytic conductivity detector (ELCD)	Compounds are mixed with a reaction gas and pass through a high temperature reaction tube. Specific reaction products are created, which mix with a solvent and pass through an electrolytic conductivity cell, changing the electrolytic conductivity of the solvent. Reaction tube temperature and solvent determine which types of compounds are detected
Nitrogen–phosphorus detector (NPD)	Compounds are burned in a plasma surrounding a Rb bead supplied with H and air. N- and P-containing compounds produce ions that are attracted to the collector. The number of ions hitting the collector is measured
Mass spectrometer (MS)	The detector is maintained under vacuum. Compounds are bombarded with electrons or gas molecules. Compounds fragment into characteristic charged ions or fragments. The resulting ions are focussed and accelerated towards a mass filter. The mass filter selectively allows all ions of a specific mass to pass through to the electron multiplier and will be detected. The mass filter then allows the next mass to pass through while excluding all others. The mass filter scans stepwise through the designated range of masses several times per second. The total number of ions is counted for each scan. The abundance or number of ions per scan is plotted versus time to obtain the chromatogram. A mass spectrum is obtained for each scan, which plots the various ion masses versus their abundance or number

Source: From [63].

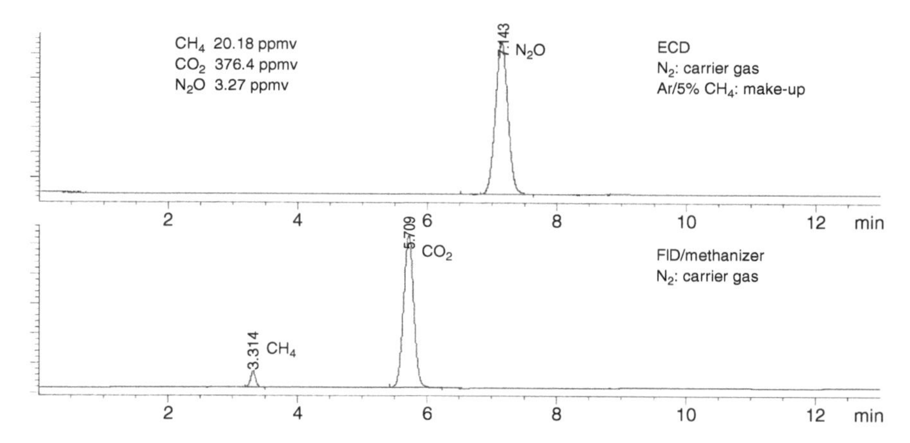

Figure 3.15 Chromatograms of a greenhouse gas sample. CO_2 is methanized and measured by an FID, whereas N_2O is sensed by an electron capture detector. Source: From [64].

applications provided by column manufacturers and suppliers and GC manufacturers and in published literature [63].

3.2.3.6 Detectors

After eluting from the column, solutes pass the detector that measures either its mass flow or the concentration. The detector delivers an electronic signal as a plot of magnitude over retention time, the so-called chromatogram (cp. Figure 3.15).

There are a huge variety of detectors (Table 3.7) that either are sensitive to any solute exiting the column or – in the case of selective detectors – respond only to solutes with specific structures, functional groups, or atoms.

Detectors usually require one or more gases to function properly, e.g. reagent, carrier, and make-up gases. Gases may also serve multiple purposes, e.g. as carrier and make-up gas. The flow rates of the relevant gases determine sensitivity, selectivity, and linear range of the particular detector. For that reasons, manufacturers recommend optimum values for specific chromatographs and detectors.

3.2.3.7 Data Analysis

The chromatogram provides information on the components of the measured sample by the separate peaks (cp. Figure 3.15). The concentration for each analyte can then be determined by calculating the area of the peak, which is proportional to the amount of analyte. In modern GC systems, computer software is used to draw and integrate peaks and to match measured MS spectra to library spectra [65].

The determination of absolute concentration values needs calibration procedures based on calibration curves. One can use internal or external standards with a known amount of analyte and a constant amount of internal standard, which is a chemical added to the sample at a constant concentration.

3.2.4 Gas Chromatography of Gaseous CO_2

GC is often used for CO_2 detection when CO_2 is part of a gas mixture. Examples are:

- Continuous greenhouse gas sensing in ambient air to measure, for instance, carbon dioxide (CO_2), methane (CH_4), and nitrous oxide (N_2O) simultaneously (Figure 3.15) [66–68],
- Trace gas analysis in gas compounds, e.g. in natural gas [69],
- Determination of ancient atmospheric CO_2 concentrations (see Chapter 13), and
- CO_2 in food production, e.g. dairy products [70].

Resolution and accuracy of GC is comparable with Fourier-transform infrared (FTIR) analysis [70]. However, different gases of the gas compound might need specific detectors in GC analysis (Figure 3.15). For CO_2, FID [64, 71] and thermal conductivity detectors [72] are preferred. Stationary phases of columns for CO_2 detection are polystyrene/divinylbenzene, divinylbenzene/ethylene glycol dimethacrylate, and bonded monolithic carbon layers [63].

3.3 Analytical Determination of CO_2 in Liquids

The actual content of free carbon dioxide in an aqueous solution depends significantly on its pH value. In Figure 3.16 the characteristic relationship between the CO_2 equilibrium concentration and the pH value in water is shown. Apart from the pH value, the curves depend slightly on temperature and on the alkalinity and salinity of the solution.

As can be seen in the diagram, the fractional amount of the individual carbonate species changes with the pH value. At pH values below 4.3, nearly the total carbonate exists as freely dissolved CO_2 in the solution. With increasing pH value its percentage decreases to almost zero at pH = 8.2. In the pH range 7–10,

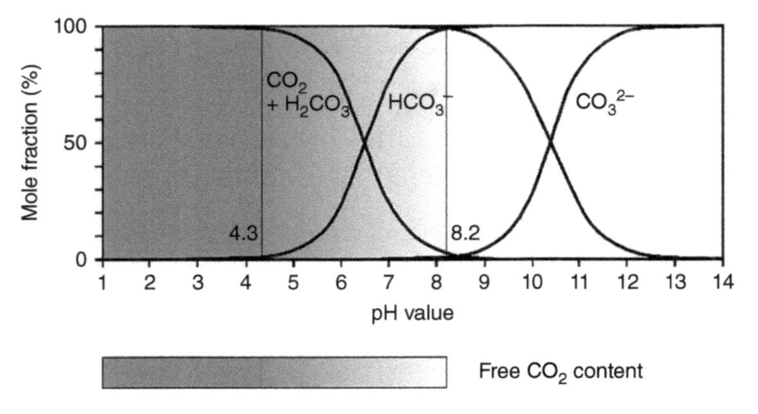

Figure 3.16 pH dependence of the carbonate system.

hydrogen carbonate HCO_3^- is the predominant species. At higher pH values the carbonate equilibrium shifts increasingly to CO_3^{2-} ions. The conventional analytical determination of carbon dioxide by titration procedures and the evaluation of the so-called *p*- and *m*-values are based on these facts.

In the ASTM standard D 513-02 [73], two test methods (A and B) are proposed, providing the measurement of total or dissolved carbon dioxide present as carbon dioxide (CO_2), carbonic acid (H_2CO_3), bicarbonate ions (HCO_3^-), and carbonate ions (CO_3^{2-}) in water. Test method A is applicable to various natural waters and brines; test method B is applicable to natural waters, brines, and various industrial waters.

Test method A (gas sensing electrode test method) stipulates that carbon dioxide is liberated by acidification of the sample. For this purpose, samples are treated prior to measurement with a buffer solution that sets the pH between 4.8 and 5.2. At this pH, interferences are minimized, and according to Figure 3.16, the various ionic forms are converted to CO_2. The expelled CO_2 is measured by a gas sensing electrode like a Severinghaus CO_2 sensor. Volatile weak acids as well as water vapour are potential interferences and can cause deviations. Own experiments with ammonium formate ($HCOONH_4$) and sodium sulphite (Na_2SO_3) added as an interfering substance confirmed this quantitatively. On the other hand, it could be demonstrated that by lowering the pH value to pH = 4.2, the deviation was considerably reduced.

Test method B (CO_2 evolution, coulometric titration test method) allows to analyse samples that contain between 5 and 800 mg l^{-1} of total CO_2. Carbon dioxide is liberated by acidification and heating the samples. The liberated CO_2 is swept through a scrubber by carbon dioxide-free air into an absorption cell where it is automatically coulometrically titrated. Individual concentrations of the several carbonate species are determined from the pH and total CO_2 values.

According to the German industrial standard DEV D8 [74], carbon dioxide dissolved in water and the anions of the carbonic acid can be quantified indirectly only. The application of this method requires:

- The measurement of pH value, temperature, and conductivity and
- The analytical determination of acid capacity (K_a) and base capacity (K_b) or the formerly used *m*- and *p*-values.

In DIN 38409 (H7) [75], the acid capacity K_a and base capacity K_b are defined as

$$K_a = n(H_3O^+)/V(H_2O)\ (\text{mmol l}^{-1}) \tag{3.27}$$

$$K_b = n(OH^-)/V(H_2O)\ (\text{mmol l}^{-1}) \tag{3.28}$$

where n is the amount of substance of the specified ions that the water (volume V) can take up until defined pH values are reached. K_a and K_b or the *m*- and *p*-values are determined by titration with HCl and NaOH to the characteristic pH values 8.2 and 4.3 or with indicator substances according to Table 3.8. It is obvious that in the pH range 4.5–7.8, the +*m*-value (alkalinity) and the −*p*-values are determined. The main components of the sample can be roughly estimated from the *m*- and *p*-values according Table 3.9.

Table 3.8 Analytical determination of K_a and K_b and *m*-value and *p*-value.

Determination	Electrometric	With indicator
Value (at pH)	Titration with (0.1 or 0.02 mol l^{-1})	Phenolphthalein (*p*), methyl orange (*m*)
K_a (8.2)	HCl	$= +p$-value
K_a (4.3)	HCl	$\approx +m$-value
K_b (8.2)	NaOH	$\approx -p$-value
K_b (4.3)	NaOH	$= -m$-value

Table 3.9 Estimation of the main components of the sample from the *m*- and *p*-values.

Determined *m*- and *p*-values	Sample contains
$m > 2p$	CO_3^{2-}, HCO_3^-
$m = 2p$	CO_3^{2-}
$p < m < 2p$	CO_3^{2-}, OH^-
$m = p$	OH^-
$p = 0$	CO_2

The total amount of inorganic carbon (TIC), referred to as Q_c, can be calculated from the *m*- and *p*-values according to

$$Q_c = \Sigma\, c(CO_2; HCO_3^-; CO_3^{2-}) = m - p \tag{3.29}$$

Provided that the values of pH, temperature, conductivity, and the analytical parameters *m*, *p*, and Q_c are known, the CO_2 concentration can be calculated by two methods. While the range of application of the first method, which evaluates the *p*- and *m*-values, is restricted, the second method, which uses the Q_c value, is more generally applicable. The CO_2 concentration can be determined as follows:

Calculation according to method 1:

- Calculation of $m_{relative} = m/(m - p)$,
- Determination of the corresponding pH value from the titration curve shown in Figure 3.17, and
- Comparison with the measured pH value.

  In the case of conformity of both pH values, the CO_2 concentration yields

$$c(CO_2) = -p\,(\text{mmol}\,l^{-1}) - [H_3O^+] \quad \text{if pH} < 4.5 \tag{3.30}$$

$$c(CO_2) = -p\,(\text{mmol}\,l^{-1}) \quad \text{if } 4.5 < \text{pH} < 78 \tag{3.31}$$

Calculation according to method 2:

- Calculation of $Q_c = m - p$ (mmol l^{-1}) and
- Determination of the percent content Q_c of CO_2 of total carbonic acid from [74], Eq. (3.31) or from [76].

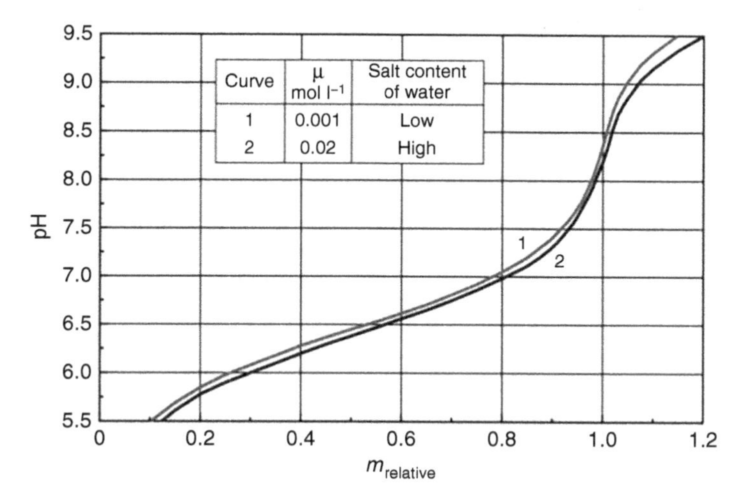

Figure 3.17 General titration curve of carbonic acid at 10 °C according to [74].

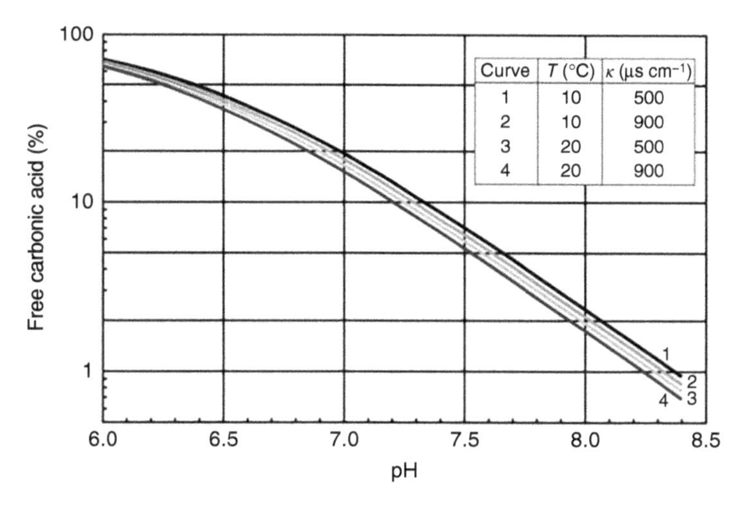

Figure 3.18 Percentage Q_c of free carbonic acid from total carbonic acid versus pH value, temperature, and conductivity. Source: From [76].

Numerical values for the thermodynamic constants $k_1(T)$, $k_2(T)$ and $f_1(\kappa)$, $f_2(\kappa)$ as functions of temperature T and specific conductivity κ of the water may be taken from the tables "Tabelle 1" and "Tabelle 2" in the German standard DEV D8 for the calculation of dissolved CO_2 [74].

For illustration, for some selected values of temperature T and conductivity κ, Figure 3.18 depicts the numerical values of [76] for the percentage of free carbonic acid as function of the total carbonic acid Q_c.

In practice, the CO_2 concentration is usually taken from relevant tables. For that, it is generally necessary to determine the total alkalinity analytically by titration. Rebsdorf [77] presented a set of tables for easy calculation of total carbon dioxide, the partial pressure of free CO_2, and other components of the carbon dioxide system in freshwater. The tables are based on measured values of pH, temperature, total alkalinity, and electrolytic conductivity, from which an approximate figure of the ionic strength can be determined. They are valid only in aqueous solutions where the overall dominating buffer capacity is caused by the carbonate system and where the Debye–Hückel formula for the mean activity coefficients in diluted solutions can be applied without great deviation. Tables in [78] show influencing parameters. The accuracy of the table values decreases for higher ionic strengths, e.g. Ca^{++} and Mg^{++} ions can affect the calculation of CO_2 from titration alkalinity.

Ca^{++} and Mg^{++} ions can also influence the calculation of CO_2 from titration alkalinity. Neglecting this influence can cause considerable deviations. General relationships among the influencing factors are given by tables too [77]. Figure 3.19 illustrates the dependence of the CO_2 concentration on pH value and alkalinity of two solutions with different alkalinities at selected values of temperature and conductivity.

Of course, it is possible to store the table values in a microprocessor as it was realized by the microprocessor-based handheld pH/CO_2 analyser (Model 503, Royce Instrument Corporation, USA). This device uses a conventional combination pH glass electrode and calculates the CO_2 concentration from the measured values of both the current pH and the temperature and the analytically determined alkalinity (*m*-value) and salinity of the water being tested. If the water contains substances that affect the ionic strength, this may be compensated by adjusting a salinity factor.

For an experimental comparison of different measuring methods, CO_2 measurements were performed in mineral waters with varying dilution by using the different analytical methods described above and with a conventional electrochemical CO_2 sensor (Figure 3.20) [79]. Despite high experimental care, noticeable differences occurred, which seem not to be systematically and increase with CO_2 concentration. One source of deviation is the essential difference between the kinds of CO_2 quantities to be measured. The analytical methods determine the CO_2 concentration, whereas the electrochemical CO_2 sensor primarily measures the CO_2 partial pressure in the solution. This fact plays a role in many of the application examples described in this book where

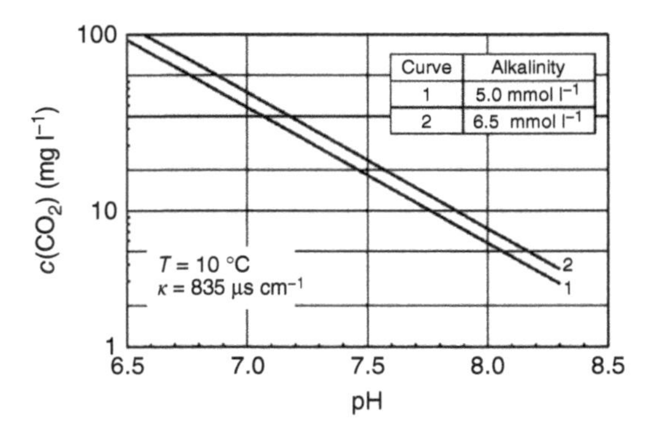

Figure 3.19 Dependence of the CO_2 concentration on pH value and alkalinity of the solution.

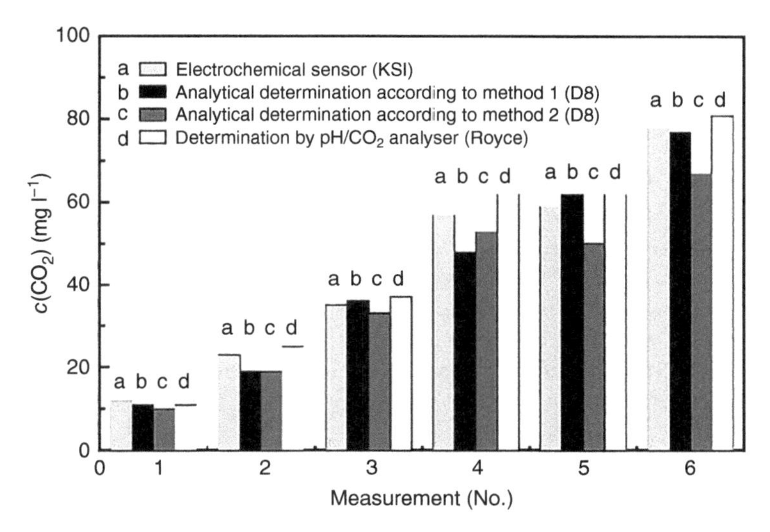

Figure 3.20 Determination of CO_2 concentrations in various samples of diluted mineral waters by using different methods of measurement [79]. KSI means Kurt Schwabe Institute, Meinsberg, Germany, and Royce means Royce Technologies, USA.

CO_2 is measured in natural waters containing various organic or inorganic substances, which can affect the titration.

References

1 Skoog, D.A., Holler, F.J., and Crouch, S.R. (2007). *Principles of Instrumental Analysis*, 6e. Belmont, CA: Thomson Brooks/Cole.
2 Christian, G.D., Dasgupta, P.K., and Schug, K.A. (2013). *Analytical Chemistry*, 7e. Chichester: Wiley.
3 Rouessac, F. and Rouessac, A. (2007). *Chemical Analysis: Modern Instrumentation Methods and Techniques*, 2e. Chichester: Wiley.
4 Danzer, K. (2007). *Analytical Chemistry*. Berlin, Heidelberg: Springer-Verlag.
5 Valcarcel, M. (2000). *Principles of Analytical Chemistry*. Berlin, Heidelberg: Springer-Verlag.
6 Budzier, H. and Gerlach, G. (2011). *Thermal Infrared Sensors – Theory, Optimisation and Practice*. Chichester: Wiley.
7 Shimanouchi, T. (1972). *Tables of Molecular Vibrational Frequencies, Consolidated Volume*, NSRDS-NBS-39. Washington, DC: National Bureau of Standards.
8 Linstrom, P.J. and Mallard, W.G. (eds.) (2018). *NIST Chemistry WebBook*, NIST Standard Reference Database Number 69. Gaithersburg, MD: National Institute of Standards and Technology, 20899. http://webbook.nist.gov (retrieved 15 August 2016).
9 Banwell, C.N. and McCash, E.M. (2008). *Fundamentals of Molecular Spectroscopy*, 4e. New York: McGraw-Hill Education.

10 Love, T.J. (1989). Environmental effects on radiation thermometry. In: *Theory and Practice of Radiation Thermometry*, Chapter 3 (ed. D.P. DeWitt and G.D. Nutter), 189–229. New York: Wiley.
11 Bouguer, M. (1729). *Essai d'optique, Sur la gradation de la lumière* (essay about the gradation of light). Paris: Claude Jombert (in French).
12 Lambert, J.H. (1760). *Photometria, sive de mensura et gradibus luminis, colorum et umbrae* (photometry, or, on the measure and gradation of light, colors, and shade). Augsburg: Sumptibus Vidae Eberhardi Klett (in Latin).
13 Beer, A. (1852). Bestimmung der Absorption des rothen Lichts in farbigen Flüssigkeiten (determination of the absorption of red light in coloured liquids). *Ann. Phys. Chem.* 86 Erstes Stück: 78–88 (in German).
14 Mayerhöfer, T.G., Mutschke, H., and Popp, J. (2016). Employing theories far beyond their limits – the case of the (Bouguer-) Beer-Lambert law. *ChemPhysChem* 17 (13): 1948–1955.
15 Rothman, L.S., Gordon, I.E., Barbe, A. et al. (2009). The HITRAN 2008 molecular spectroscopic database. *J. Quant. Spectrosc. Radiat. Transfer* 110 (9–10): 533–572.
16 Rothman, L.S., Gordon, I.E., Babikov, Y. et al. (2013). The HITRAN 2012 molecular spectroscopic database. *J. Quant. Spectrosc. Radiat. Transfer* 130 (1): 4–50.
17 http://hitran.org/docs/hitran-papers (retrieved 15 August 2016).
18 www.cfa.harvard.edu/hitran/molecules.html (retrieved 15 August 2016).
19 Bradley, M. (2007). *Curve Fitting in Raman and IR Spectroscopy: Basic Theory of Line Shapes and Applications*, Application Note 50733. Thermo Scientific.
20 De Bievre, P., Gallet, M., Holden, N.E., and Barnes, I.L. (1984). Isotopic abundances and atomic weights of the elements. *J. Phys. Chem. Ref. Data* 13 (3): 809–891.
21 http://spectralcalc.com/spectral_browser/db_intensity.php (retrieved 16 August 2016).
22 http://www.toptica.com (retrieved 23 January 2017).
23 http://www.daylightsolutions.com (retrieved 23 January 2017).
24 Diddams, S.A. (2010). The evolving optical frequency comb. *J. Opt. Soc. Am. B* 27: B51–B62.
25 Coddington, I., Newbury, N., and Swann, W. (2016). Dual-comb spectroscopy. *Optica* 3: 414–426.
26 Millot, G., Pitois, S., Yan, M. et al. (2016). Frequency-agile dual-comb spectroscopy. *Nat. Photonics* 10: 27–37.
27 http://nanoplus.com (retrieved 09 January 2017).
28 http://www.vertilas.com (retrieved 09 January 2017).
29 http://www.eblanaphotonics.com (retrieved 09 January 2017).
30 http://www.alpeslasers.ch (retrieved 09 January 2017).
31 http://www.hamamatsu.com (retrieved 09 January 2017).
32 http://www.atoptics.com (retrieved 09 January 2017).
33 Beyer, T., Braun, M., and Lambrecht, A. (2003). Fast gas spectroscopy using pulsed quantum cascade lasers. *J. Appl. Phys.* 93: 3158–3160.

34 Normand, E., McCulloch, M., Duxbury, G., and Langford, N. (2003). Fast, real-time spectrometer based on a pulsed quantum-cascade laser. *Opt. Lett.* 28: 16–18.

35 Francis, D., Hodgkinson, J., Livingstone, B. et al. (2016). Low-volume, fast response-time hollow silica waveguide gas cells for mid-IR spectroscopy. *Appl. Opt.* 55: 6797–6806.

36 Pogany, A., Ott, O., Werhahn, O., and Ebert, V. (2013). Towards traceability in CO_2 line strength measurements by TDLAS at 2.7 μm. *J. Quant. Spectrosc. Radiat. Transfer* 130: 147–157.

37 Reid, J. and Labrie, D. (1981). Second-harmonic detection with tunable diode lasers – comparison of experiment and theory. *Appl. Phys. B* 26: 203–210.

38 Rieker, G.B., Jeffries, J.B., and Hanson, R.K. (2009). Calibration-free wavelength-modulation spectroscopy for measurements of gas temperature and concentration in harsh environments. *Appl. Opt.* 48: 5546–5560.

39 Hayden, T.R.S. and Rieker, G.B. (2016). Large amplitude wavelength modulation spectroscopy for sensitive measurements of broad absorbers. *Opt. Express* 24: 27910–27921.

40 Li, H.J., Rieker, G.B., Liu, X. et al. (2006). Extension of wavelength-modulation spectroscopy to large modulation depth for diode laser absorption measurements in high-pressure gases. *Appl. Opt.* 45: 1052–1061.

41 Spearrin, R.M., Goldenstein, C.S., Schultz, I.A. et al. (2014). Simultaneous sensing of temperature, CO, and CO_2 in a scramjet combustor using quantum cascade laser absorption spectroscopy. *Appl. Phys. B* 117: 689–698.

42 Brooke, J.S.A., Bernath, P.F., Kirchengast, G. et al. (2012). Greenhouse gas measurements over a 144 km open path in the Canary Islands. *Atmos. Meas. Tech.* 5: 2309–2319.

43 Ebert, V., Fernholz, T., Giesemann, C. et al. (2001). Diode-laser based in-situ detection of multiple gas species in a power plant. *Tech. Mess.* 68: 406–414.

44 White, J.U. (1942). Long optical paths of large aperture. *J. Opt. Soc. Am.* 32: 285–288.

45 Herriott, D.R. and Schulte, H.J. (1965). Folded optical delay lines. *Appl. Opt.* 4: 883–889.

46 O'Keefe, A. and Deacon, D.A.G. (1988). Cavity ring-down optical spectrometer for absorption measurements using pulsed laser sources. *Rev. Sci. Instrum.* 59: 2544–2551.

47 http://www.picarro.com (retrieved 16 January 2017).

48 Munksgaard, N.C., Davies, K., Wurster, C.M. et al. (2013). Field-based cavity ring-down spectrometry of $\delta^{13}C$ in soil-respired CO_2. *Isot. Environ. Health Stud.* 49 (2): 232–242.

49 Berden, G., Peeters, R., and Meijer, G. (2000). Cavity ring-down spectroscopy: experimental schemes and applications. *Int. Rev. Phys. Chem.* 19: 565–607.

50 Baran, S.G., Hancock, G., Peverall, R. et al. (2009). Optical feedback cavity enhanced absorption spectroscopy with diode lasers. *Analyst* 134: 243–249.

51 Baer, D.S., Paul, J.B., Gupta, M., and O'Keefe, A. (2002). Sensitive absorption measurements in the near-infrared region using off-axis integrated-cavity-output spectroscopy. *Appl. Phys. B* 75: 261–265.

52 http://www.lgrinc.com (retrieved 16 January 2017).
53 Dettmer-Wilde, K. and Engewald, W. (eds.) (2014). *Practical Gas Chromatography – A Comprehensive Reference*. Berlin: Springer-Verlag.
54 Poole, C. (2012). *Gas Chromatography*. Amsterdam: Elsevier.
55 McNair, H.M. and Miller, J.M. (2009). *Basic Gas Chromatography*, 2e. Hoboken, NJ: Wiley.
56 Jennings, W., Mittlefehldt, E., and Stremple, P. (1997). *Analytical Gas Chromatography*, 2e. San Diego, CA: Academic Press.
57 Grob, R.L. and Barry, E.F. (eds.) (2004). *Modern Practice of Gas Chromatography*, 4e. Hoboken, NJ: Wiley.
58 Tsvett, M.S. (1905). On a new category of adsorption phenomena and on its application to biochemical analysis. *Proc. Warsaw Soc. Natur., Biol. Sect.* 14 (6): 20–39 (in Russian).
59 Tswett, M. (1906). Adsorptionsanalyse und chromatographische Methode. Anwendung auf die Chemie des Chlorophylls (Adsorption analysis and chromatographic method. Application to the chemistry of chlorophyll). *Ber. Deut. Bot. Ges.* 24 (7): 384–393 (in German).
60 Pawliszyn, J. (2009). *Handbook of Solid Phase Microextraction*. Beijing: Chemical Industry Press.
61 Seader, J.D. and Henley, E.J. (2006). *Separation Process Principles*, 2e. Hoboken, NJ: Wiley.
62 Barry, E.F. and Grob, R.L. (eds.) (2007). *Columns for Gas Chromatography: Performance and Selection*. Hoboken, NJ: Wiley.
63 Agilent J&W GC Column Selection Guide (2007). Agilent Technologies. http://selectgc.chem.agilent.com, http://www.team-cag.com/products/chrom/gc/brosh/5989-6159.pdf (retrieved 24 August 2016).
64 Wang, C. (2010). *Simultaneous Analysis of Greenhouse Gases by Gas Chromatography*, Application note. Agilent Technologies. https://www.agilent.com/cs/library/applications/5990-5129EN.pdf (retrieved 25 August 2016).
65 Kromidas, S. and Kuss, H.-J. (eds.) (2008). *Chromatogramme richtig integrieren und bewerten: Ein Praxishandbuch für die HPLC und GC (correct integration and evaluation of chromatograms: a practical guide for HPLC and GC)*. Weinheim: Wiley-VCH (in German).
66 Wang, Y.H., Wang, Y.S., Sun, Y. et al. (2006). An improved gas chromatography for rapid measurement of CO_2, CH_4 and N_2O. *J. Environ. Sci.* 18 (1): 162–169.
67 Yuesi, W. and Yinghong, W. (2003). Quick measurement of CH_4, CO_2 and N_2O emissions from a short-plant ecosystem. *Adv. Atmos. Sci.* 20 (5): 842–844.
68 SRI Instruments (2008). Greenhouse gas GC configuration. http://srigc.com/cn/downloads/9/GreenHouseGasGC112008.pdf (retrieved 25 August 2016).
69 SRI Instruments. Multi gas analyzer #1. http://srigc.com/cn/downloads/8/SRImultigas1GC.pdf (retrieved 25 August 2016).
70 Mohr, S.A., Zottola, E.A., and Reineccius, G.A. (1993). The use of gas chromatography to measure carbon dioxide production by dairy starter cultures. *J. Dairy Sci.* 76 (11): 3350–3353.

71 Esler, M.B., Griffith, D.W.T., Wilson, S.R., and Steele, L.P. (2000). Precision trace gas analysis by FT-IR spectroscopy. 1. Simultaneous analysis of CO_2, CH_4, N_2O, and CO in air. *Anal. Chem.* 72 (1): 206–215.

72 Woebkenberg, M.L. (1994). Carbon dioxide. In: *NIOSH Manual of Analytical Methods (NMAM)*, 4e. The National Institute for Occupational Safety and Health, https://www.cdc.gov/niosh/docs/2003-154/pdfs/6603.pdf (retrieved 25 August 2016).

73 ASTM Designation D 513-02 (2002). Standard test methods for total and dissolved carbon dioxide in water. ASTM.

74 Deutsches Institut für Normung e.V. DEV D8, 6 (1971). Deutsche Einheitsverfahren zur Wasser-, Abwasser- und Schlammuntersuchung. Die Berechnung des gelösten Kohlendioxids (der freien Kohlensäure), des Carbonat- und Hydrogencarbonat-Ions (German standards for Water, Waste water and Sludge investigations. The calculation of dissolved CO_2 [free carbonic acid], of carbonate and hydrocarbonate). Lieferung 1971. Weinheim: Verlag Chemie (in German).

75 Deutsches Institut für Normung e.V. DIN 38409-7 (2005). Deutsche Einheitsverfahren zur Wasser- Abwasser- und Schlammuntersuchung – Summarische Wirkungs- und Stoffkenngrößen (Gruppe H) – Teil 7: Bestimmung der Säure- und Basekapazität (H 7) (German standards for Water, Waste water and sludge investigations. Determination of the acid-base capacity), 1–28. Berlin: Beuth-Verlag (in German).

76 Bauer, K. (1981). Zur Bedeutung der freien Kohlensäure in Forellenzuchtbetrieben (To the relevance of free carbonic acid for the fish farming of trout). *Z. Binnenfischerei* 31: 1–5 (in German).

77 Rebsdorf, A. (1972). *The Carbon Dioxide System in Freshwater*. A set of tables for easy computation of total carbon dioxide and other components of the carbon dioxide system. Hillerod, DK: Freshwater Biological Laboratory.

78 Gelbrecht, J., Henrion, G., and Henrion, R. (1987). Zur Bestimmung des gesamten anorganischen Kohlenstoffes in natürlichen Gewässern durch Titration mit Salzsäure (Determination of total inorganic carbon in natural waters by titration with hydrochloric acid). *Acta Hydrochim. Hydrobiol.* 15 (1): 19–28 (in German).

79 Zosel, J., Oelßner, W., Decker, M. et al. (2011). The measurement of dissolved and gaseous carbon dioxide concentration. *Meas. Sci. Technol.* 22: 072001.

4

Electrochemical CO_2 Sensors with Liquid or Pasty Electrolyte

Manfred Decker, Wolfram Oelßner, and Jens Zosel

Kurt-Schwabe-Institut für Mess- und Sensortechnik e.V. Meinsberg, Kurt-Schwabe-Straße 4, 04736 Waldheim, Germany

4.1 Severinghaus-Type Membrane-Covered Carbon Dioxide Sensors

4.1.1 The Severinghaus Principle

The basic and at that time new idea of a membrane-covered carbon dioxide electrode was first published by Richard W. Stow and Barbara F. Randall at the Fall Meeting of the American Physiological Society (APS) in 1954 [1]. It consisted essentially in measuring the pH value of a film of water separated from the measuring medium (in this case blood) by a thin rubber membrane. The early development work on the electrode was made by Stow et al. alone at the Ohio State University, Columbus, United States. Later, he was assisted by B.F. Randall as well as R.F. Baer who did his master's thesis on the electrode. In their fundamental paper [2], they reported on the new method for measuring the partial pressure pCO_2 of blood by means of an electrochemical electrode. It was presented at the annual meeting of the American Academy of Physical Medicine and Rehabilitation in 1956, but it could be published only in 1957 because the manuscript was lost initially in the editorial office [3]. Stow and Randall had decided not to patent the electrode. He had offered his idea to Beckman Company before his APS lecture in 1954, but they refused and rewrote a patent by Clark to apply it to all membrane-covered electrochemical analysers including the Stow electrode, which resulted in nearly 15 years of litigation [4].

Figure 4.1 illustrates schematically the essential components of Stow's original CO_2 electrode shown in [2]. A rubber membrane (finger cot) (B) was pulled tightly over the sensitive end (E) of a standard Beckman general-purpose glass electrode (A) and the tip of the reference electrode (K), and fixed there with a rubber band (C). This membrane has been shown to be highly impermeable to water, hydrogen carbonate, chloride, and hydrogen ions in particular and to the reactive constituents of blood in general. The silver–silver chloride reference electrode was prepared on the outer wall of the glass electrode on a strip of Tygon™ enamel (H). A strip of 'silver paint' (J) was painted on Tygon™ and covered by

Carbon Dioxide Sensing: Fundamentals, Principles, and Applications,
First Edition. Edited by Gerald Gerlach, Ulrich Guth, and Wolfram Oelßner.

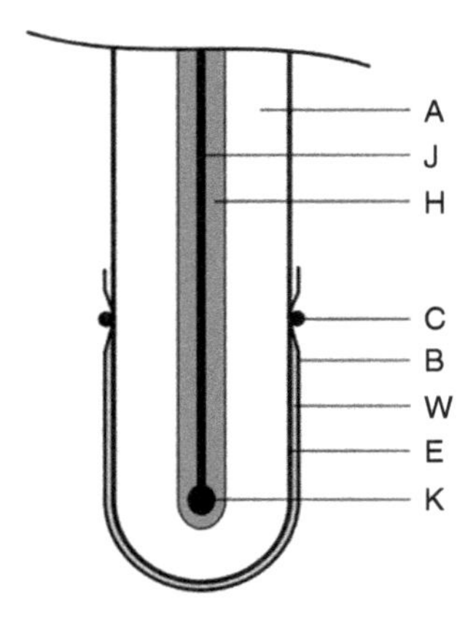

Figure 4.1 Top of Stow's pCO_2 cell according to [2]. A, Beckman glass electrode; B, rubber finger cot; C, rubber band; E, sensitive glass surface; H, Tygon™ enamel strip covering silver except at lower tip (not shown); J, silver strip; K, exposed tip of silver chloride layer; W, thin water film.

another strip of Tygon™ except at the lower tip (K) upon which silver chloride is deposited electrolytically. A very thin film of distilled water (W) was trapped between the sensitive surface (G) of the glass electrode and the membrane.

Calibration of the CO_2 electrode was accomplished by bringing the electrode system in contact with water, which has been equilibrated with a given pCO_2. It was found that the calibration curves closely follow the theoretical prediction. Within the precision of measurement (0.01 pH unit), blood and water of the same pCO_2 produced identical results.

In some respects, this new method for determining the pCO_2 of blood had advantages over the two classical chemical methods that were available for this purpose and common in the laboratory up to this time: with the indirect method, using the manometric extraction apparatus of Van Slyke and Sendroy [5], the pCO_2 of blood was calculated from measuring values of pH and plasma CO_2 content with the Henderson–Hasselbalch equation or determined by interpolation on a plot of CO_2 content versus the pCO_2 of an equilibration gas, while with the Riley et al. or direct method [6, 7] a blood sample was equilibrated with a bubble of gas followed by chemical analysis of the gas for its CO_2 content [2]. The accuracy of the results obtained with Stow's electrode rivalled those obtained under the best conditions with either of the chemical methods, and the speed of analysis was modestly improved.

However, the electrode drifted, and Stow himself was sceptical of ever making it stable. John Wendell Severinghaus contacted Stow and Randall at the APS meeting in 1954 and suggested adding hydrogen carbonate to the film of water between the membrane and the surface of the pH glass electrode to stabilize the electrode signal [4]. At the Anesthesia Research Laboratory, University of California, San Francisco, he immediately began to build Stow-type CO_2 electrodes with the following modifications:

- A layer of cellophane is inserted as a spacer between the membrane and the glass electrode instead of a thin water film.
- A solution of $NaHCO_3$ and NaCl is used instead of distilled water as cell electrolyte.
- A Teflon™ membrane is used instead of a rubber membrane.

Compared with the original Stow electrode, these modified Severinghaus CO_2 electrodes exhibit the following benefits:

- They are twice as sensitive as the Stow system.
- They respond faster to changes of the CO_2 partial pressure.
- They are more stable and drift much less due to stabilizing effect of the NaCl on the silver reference electrode and the greater conductivity of the solution.

Thus, by using hydrogen carbonate solution instead of pure water as sensor electrolyte, Severinghaus increased the sensitivity and stability of the CO_2 sensor significantly. For this reason, membrane-covered carbon dioxide sensors are commonly called Severinghaus electrodes, despite the basic idea coming from Stow.

As the medical definition of the Severinghaus-type electrode indicates, this electrode was intended in particular for the analysis of blood [8]: 'A glass electrode in a film of hydrogen carbonate solution covered by a thin plastic membrane permeable to carbon dioxide but impermeable to water and electrolytes; the carbon dioxide pressure of a gas or liquid sample quickly equilibrates through the membrane and is measured in terms of the resulting pH of the hydrogen carbonate solution, as sensed by the glass electrode; commonly used to analyse arterial blood samples for CO_2'. Severinghaus utilized oxygen and carbon dioxide electrodes invented by others to make the first useful blood gas analysis apparatus and assisted various manufacturers in marketing such devices [9–12]. To permit rapid and accurate analysis of oxygen and carbon dioxide tensions in gas, blood, or any liquid mixture, he developed a thermoregulated apparatus containing essentially a membrane-covered amperometric pO_2 electrode, just developed by L.C. Clark [13, 14], and a pCO_2 electrode as described in Figure 4.2. At this improved Stow electrode, a piece of wet cellophane film (thickness 25 μm when wet) is stretched as spacer over the glass electrode and held in place with an O-ring. As permeation membrane a 25 μm thick Teflon™ film is mounted on the end of the thin-walled Lucite™ tube with an O-ring using silicone grease to seal the Lucite™ to the Teflon™. This tube is filled with the electrolyte ($NaHCO_3$ + NaCl). The glass electrode is inserted into the Lucite™ until its tip indents the Teflon™ membrane. The assembly is slid into the chamber until the shoulder of the glass electrode seats on

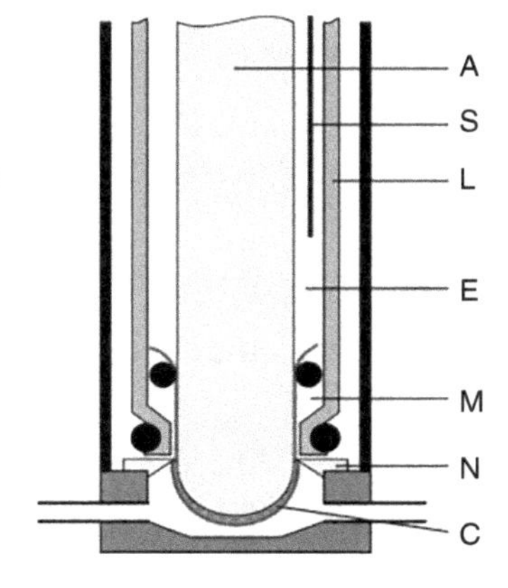

Figure 4.2 Schematic drawing of the Severinghaus CO_2 electrode. A, glass electrode; S, silver reference electrode; L, lucite™ tube; E, electrolyte; M, Teflon™ membrane; N, Nylon™ orifice; C, cellophane layer. Source: Severinghaus and Bradley 1958 [9]. Reproduced with permission of The American Physiological Society.

the Nylon orifice of the cuvette. The silver reference electrode is a sterling silver wire that has been provided with an AgCl coating in 0.1 N HCl [9].

Severinghaus investigated the response behaviour of his CO_2 electrodes in dependence of the hydrogen carbonate concentration in the sensor electrolyte and the material and thickness of the spacer and the permeation membrane. The relationship of pH to log(pCO_2) was linear over the range tested, from 1.38% to 11.37% CO_2. He found that at any pCO_2 a 0.01 pH change is about equal to 2.5% change in pCO_2. The response time was about two minutes for a fourfold rise in CO_2 concentration and about four minutes for a fourfold fall in CO_2 concentration. This was believed due to the logarithmic nature of the response, 'magnifying light changes at low concentrations and making the tail more prominent' [9].

In 1958, obviously unaware of the work of Severinghaus, Gertz, and Loeschke developed also a method to determine the CO_2 pressure in liquids and gases with a membrane-covered glass electrode. Their electrode used simple polyethylene sheathing with a thickness of a few micrometres around the electrode and was filled with 0.01 M $NaHCO_3$ solution. The reference electrode was located in the capillary gap between the glass electrode and the polyethylene sheathing. At a CO_2 pressure of about 53 mbar, the CO_2 pressure could be determined with an uncertainty of ±1.3 mbar. The equilibrium was reached within 5–10 minutes [15, 16].

Despite many efforts in research and development, the Severinghaus basic principle for potentiometric CO_2 sensors has remained almost unchanged since its introduction in 1954. Figure 4.3 shows schematically the Severinghaus CO_2 sensor (also called CO_2 electrode) and illustrates its mode of operation. Main constituents of the sensor are:

- A pH electrode, usually a glass electrode with a flat membrane;
- A sensor electrolyte containing the hydrogen carbonate and KCl;
- A reference electrode in the sensor electrolyte, mostly an $Ag/AgCl/Cl^-$ electrode;
- A thin hydrophilic spacer sheet soaked with the sensor electrolyte; and
- A thin polymer membrane.

As indicated in Figure 4.3, CO_2 permeates from the gaseous or liquid test specimen through the membrane into the electrolyte film in the spacer until

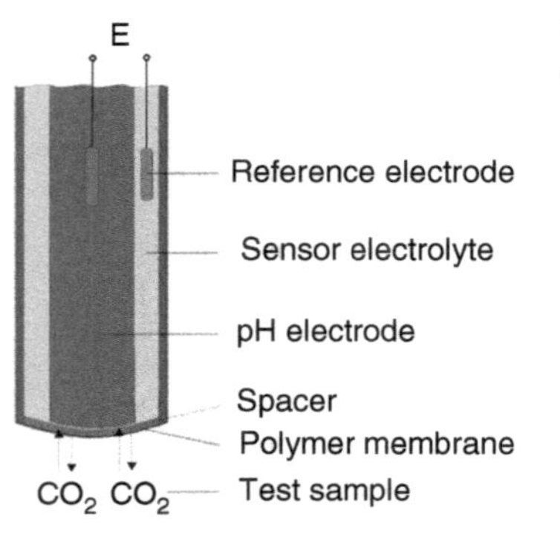

Figure 4.3 Main components and mode of operation of the Severinghaus carbon dioxide sensor.

equilibrium between the CO_2 partial pressures on both sides of the membrane has been established. According to the potentiometric sensor principle, the potential difference E between the inner reference electrode of the glass electrode and the outer reference electrode in the sensor electrode is measured as sensor signal. As will be shown below, E is proportional to the logarithm of the CO_2 partial pressure pCO_2 in the measuring medium. The main advantages of such an equilibrium sensor are that virtually no CO_2 from the measuring medium is consumed and that the CO_2 partial pressure can be measured over a range of several orders of magnitude with the same relative accuracy.

In the sensor electrolyte a series of equilibrium reactions take place, resulting in the formation or diminishment of H^+ ions and, thus, in a CO_2-dependent shift of the pH value that is measured by means of the integrated pH electrode:

$$CO_2 + H_2O \rightleftarrows H_2CO_3 \text{ (carbonic acid)} \tag{4.1}$$

$$H_2CO_3 \rightleftarrows H^+ + HCO_3^- \text{ (hydrogen carbonate ion)} \tag{4.2}$$

$$HCO_3^- \rightleftarrows H^+ + CO_3^{2-} \text{ (carbonate ion)} \tag{4.3}$$

Equations (4.1) and (4.2) may be summarized as

$$CO_2 + H_2O \rightleftarrows H^+ + HCO_3^- \tag{4.4}$$

with the first and second dissociation constants

$$K_1 = \frac{[H^+][HCO_3^-]}{[CO_2]} = 4.4 \times 10^{-7} \text{ mol l}^{-1} \text{ (at 25°C)} \tag{4.5}$$

and

$$K_2 = \frac{[H^+][CO_3^{2-}]}{[HCO_3^-]} = 5.6 \times 10^{-11} \text{ mol l}^{-1} \text{ (at 25°C)} \tag{4.6}$$

respectively. The dissociation constant of water is

$$K_W = [H^+][OH^-] = 1.01 \times 10^{-14} \text{ (mol l}^{-1})^2 \text{ (at 25°C)} \tag{4.7}$$

According to Henry's law in the case of equilibrium between the gaseous and liquid phases, the concentration of dissolved carbon dioxide $[CO_2]$ in the sensor electrolyte is

$$[CO_2] = K_H\, pCO_2 \tag{4.8}$$

with Henry's coefficient

$$K_H = 3.3 \times 10^{-4} \text{ (4.1 mol l}^{-1})/\text{kPa (at 25°C)} \tag{4.9}$$

where pCO_2 is the partial pressure of CO_2 in the sample (and finally also in the sensor electrolyte) and K_H is the solubility coefficient of CO_2.

As explained above, a certain amount of $NaHCO_3$ has to be added to the sensor electrolyte in order to achieve the required sensitivity and stability of the Severinghaus carbon dioxide sensor.

The equation for charge neutrality in the sensor electrolyte is then

$$[H^+] = [OH^-] + [HCO_3^-] + 2[CO_3^{2-}] - [Na^+] \tag{4.10}$$

Inserting Eqs. (4.5)–(4.8) in Eq. (4.10) results in

$$[H^+] = \frac{K_W}{[H^+]} + \frac{K_1 K_H pCO_2}{[H^+]} + \frac{2K_1 K_2 K_H pCO_2}{[H^+]^2} - [Na^+] \quad (4.11)$$

which can be written as cubic equation for $[H^+]$:

$$[H^+]^3 + [Na^+][H^+]^2 - (K_W + K_1 K_H pCO_2)[H^+] - 2K_1 K_2 K_H\, pCO_2 = 0 \quad (4.12)$$

or solved for pCO_2:

$$pCO_2 = \frac{[H^+]^3 + [Na^+][H^+]^2 - K_W[H^+]}{K_1 K_H [H^+] + 2K_1 K_2 K_H} \quad (4.13)$$

By definition, the pH value is

$$\text{pH} = -\log[H^+] \quad (4.14)$$

Finally, the output voltage signal E of the pH electrode in Figure 4.3 can be written as

$$\begin{aligned} E = E_1 - S_{\text{rel}} \cdot F_N \cdot \text{pH} &= E_2 + S_{\text{rel}} \cdot F_N \cdot \log(pCO_2) \\ &= E_0 + S_{\text{rel}} \cdot F_N \cdot \log[CO_2] \end{aligned} \quad (4.15)$$

Here, F_N is the Nernst factor (59.16 mV dec^{-1} at 25 °C) and S_{rel} the relative sensitivity of the carbon dioxide sensor, ranging practically between 0.85 and 0.98. The values of E_1, E_2, and E_0 result from the parameters in Eq. (4.13) and also from the reference electrode, from the pH value of the buffer solution and the internal reference electrode of the pH electrode in Figure 4.3. All of these parameters are temperature dependent.

The interesting sensitivity S of the CO_2 sensor is generally characterized by

$$S = \text{dpH}/\text{d}(\log(pCO_2)) = \text{d}E/\text{d}(\log(pCO_2)) \quad (4.16)$$

To determine $[H^+]$ as a function of pCO_2, the cubic Eq. (4.12) can be solved, e.g. by means of Cardano's formula as shown in [17]. However, this method is somewhat cumbersome. Therefore, it is more appropriate to simplify Eq. (4.13) with some very reasonable assumptions. In detail, two cases will be considered [18].

4.1.1.1 Stow's Electrode with $[Na^+] = 0$ mol l^{-1}

In the range $[H^+] > 10^{-6}$ (pH < 6), the terms $K_W[H^+]$ and $2K_1K_2K_H$ in Eq. (4.13) can be neglected. Then, only the terms $[H^+]^3$ in the numerator and $K_1K_H[H^+]$ in the denominator remain in Eq. (4.13), and it reduces to

$$pCO_2 = [H^+]^2/(K_1 K_H) \quad (4.17)$$

$$\log(pCO_2) = 2 \cdot \log[H^+] - \log(K_1 K_H) = -2\,\text{pH} - \log(K_1 K_H), \quad (4.18)$$

$$\text{pH} = -\frac{\log(pCO_2) + \log(K_1 K_H)}{2} \quad (4.19)$$

According to Eq. (4.16), the sensitivity S_{St} of Stow's CO_2 sensor with $[Na^+] = 0$ mol l^{-1} becomes

$$S_{St} = \text{dpH}/\text{d}(\log(pCO_2)) = -1/2 \quad (4.20)$$

4.1.1.2 Severinghaus Electrode with $[Na^+] > 0.001\ mol\ l^{-1}$

In the range $[H^+] > 10^{-6}$ (pH < 6), the terms $[H^+]^3$, $K_W[H^+]$, and $2K_1K_2K_H$ in Eq. (4.13) can be neglected. Then, only the terms $[Na^+][H^+]^2$ in the numerator and $K_1K_H[H^+]$ in the denominator remain, and the Eq. (4.13) reduces to

$$pCO_2 = [Na^+][H^+]/(K_1K_H) \tag{4.21}$$

Analogously to Eqs. (4.18)–(4.20) it follows

$$pH = -\log(pCO_2) - \log[Na^+] + \log(K_1K_H) \tag{4.22}$$

and for the sensitivity S_{Se} of the Severinghaus sensor with $[Na^+] > 0.001\ mol\ l^{-1}$:

$$S_{Se} = dpH/d(\log(pCO_2)) = -1 \tag{4.23}$$

which is twice as high as that one of Stow's sensor in Eq. (4.20).

Figure 4.4 shows the influence of the $NaHCO_3$ concentration and the resulting Na^+ concentration of the sensor electrolyte on the $[H^+]/pCO_2$ function of the Severinghaus carbon dioxide sensor. The curves calculated from Eqs. (4.13) and (4.14) are in good accordance with experimental results (Figure 4.4) [18]. They were helpful to find an optimum sensor design. In principle, the sensor works even if pure water would be used as sensor electrolyte. However, in this case, the sensor function would be less stable, and the CO_2 sensitivity is only $\Delta pH = -0.5$/dec pCO_2 (curve 0 in Figure 4.4). By adding comparatively small amounts of $NaHCO_3$, as proposed in [9], the sensitivity can be considerably raised to about $\Delta pH = -1$/dec pCO_2. The parallel slope of all of the curves with $[Na^+] > 10^{-4}\ mol\ l^{-1}$ in Figure 4.4 indicates that the sensitivity cannot be further increased by adding more $NaHCO_3$. For reasons of sensor stability, typically $10^{-2}\ mol\ l^{-1}$ $NaHCO_3$ is used.

In the range $[H^+] < 10^{-9}\ mol\ l^{-1}$ (pH > 9), the curves in Figure 4.4 are no longer linear, and the sensitivity $dpH/d(\log(pCO_2))$ decreases. For the curve with

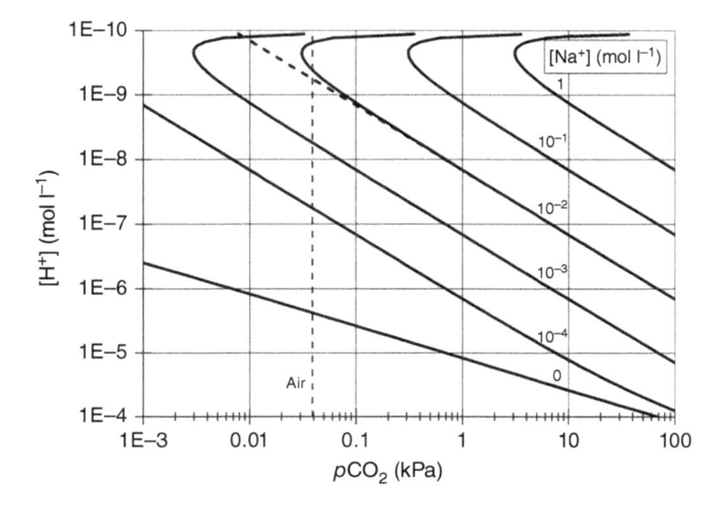

Figure 4.4 Dependence of the sensitivity of the Severinghaus carbon dioxide sensor on the $NaHCO_3$ concentration in the sensor electrolyte, calculated from Eq. (4.13).

$Na^+ = 10^{-2}\,mol\,l^{-1}$, the deviation from the straight line becomes visible at CO_2 partial pressures below c. 0.7 kPa and is already considerable at the partial pressure of CO_2 in the atmosphere of c. 0.039 kPa. According to Eqs. (4.8) and (4.9), the CO_2 partial pressure of 0.7 kPa corresponds to a CO_2 concentration of about $10\,mg\,l^{-1}$.

Figure 4.5 shows the shape, the outer dimensions, and the internal structure of a robust Severinghaus-type electrochemical CO_2 sensor that has been used in several of the application examples described in Chapters 12–15. It can be used both for measurements in liquids and in gases [19].

A calibration curve of such a sensor is shown in Figure 4.6. The CO_2 sensitivity of this sensor decreases at CO_2 concentrations below about $10\,mg\,l^{-1}$, which is in good agreement with Figure 4.4. Consequently, this sensor is not suitable

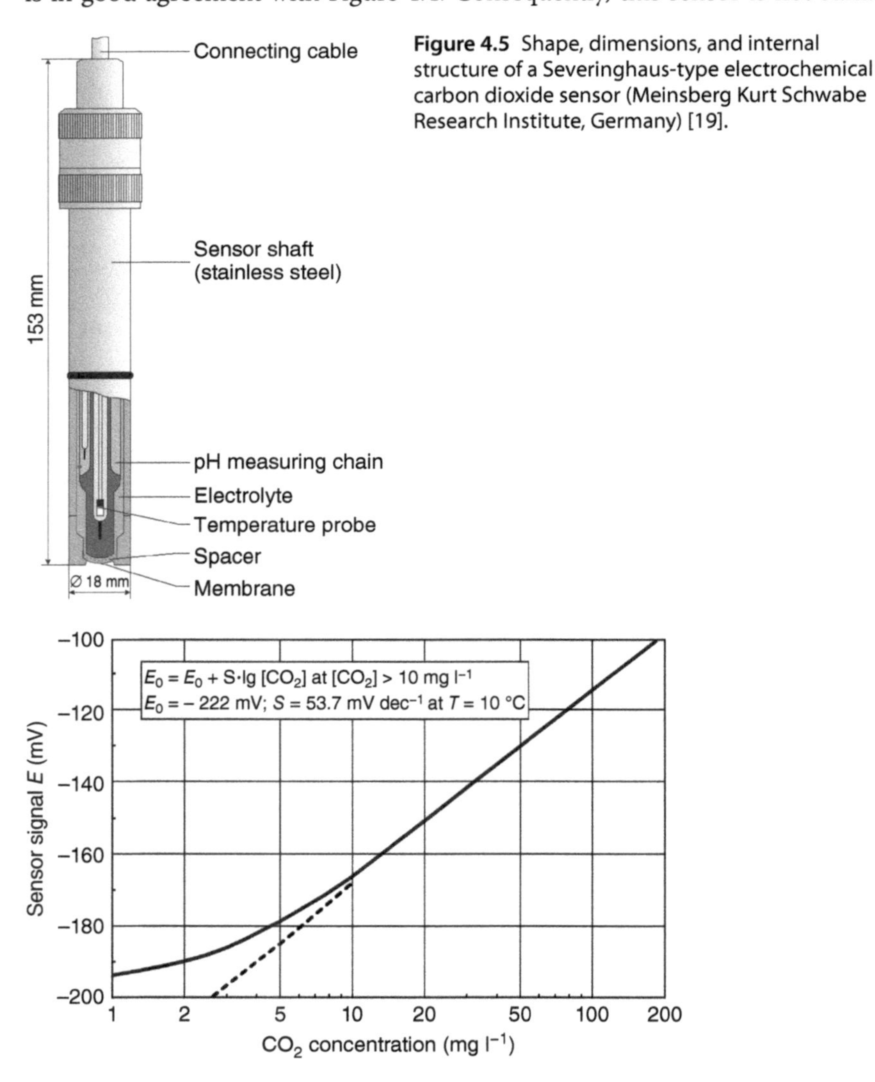

Figure 4.5 Shape, dimensions, and internal structure of a Severinghaus-type electrochemical carbon dioxide sensor (Meinsberg Kurt Schwabe Research Institute, Germany) [19].

Figure 4.6 Calibration curve of the Severinghaus carbon dioxide sensor as shown in Figure 4.5.

particularly for the direct measurement of CO_2 in the normal atmosphere, where the resulting CO_2 concentration is somewhat below 1 mg l^{-1}, but it can be used favourably to measure higher concentrations, for example, with CO_2 warning devices.

4.1.2 Sensor Electrolyte

The sensor characteristics of the Severinghaus electrode are substantially influenced by the hydrogen carbonate concentration of the internal electrolyte. A profound graphical representation of the sensitivity and of the measuring range of the sensor in dependency of the dissolved concentration of hydrogen carbonate is shown in Figure 4.4. A tailored fit of the $NaHCO_3$ amount in the inner electrolyte in relation to the expected carbon dioxide concentrations enables – in case of a sufficient small inner solution volume – a fast-responding sensor with appropriate measuring range and allows the reduction of the influences of interfering species [20].

For a long-lasting stability of the sensor characteristics and a minimum of calibration and maintenance efforts, the composition of the internal electrolyte has to be kept constant during the utilization of the electrode. This is still a challenge when the sensors are stored in dry ambient air or the probes are used for long-term measurements in gaseous phases. The vapour pressure of H_2O causes a decrease of the water volume by permeation through the covering membrane resulting at least in a rising drift and an increased demand of calibration measures. The addition of ethylene glycol to the electrolyte offers the possibility to decrease the vapour pressure of the water without a significant deterioration of the essential electrochemical properties of the Severinghaus electrode. If the ambient conditions for the long-term measurements are known and constant, the ratio of water to ethylene glycol can be adapted to the relative humidity and the mean temperature of the surrounding environment of the probe. The resulting H_2O pressure p_w above a water/ethylene glycol mixture in dependency of the volumetric water content c_w can be calculated by the following empirical equation:

$$p_w(\text{Pa}) = \sum_{i=0}^{3} a_i \cdot c_w^i(\text{vol}\%) \tag{4.24}$$

At 25 °C the values of the constants have been determined as $a_0 = 95.97$, $a_1 = 80.19$, $a_2 = -0.905$, and $a_3 = 4.4 \times 10^{-3}$. If, for instance, the measured gas matrix at room temperature would have a relative humidity of 50%, an admixture of nearly 75 vol% of ethylene glycol in the internal electrolyte would establish a stable equilibrium between the inner and the outer water vapour pressure. This composition would allow a calibration-free monitoring of CO_2 with these probes for more than six months.

A comparable way has been chosen for the transcutaneous determination of arterial carbon dioxide partial pressure in newborn children with Severinghaus-type sensors. The determination of the measurand requires an elevated temperature of 44 °C at the sensor surface for the establishment of a sufficient CO_2 transport through the skin. For the stabilization of the inner electrolyte of the carbon dioxide sensor, an addition of ethylene glycol in the internal hydrogen carbonate electrolyte has been applied [21].

If the Severinghaus sensors are stored under wet condition and are only introduced in aqueous solution, an addition of water vapour suppressors like ethylene glycol is not required.

An interesting attempt to improve the response time of the carbon dioxide sensors has been undertaken by the addition of carbonic anhydrase (CA). This enzyme is known to catalyse the hydration of CO_2 and was expected to accelerate the measuring process of Severinghaus-type CO_2 sensors when added to the internal electrolyte. Although the research results showed improved response times and reduced hysteresis, this additive has not been established in general for potentiometric carbon dioxide probes [22–26].

4.1.3 Membrane Materials

The gas-permeable membrane is an essential part of the Severinghaus-type carbon dioxide electrode. The necessity and the advantages of the membrane coverage of the inner electrolyte for this kind of sensor have been addressed in [27] and can shortly be summarized as follows:

- The membrane encloses continuously a small and defined volume of the inner electrolyte. This ensures a stable geometry of the liquid layer in front of the pH electrode.
- The need for an exchange of this inner hydrogen carbonate solution for maintenance can be reduced, because any passing of ionic components between inner and outer phases is prevented. Only the required permeation of carbon dioxide (and other gaseous or non-ionic compounds) is still possible.
- The evaporation of water is restricted as far as possible, and a drying-out of the electrolyte with grave impact on the sensor functions and the quality of the pH-sensitive layer of the glass electrode can be reduced.
- The stabilizing membrane enables the utilization of the sensor system in continuous flow analysis.

For the sensor membranes, hydrophobic materials, e.g. silicone rubber, polymethylpentene (TPX), polypropylene (PP), polyvinyl chloride (PVC), and polytetrafluoroethylene (PTFE), have been introduced. While Stow et al. used a stretched rubber membrane for his CO_2 probe [2], Severinghaus employed a Teflon™ sheet for his sensor assembly [4]. These materials represent two different kinds of membranes available for the gas transport. On one hand side, inert microporous membranes are applied for the CO_2 transport, which is established by the diffusion of the gas through the electrolyte-free but air-filled pores and channels [28]. On the other hand, passing of the analyte through the homogeneous membrane is induced by a first dissolving step of CO_2 in the bulk of the polymer [28, 29].

Microporous materials (e.g. Teflon™, cellulose acetate, polypropylene, polyethylene) show faster responses because of the high diffusion coefficient D of carbon dioxide in the gas phase. Pore sizes for PTFE polymers of 1 μm are estimated in potentiometric gas sensors for membranes with a thickness of about 100 μm. A slight increase of the pore size can be achieved by a prolongation of the membrane material. In contrast to the microporous membranes, the CO_2

transport through the homogeneous polymer bulk suffers from considerable smaller diffusion coefficients, reducing the response speed of these sensor systems. The values for D in this matrix are estimated to 4 orders of magnitude smaller than in air. In order to decrease the time for the passage between the liquid phases, the thickness of the polymer has to be reduced. Mechanically stabilized membranes with thicknesses of about 10–25 μm have been described.

The impact of the different membrane types on the time response as well as on the selectivity has been investigated in comparing studies. Especially the influence of microporous or homogeneous membranes on the suppression of interfering organic and inorganic acids has been in the focus of research. A detailed investigation has been executed by Kobos et al. [30] who have determined the influence of NO_2 and SO_2 and of different organic acids such as formic acid, acetic acid, ascorbic acid, benzoic acid, and cinnamic acid (interfering substances dissolved in slightly acidic sodium citrate buffer with a pH of 4.45) on the electrode potential within 20 minutes. The Severinghaus-type sensors have been covered with microporous Teflon™ or polypropylene membranes or with homogeneous Teflon™ or silicone rubber films, respectively. The probes with microporous coverage showed a remarkable response on volatile organic acids such as formic acid and acetic acid as well as on NO_2. In contrary, the silicone rubber suppressed the influence of the short-chained carboxylic acids and decreased the interference of NO_2, but the influence of cinnamic acid or benzoic acid was considerably increased. These observations have been explained by the different transport mechanisms through the gas-permeable membrane. Compounds with high volatility cross the micropores of the membrane in higher amounts than substances with lower vapour pressures. Meanwhile, the first step for the transport through homogeneous membrane is the dissolution of the interfering substances in the polymer film, followed by the diffusion-driven transport through the bulk. Obviously, aromatic acids such as cinnamic acid, benzoic acid, or salicylic acid show a better solubility in the membrane polymer than the short-chained carboxylic acids, resulting in an explicitly faster response of the sensor on these compounds.

The best selectivities have been obtained with a Tefzel™ membrane, a polyethylene/PTFE-based copolymer establishing a homogeneous Teflon-like material. The determined characteristics are comparable with the homogeneous silicone rubber membrane with respect to volatile compounds. Aromatic acids crossed the fluoropolymer remarkably slower due to their lower solubility. But the improved selectivity of the Tefzel™ polymer film was accompanied with a longer response time. In addition, a double-layer membrane as a combination of microporous and homogeneous materials was tested to resemble the positive effects of both materials. However, the response time was prolonged by a factor of two, and the selectivity did not reach the quality of the Tefzel™ polymer films.

While the studies in [30] were mainly focussed on the rate of the passage of interfering compounds through the different types of membranes during a defined period, the test series in [31] dealt with the selectivity results after reaching the steady state of the electrode potential. Microporous Teflon™ with a polyethylene support and a silicone rubber membrane based on polymethylvinyl-siloxane had been applied for the investigation of clinically important organic acids present in blood. Use of both materials showed no significant response to pyruvic, DL-lactic, oxalic, or succinic acid. In addition, the theoretical selectivity coefficient of the potentiometric output of the Severinghaus-type electrode has been calculated for numerous organic compounds under consideration of their acidity constants related to carbon dioxide. A closer look on the derivation of the selectivity values has been executed in [31]. The calculated results based on the selectivity coefficients differed only slightly from the experimental data. This allows the conclusion that only the chemical equilibrium in the inner electrolyte is decisive for the steady-state response of the sensor system. A higher volatility is accelerating the establishment of the equilibrium and the sensor response of Teflon-based CO_2 probes, respectively, while low vapour pressure reduces the exchange and the response time. The different behaviour of silicone membrane-covered sensors on aromatic acids and short-chained carboxylic acids can clearly been addressed to the widespread solubility of the different compounds in the polymer film. While acetic and formic acids are nearly insoluble, benzoic acid is remarkably dissolved in the membrane bulk and is available for the transport in the inner electrolyte.

A further development step in the calculation of the sensor response was described in [32], where the permeability coefficients as well as the acidity constants were taken into account in a theoretical consideration of the progress of the CO_2 sensor signal of Severinghaus electrodes. A good agreement of the calculated and the observed potential curves in dependency from the thickness of plasticized PVC membranes has been achieved by the theoretical model introduced there.

4.1.4 Temperature Dependence

The temperature dependence of the permeability P can be calculated from the activation energy E_a:

$$P = P_0 \cdot \exp\left(-\frac{E_a}{RT}\right) \tag{4.25}$$

The ratio $p(H_2O)/p(CO_2)$ of the membrane permeabilities for H_2O and CO_2 is of special importance regarding to the long-term stability of equilibrium-type carbon dioxide sensors. This parameter depends on the membrane material and amounts to c. 13 and for less than unity for TPX. The latter is therefore especially suited as membrane material for electrochemical CO_2 gas sensors.

4.1.5 Response Behaviour

The time response of a potentiometric membrane-covered CO_2 probe on concentration changes of the analyte is influenced by the characteristics of the different sensor components themselves and by their kind of assembly in the Severinghaus-type electrode. Profound depictions of the influencing factors on the response behaviour of potentiometric gas sensors covered by gas-permeable membranes are topics in various publications mainly focussed on the determination of carbon dioxide and ammonia [29, 32–37].

The following parameters have been specified in [27] to decisively influence the response speed of Severinghaus-type potentiometric probes:

- Diffusion coefficient of the gaseous analyte in the separating membrane,
- Partition coefficient of the gas between separating membrane and analysed matrix,
- Thickness of the separating membrane,
- Geometry and thickness of the electrolyte film between pH sensor probe and separating membrane,
- Response characteristics of the pH electrode,
- Concentration of the measurand, and
- Temperature of the solution.

The central and most decisive component of the CO_2 sensor is the pH electrode. Neither improvements of the complete sensor assembly nor changes of the gas-permeable membrane can accelerate the dynamics of the pivotal pH glass probe. But investigations have shown that the response characteristic of the glass electrode is much faster than the time needed for the establishment of the chemical equilibrium in the $NaHCO_3$ solution placed between the membrane and the sensory pH glass layer [27]. Nevertheless, degradation of both the response time and the slope of the electrode, especially in slightly buffered electrolyte solutions, during continuous operation within several months must be taken into consideration.

Other main influences are the geometry of the sensing tip of the Severinghaus-type electrode and the properties of the separating membrane. For a rapid pH adjustment after changing of the CO_2 concentration in the sample, a reduction of the thickness of the hydrogen carbonate electrolyte film between the membrane and the sensing area of the glass electrode is essential. In addition, the thickness and kind of the gas-permeable polymer plays a central role for the sensor response.

A comprehensive theoretical description of the electrode response characteristics of potentiometric gas sensing electrodes has been given in [28]. Main aspects of the theoretical paper can be shortly summarized in the following passages.

A basic assumption is that the transport of the analyte through the gas-permeable membrane is the time-determining process for the establishment of the equilibrium between the analyte outside and the CO_2/HCO_3^- system in the internal electrolyte. In addition, the buffer film placed between the electrode surface and the polymer should be presumed to be thin in comparison with the thickness of the membrane. In this case, the diffusion processes and the pH gradient in this inner electrolyte can be neglected. Another assumption concerns the fast achievement of the distribution of carbon dioxide across the two interfaces between polymeric phase and sample matrix and polymeric phase and inner buffer solution.

Figure 4.7 illustrates the concentration levels and gradients of CO_2 in the involved phases shortly after an increase of the analyte concentration from C_1 to C_2. The thickness m of the membrane is large compared with that of the space needed for the inner electrolyte. The rise of carbon dioxide in the analysed matrix causes a transport through the membrane to adapt finally the inner

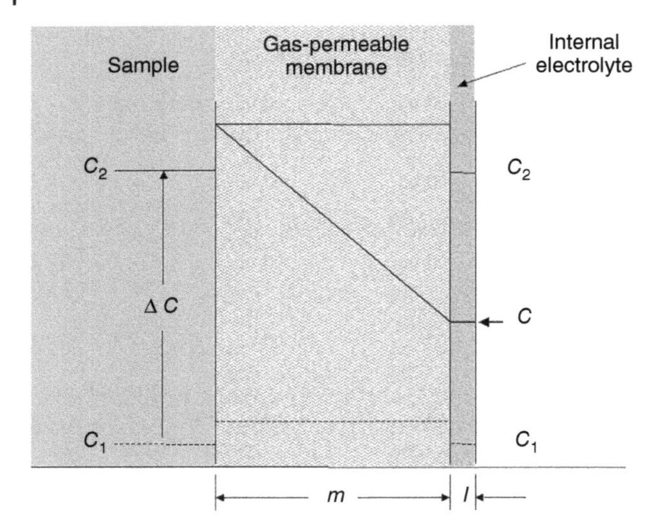

Figure 4.7 Steady-state model of the electrode response of the CO_2 electrode according to [28]. Source: Ross et al. 1973 [28]. Reproduced with permission of Elsevier.

concentration to the ambient one. The flux of CO_2 through the polymer film stops when the equilibrium state is achieved.

The fractional approach

$$\varepsilon = \left| \frac{C_2 - C}{C_2} \right| \tag{4.26}$$

during this process is dependent on the end concentration C_2 and can be calculated for an intermediate concentration C at the inner interface. Considering the diffusion coefficient D of CO_2 in the gas-permeable membrane and its partition coefficient k between the solutions and the bulk of the polymer material, the following relationship applies [28]:

$$t = \frac{lm}{Dk} \left[1 + \frac{dC_B}{dC} \right] \ln \frac{\Delta C}{\varepsilon C_2} \tag{4.27}$$

where ΔC represents the difference between the starting concentration of the membrane-passing CO_2 in the sample and its amount after the contemplated change. The term dC_B/dC reflects a concentration-dependent change of the ratio of the ionic species such as HCO_3^- and CO_3^{2-} in the inner electrolyte towards H_2CO_3 and CO_2:

$$\frac{dC_B}{dC} = \frac{\Delta[HCO_3^-] + \Delta[CO_3^{2-}]}{\Delta[H_2CO_3]} \tag{4.28}$$

For a simplified estimation of the sensor response, the relationship of dC_B/dC can be assumed as constant and is therewith negligible for the calculation of the concentration dependency of the response time t. The latter can directly be linked to the concentration change:

$$t \approx \ln \frac{\Delta C}{\varepsilon C_2} \tag{4.29}$$

Table 4.1 $\ln \Delta C/\varepsilon C_2$ ($\varepsilon = 0.01$) versus analyte concentration change reflecting the relative response times.

Start concentration, C_1 (mol l^{-1})	End concentration, C_2 (mol l^{-1})	$\ln \Delta C/\varepsilon C_2$
10^{-3}	10^{-1}	4.6
10^{-1}	10^{-3}	9.2
10^{-5}	10^{-1}	4.6
10^{-1}	10^{-5}	13.8

The time required to establish either 90% (t_{90}) or 99% (t_{99}) of the end signal is a practicable and meaningful parameter for the characterization of chemical sensors. The t_{99} time is equivalent with a fractional approach factor ε of 0.01. Assuming that l, m, k, and D stay constant during the measuring process, Eq. (4.27) allows the calculation of the temporal progress of the equilibrium and the analytical signal in dependency of the chosen boundary concentrations C_1 and C_2. It is worth mentioning that the response times of a sensor differ if the content of the analyte is changing from a lower value to a higher value or vice versa. If the starting concentration is small compared with the final one, then the ratio of $\Delta C/C_2$ approximates unity and the response time is nearly independent of the extent of the analyte change. In contrast, for a change from a higher concentration to a distinct lower concentration, the response time varies strongly on the magnitude of this change because ΔC is much larger than the final value C_2 (Table 4.1). It can be seen that the response time t_{99} needed for the reduction of the analyte content to 1% is twice the value than for the corresponding concentration increase.

In [38] it was shown that the assumption of a constant ratio of dC_B/dC is no longer valid for small H_2CO_3 concentrations. If the concentration of H_2CO_3 exceeds 2 mM in the analyte, then the response time is nearly independent of the solved hydrogen carbonate content in the internal electrolyte film of the Severinghaus-type sensor. However, for lower CO_2 values, the time characteristics were strongly dependent on the amount of HCO_3^- solved in the inner electrolyte. An increase of hydrogen carbonate resulted in a distinct extension of the response time for the measurement of low carbon dioxide values due to the higher need of CO_2 permeating through the membrane to establish the equilibrium in front of the pH electrode.

All these considerations were made for Severinghaus-type electrodes and based on the assumption that the thickness of the electrolyte film placed between the pH glass electrodes is small and does not affect the response time. However, especially electrolyte located at the rim of the sensing electrode placed beside the bulk electrolyte, which has to be measured by the pH electrode, cannot be neglected. This additional electrolyte reservoir damps the establishment of the sensor potential at the pH electrode surface. A more detailed consideration of these effects on the response time of different designs of Severinghaus-type electrodes can be found in [29].

4.1.6 Calibration of Electrochemical CO_2 Sensors

As it is generally required for electrochemical sensors, Severinghaus-type CO_2 sensors have to be calibrated regularly according to given temporal or user-specific regimes to compensate for unavoidable changes of the sensor parameters and to maintain high measuring accuracy over longer periods. Whereas CO_2 sensors for measurements in gases can simply be calibrated by using commercially available test gases, the calibration of electrochemical CO_2 sensors for measurements in liquids is not trivial [39]. Contrary to standardized, long-term stable buffer solutions with defined pH values, which are generally used in pH measuring techniques, solutions with defined CO_2 content are not stable over longer periods of time and, therefore, commercially not available. For this reason, calibration solutions with defined CO_2 concentration must be prepared immediately before starting the calibration procedure. This can be made as follows.

At first, solutions with the required $NaHCO_3$ concentrations are prepared. Then, by adding an organic acid (e.g. succinic or oxalic acid), the pH value of the solution is decreased to $pH \approx 3$, changing the hydrogen carbonate in the solution completely into dissolved free CO_2 in the desired concentration according to reaction equation

$$2NaHCO_3 + C_4H_6O_4 \rightleftarrows Na_2C_4H_4O_4 + 2CO_2 + 2H_2O \tag{4.30}$$

A special calibrating vessel [19] and high experimental care are necessary to prevent the escape of CO_2 from the calibrating solution as it happens when opening a mineral water bottle. Especially the user has to avoid any airspace in the vessel to prevent a change of dissolved CO_2 in the gas phase. Nevertheless, the calibrating solutions cannot be stored over long time.

Since the sensitivity of electrochemical CO_2 sensors changes only slightly in the course of time, it is sufficient to carry out regularly one-point calibrations in order to correct the potential drift of the sensor. To consider temporal changes of the sensitivity, two-point calibrations are only necessary in longer time intervals or after the sensor had been out of use for a longer period. If the sensor is continuously in use, as a rule, then one-point calibrations at least once per week and two-point calibrations at least once per month are recommended. For two-point calibrations, two calibration solutions with different CO_2 concentrations are needed. Generally, the expected CO_2 concentration of the measuring liquid should be within these concentration values. To avoid larger temperature-dependent errors, the temperature of the calibration solution should deviate by no more than 5 K from that of the measuring solution. Before starting the calibration procedure, the electrode should be inserted at least for 15 minutes into the solution.

However, the calibration of electrochemical CO_2 sensors is particularly difficult if measurements under field conditions and in solutions of unknown and changing composition must be carried out. It was found that considerable deviations might occur between the analytically determined CO_2 concentrations and those measured with the sensor if the measuring and the calibrating solutions differ substantially in their composition. For this reason, it was proposed to check the electrochemical CO_2 sensor in the actual measuring solution with a multi-parameter measuring instrument also containing sensors to determine pH value, conductivity, and temperature [40]. During measurement and calibration, an integrated microcomputer compares the measured values of the CO_2 sensor with the values measured with the other sensors according to given functions and limit values.

4.2 Coulometric and Amperometric CO_2 Sensors

4.2.1 Operation Principle

Since carbon dioxide exhibits a highly negative reduction potential, the coulometric determination of CO_2 is difficult. Instead, the detection of reaction products of CO_2 that are formed in aqueous solutions can be used to overcome this disadvantage.

The first one-electron reduction of CO_2 results in the formation of radical anions $CO_2^{\bullet -}$ according to

$$CO_2 + e^- \rightleftarrows CO_2^{\bullet -} \tag{4.31}$$

The strong negative reduction potential ($E^0 = -1.8\,V$ [41] and $-1.85\,V$ [42] versus normal hydrogen electrode (NHE) has been explained by the required bending of the sp orbital in the sp^2 state yielding in a slow reduction kinetics and a high overpotential [43, 44]. The next following one-electron reduction

$$CO_2^{\bullet -} + e^- \rightleftarrows CO_2^{2-} \tag{4.32}$$

leads to the formation of carbon dioxide dianions. The corresponding reduction potential amounts to $E^0 = -\ 1.2\,V$ versus NHE [45]. Finally, formic acid HCO_2H is formed.

Previous research has been targeted on the electrochemical titration of the protons evolving during the solvation of carbon dioxide. Here, the required hydroxide anions are generated by a cathodic water electrolysis. Therefore, these sensors consist both of a pH electrode for pH control and of a cathode for the evolution of hydrogen, resulting in the formation of titrating hydroxide anions.

Alberty and Uttamlal have described the working principle of a coulometric CO_2 sensor [46] and its design [47] as well as the application of the probes for monitoring the fermentation process of beer [48]. In their sensor design, a pH-sensitive IrO_2 electrode was placed in the centre of the device closely surrounded by a platinum ring used as cathode. A further platinum ring served as counter electrode. Potential measurements were controlled by an $Ag/AgCl/Cl^-$ reference electrode. The electrolyte between the sensor element and a gas-permeable membrane consisted of 1 M $NaClO_4$ with an additional 0.1 M amount of chloride to establish a constant reference potential. Fluidic parts allowed a change of the electrolyte after each measurement to guarantee comparable starting conditions. For the generation of hydroxide according to

$$H_2O + 2e^- \rightarrow 2OH^- + H_2 \tag{4.33}$$

a constant current of 15 μA had been applied for 100 seconds. After electrolysis the permeation of carbon dioxide into the electrolyte establishes a hydrogen carbonate system and results in a slow pH drift into the acidic direction monitored by the IrO_2 electrode. Theoretical considerations showed that assuming a constant permeation of carbon dioxide into the electrolyte, the time-dependent increase $d(\ln[H^+])/dt$ of the proton concentration reaches a maximum at a pH value close to 8. In doing so, the inflexion point of the potential versus time curve of the pH sensor could be used for the definition of an endpoint of the titration of hydroxide with the carbon dioxide to be analysed. The concentration of carbon dioxide is inversely proportional to the time required to complete the titration. After changing the consumed electrolyte, the measuring cycle could be started again. The measuring time could be affected by the choice of the gas-permeable membrane influencing the CO_2 uptake and by the variation of the charge used for the hydroxide evolution.

This suitability of such sensors has been shown for the determination of CO_2 in greenhouses and for an alcoholic fermentation process.

One disadvantage of coulometric CO_2 sensors is the necessary exchange of the inner electrolyte, which requires a fluidic component in the sensor assembly. To overcome this drawback, an alternative set-up with a long-term stable inner electrolyte as shown in Figure 4.8 was proposed by Trapp et al. [49].

Two measures have been implemented to increase the sensor lifetime without maintenance, to reduce the size of the probe, and to minimize the influence of the electrochemical side reaction at the anode on the measuring process: (i) For separating the anolyte from the working catholyte at the sensing element, a diaphragm has been introduced. This reduces the impact of the counter reaction at the anode on the pH measurement during the evolution of hydroxide anions at the working electrode. (ii) The pH sensing IrO_2 electrode has been used in parallel as the essential cathode for the production of the titrating hydroxide anions. The electrical current pulses for the evolution of hydrogen will be shortly interrupted by potentiometric measuring cycles to monitor the pH development at the IrO_2 electrode. These sequences are repeated until the designated pH value is reached. After the pH-adjusting process, the intruding carbon dioxide acidifies the electrolyte until a defined threshold pH value is reached and the cathodic formation of hydroxide for the titration starts again. The time for reaching the acidically limiting value represents the time-averaged partial pressure of CO_2. Figure 4.9 shows in principle the pH course during two complete measurement cycles.

In sections I and III, the pH value changes by CO_2 entering the measuring electrolyte; in sections II and IV, the starting conditions are re-established by a

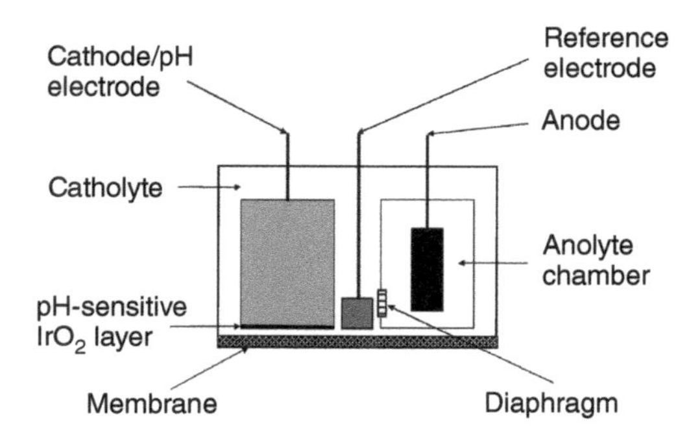

Figure 4.8 Scheme of the cell design used for coulometric CO_2 sensors without fluidic components. Source: Trapp et al. 1998 [49]. Reproduced with permission of Elsevier.

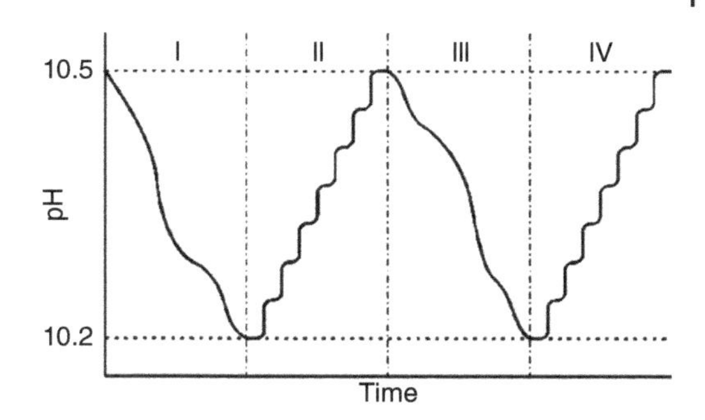

Figure 4.9 Course of the pH value during two measuring cycles for the measuring set-up of Figure 4.8. Source: Trapp et al. 1998 [49]. Reproduced with permission of Elsevier.

stepwise coulometric titration and subsequent potentiometric monitoring of the pH value during the breaks between the cathodic pulses.

4.2.2 IrO_2 Electrode

The preparation of the IrO_2 electrode required for the pH measurement as well as for the hydrogen evolution is of central importance. It has been executed by thermal treatment of $IrCl_3$ as well as by reactive sputtering of Ir in oxygen atmosphere. The cathode was surrounded by an Ag/AgCl ring as referencing element. The electrolyte had been adjusted to a starting pH of 11.5, ensuring a rapid reaction of entering carbon dioxide with dissolved hydroxide. The electrolyte additionally included a 1 M amount of potassium chloride and a 1 M concentration of $BaCl_2$. The Ba^{2+} salt was chosen to eliminate carbonate from the alkaline by precipitating hardly soluble $BaCO_3$. The measuring process was switched between the pH values of 11.2 and 11.5, guaranteeing a rapid solvation of carbon dioxide.

The lifetime has been improved by the addition of hygroscopic compounds like glycols to reduce the drying of the electrolyte. The sensors have been successfully tested in air with CO_2 concentrations between 200 and 20 000 ppm When the electrolyte reservoir was equipped with a 12.5 μm thick Teflon™ membrane, the response time varied between a few seconds for carbon dioxide amounts of more than 10 000 ppm and several minutes for concentrations below 1000 ppm The coulometric probes have been applied for the monitoring of carbon dioxide in marine waters [50].

4.2.3 Amperometric CO_2 Sensors

An alternative electrochemical sensor for the detection of CO_2 in seawater is described in [51, 52]. Here, the pH-dependent reduction of platinum oxide [51] and ruthenium oxide is used. Driving force of the sensor is the interplay of the protons, evolved by the dissociation of carbonic acid with the electrochemical reduction process of the noble metal oxide. The essential reaction equations for platinum and for ruthenium are initiated by a proton-generating step:

$$CO_2 + H_2O \rightleftarrows HCO_3^- + H^+ \quad (4.34)$$

It is followed by the proton-assisted reduction of the noble metal oxide for Pt and Ru, respectively:

$$PtO + 2H^+ + 2e^- \rightleftarrows Pt + H_2O \quad (4.35)$$

$$RuO_2 + 4H^+ + 4e^- \rightleftarrows Ru + 2H_2O \quad (4.36)$$

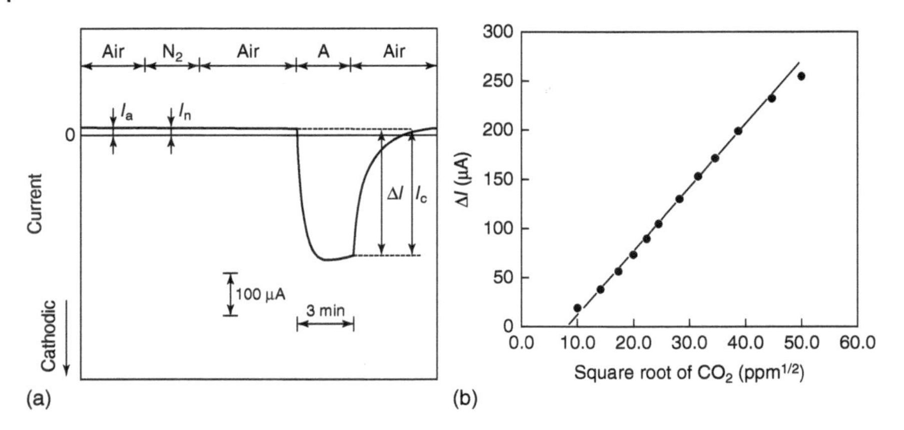

Figure 4.10 Current response of the RuO_2 electrode on pure N_2, air, and air with an amount of 2500 ppm carbon dioxide (a). Linearity of the current as function of the square root of the carbon dioxide concentration in the range between 100 and 2500 ppm CO_2 (b). Source: Ishiji et al. 2001 [52]. Reproduced with permission of Elsevier.

The reduction process is promoted by the released protons, and the resulting current is proportional to the square root of the CO_2 concentration [51]. Figure 4.10 shows the sensor response of the RuO_2 electrode and the dependency between current I and $(pCO_2)^{1/2}$.

The Pt/PtO working electrodes (thickness of 0.5 μm) have been prepared on porous polytetrafluoroethylene (PTFE) membranes by sputtering processes with a Pt target in reactive oxygen–argon atmosphere. RuO_2 layers were deposited via screen printing methods by application of RuO_2 powder on gas-permeable PTFE. Due to the spatial proximity of porous membrane, noble metal coating, and inner electrolyte, such a set-up guarantees the shortest distance between permeating and solving carbon dioxide and electrode surface, finally resulting in a fast response of the sensor. For both metals the best results had been achieved for a working potential of 0.2 V versus a saturated Ag/AgCl/Cl^- electrode. Higher potentials decreased the sensitivity of the device, while values below 0.2 V resulted in remarkably rising interfering currents caused by oxygen reduction.

Another approach for the detection of carbon dioxide has been demonstrated in [53] for a breath-by-breath determination of CO_2 during respiration. The sensor is based on a pulsed titration of carbon dioxide dissolved in dimethyl sulfoxide (DMSO). A direct electrochemical reduction of carbon dioxide, leading finally to the generation of formate, can be executed in wet DMSO at silver, gold, or mercury electrodes and at working potentials of c. −2.0 V versus saturated calomel electrode (SCE) [54]. This reaction is interfered by solved oxygen, which is already reduced at less negative potentials. To overcome this problem, a membrane-shaped cathode was placed in front of the CO_2 electrode. This additional electrode was introduced for a reductive removal of oxygen, decreasing the interference on the CO_2 measurement. It can also be applied for the amperometric measurement of O_2 [55]. For a rapid breath-by-breath monitoring of exhaled carbon dioxide, the double-membrane system proved to be insufficient. Instead of that, the reversible one-electron reduction of O_2 to an oxygen radical anion $O_2^{\bullet-}$ in dry DMSO was used:

$$O_2 + e^- \rightleftarrows O_2^{\bullet-} \quad (4.37)$$

The formation of $O_2^{\bullet-}$ was executed in pure DMSO with a 0.20 M content of tetraethylammonium perchlorate as conducting salt. This intermediate oxygen compound reacts in DMSO in several steps by combination with dissolved carbon dioxide according to the following reaction scheme:

$$2CO_2 + 2O_2^{\bullet-} \rightarrow C_2O_6^{2-} + O_2 \quad (4.38)$$

forming at least peroxodicarbonate and oxygen. This titration step for CO_2 is followed during the CO_2 quantification by a reoxidation of the remaining oxygen radical anions. The extent of the decrease of the anodic current compared with the previous cathodic step is a measure for the amount of dissolved carbon dioxide. A more detailed description including theoretical considerations concerning transport process and electrode kinetics can be found in [56].

The sensor was designed similar to a Clark-type oxygen electrode with a gold working electrode, a platinum counter electrode, and a saturated $Ag/AgCl/Cl^-$ reference electrode. Typical pulse times for the oxygen reduction resulting in the titration of dissolved CO_2 lasted 200 ms, while the reverse pulse for the reoxidation of the radical anion took 350 ms. This time sequence of measurement cycles enabled the monitoring of expired CO_2 during breathing, breath by breath (Figure 4.11).

Further improvements of this sensor principle for the CO_2 detection including deep insights in the electrochemical steps and the reaction sequence for the formation of peroxodicarbonate [57–61] and an extension of the operational area of the system on the determination of excessive N_2O in anaesthetic gases have been described in [62].

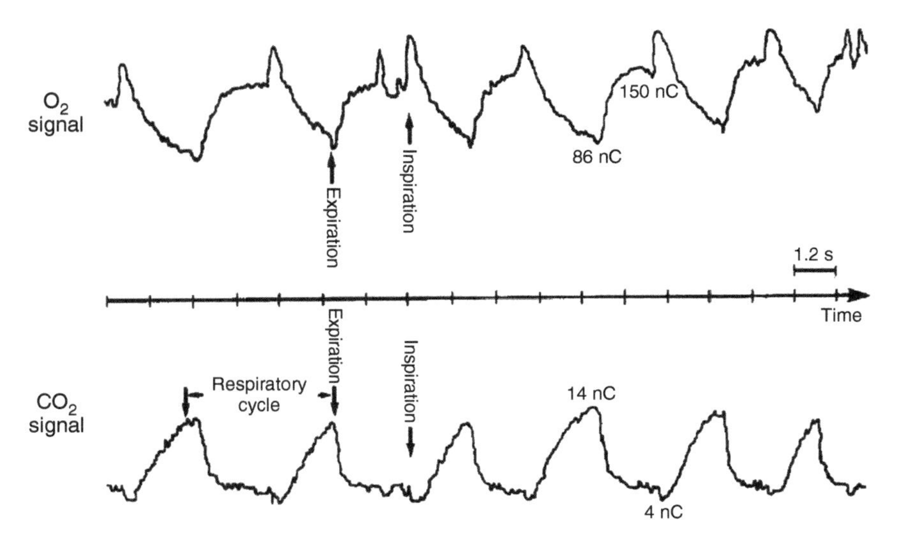

Figure 4.11 CO_2 and O_2 monitoring of human breath with a pulsed titration sensor according to [53]. Source: Albery et al. 1992 [53]. Reproduced with permission of Elsevier.

4.3 Conductometric CO_2 Sensors

Conductometric carbon dioxide sensors use a membrane-covered set-up as shown in Section 4.1. One of the first publications on CO_2 sensors working on conductometry [63] outlines that the motivation behind this development was the significant decrease of the relatively long response times of equilibrium sensors to meet the requirements of medical applications, e.g. breathing gas monitoring. The idea behind this approach is based on the fact that the relatively slow dissociation reaction of CO_2 in liquid electrolytes can be excluded from the sensor response by its dislocation into a flowing electrolyte. Additionally, the flowing electrolyte causes maximum partial pressure difference on both sides of the diffusion membrane, which promotes short response times.

Conductometric sensors usually contain two conductometric flow-through cells upstream and downstream of the membrane diffusion zone, respectively (Figure 4.12). The sensor signal is generated by conductivity changes in the electrolyte due to CO_2 dissociation and measured as conductivity difference between the upstream and the downstream cell. The molar conductivity difference $\Delta\Lambda$ depends on the CO_2 flow across the membrane and the electrolyte flow rate and can be calculated roughly by

$$\Delta\Lambda = 394.5\ \mathrm{S\,cm^2\,mol^{-1}}\ c(\mathrm{H_2CO_3}) \tag{4.39}$$

During the last four decades different attempts have been made to adapt this sensor principle to the requirements of different applications [64–66]. Especially the use of planar electrode structures for conductivity measurements has been proposed in several publications [67–70]. Such electrodes are easily miniaturizable and cost effective and, therefore, have decisive advantages over conventional conductivity electrodes. They can be produced using standardized thick- or thin-film processes and, hence, are suited for mass production.

The approach of Figure 4.12 utilizes the multilayer printed-circuit-board (PCB) technology as used today for highly integrated electronic devices such as mobile phones and portable computers [71].

All electrodes were made from Pt foils with a thickness of 20 μm and with a distance of 100 μm between the two electrodes of every cell. The microchannel with a cross section of $300 \times 30\ \mu m^2$ and a length of 10 mm was engraved in the top layer by milling. Silicone membranes of different thicknesses were glued onto the surface of the substrate. Due to the relatively low H_2CO_3 concentrations in the flow at the downstream cell, the sensor was fed with deionized water with conductivities below $1\ \mu S\ cm^{-1}$ at different flow rates.

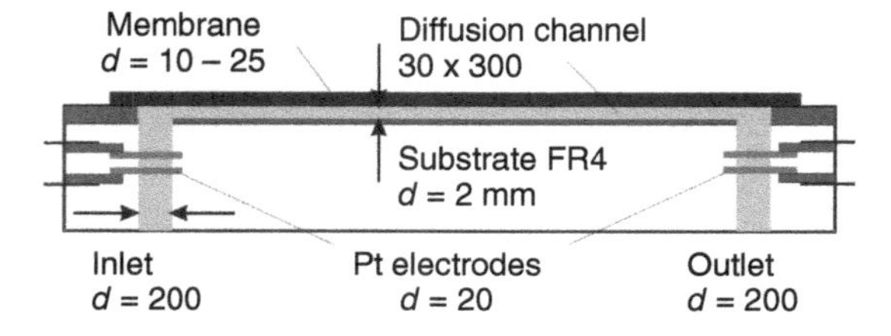

Figure 4.12 Schematic drawing of the cross section of a planar conductivity-type sensor manufactured by printed-circuit-board technology as a four-layer structure (from [70]), all dimensions in μm.

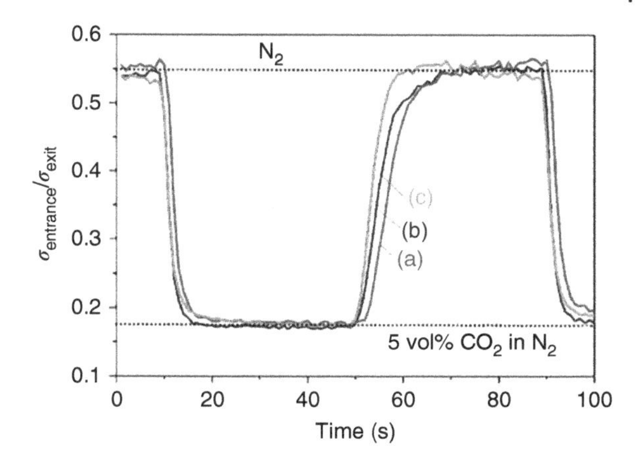

Figure 4.13 Response of the planar CO_2 sensor from Figure 4.12 to fast changes of the CO_2 concentration. Membrane thickness, 25 µm; membrane material, silicone; temperature, 22 °C; electrolyte, deionized water. Flow rates: (a) 20, (b) 40, and (c) 80 µl min^{-i}.

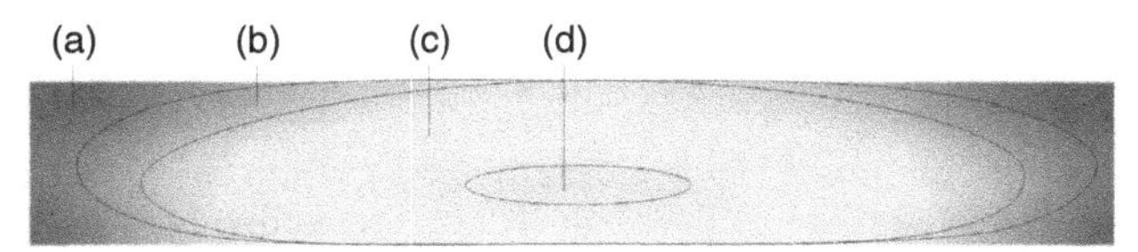

Figure 4.14 Simulated concentration profile in the diffusion channel at the position 7 mm downstream of the electrolyte inlet, H_2CO_3 concentration: (a) 3.6–4 × 10^{-4} mg l^{-1}, (b) 3.0–3.6 × 10^{-4} mg l^{-1}, (c) 2.8–3.0 × 10^{-4} mg l^{-1}, and (d) 2.6–2.8 × 10^{-4} mg l^{-1}.

As illustrated in Figure 4.13, the response time decreases slightly with increasing flow. The sensor responds rapidly to 5 vol% CO_2 compared with an equilibrium-type sensor as described in Section 4.1.1. The increase of the electrolyte conductivity during the passage of the diffusion zone during the exposition of the sensor to N_2 atmosphere is mainly caused by copper corrosion in the flow channel, which can be prevented by covering the channel walls with a thin gold layer.

To understand the sensor behaviour, the concentrations of dissolved CO_2 and H_2CO_3 in the electrolyte were studied by numerical simulation based on the software CFX solver (ANSYS, Germany). The calculated H_2CO_3 concentration profile at a CO_2 concentration of 5 vol% outside the membrane, as shown in Figure 4.14, indicates that the regions of higher concentrations are located at the edges of the channel. Here, the flow velocity is lower, and, therefore, more CO_2 can dissociate than in the channel centre.

The temporal change of the concentration profile in the downstream conductivity cell was also simulated after a steep CO_2 concentration change outside the membrane. Figure 4.15 shows the comparison between measured and simulated results for a real sensor.

Based on previous studies [72–74], a conductivity pCO_2 sensor has been proposed for clinical applications [75]. It allows the fabrication of miniaturized sensors to detect ischemia in organs. Based on a bridge design with two cavities, both a planar and a cylindrical set-up were studied. The results showed that both geometries have advantages and disadvantages with respect to future miniaturization. In general, the cylindrical set-up is favoured because of its

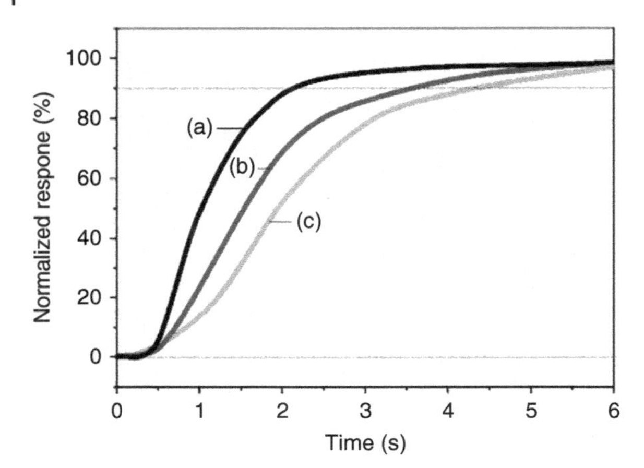

Figure 4.15 Comparison between experimental and simulation results of the sensor response after a fast concentration change from 0 to 5 vol% CO_2: (a) experimental, $d = 15\,\mu m$, $t_{90} = 1.9$ seconds; (b) simulation, $d = 30\,\mu m$, $t_{90} = 3.7$ seconds; and (c) experimental, $d = 25\,\mu m$, $t_{90} = 4.9$ seconds.

large contact area and simpler miniaturization. However, the planar sensor is advantageous for surface measurements, while the cylindrical sensor is preferred for tissue insertion. In a further study [76], a miniaturized cylindrical design of the sensor according to [75] was used to perform *in vivo* measurements in a small muscle flap of a pig. The measured values of the sensor were comparable with those obtained with a commercial Severinghaus microsensor.

In [77] the development and characterization of conductivity-based CO_2 sensors for the determination of CO_2 elimination during extracorporeal membrane oxygenation is described. The sensors measure the CO_2-dependent conductivity change in a thin water film, which is separated from the gas volume by a semipermeable Teflon™ membrane. The use of a commercially available thin-film interdigital electrode (IDE) enables a simple and practical sensor configuration. In this work the relevant chemical equilibrium reactions, the electrical model, and equivalent circuit diagram as well as the electrode materials and temperature compensation of the conductivity measuring cell are explained in detail. Conductivity-based CO_2 sensors were compared with ultrasonic CO_2 sensors with regard to pCO_2 measurement on a membrane oxygenator. While the ultrasonic sensor had significant advantages in signal stability, the conductivity sensor was superior in sensor resolution, cost efficiency, and integrability.

4.4 Quinhydrone CO_2 Electrode

To simplify the relatively complex integration of the pH sensor inside an equilibrium-type CO_2 sensor, an easily fabricable pH electrode with a quinhydrone system was suggested [63]. Quinhydrone is a 1 : 1 mixture of quinone (Q) and hydroquinone (H_2Q). The redox potential of this electrode (Pt| quinone, H^+) can be utilized in the pH range between 1 and 8 [78]. The detection principle is based on the oxidation and reduction of hydroquinone and quinone, respectively. The corresponding potential of the redox electrode is governed by

$$E = E_0 + \frac{RT}{2F} \ln \frac{a_Q}{a_{H_2Q}} + \frac{RT}{F} \ln a_{H^+} \tag{4.40}$$

For equal activities of quinone and hydroquinone (the quinhydrone system), the electrode potential depends only on temperature and the hydrogen ion activity. The performance of this electrode is comparable with that of conventional CO_2 electrodes. It has the advantage of being an inexpensive low-impedance electrode with a high potential for miniaturization.

Further developments that were directed on the utilization of this electrochemical system for CO_2 sensors are described in [79–82].

References

1 Stow, R.W. and Randall, B.F. (1954). Electrical measurement of the pCO_2 of blood. *Am. J. Physiol.* 179 (3): 678.
2 Stow, R.W., Baer, R.F., and Randall, B.F. (1957). Rapid measurement or the tension of carbon dioxide in blood. *Arch. Phys. Med. Rehabil.* 38 (10): 646–650.
3 Stow, R.W. (1987). This week's citation-classic – rapid measurement of the tension of carbon-dioxide in blood. *Curr. Cont./Life Sci.* 47: 21.
4 Severinghaus, J.W. (2004). First electrodes for blood P_{O2} and P_{CO2} determination. *J. Appl. Phys.* 97 (5): 1599–1600.
5 Van Slyke, D.D. and Sendroy, J. Jr. (1927). Carbon dioxide factors for the manometric blood gas apparatus. *J. Biol. Chem.* 73 (1): 127–144.
6 Riley, R.L., Proemmel, D.D., and Franke, R.E. (1945). A direct method for determination of oxygen and carbon dioxide tensions in blood. *J. Biol. Chem.* 161 (2): 621–633.
7 Riley, R.L., Campbell, E.J.M., and Shepard, R.H. (1957). A bubble method for estimation of P_{O2} and P_{CO2} in whole blood. *J. Appl. Phys.* 11 (2): 245–249.
8 Venes, D. (2005). *Taber's Cyclopedic Medical Dictionary*, 20e, 1981. Philadelphia, PA: F. A. Davis Company.
9 Severinghaus, J.W. and Bradley, A.F. (1958). Electrodes for blood pO_2 and pCO_2 determination. *J. Appl. Phys.* 13 (3): 515–520.
10 Severinghaus, J.W. (1968). Measurements of blood gases P_{O2} and P_{CO2}. *Ann. N.Y. Acad. Sci.* 148 (1): 115–132.
11 Severinghaus, J.W. and Astrup, P.B. (1986). History of blood gas analysis. III. Carbon dioxide tension. *J. Clin. Monit.* 2 (1): 60–73.
12 Severinghaus, J.W. (2002). The invention and development of blood gas analysis apparatus. *Anesthesiology* 97 (1): 253–256.
13 Clark, L.C. Jr. (1956). Electrochemical device for chemical analysis. US Patent 2,913,386 A, filed 21 March 1956 and issued 17 November 1959.
14 Clark, L.C. Jr. (1956). Monitor and control of blood and tissue oxygenation. *Trans. Am. Soc. Art. Int. Org.* 2 (1): 41–48.
15 Gertz, K.H. and Loeschcke, H.H. (1958). Elektrode zur Bestimmung des CO_2-drucks (electrode for the determination of the CO_2 pressure). *Naturwissenschaften* 45 (7): 160–161 (in German).
16 Gertz, K.H. and Loeschcke, H.H. (1958). Eine Methode zur Bestimmung der CO_2-Drucke in Flüssigkeiten und Gasen (a method for the determination of the CO_2 pressures in liquids and gases). *Pflügers Arch.* 268 (1): 70 (in German).

17 Zosel, J., Oelßner, W., Decker, M. et al. (2011). The measurement of dissolved and gaseous carbon dioxide concentration. *Meas. Sci. Technol.* 22: 072011, 45pp, doi: https://doi.org/10.1088/0957-0233/22/7/072001.
18 Herber, S. (2005). Development of a hydrogel based carbon dioxide sensor. PhD thesis. Enschede, The Netherlands: University of Twente.
19 Schindler, W. (1983). Wirkungsweise, Aufbau und Anwendung von CO_2-Sensoren (mode of operation, assembly and utilization of CO_2 sensors). In: *Informationen aus dem Forschungsinstitut Meinsberg*, 55–72. Zentralinstitut für Information und Dokumentation der DDR (ZIID) (in German).
20 Mascini, M. and Cremisini, C. (1978). Evaluation of measuring range and interferences for gas-sensing potentiometric probes. *Anal. Chim. Acta* 97 (2): 237–244.
21 Whitehead, M.D., Halsall, D., Pollitzer, M.J. et al. (1980). Transcutaneous estimation of arterial pO_2 and pCO_2 in newborn infants with a single electrochemical sensor. *Lancet* 315 (8178): 1111–1114.
22 Donaldson, T.L. and Palmer, H.J. (1979). Dynamic response of the carbon dioxide electrode. *AlChE J.* 25 (1): 143–151.
23 Donaldson, T.L. and Ho, S.P. (1985). Electrokinetic effects: carbonic anhydrase and CO_2 electrode dynamics. *Chem. Eng. Commun.* 37 (1-6): 223–231.
24 Cammaroto, C., Diliberto, L., Ferralis, M. et al. (1998). Use of carbonic anhydrase in electrochemical biosensors for dissolved CO_2. *Sens. Actuators, B* 48 (1–3): 439–447.
25 Voipio, J. and Kaila, K. (1993). Interstitial PCO_2 and pH in rat hippocampal slices measured by means of a novel fast CO_2/H^+-sensitive microelectrode base on a PVC-gelled membrane. *Pflügers Arch.* 423 (3–4): 193–201.
26 Zhao, P. and Cai, W.-J. (1997). An improved potentiometric pCO_2 microelectrode. *Anal. Chem.* 69 (24): 5052–5058.
27 Bailey, P.L. and Riley, M. (1975). Performance characteristics of gas-sensing membrane probes. *Analyst* 100 (1188): 145–156.
28 Ross, J.W., Riseman, J.H., and Krueger, J.A. (1973). Potentiometric gas sensing electrodes. *Pure Appl. Chem.* 36 (4): 473–487.
29 Lindner, E., Gyurcsányi, R.E., and Buck, R.P. (2002). Membranes in electroanalytical chemistry: membrane-based chemical and biosensors. In: *Encyclopedia of Surface & Colloid Science* (ed. P. Somasundaran and founding editor A.T. Hubbard), 3239–3264. New York: Marcel Dekker.
30 Kobos, R.K., Parks, S.J., and Meyerhoff, M.E. (1982). Selectivity characteristics of potentiometric carbon dioxide sensors with various gas membrane materials. *Anal. Chem.* 54 (12): 1976–1980.
31 Lopez, M.E. (1984). Selectivity of the potentiometric carbon dioxide gas-sensing electrode. *Anal. Chem.* 56 (13): 2360–2366.
32 Morf, W.E., Mostert, I.A., and Simon, W. (1985). Time response of potentiometric gas sensors to primary and interfering species. *Anal. Chem.* 57 (6): 1122–1126.
33 Samukawa, T., Ohta, K., Onitsuka, M. et al. (1995). Numerical approach to the explanation of the response time of the Severinghaus type electrode. *Anal. Chim. Acta* 316 (1): 83–92.

34 Roditaki, A., Nikolelis, D.P., and Papastathopoulos, D.S. (1993). Studies on operational and dynamic response characteristics of the potentiometric carbon dioxide gas-sensing probe by using a Teflon membrane. *Anal. Lett.* 26 (3): 393–414.

35 Nikolelis, D.P. and Krull, U.J. (1990). Dynamic response characteristics of the potentiometric carbon dioxide sensor for the determination of aspartame. *Analyst* 115 (7): 883–888.

36 Meyerhoff, M.E., Fraticelli, Y.M., Opdycke, W.N. et al. (1983). Theoretical predictions on the response properties of potentiometric gas sensors based on internal polymer membrane electrodes. *Anal. Chim. Acta* 154: 17–31.

37 Hato, M., Masuoka, T., and Shimura, Y. (1983). Kinetic assay of enzymes with an ammonia gas-sensing electrode. Part 1. Dynamic responses of the sensor. *Anal. Chim. Acta* 149: 193–202.

38 Jensen, M.A. and Rechnitz, G.A. (1979). Response time characteristics of the pCO_2 electrode. *Anal. Chem.* 51 (12): 1972–1977.

39 Gumbrecht, W. and Stanzel, M. (1996). Verfahren zur Eichung von Gasmesssensoren für gelöste Gase und Verfahren zur Konzentrationsmessung von CO_2 in Blut mit Hilfe eines solchen Eichverfahrens. (Procedure for calibration of gas sensors of dissolved gases and procedure for measurement of CO_2 concentration in blood by means of such calibration procedure). DE Patent 19,605,246 A 1, filed 13 February 1996 and issued 14 August 1997 (in German).

40 Domanowski, A., Gerlach, F., Oelßner, W., and Sauer, R. (2002). Vorrichtung und Verfahren zur Messung der CO_2-Konzentration in Flüssigkeiten (Equipment and procedure for measurement of the CO_2 concentration in liquids). DE Patent 10,251,183 A1, filed 12 January 2002 and issued 24 June 2003 (in German).

41 Schwarz, H.A. and Dodson, R.W. (1989). Reduction potentials of CO_2^- and the alcohol radicals. *J. Phys. Chem.* 93 (1): 409–414.

42 Surdhar, P.S., Mezyk, S.P., and Armstrong, D.A. (1989). Reduction potential of the $^{\bullet}CO_2^-$ radical anion in aqueous solutions. *J. Phys. Chem.* 93 (8): 3360–3363.

43 Schwarz, H.A., Creutz, C., and Sutin, N. (1985). Cobalt(I) polypyridine complexes. Redox and substitutional kinetics and thermodynamics in the aqueous 2,2'-bipyridine and 4,4′-dimethyl-2,2′- bipyridine series studied by the pulse-radiolysis technique. *Inorg. Chem.* 24 (3): 433–439.

44 Amatore, C. and Savéant, J.-M. (1981). Mechanism and kinetic characteristics of the electrochemical reduction of carbon dioxide in media of low proton availability. *J. Am. Chem. Soc.* 103 (17): 5021–5023.

45 Lilie, J., Beck, G., and Henglein, A. (1971). Pulsradiolyse und Polarographie: Halbstufenpotentiale für die Oxydation und Reduktion von kurzlebigen organischen Radikalen an der Hg-Elektrode (pulse-radiolysis and polarography: half-wave potentials for the oxidation and reduction of short-lived organic radicals at the Hg-electrode). *Ber. Bunsen. Phys. Chem.* 75 (5): 458–465 (in German).

46 Albery, W.J. and Uttamlal, M. (1994). A CO_2 titration electrode Part I: Theoretical description. *J. Appl. Electrochem.* 24 (1): 8–13.

47 Albery, W.J., Uttamlal, M., Appleton, M.S. et al. (1994). A CO_2 titration electrode Part II: Development of the sensor. *J. Appl. Electrochem.* 24 (1): 14–17.
48 Albery, W.J., Appleton, M.S., Pragnell, T.R.D. et al. (1994). Development of an electrochemical workstation for monitoring the fermentation of beer. *J. Appl. Electrochem.* 24 (6): 521–524.
49 Trapp, T., Ross, B., Cammann, K. et al. (1998). Development of a coulometric CO_2 gas sensor. *Sens. Actuators, B* 50 (2): 97–103.
50 Wiegran, K., Trapp, T., and Cammann, K. (1999). Development of a dissolved carbon dioxide sensor based on a coulometric titration. *Sens. Actuators, B* 57 (1–3): 120–124.
51 Ishiji, T., Takahashi, K., and Kira, A. (1993). Amperometric carbon dioxide gas sensor based on electrode reduction of platinum oxide. *Anal. Chem.* 65 (20): 2736–2739.
52 Ishiji, T., Chipman, D.W., Takahashi, T., and Takahashi, K. (2001). Amperometric sensor for monitoring of dissolved carbon dioxide in seawater. *Sens. Actuators, B* 76 (1–3): 265–269.
53 Albery, W.J., Clark, D., Drummond, H.J.J. et al. (1992). Pulsed titration sensors Part 1: A breath-by-breath CO_2 sensor. *J. Electroanal. Chem.* 340 (1–2): 99–110.
54 Haynes, L.V. and Sawyer, D.T. (1967). Electrochemistry of carbon dioxide in dimethyl sulfoxide at gold and mercury electrodes. *Anal. Chem.* 39 (3): 332–338.
55 Albery, W.J. and Barron, P. (1982). A membrane electrode for the determination of CO_2 and O_2. *J. Electroanal. Chem.* 138 (1): 79–87.
56 Albery, W.J., Clark, D., Young, W.K., and Hahn, C.E.W. (1992). Pulsed titration sensors Part 2: A general theoretical model for the breath-by-breath CO_2 sensor. *J. Electroanal. Chem.* 340 (1–2): 111–126.
57 Hahn, C.E.W., McPeak, H., Bond, A.M., and Clark, D. (1995). The development of new microelectrode gas sensors; an odyssey. Part 1. O_2 and CO_2 reduction at unshielded gold microdisc electrodes. *J. Electroanal. Chem.* 393 (1–2): 61–68.
58 Hahn, C.E.W., McPeak, H., and Bond, A.M. (1995). The development of new microelectrode gas sensors: an odyssey. Part 2. O_2 and CO_2 reduction at membrane-covered gold microdisc electrodes. *J. Electroanal. Chem.* 393 (1–2): 69–74.
59 Welford, P.J., Brookes, B.A., Wadhawan, J.D. et al. (2001). The electro-reduction of carbon dioxide in dimethyl sulfoxide at gold microdisk electrodes: current | voltage waveshape analysis. *J. Phys. Chem. B* 105 (22): 5253–5261.
60 Wadhawan, J.D., Welford, P.J., Maisonhaute, E. et al. (2001). Microelectrode studies of the reaction of superoxide with carbon dioxide in dimethyl sulfoxide. *J. Phys. Chem. B* 105 (43): 10659–10668.
61 Floate, S. and Hahn, C.E.W. (2005). Simultaneous electrochemical determination of oxygen and carbon dioxide gas mixtures in a non-aqueous solvent using unshielded and membrane shielded gold microelectrodes. *Sens. Actuators, B* 110 (1): 137–147.

62 McPeak, H., Bond, A.M., and Hahn, C.E.W. (2000). The development of new microelectrode gas sensors: an odyssey. Part IV. O_2, CO_2 and N_2O reduction at unshielded gold microdisc electrodes. *J. Electroanal. Chem.* 487 (1): 25–30.
63 van Kempen, L.H.J. and Kreuzer, F. (1977). Alternative methods of CO_2 measurement, with particular reference to continuous recording. In: *Blood pH, Gases, and Electrolytes* (ed. R.A. Durst), NBS Special Publication 450, 239–246. Washington, DC: National Bureau of Standards.
64 Bowman, G.E. (1966). Apparatus for supplying carbon dioxide to growing plants, patent. Patent GB 1,143,403, filed 4 May 1966 and issued 19 February 1969.
65 Moyat, P.E. (1971). Kohlendioxid-Warngerät (carbon dioxide warning device). DE Patent 2,147,718 A, filed 24 September 1971 and issued 12 April 1973 (in German).
66 Symanski, J.S., Martinchek, G.A., and Bruckenstein, S. (1983). Conductometric sensor for atmospheric carbon dioxide determination. *Anal. Chem.* 55 (7): 1152–1156.
67 Sheppard, N.F. Jr., Tucker, R.C., and Wu, C. (1993). Electrical conductivity measurements using microfabricated interdigitated electrodes. *Anal. Chem.* 65 (9): 1199–1202.
68 Timmer, B., Sparreboom, W., Olthuis, W. et al. (2002). Optimization of an electrolyte conductivity detector for measuring low ion concentrations. *Lab Chip* 2 (2): 121–124.
69 Lvovich, V.F., Liu, C.C., and Smiechowski, M.F. (2006). Optimization and fabrication of planar interdigitated impedance sensors for highly resistive non-aqueous industrial fluids. *Sens. Actuators, B* 119 (2): 490–496.
70 Olthuis, W., Streekstra, W., and Bergveld, P. (1995). Theoretical and experimental determination of cell constants of planar-interdigitated electrolyte conductivity sensors. *Sens. Actuators, B* 24 (1–3): 252–256.
71 Gambert, R., Oelßner, W., Wex, K., and Zosel, J. (2004). Kohlendioxidsensor (carbon dioxide sensor). DE Patent 102,004,058,135 A1, filed 2 December 2004 and issued 8 June 2006 (in German).
72 Baker, J.M., Spaans, E.J.A., and Reece, C.R. (1996). Conductometric measurement of CO_2 concentration: theoretical basis and its verification. *Agron. J.* 88 (4): 675–682.
73 Lis, K., Acker, H., Lübbers, D.W., and Halle, M. (1979). A PCO_2 surface electrode working on the principle of electrical conductivity. *Pflügers Arch.* 381 (3): 289–291.
74 Varlan, A.R. and Sansen, W. (1997). Micromachined conductometric $p(CO_2)$ sensor. *Sens. Actuators, B* 44 (1–3): 309–315.
75 Mirtaheri, P., Grimnes, S., Martinsen, O.G., and Tønnessen, T.I. (2004). A new biomedical sensor for measuring PCO_2. *Physiol. Meas.* 25 (2): 421–436.
76 Mirtaheri, P., Omtveit, T., Klotzbuecher, T. et al. (2004). Miniaturization of a biomedical gas sensor. *Physiol. Meas.* 25 (6): 1511–1522.
77 van der Weerd, B. (2016). Entwicklung und Charakterisierung von CO_2-Sensoren für die Bestimmung der CO_2-Eliminierung während extrakorporaler Membranoxygenierung (Development and characterization of CO_2 sensors for the determination of CO_2 elimination during extracorporeal

membrane oxygenation). PhD thesis. Germany: Universität Regensburg (in German).

78 Galster, H. (1990). *pH-Messung (pH-measurement)*, 163–164. Weinheim: VCH-Verlag (in German).

79 Gambert, R. (2001). Vorrichtung zur Messung des Partialdrucks von Kohlendioxid (Equipment for measurement of the CO_2 partial pressure). DE Patent 10,122,150 A1, filed 8 May 2001 and issued 21 November 2002 (in German).

80 Scholz, F., Düssel, H., and Meyer, B. (1993). A new pH-sensor based on quinhydrone. *Fresenius J. Anal. Chem.* 347 (10–11): 458–459.

81 Düssel, H., Komorsky-Lovric, Š., and Scholz, F. (1995). A solid composite pH sensor based on quinhydrone. *Electroanalysis* 7 (9): 889–894.

82 Aquino-Binag, C.N., Kumar, N., Lamb, R.N., and Pigram, P.J. (1996). Fabrication and characterization of a hydroquinone-functionalized polypyrrole thin-film pH sensor. *Chem. Mater.* 8 (11): 2579–2585.

5

Potentiometric CO_2 Sensors with Solid Electrolyte

Hans Ulrich Guth

Technische Universität Dresden, Faculty of Chemistry and Food Chemistry, 01062 Dresden, Germany

5.1 Indirect Measurement of CO_2 in Hot Water Gas

Electrochemical cells based on solid electrolytes are suitable to determine directly oxygen and dissolved oxygen in the temperature range between 350 and 1400 °C. Such cells exhibit many advantages: they work potentiometrically according to Nernst's equation or in coulometric mode according to Faraday's law over a long time without calibration and with response times of less than one second. The solid electrolyte used for such cells is an oxide ion conductor. Mostly, yttria-stabilized zirconia (YSZ) as tubes, screen-printed layers, or discs is utilized for oxygen sensors [1]. Solid electrolyte sensors are widely applied to control high temperature processes like combustions for industrial and automotive application (Figure 5.1a).

In most cases the measuring electrode, consisting of platinum, is catalytically highly active so that in its vicinity the gas mixture is in the state of chemical equilibrium. Reference and measuring electrode are separated gas-tightly from each other (Figure 5.1b).

The measurement of CO_2 concentration at high temperature can be performed *in situ* as well. High temperature processes are suitable to convert CO_2 into syngas, a fuel gas mixture consisting primarily of hydrogen, carbon monoxide, and very often some carbon dioxide, in order to prevent the increase of carbon dioxide in the atmosphere. For this purpose, the *in situ* determination of CO_2 is of large interest. Potentiometrically working solid electrolyte cells can be applied to measure the CO_2/CO ratio in gas mixtures or the redox (product) quotient $\frac{p_{CO_2} p_{H_2O}}{p_{CO} p_{H_2}}$ of burnt and unburnt gas species [2]. In both cases, an equilibrium oxygen partial pressure is obtained, which is given by

$$U(CO_2, H_2, CO, H_2O)/\text{mV} = 0.021\,54(T/[\text{K}]) \ln \left(\frac{p_{O_2}}{p_t} \right) \tag{5.1}$$

Carbon Dioxide Sensing: Fundamentals, Principles, and Applications,
First Edition. Edited by Gerald Gerlach, Ulrich Guth, and Wolfram Oelßner.

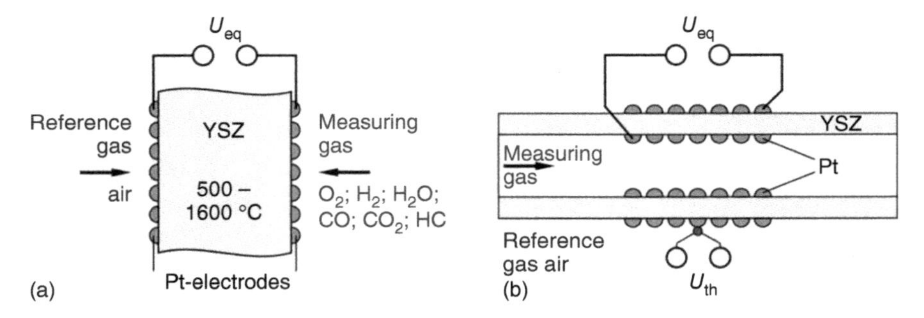

Figure 5.1 Potentiometric sensors: (a) basic principle of thermodynamically controlled potentiometric solid electrolyte sensors and (b) tube-like potentiometric sensor for complex gas analysis.

where U is the equilibrium voltage, and with p_t = 101 325 Pa for CO/CO_2 gas mixtures, it yields [2]

$$U(\mathrm{CO_2}, \mathrm{CO})/\mathrm{mV} = -1458.4 + \left[0.447 + 0.043\,08 \ln\left(\frac{p_{\mathrm{CO_2}}}{p_{\mathrm{CO}}}\right)\right] T/[\mathrm{K}] \tag{5.2}$$

The redox quotient can be determined by

$$U(\mathrm{CO_2}, \mathrm{H_2}, \mathrm{CO}, \mathrm{H_2O})/\mathrm{mV} = -1374.5 + \left[0.3697 + 0.021\,54 \ln\left(\frac{p_{\mathrm{CO_2}} p_{\mathrm{H_2O}}}{p_{\mathrm{CO}} p_{\mathrm{H_2}}}\right)\right] T/[\mathrm{K}] \tag{5.3}$$

The use of Eqs. (5.1)–(5.3) allows to measure only the ratios of burnt and unburnt species. In a simple way, the concentration of the single components cannot be determined due to the temperature dependence of the chemical equilibria. However, it becomes possible if additional relations are used:

$$A_b = \frac{p_b}{p_{\mathrm{CO_2}} + p_{\mathrm{H_2}} + p_{\mathrm{CO}} + p_{\mathrm{H_2O}}} = x_b \tag{5.4}$$

where A_b is the mole fraction with $A_b = x_b$ if the gas is not diluted. Then the carbon/hydrogen relation V

$$V = \frac{A(\mathrm{CO}) + A(\mathrm{CO_2})}{A(\mathrm{H_2}) + A(\mathrm{H_2O})} \tag{5.5}$$

and the redox (sum) quotient Q

$$Q = \frac{A(\mathrm{H_2O}) + A(\mathrm{CO_2})}{A(\mathrm{H_2}) + A(\mathrm{CO})} \tag{5.6}$$

which depend on temperature, can be calculated if the voltage is measured at two temperatures:

$$V = \left(\frac{1}{1 + (\sqrt{PO}/KC)_2} + \frac{1}{1 + (\sqrt{PO}/KH)_1}\right) \Big/ \left(\frac{1}{1 + (\sqrt{PO}/KC)_1} + \frac{1}{1 + (\sqrt{PO}/KC)_2}\right) \tag{5.7}$$

$$Q = (1+V)/\left(\frac{V}{1+(\sqrt{PO}/KC)_1} + \frac{1}{1+(\sqrt{PO}/KH)_1}\right) - 1 \tag{5.8}$$

$PO = p_{O_2}/p_t$ is the oxygen partial pressure, and KC, KH, and K are the mass action constants for the equilibrium of CO_2, the H_2O decomposition, and the water gas shift reaction, respectively. The indices 1 and 2 denote the measurement at temperature 1 and temperature 2, respectively. In such a way, the concentrations of all components in a hot gas mixture can be determined:

$$A(CO) = \frac{1}{2}\left(\frac{1}{1+Q} - \frac{1+VK}{(1+V)(1-K)} \pm \sqrt{\left(\frac{1}{1+Q} - \frac{1+VK}{(1+V)(1-K)}\right)^2 + \frac{4VK}{(1+Q)(1+V)(1-K)}}\right) \tag{5.9}$$

$$A(CO_2) = \frac{V}{V+1} - A(CO) \tag{5.10}$$

$$A(H_2) = \frac{1}{1+Q} - A(CO) \tag{5.11}$$

$$A(H_2O) = 1 - A(CO) - A(H_2) - A(CO_2) \tag{5.12}$$

In few cases, e.g. when CO_2 is formed by fermentation processes, it can be measured indirectly by means of such oxygen sensors in binary CO_2/O_2 gas mixtures. Oxygen-free CO_2 is necessary for carbonizing of beer and non-alcoholic beverages.

5.2 Direct CO_2 Measurement with Solid Electrolyte Cells

5.2.1 Functional Principles of Solid Electrolyte CO_2 Cells

Although no solid electrolyte is known, which conducts carbonate anions, CO_2 can be measured with potentiometric solid electrolyte sensors. For this purpose, an electrochemical equilibrium has to be established between the CO_2 to be measured, the ions of the solid electrolyte in use, and the electrons in the electrodes.

Solid alkali carbonates are solid electrolytes at temperatures of more than 300 °C because the charge carriers, the alkaline ions, are mobile via alkali ion vacancies V'_{Na}:

$$Na_2CO_3 \rightarrow Na_i^{\circ} + V'_{Na} + CO^x_{3CO_3} \tag{5.13}$$

That is due to the intrinsic disorder of Frenkel type. The mobility and, therefore, the electrical conductivity can be enhanced by doping with divalent or threevalent ions, respectively. In that case the doping leads to alkaline ion vacancies [3]:

$$SrCO_3 \xrightarrow{Na_2SO_4} Sr^{\circ}_{Na} + V'_{Na} + CO^x_{3CO_3} \tag{5.14}$$

For Eqs. (5.13) and (5.14), Kröger–Vink's relative notation is used. With such solid carbonates Nernstian gas concentration cells can be built up. Those cells consist of sintered gas-tight sodium carbonate (or doped sodium carbonate) covered on both sides with gold layers that separate the two gas chambers (Figure 5.2a).

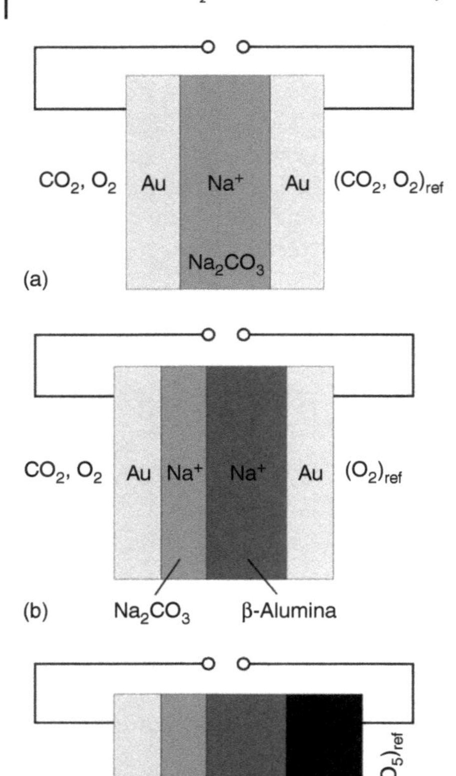

Figure 5.2 Basic principles of solid electrolyte CO_2 sensors. (a) Sodium carbonate, covered on both sides with gold layers; (b) alkali carbonate with a stable alkali ion conductor, covered on both sides with gold layers; and (c) alkali carbonate with a stable alkali ion conductor, covered with a gold layer and a $Na_2Si_2O_5$ reference electrode.

The cell voltage U (−E) depends on both concentrations CO_2 and O_2 [4]:

$$CO_2(g) + \frac{1}{2}O_2(g) + 2e^-(Au) \rightleftarrows CO_3^{2-}(s) \quad \text{(Measuring electrode)} \tag{5.15}$$

$$CO_3^{2-} \rightleftarrows CO_2(g) + \frac{1}{2}O_2(g) + 2e^-(Au) \quad \text{(Reference electrode)} \tag{5.16}$$

$$CO_2'(g) + \frac{1}{2}O_2'(g) \rightleftarrows CO_2''(g) + \frac{1}{2}O_2''(g) \quad \text{(Total cell reaction)} \tag{5.17}$$

$$U(CO_2, O_2)/\mathrm{mV} = \frac{RT}{2F}\ln\left(\frac{p'_{CO_2}p'_{O_2}}{(p''_{CO_2}p''_{O_2})_{\mathrm{ref}}}\right) \tag{5.18}$$

The charge compensation occurs by migration of sodium ions from the side having a lower CO_2 and O_2 concentration to the other one with higher concentration. If a current flows, sodium carbonate is formed on the side with higher gas concentration and disappears on the other side. That behaviour is different to that one of oxygen concentration cells using stabilized zirconia. This is the reason why the cell voltage U should be measured with an instrument having a high input impedance. Carbonate cells cannot be used stably in amperometric or coulometric mode. Nevertheless, the sintered solid carbonates remain gas-tight over a long time. Using those cells, CO_2 can be measured continuously from ppm to the percentage level over more than one year in the cell temperature range of 450–700 °C according to Nernst's equation [3, 4] provided that the cell is continuously heated.

There are two main issues that had to be solved for the practical applications. Pressed alkali carbonates are often hygroscopic and tend to sublimate so that electrolytes become porous. Beside this effect in changing the electrolyte, the realization of a stable reference system is a decisive topic. To establish a gas reference, CO_2 and O_2 concentrations have to be fixed. It is not possible using air as in the case of oxygen concentration cells because the CO_2 content in the air is normally not fixed (see below).

To overcome these problems of the solid electrolyte, various admixtures of aliovalent carbonates (homogeneous doping) and of γ-alumina (heterogeneous doping) were proposed to enhance the mechanical strength of the carbonates to increase the electrical conductivity [5, 6] and to diminish the cross sensitivity to water vapour [7, 8]. For that purpose, alkali carbonates are usually combined with a stable alkali ion conductor such as NASICON, β′-alumina, or β″-alumina as an electrolyte (Figure 5.2b) [9]:

- NASICON is a three-dimensional sodium ion conductor having the general chemical composition $Na_{1-x}Zr_2Si_xP_{3-x}O_{12}$. The sodium concentration can vary up to three. With $Na^+ = 2.2$ the compound exhibits the highest electrical conductivity.
- β′-Alumina is a two-dimensional sodium ion conductor of the general composition $Na_2O_{11}\ Al_2O_3$.
- β″-Alumina contains only 5–7 Al_2O_3 per unit Na_2O and is often stabilized by MgO.

β,β″-Alumina can be formed as a thin layer on an α-alumina substrate [10, 11]. These electrolytes are gas-tight so that two separate compartments can be realized.

On the measuring electrode CO_2 and O_2 are electrochemically transformed, whereas on the reference side oxygen is reduced:

$$Na_2CO_3 \rightleftarrows 2\,Na^+ + \tfrac{1}{2}\,O_2\,(g) + CO_2\,(g) + 2\,e^-(Au) \quad \text{(Measuring electrode)} \tag{5.19}$$

$$2\,Na^+ + 1/2\,O_2\,(g) + 2e^- \rightleftarrows Na_2O\,(s) \quad \text{(Reference electrode)} \tag{5.20}$$

$$Na_2CO_3\,(s) \rightleftarrows Na_2O\,(\text{dissolved}) + CO_2\,(g) \quad \text{(Total cell reaction)} \tag{5.21}$$

At the phase boundary between the sodium ion-conducting carbonate and the sodium ion conductor, e.g. of β-alumina, only sodium ions can be exchanged. The sensitive electrode consisting of Na_2CO_3/Au should be porous. After [12] this type of sensor is called potentiometric sensor type III. The loss in carbonate leads to an increase of porosity and can diminish the three-phase contacts between gold, carbonate, and gas.

The reference electrode potential can be established by fixing the sodium oxide activity using $Na_x(Hg)$ and Na_xCoO_2, respectively [13]. It is also possible to cover the gold or platinum layer by glass. Under these circumstances, the oxygen surface concentration and, hence, the oxide ion activity are fixed provided that no current flows. An elegant solution is to fix the O_2 by a solid thermodynamic system such as SiO_2 and $Na_2Si_2O_5$ (Figure 5.2c) [14, 15]. Then the reaction on the reference electrode is

$$2\,SiO_2\,(s) + 2\,Na^+ + ½\,O_2\,(g) + 2e^-\,(Au) \rightleftarrows Na_2Si_2O_5\,(s) \tag{5.22}$$

and, hence, the cell reaction is

$$Na_2CO_3\,(s) + 2\,SiO_2\,(s) \rightleftarrows Na_2Si_2O_5\,(s) + CO_2\,(g) \tag{5.23}$$

If the solid components are pure, then the cell electromotive force E depends only on the CO_2 partial pressure:

$$-U = E = \text{const.} + \frac{RT}{2F} \ln p_{CO_2} \tag{5.24}$$

Instead of the system silica/silicate [15], also other systems such as titania/titanate, zirconia/zirconate [16], or mixed pure oxides such as $Na_2Ti_6O_{13}$/$Na_2Ti_3O_7$ can be used as reference system [17]. In such cases no gas separation is necessary because only one electrode is gas sensitive. Other reference systems based on contacts with an additional solid electrolyte (YSZ) were also proposed [18].

The response time of freshly fabricated thick-film sensors based on β-alumina thin films is very short (Figure 5.3, about 11 ms at 650 °C).

After several weeks of operation, this time increases strongly (150 ms) [19]. Solid electrolyte CO_2 sensors using a Ni/carbonate composite as measuring electrode are suited for measuring of CO_2 in equilibrated water gases [20]. A few authors reported that NASICON-based CO_2 sensors using mixtures of semiconducting oxides and carbonates such as SnO_2 or ITO (indium tin oxide) [21] and In_2O_3 [22] as electrodes are able to measure at room temperature. However, all solid electrolyte-based sensors, which are commercially available, operate at high temperature.

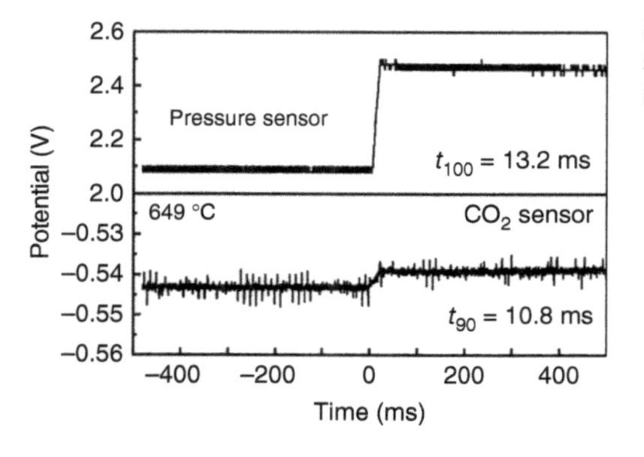

Figure 5.3 Response behaviour of a solid electrolyte CO_2 sensor during a pressure jump (own results).

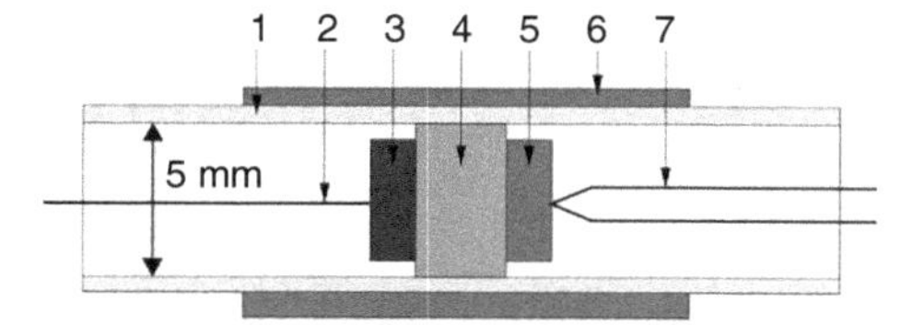

Figure 5.4 Schematic cross section of a CO_2 sensor using the pellet design. 1, quartz tube; 2, contact; 3, Au measuring electrode; 4, auxiliary solid electrolyte; 5, reference electrode; 6, heater; and 7, spring load mechanism [23]. Source: With Courtesy of Zirox GmbH, Greifswald, Germany.

5.2.2 General Setup

Commercially available CO_2 solid electrolyte sensors can be divided into two main types: (i) sensors with pellet shape and (ii) thick-film sensors.

5.2.2.1 Pellet Sensors

Figure 5.4 shows the setup of an electrochemical cell according to the principle from Figure 5.2c. Three sintered pellets are arranged by a spring load mechanism in a heatable quartz tube as measuring electrode, auxiliary solid electrolyte, and reference electrode, respectively. The measuring electrode consists of an alkaline carbonate, mostly sodium carbonate, mixed with small gold particles. The pellets are produced by uniaxial pressing of corresponding powders and subsequent sintering. Specifics of the fabrication are described in literature in more detail [2, 15, 16].

The long-term stability of such sensors is influenced by the sublimation of alkaline carbonates. This phenomenon can be easily observed when carbonate dendrites form in the colder part of the oven. Carbonates can move via the gas phase onto the reference side and influence the cell voltage. Additionally, the loss of carbonates leads to a loss of the three-phase contacts on the measuring electrode, which in turn is connected with an increase of the response time. Impedance studies are useful to investigate kinetic and ageing effects of sensors [24].

Commercially available sensors according to this design operate at moderate temperatures (450 °C) over a very long time (several years) in clean air without calibration (Figure 5.4).

5.2.2.2 Thick-Film Sensors

Thick-film sensors can be produced more economically in a large scale [25–28]. As an electrolyte a sheet of pure β,β''-alumina or NASICON may be used. The drawback of such conducting materials is that the metallic heater structures cannot be printed directly on this without an additional insulating layer. An elegant method is to modify a thick-film substrate, e.g. an α-alumina substrate, by a thin film of β,β''-alumina by topotactical growing via a high temperature process. By this means the mechanical strength of the α-alumina and the sodium ion conductivity are combined. Moreover, on the backside of the substrate (after removing the β-alumina layer), a platinum heater can be printed so that there is no influence

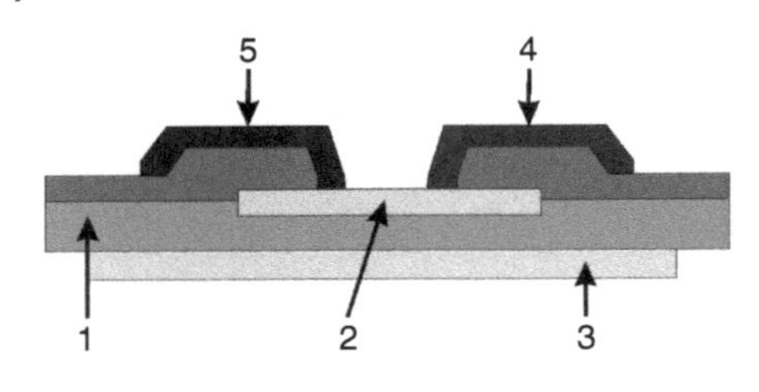

Figure 5.5 Cross section of a thick-film sensor using β-alumina layer on an alumina substrate. 1, ceramic substrate; 2, β″-alumina, solid electrolyte thin film; 3, heater; 4, $Au/SiO_2/Na_2Si_2O_5$ mix (reference electrode); and 5, $Au/Na_2CO_3/BaCO_3$ mix (measuring electrode).

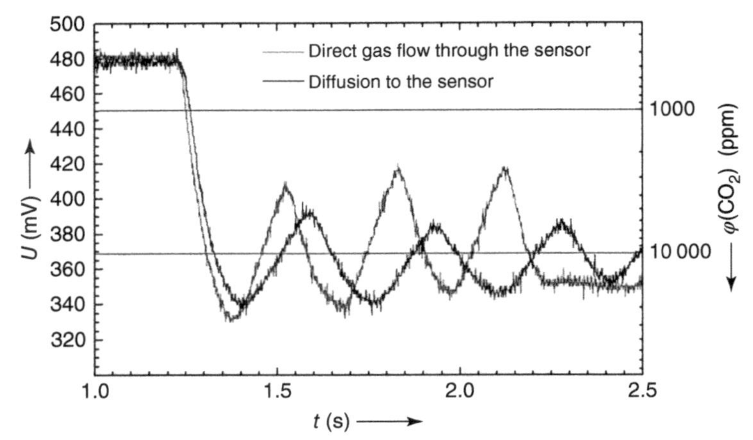

Figure 5.6 Measurement in breath gas. Due to the fast response of the thick-film sensor, breath-by-breath measurement is possible directly (light gray) or over a tube (dark gray).

of the cell voltage by the heater voltage. The setup of such a sensor is shown schematically in Figure 5.5.

Sensors made of this design are suitable to measure CO_2 breath by breath (Figure 5.6).

The preparation of layered alumina is shown in Figure 5.7. A thin layer of β-alumina with a thickness of a few micrometres is formed by tempering of α-alumina substrates in a bed of β-alumina powder at 1000 °C for 12 hours. The alkaline ion can be easily exchanged by other mono- or divalent ions. The conditions for the exchange process and the results are given in Figure 5.7. The excess of ion exchange could be observed by XRD and EDX investigations (Figure 5.7).

The conductivity, shown in Figure 5.8 also for the divalent-doped material, is high enough for being used in potentiometric cells [28–30].

For the fabrication of thick-film sensors, mostly screen-printing is used. Although the sensor design looks very easy and the fabrication process also seems to be simple, the window for the process parameters is small. In particular, the firing temperature for the layers is very sensitive. It should be high enough to achieve a sufficiently good adhesion.

On the other hand, the higher the temperature is, the more material of the layers can react with each other, leading to the formation of a new phase in this region. This in turn leads to both a deviation in the cell voltage and a higher response time as can be seen in Figure 5.9.

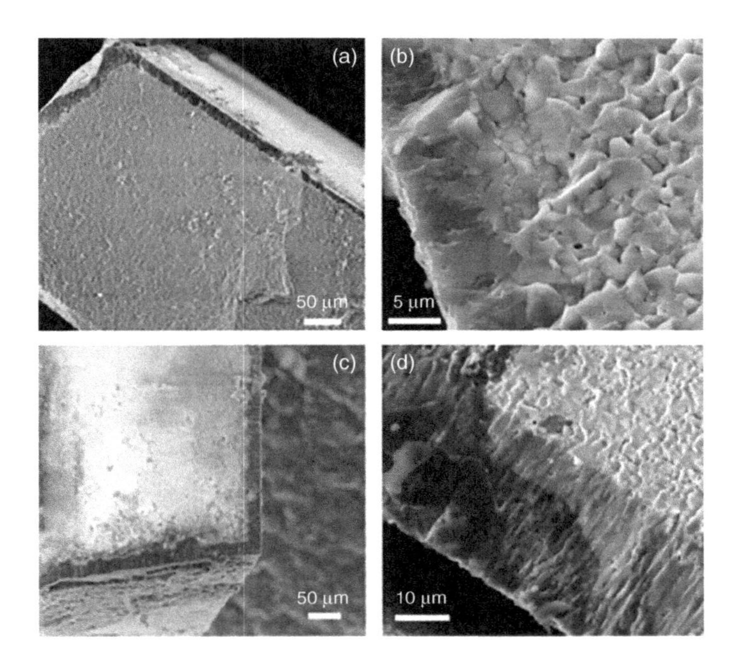

Figure 5.7 SEM images of layered alumina substrates. (a) Na^+-β′-alumina, (b) K^+-β″-alumina, (c) Li^+-β″-alumina, and (d) H_3O^+-β″-alumina, conditions for ion exchange process, see Table 5.1.

Table 5.1 Conditions for the ion exchange process in layered β″-alumina.

Basis material	Modified β″-alumina	Conditions for ionic exchange
Na^+-β″-alumina layer on α-alumina sheet	Li^+-β″-alumina	Molten $LiNO_3$; 300 °C, 168 h
	Ca^{2+}-β″-alumina	Molten $CaCl_2$; 800 °C, 48 h
	Sr^{2+}-β″-alumina	Molten $SrCl_2$; 800 °C, 48 h
K^+-β″-alumina layer on α-alumina sheet	H^+-β″-alumina	H_2O; 95 °C, 48 h

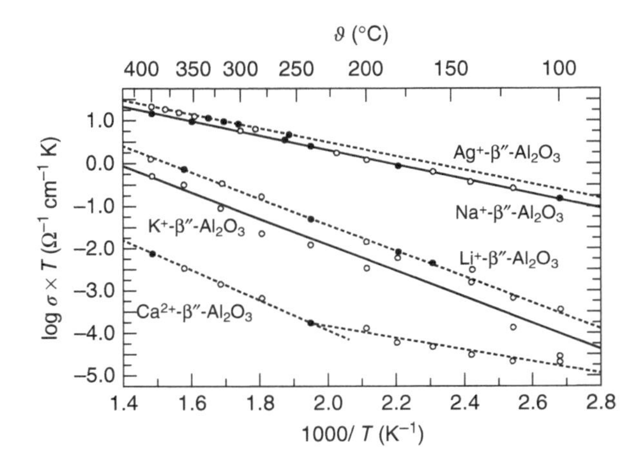

Figure 5.8 Arrhenius plots for the temperature dependence of ion-exchanged β″-alumina.

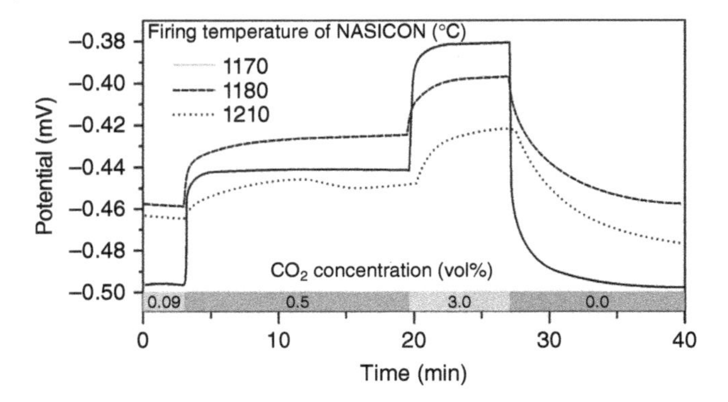

Figure 5.9 Influence of the firing temperature of NASICON on the CO_2 signal.

Influence of the Electrolyte The sensor consists of two electrolytes: the carbonate electrolyte, which is necessary for the electrochemical (sensing) process, and the auxiliary electrolyte, in which no electrochemical reaction with CO_2 takes place but a reaction with oxygen and a subsequent alkaline ion transport. Sometimes, the carbonate/Au phase is called an auxiliary phase, but this phase is essential for the operation, whereas the alkali ion electrolyte is not. Both electrolytes were modified mainly in order to improve the sensor stability and its selectivity and to reduce the cross sensitivity versus water vapour and combustibles [7, 8].

The auxiliary electrolyte has no influence on the cell performance. Different compositions of NASICON and β-alumina were used successfully [9, 14, 29, 30]. Own experiments showed that NASICON seems to be less stable against alkali carbonate than β-alumina. Also thin layers of Li_3PO_4 prepared by radio frequency (RF) magnetron sputtering together with Li_2CO_3 as an electrolyte and a mixture Li_2TiO_3 and TiO_2 as reference were used as stable sensor arrangement. In this case the sensor response was smaller than the expected Nernst voltage [31]. In the thick-film stack, CO_2 can be measured with a sodium carbonate electrolyte mixed with gold particles over a temperature range of 250–700 °C according to the Nernst's equation (Figure 5.10). In contrast, with lithium carbonate it is only possible at temperatures above 500 °C.

The doping of Na_2CO_3 with a small amount of $BaCO_3$ has only a small influence on the cell voltage (Figure 5.11a). With pure $SrCO_3$ as an electrolyte nearly the same results can be obtained (Figure 5.11b).

Sensitivity on Water Vapour The cross sensitivity versus water vapour behaviour is displayed in Figure 5.12. The influence on the signal can be explained by a different heat conductivity of gas with and without water vapour. Taking into account results of impedance measurements shown as Nyquist plots (Figure 5.12a), one can conclude that water reduces the total cell resistance in fresh cells. This result suggests that water plays a role in electrode kinetics. However, in general, the water vapour sensitivity is low.

Operating Temperature Higher concentrations of combustibles in air disturb the signal. If the temperature and the residence time are high enough, then the

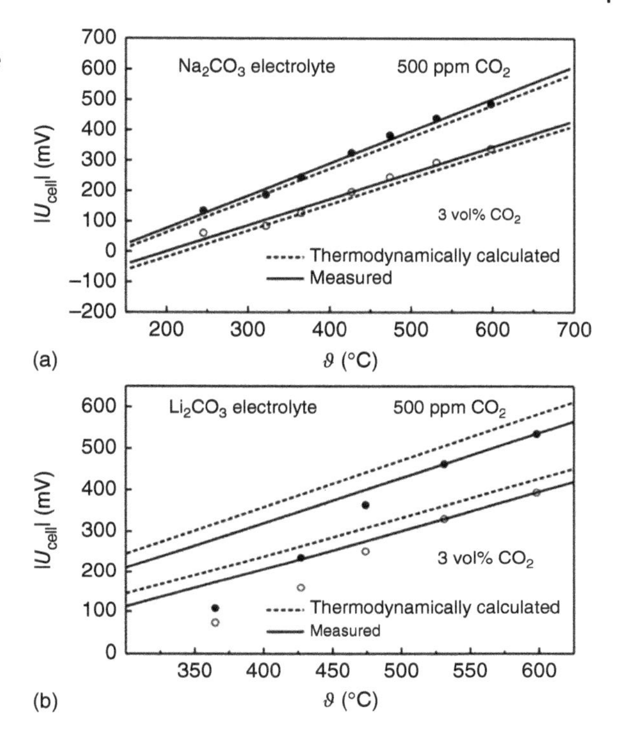

Figure 5.10 Temperature dependence of the cell voltage of (a) Na_2CO_3 and (b) Li_2CO_3 cells.

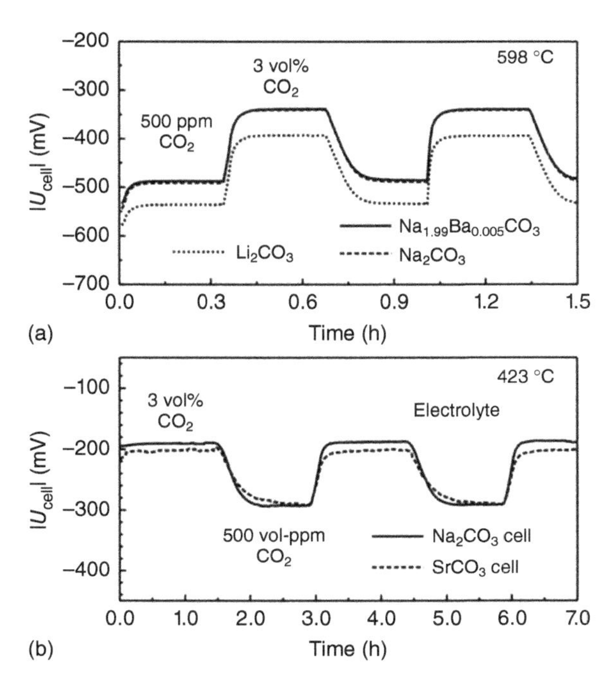

Figure 5.11 Influence of the electrolyte on cell voltage at 598 °C (a) and 423 °C (b).

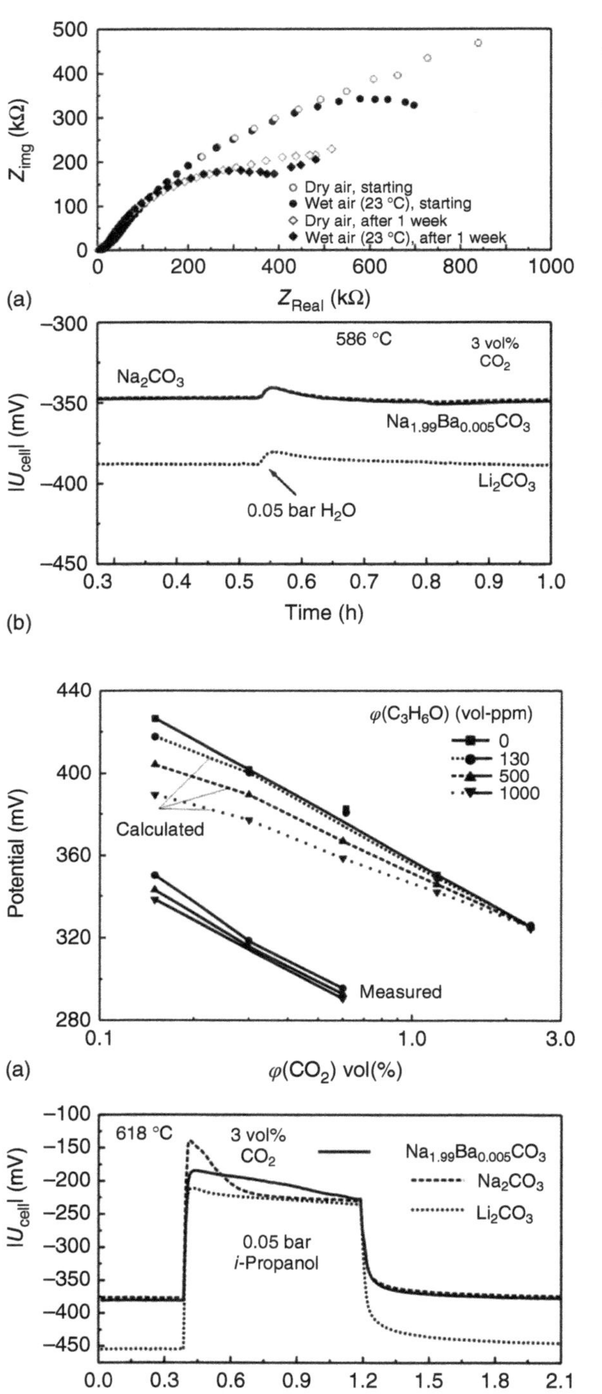

Figure 5.12 Influence of water vapour on (a) impedance and (b) cell voltage *U*.

Figure 5.13 Influence of combustibles on (a) 2-propanol and (b) acetone. The dotted lines are calculated assuming complete combustion.

organic components are burnt completely so that a higher CO_2 value (i.e. a lower cell voltage U) is displayed. At temperatures below 500 °C, the gold electrode is also sensitive to combustibles, generating a mixed potential, similar to that observed on YSZ cells at low temperatures, with lower values (Figure 5.13). To avoid this influence the combustible components have to be removed by an activated charcoal filter.

Main drawback of a solid electrolyte sensor is the sensitivity to acidic gases such as SO_2 or chlorine, which react with the sensitive carbonate layer. In this case the sensors become sensitive to SO_2 or Cl_2. With sensors operating at much higher temperatures, the values obtained are not comparable to such observed at room temperature because in the latter case there is usually a non-equilibrium state. Here, it has to be taken into account to what extent high temperature data are applicable to room temperature measurements.

As compared with other CO_2 sensors, high temperature sensors have to be heated electrically up to 500–700 °C. However, only a small electric heating power of usually c. 2–3 W is necessary for that.

5.3 Solid-State Sensors Based on Changes in Capacity and Resistivity

Apart from the physical adsorption, there exist also methods based on the interaction of carbon dioxide as an acid gas with solid surfaces that exhibit basic properties, i.e. acid–base interactions. The basicity can be improved by doping. From the point of solid-state theory, this phenomenon is often called chemisorption. As a result, the chemisorbed species cause a change in electrical properties of the semiconducting solid, they vary the charge carrier concentration, and this in turn leads to a change in the resistivity, impedance, or capacity. The electrical values can be measured very easily. But the interaction of other acidic or alkaline gases cannot be avoided so that a cross sensitivity to other acidic gases and water vapour has to be expected. Furthermore, compared among physisorbed species, the stronger interaction between surface and CO_2 leads to longer recovery times of the sensors. In the literature a lot of pure and mixed oxides such as $CuO–BaTiO_3$, La_2O_3 or $LaCl_3$-doped SnO_2, $CuO–SnO_2$, $BaCeO_3$, LaOCl, $LaOCl–SnO_2$, $SmCoO_3$, and $GdCoO_3$ were investigated [32–38]. Mostly the resistivity is enhanced if CO_2 is present.

Base-type poly(anthranilic acid) (PANA) and poly(vinylalcohol) (PVA) were used to modify the surface. The AC impedance was enhanced with increasing CO_2 concentration in the frequency region above 100 Hz [39]. SnO_2 doped with different contents of LaOCl were prepared by electrospinning as a sensing material. Such sensors showed a typical response of $R_{gas}/R_{air} = 3.7$ at a concentration of 1000 ppm CO_2 with response/recovery times of 24 and 92 seconds, respectively. A wide range of CO_2 concentrations (100–20 000 ppm) can be detected without saturation [40]. However, there was no information about cross sensitivities provided. Furthermore, sensors based on the measurement of the work function are known [41].

References

1 Möbius, H.-H., Sandow, H., Hartung, H. et al. (1992). Entwicklung neuer Sensorsysteme mit galvanischen Hochtemperatur-Festelektrolytzellen (Development of new sensor systems with galvanic high-temperature solid electrolyte cells). In: *Dechema-Monographien*, vol. 126 (ed. W. Göpel), 329–344. Weinheim: Wiley-VCH (in German).

2 Möbius, H.-H. (1992). Solid-state electrochemical potentiometric sensors for gas analysis. In: *Sensors: A Comprehensive Survey*, vol. 3 (ed. W. Göpel, J. Hesse and J.N. Zemel), 1106–1151. Weinheim: Wiley-VCH.

3 Guth, U., Barwisch, F., Wulff, H. et al. (1987). Electrical conductivity and crystal structure of pure and $SrCO_3$-doped Na_2CO_3. *Cryst. Res. Technol.* 22 (1): 141–145.

4 Gauthier, M. and Chamberland, A. (1977). Solid-state detectors for the potentiometric determination of gaseous oxides. *J. Electrochem. Soc.* 124 (10): 1579–1583.

5 Brosda, S., Bouwmeester, H.J.M., and Guth, U. (1997). Electrical conductivity and thermal behavior of solid electrolytes based on alkali carbonates and sulfates. *Solid State Ionics* 101–103 (Part 2): 1201–1205.

6 Guth, U., Brosda, S., Löscher, B. et al. (1991). Composite based on oxoanionic solid electrolytes. *Mater. Sci. Forum (Zürich)* 76: 137–140.

7 Dubbe, A., Wiemhöfer, H.-D., Sadaoka, Y., and Göpel, W. (1995). Microstructure and response behaviour of electrodes for CO_2 gas sensors based on solid electrolytes. *Sens. Actuators, B* 24-25: 600–602.

8 Lee, I., Akbar, S.A., and Dutta, P.K. (2009). High temperature potentiometric carbon dioxide sensor with minimal interference to humidity. *Sens. Actuators, B* 142 (1): 337–341.

9 Hötzel, G. and Weppner, W. (1986). Application of fast ionic conductors in solid state galvanic cells for gas sensors. *Solid State Ionics* 18–19 (Part 2): 1223–1227.

10 Chu, W.F., Fischer, D., Erdmann, H. et al. (1992). Thin and thick film electrochemical CO_2 sensors. *Solid State Ionics* 53–56 (Part 1): 80–84.

11 Chu, W.F., Tsagarakis, E.D., Metzing, T., and Weppner, W. (2003). Fundamental and practical aspects of CO_2 sensors based on NASICON electrolytes. *Ionics* 9 (5-6): 321–328.

12 Weppner, W. (1987). Solid-state electrochemical gas sensors. *Sens. Actuators* 12 (2): 107–119.

13 Guth, U. (2008). Gas sensors. In: *Electrochemical Dictionary* (ed. A.J. Bard, G. Inzelt and F. Scholz), 294–299. Berlin, Heidelberg: Springer.

14 Maier, J. and Warhus, U. (1986). Thermodynamic investigations of Na_2ZrO_3 by electrochemical means. *J. Chem. Thermodyn.* 18 (4): 309–316.

15 Möbius, H.-H. (2004). Galvanic solid electrolyte cells for the measurement of CO_2 concentrations. *J. Solid State Electrochem.* 8 (2): 94–109.

16 Fergus, J.W. (2008). A review of electrolyte and electrode materials for high temperature electrochemical CO_2 and SO_2 gas sensors. *Sens. Actuators, B* 134 (2): 1034–1041.

17 Ramirez, J. and Fabry, P. (2001). Investigation of a reference electrode based on perovskite oxide for second kind potentiometric gas sensor in open systems. *Sens. Actuators, B* 77 (1-2): 339–345.

18 Okamoto, T., Shimamoto, Y., Tsumura, N. et al. (2005). Drift phenomena of electrochemical CO_2 sensor with Pt,Na_2CO_3/Na^+-electrolyte//YSZ/Pt structure. *Sens. Actuators, B* 108 (1-2): 346–351.

19 Widmer, T., Brüser, V., Schäf, O., and Guth, U. (1999). CO_2/SO_x-sensors with different β''-aluminas as solid electrolytes. *Ionics* 5 (1): 86–90.

20 Möbius, H.-H., Shuk, P., and Zastrow, W. (1996). Solid state systems for the potentiometric determination of CO_2. *Fresenius J. Anal. Chem.* 356 (3): 221–227.

21 Bredikhin, S., Liu, J., and Weppner, W. (1993). Solid ionic conductor/semiconductor junctions for chemical sensors. *Appl. Phys. A* 57: 37–43.

22 Obata, K., Shimanoe, K., Miura, N., and Yamazoe, N. (2003). Influences of water vapor on NASICON-based CO_2 sensor operative at room temperature. *Sens. Actuators, B* 93 (1–3): 243–249.

23 http://www.zirox.de/produkte/analysatoren/spezielle-messsysteme.html (retrieved 07 July 2016).

24 Guth, U., Schmidt, P., Jahn, R. et al. (1989). Impedance studies on galvanic cells using oxoanionic solid electrolytes. *Solid State Ionics* 36 (1-2): 127–128.

25 Brüser, V., Klingner, W., Möbius, H.-H., and Guth, U. (1997). Galvanic solid state sensors for potentiometric determination of CO_2. In: *8th International Conference of Sensors Transducers and Systems*, vol. 3, 209–214. Nuremberg.

26 http://www.ionic-systems.de/Sensoren/CDS_02.html (retrieved 09 January 2019).

27 Belda, C., Fritsch, M., Feller, C. et al. (2009). Stability of solid electrolyte based thick-film CO_2 sensors. *Microelectron. Reliab.* 49 (6): 614–620.

28 www.nova1to1.com/sensors.php#co2 (retrieved 07 July 2016).

29 Liu, J. and Weppner, W. (1991). Potentiometric CO_2 gas sensor based on Na-β, β''-alumina solid electrolytes at 450 °C. *Eur. J. Solid State Inorg. Chem.* 28: 1151–1160.

30 Schäf, O., Widmer, T., and Guth, U. (1997). In-situ formation of thin-film like β''-alumina on α-alumina substrates. *Ionics* 3 (3): 277–281.

31 Lee, I. and Akbar, S.A. (2014). Potentiometric carbon dioxide sensor based on thin Li_3PO_4 electrolyte and Li_2CO_3 sensing electrode. *Ionics* 20 (4): 563–569.

32 Fan, K., Qin, H., Wang, L. et al. (2013). CO_2 gas sensors based on $La_{1-x}Sr_xFeO_3$ nanocrystalline powders. *Sens. Actuators, B* 177: 265–269.

33 Michel, C.R., Martínez-Preciado, A.H., and Rivera-Tello, C.D. (2015). CO_2 gas sensing response of YPO_4 nanobelts produced by a colloidal method. *Sens. Actuators, B* 221: 499–506.

34 Ong, K.G. and Grimes, C.A. (2001). A carbon nanotube-based sensor for CO_2 monitoring. *Sensors* 1 (6): 193–205.

35 Jiao, Z., Chen, F., Su, R. et al. (2002). Study on the characteristics of Ag doped CuO-$BaTiO_3$ CO_2 sensors. *Sensors* 2 (9): 366–373.

36 Anderson, T., Ren, F., Pearton, S. et al. (2009). Advances in hydrogen, carbon dioxide, and hydrocarbon gas sensor technology using GaN and ZnO-based devices. *Sensors* 9 (6): 4669–4694.

37 Mandayo, G.G., Herrán, J., Castro-Hurtado, I., and Castaño, E. (2011). Performance of a CO_2 impedimetric sensor prototype for air quality monitoring. *Sensors* 11 (4): 5047–5057.
38 Bezerra, T., Terencio, C., Bavastrello, V., and Nicolini, C. (2012). Calcium oxide matrices and carbon dioxide sensors. *Sensors* 12 (5): 5896–5905.
39 Oho, T., Tonosaki, T., Isomura, K., and Ogura, K. (2002). A CO_2 sensor operating under high humidity. *J. Electroanalyt. Chem.* 522 (2): 173–178.
40 Xiong, Y., Xue, Q., Ling, C. et al. (2017). Effective CO_2 detection based on LaOCl-doped SnO_2 nanofibers: Insight into the role of oxygen in carrier gas. *Sens. Actuators, B* 241: 725–734.
41 Stegmeier, S., Fleischer, M., Tawil, A. et al. (2011). Sensing of CO_2 at room temperature using work function readout of (hetero-) polysiloxanes sensing layers. *Sens. Actuators, B* 154 (2): 206–212.

6

Opto-Chemical CO_2 Sensors

Gerald Gerlach[1] *and Wolfram Oelßner*[2]

[1] *Technische Universität Dresden, Faculty of Electrical and Computer Engineering, Institute of Solid-State Electronics, 01062 Dresden, Germany*
[2] *Kurt-Schwabe-Institut für Mess- und Sensortechnik e.V.Meinsberg, Kurt-Schwabe-Straße 4, 04736 Waldheim, Germany*

6.1 Liquid Reagent-Based Opto-Chemical CO_2 Sensors

In contrast to optical CO_2 sensors that directly evaluate spectrochemical properties of the analyte (cp. Section 6.3), CO_2-caused changes in the optical properties of certain materials, i.e. usually electrolytes, can also be used to measure the concentration of CO_2 [1–5]. Such sensors are called reagent-based opto-chemical sensors (ROCS). They utilize a CO_2-selective reagent to transduce the CO_2 concentration of the analyte into an optical signal. For this purpose, light is directed via fibre optics into a volume of a colorimetric or fluorometric reagent that is entrapped within a small membrane [6].

Opto-chemical sensors are usually very sensitive, robust, fast, and inexpensive and can be easily miniaturized. In [7] the basic concepts of the different colorimetric and luminescent optical sensors for the detection and quantitative analysis of carbon dioxide are reviewed, the major applications of these sensors are discussed, and their strengths and weaknesses are highlighted.

[8] describes in detail an invention for an opto-chemical approach of sensing CO_2, which 'comprises a method for determining an analyte in which a dye solution is illuminated to induce a first and a second output light at a first and second wavelength, respectively, and the analyte concentration is determined from the measured first and second output intensities'. Based on this patent, a CO_2 monitor for the *in situ* measurement of dissolved CO_2 has been developed and commercialized (Figure 6.1) [9]. Here, CO_2 diffuses through a polymer membrane into a buffer solution and changes its pH value. The thin membrane is mechanically protected by a perforated stainless steel foil.

The disposable stainless steel sensor capsule of the CO_2 monitor contains essentially a small reservoir of bicarbonate buffer with an admixture of HPTS (1-hydroxypyrene-3,6,8-trisulfonate), a pH-sensitive fluorescent luminophore. HPTS has been used extensively for opto-chemical carbon dioxide and pH sensors. It shows a pK_A value of c. 7.3, a large Stokes shift, good photo stability, high

Carbon Dioxide Sensing: Fundamentals, Principles, and Applications,
First Edition. Edited by Gerald Gerlach, Ulrich Guth, and Wolfram Oelßner.

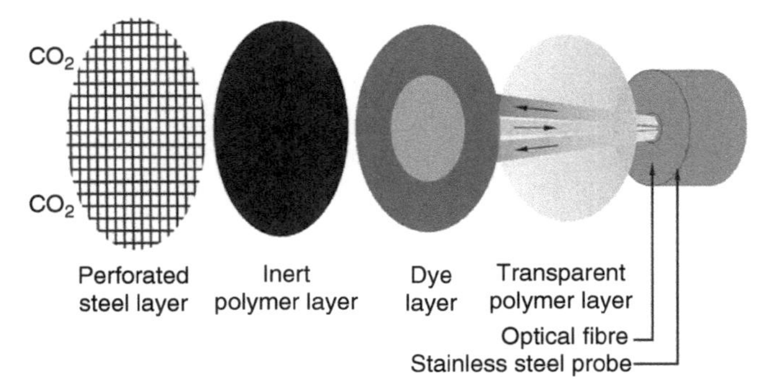

Figure 6.1 Sensor set-up of the YSI 8500 CO_2 monitor. (YSI Inc., USA, according to [9]).

quantum yield, and spectral compatibility with blue LEDs [10]. As explained in [9], CO_2 that permeates into the capsule's dye layer causes a chemical reaction, which changes the pH value and, consequently, the fluorescence of the dye according to

$$CO_2 + H_2O \rightleftarrows H_2CO_3 \rightleftarrows H^+ + HCO_3 \tag{6.1}$$

$$PTS^- + H^+ \rightleftarrows HPTS \tag{6.2}$$

A fibre-optic fibre transfers light into the sensor capsule with the fluorescent dye. To determine the CO_2 concentration of the sample medium, the instrument compares the pH-dependent fluorescence of the dye at two different wavelengths. The resulting light emission from the dye is transferred back through the fibre to the monitor, which calculates the level of dissolved CO_2 based on ratiometric analysis of the dye's fluorescence.

The sensor capsule is easy to replace and can be autoclaved multiple times. It measures dissolved CO_2 in the range of 1–25% with an accuracy of 5% of the reading or 0.2% absolutely. Over a seven-day period, the sensor drift is 2% of the reading; the 90% response time amounts to seven minutes. The sensor was found to show good shelf life for over 45 days of continuous use, depending upon the special application. Applications of this opto-chemical CO_2 sensor are reported particularly with regard to measurement and control of dissolved carbon dioxide in mammalian cell culture processes and microbial fermentations [11].

Figure 6.2 shows a different set-up for opto-chemical sensors. The SAMI-CO_2 sensor (Submersible Autonomous Moored Instrument, Sunburst Sensors, Missoula, MT, USA) is intended for studying seawater pCO_2 dynamics, preferably for the long-term deployment on moorings [12, 14] (see also Section 12.3).

The heart of the sensor is a colorimetric (sulfonephthalein-type) pH indicator contained in a tubular gas-permeable membrane. The latter is exposed directly to the ambient seawater, allowing diffusion of CO_2 into the indicator solution. To enhance stability and sensitivity of the sensor, the indicator solution is periodically renewed. A small volume of solution is pumped by a solenoid pump into the gas-permeable membrane, and the solution equilibrates with the analyte pCO_2 for a set time period. The equilibrator membrane consists of a coiled silicone rubber tube (coil length of 70 cm, outer diameter of 630 μm, and inner diameter of 300 μm). The total membrane internal volume is c. 50 μl. Blank

(a)

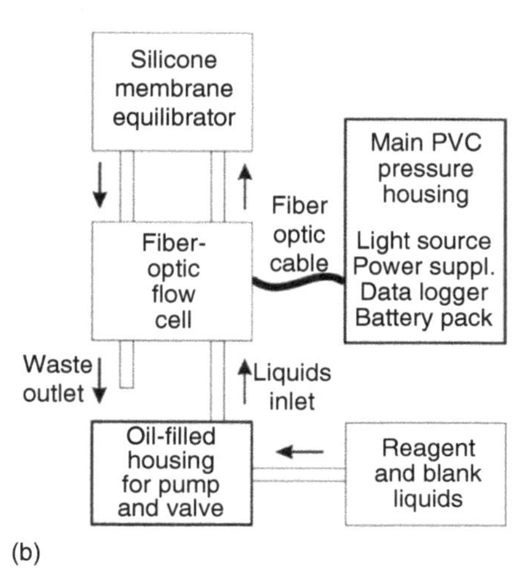

(b)

Figure 6.2 SAMI-CO_2 sensor (Sunburst Sensors, Missoula MT, USA) based on a membrane equilibrator and a fibre-optic flow cell, (a) photograph and (b) functional principle. Source: Panel (a): DeGrandpre et al. 1999 [12]. Reproduced with permission of American Chemical Society. Panel (b): DeGrandpre et al. 1995 [13]. Reproduced with permission of John Wiley & Sons.

solutions are introduced through a 3-way solenoid valve. Both the pump and the valve are contained in a silicone oil-filled housing that permits equilibration with the ambient hydrostatic pressure.

Since pump and valve are driven in pulse mode, their power requirements are minimal. The fully equilibrated solution is pumped into a fibre-optic flow cell (Figure 6.2b). The light transmitted through the flow cell is measured at three wavelengths: at the absorbance maxima of the (i) acid and (ii) base forms of the indicator and (iii) at a wavelength at which the indicator does not absorb (Figure 6.3).

The sensor principle is based on the equilibrium between the reagent solution and the external sample. In this case the sensor response becomes diffusion-independent and is insensitive to changes in the diffusional boundary layer around the sensor membrane. It takes c. five minutes to achieve the equilibrium between the indicator and external solution, determining the response time of the sensor. For the intended main application in seawater, interferences with other chemical species are not expected because the concentration of other acidic (e.g. H_2S) or basic (e.g. NH_3) gases and vapours is sufficiently low. However, when measuring in other media, such interference might occur and have to be taken into account.

In summary, opto-chemical CO_2 sensors show an excellent long-term stability without the need for the commonly required periodic calibrations. This is achieved by the following measures:

- Renewing the sensing solution.
- Allowing the sensing solution to reach equilibrium with the analyte.
- Calculating the response from a ratio of the indicator solution absorbances.
- Careful solution preparation, wavelength calibration, and stray light rejection.

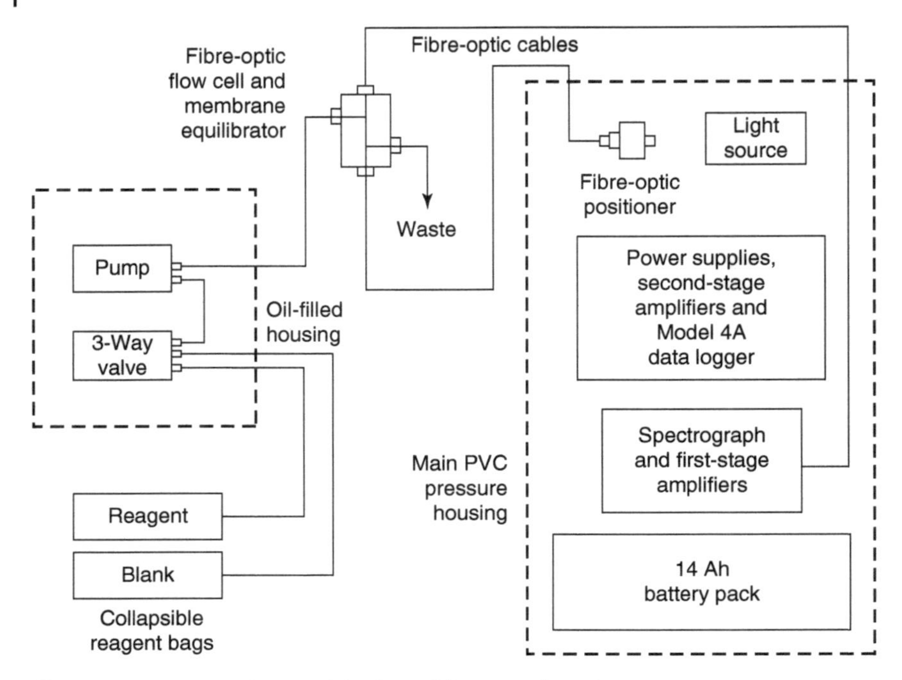

Figure 6.3 Instrument layout of the SAMI-CO_2 sensor from Figure 6.2.

6.2 CO_2 Detector Tubes

Detector tubes are thin, graduated glass tubes containing a chemically sensitive substance that reacts specifically with the gas or vapour to be measured, resulting in a clearly visible change of its colour when exposed to the gas in question. The length of the colour change produced is proportional to concentration. The point where this reaction stops is read off against graduated markings on the tube. Thus, detector tubes can also be called colorimetric chemical sensors [15] in the sense of a more general sensor definition, e.g. given in [16]. Today they belong to the classical methods used in gas analysis.

The first detector tube patent was filed in 1918 by Lamb and Hoover [17] (Figure 6.4). For the detection of the presence of carbon monoxide, they filled a mixture of iodine pentoxide and sulfuric acid in glass tubes. Their detector tubes allowed only the qualitative detection of carbon monoxide; quantitative measurement was not yet possible.

Since then, this basic principle has not changed significantly. As shown in Figure 6.5, the length of the discolouration of the reagent system is proportional to the concentration of the measured substance. A printed scale on the detector tube indicates the concentrations as quantitative values in the units ppm or volume percent. To provide long-term stability during storage, originally the tubes are hermetically sealed. The break-off tips at both ends should be opened only immediately before measurement [15].

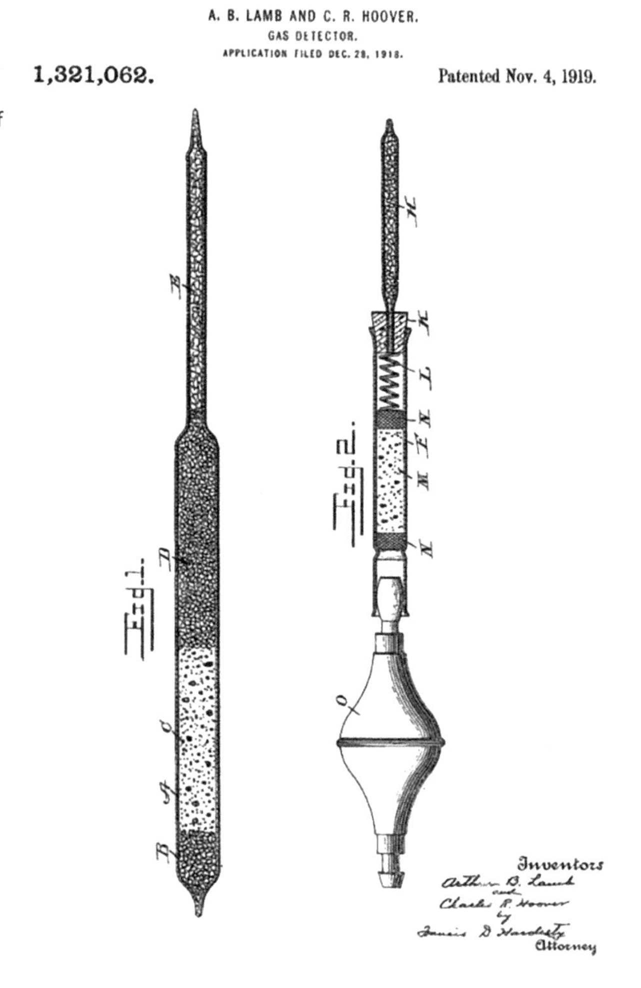

Figure 6.4 Drawings of the invented gas detector tube [17]. A, glass tube; B, granular pumice; C, granular charcoal; D, drier; E, granular mixture of iodine pentoxide and sulphuric acid; F, end of the tube; H, small glass tube of the hoolamite; K, stopper; L, spring; M, granular active charcoal; N, porous plugs; O, bulb.

The complete measurement system consists of the detector tube and a detector gas pump that sucks a defined volume of the sample strokewise through the detector tube. The necessary number n of strokes is indicated on the detector tube. It depends on the kind and concentration of the gas to be tested. For this purpose, special bellows-type pumps or piston pumps are available. Pumps from different manufacturers may have different flow patterns or deliver different volumes. Piston pumps generate a high flow initially followed by an approximately exponential decay, whereas bellows pumps provide a more steady flow initially followed by the slow decay. For example, piston pumps sometime cause a smearing of the colour stain when used on tubes originally developed for bellows pumps. Even bellows hand pumps as supplied by the companies MSA Deutschland GmbH, Berlin, Germany, and Drägerwerk AG & Co. KGaA, Lübeck, Germany, have substantially different flow patterns. The latter means

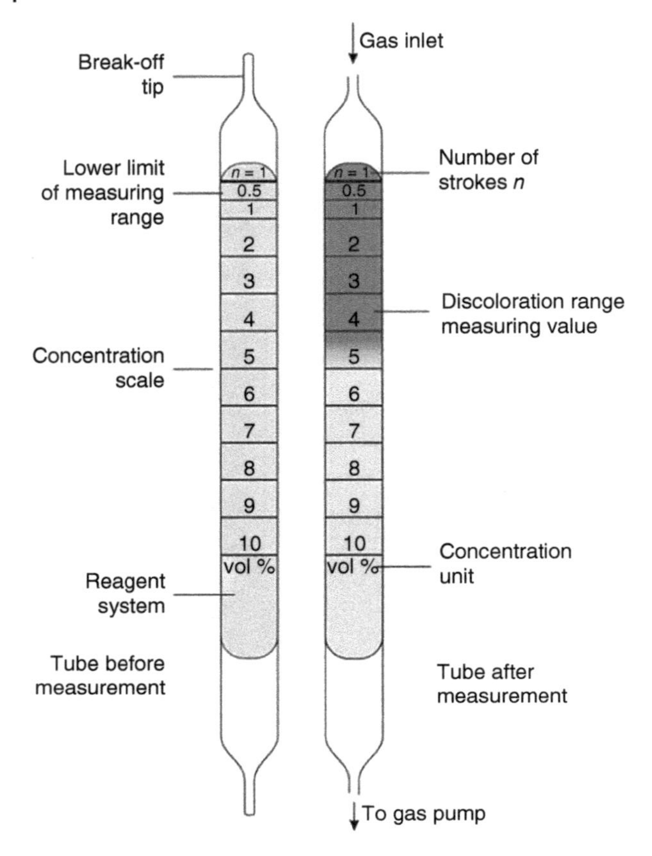

Figure 6.5 Detector tube before and after measurement. Source: Adapted from Dräger Safety AG & Co KGaA 2011 [15].

that the pumps cannot be interchanged between detector tubes from different manufacturers [18].

For long-term measurements, also diffusion tubes are offered that do not need a pump for sampling. These tubes provide integrated measurements and can be used, e.g. as personal monitors to determine the average concentration over a longer period [15]. Today short-term and long-term tubes as well as chip tube measurement systems with direct or indirect indication are available for a great number of other gases and vapours. Numerous other developments and patents are related to filters and flow-resistor elements [19, 20], chip measurement systems, and optoelectronic detectors [21, 22].

The measurement of carbon dioxide is performed by oxidation of hydrazine hydrate in the presence of crystal violet as an oxidation–reduction (redox) reaction, where the colour of the reagent in the detector tube changes from white or pale violet to blue-violet [15, 18]:

$$CO_2 + N_2H_4 \rightarrow NH_2 - NH - COOH \text{ (hydracinecarboxylic acid)} \tag{6.3}$$

As an example for a simple method of preparation of a CO_2 detector tube, in the patent [23] a composition that comprises hydrazine hydrate and crystal violet 0.05–0.5 wt% solution in absolute alcohol in the ratio of hydrazine hydrate 5–20 ml in 100 ml of crystal violet solution in absolute alcohol was used. After impregnation with silica gel, the mixture is dried by suction. It is then filled into a glass tube by means of dozer to a particular length plugged with purified white cotton and then sealed by fusion. The main objective of this set-up was the detection of carbon dioxide present in underground coal mines as well as in confined places like tunnels or caves.

Typical technical parameters of CO_2 detector tubes are:

- *Precision*: The typical relative standard deviation of CO_2 detector tubes is smaller than ±10%.
- *Selectivity*: The reaction according to Eq. (6.3) is very selective. Cross sensitivities to SO_2 or H_2S are possible but only rarely critical because in most cases carbon dioxide will be present in the measurement gas in much higher concentration than these substances.
- *Storage life*: The storage life of CO_2 detector tubes is typically two years in darkness at 5–25 °C; refrigeration is preferred.
- *Influence of humidity*: Humidity has little effect on CO_2 detector tubes. At tubes for low concentrations, the humidity has no effect in the range 5–85% RH; at tubes for higher concentrations, it is even larger (up to 100% RH).
- *Influence of temperature*: In the data sheets of CO_2 detector tubes, the temperature range is typically specified as 0–40 °C. Temperature can affect the gas tube readings because, for example, the gas density decreases or the reaction rate increases with increasing temperature. To correct the influence of temperature, correction factors CF_t are listed in the individual tube data sheets. To obtain the corrected concentration, the observed reading is multiplied by the CF_t. As it is apparent from Table 6.1, the influence of temperature can be neglected for higher CO_2 concentrations.
- *Influence of pressure*: CO_2 detector tubes are typically calibrated by the manufacturer at an atmospheric pressure of 1013 hPa. To correct the influence of pressure, the value read from the tube scale must be multiplied by a correction factor: $CF_p = 1013\,hPa/(\text{actual atmospheric pressure in hPa})$.
- *Safety aspects*: The hydrazine used in CO_2 detector tubes as a decolourizing agent has a relatively high vapour pressure and can produce toxic vapours at and above room temperature. Although it is contained in tubes only in extremely small amounts, used or expired tubes must be disposed properly and may not be simply thrown away in the domestic waste. Even CO_2 test strips prepared with hydrazine for the detection of CO_2 concentrations in small closed compartments would therefore present a health hazard to individuals. For this reason, CO_2 test strips of basic fuchsin with tetraethylene pentamine, which has a relatively low vapour pressure and is non-toxic, were used in place of hydrazine as a dye-decolourizing agent [24].

Table 6.2 gives an overview of commercially available CO_2 detector tubes for short-term measurements of the Dräger Company (Lübeck, Germany).

Table 6.1 Influence of temperature on CO_2 detector tubes.

Tube no.	CO_2 concentration (%)	Correction factors (CF_t)			
		0 °C	10 °C	21 °C	40 °C
H-10-104-30	0.03–0.5	0.90	0.95	1.0	0.95
H-10-104-40	0.05–1	1.2	1.1	1.0	0.75
H-10-104-45	0.25–3	0.85	0.95	1.0	1.05
H-10-104-50	1–20	1.0	1.0	1.0	1.0
H-10-104-60	5–40	1.0	1.0	1.0	1.0

Source: Adapted from Honeywell Analytics 2014 [18].

Table 6.2 Commercially available CO_2 detector tubes for short-term CO_2 measurements.

No.	Type	Measuring range	Number *n* of strokes	Measurement time (min)
1	100% a^{-1}	100–3000 ppm	1	4
2	0.1% a^{-1}	0.1–1.2 vol%	1	0.5
3	0.1% a^{-1}	0.5–6 vol%	5	2.5
4	0.5% a^{-1}	0.5–10 vol%	1	0.5
5	1% a^{-1}	1–20 vol%	1	0.5
6	5% a^{-1}	5–60 vol%	1	2

Source: Adapted from Dräger Safety AG & Co KGaA 2011 [15].

As Figure 6.6 shows, a CO_2 concentration range of about four orders of magnitude can be detected overlapping with these test tubes.

The main advantages of CO_2 detector tubes over other analytical methods are:

- Low costs.
- Rapid response.
- Simplicity of use.
- No calibration and very low maintenance.

For these reasons CO_2 detector tubes have been used in a broad range of applications, e.g. for analysis of technical or medical gases; for control of the concentration at the workplace and in underground coal mines, tunnels, caves, breweries, and thermal spring baths; for testing around plumbing components where a leak is suspected; and for monitoring of the breathing air quality. They are best used for one-shot evaluations when a quick and rough test is sufficient or for cursory evaluation of hazardous situations, especially when more appropriate instruments are not available. On the other hand, they are less appropriate for routine or prolonged use in the same location or if very high accuracy (standard deviation $\leq \pm 10\%$) is required.

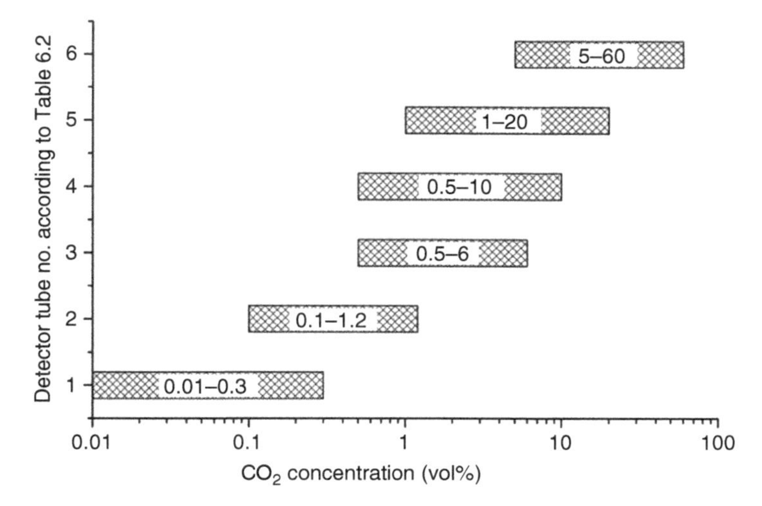

Figure 6.6 Measuring ranges of the CO_2 detector tubes from Table 6.2.

6.3 Fibre-Optic Fluorescence CO_2 Sensors

6.3.1 Fibre-Optic Sensors

6.3.1.1 Light Propagation in Optic Fibres

Optical fibres are fine strands of glass or plastic used for transmission of light from one compartment to another over a length of up to several hundreds of metres. Light itself is an electromagnetic wave of high frequency ω where both the electrical and the magnetic field strength components depend on space (x, y, z) and time t. For the electrical field strength component E_z of a planar wave travelling in x-direction (as approximately in an optical fibre), it applies

$$E_z = E_0 \exp\left\{j\left(\frac{2\pi n x}{\lambda} - \omega t + \varphi\right)\right\} \cdot \exp\left(-\frac{\omega\kappa}{c}x\right) \tag{6.4}$$

The first part describes a harmonic oscillation of amplitude E_0, frequency ω, vacuum wavelength λ, and phase shift φ. n is the refractive index of the material the wave travels in (glass fibre). The second exponential term has a real exponent, meaning that a decay of the amplitude of the electrical field strength occurs in matter. κ is called extinction coefficient.

Light can be guided in optical fibres, i.e. flexible, transparent fibres made by drawing glass (silica) or plastic. Optical fibres consist typically of a transparent core that is surrounded by a transparent cladding material with a lower index of refraction. Due to the phenomenon of total internal reflection, light is kept in the core so that the light wave can travel in the core of the fibre [25].

Light (inside or outside the fibre) interacts in many ways with matter, i.e. solids, fluids, and gases, leading to a change of the wave's properties. According to Eq. (6.4) this might be, for instance:

- Change of intensity E_z^2 by transmission, absorption, reflection, refraction, scattering, etc.
- Change of index of refraction n.
- Wavelength λ, e.g. by colour or fluorescence.
- Phase shift φ.
- Polarization.

This allows to detect the influence of certain quantities, which affect one or more of these parameters.

6.3.1.2 General Set-Up and Basic Components

Figure 6.7 shows the general set-up of a fibre-optic sensor. It comprises the following basic components:

- An optical source, e.g. a laser or a LED.
- The optical fibre for the transmission of light.
- A transducer, where the transmitted light interacts with the quantity to be measured (inside or outside the fibre; see Section 6.3.1.3).
- A receiver, i.e. a photodetector, which detects the variation in the optical signal due to the perturbation of the measurand.

6.3.1.3 Optical Fibres

To transmit light along the fibre by total internal reflection, an optic fibre (core) of refractive index n_1 must be coated with a material (cladding material) that has a refractive index n_2 smaller than that of the fibre material (Figure 6.8) [25, 28]. The acceptance cone half-angle of light that is transmitted in the fibre depends on the refractive indices of the core and the cladding as well as that one of air as the entrance medium (n_0):

$$\sin a = \frac{\sqrt{n_1^2 - n_2^2}}{n_0} \tag{6.5}$$

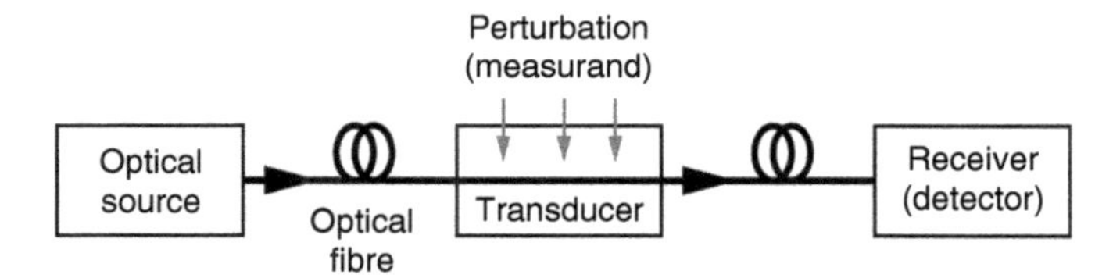

Figure 6.7 General set-up of a fibre-optic sensor [26].

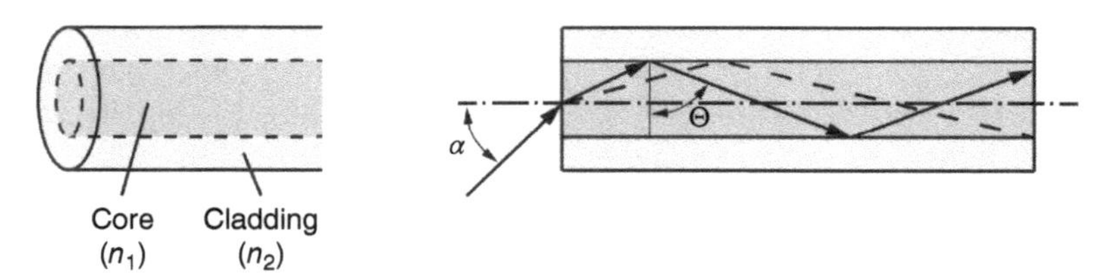

Figure 6.8 Schematic of an optical fibre. Source: Reprinted with permission from Seitz and Sepaniak [27]. Copyright 1988, American Chemical Society.

or in terms of the numerical aperture:

$$\mathrm{NA} = n_0 \sin a \tag{6.6}$$

Incident light is transmitted only if it strikes the cladding at an angle greater than the critical angle:

$$\Theta = \arcsin \frac{n_2}{n_1} \tag{6.7}$$

The transmission of light in the fibre allows only certain modes for propagation that depend on the diameter of the fibre and the wavelength of the light used. Two types of fibres are offered for a given incident wavelength:

- Monomode fibres have a narrow glass core of uniform refractive index profile and transmit only a single mode for light of a specific wavelength range and linearly polarized state. The transmitted light shows a Gaussian spatial intensity distribution over the diameter of the fibre.
- Multimode fibres have a greater core diameter and can transmit many different light modes. They have either a uniform index profile (having a so-called step index with respect to the cladding) or a parabolic (gradient) index profile. Due to their larger core size and higher numerical aperture, they can convey higher light intensities than monomode fibres. However, they show higher modal noise because disturbances such as temperature or mechanical perturbation affect different modes differently.

Even though the transmitted light is reflected totally internally, there is some penetration into the cladding at least at the boundary. This so-called evanescent wave propagates parallel to the core-cladding interface and can interact with matter, e.g. molecules, near this interface. The electrical field strength $E_z(z)$ decays exponentially in the cladding:

$$E_z(z) = E_{z0} \exp\{-z/d_\mathrm{p}\} \tag{6.8}$$

d_p describes the penetration depth of the amplitude of the evanescent wave where the field strength decreases to 1/e or 37%:

$$d_\mathrm{p} = \frac{\lambda}{2\pi n_1 \sqrt{\sin^2\theta - \left(\frac{n_2}{n_1}\right)^2}} \tag{6.9}$$

with λ as the wavelength in vacuum. Typically, penetration depths are on the order of a wavelength [27]. However, they increase – theoretically up to infinity – as the angle of incidence approaches the critical angle θ.

A comprehensive overview of the fabrication, development, and types of optic fibres is to be found, for example, in [29, 30].

The majority of applications of fibre-optic sensors operate in the visible and near-infrared spectral range (from 340 nm up to 2 μm) [31], allowing silica (SiO_2) or polymers as material for the fibres. For the visible spectral range, usually polymer fibres are used because they are mechanically more flexible and cheaper than silica fibres. Their numeric aperture NA = 0.4–0.5 is higher than of silica-based fibres (0.1–0.28), and the attenuation is c. 1 dB m^{-1} at a wavelength of 650 nm.

For higher wavelengths up to 10.6 µm – the wavelength of CO_2 lasers – chalcogenide fibres are suitable [32]. However, the application of chalcogenides for chemical and biological sensors or in medicine is limited due to the toxicity and the fragility of the fibre material. Alternatively, hollow fibres with a silica or glass capillary can be used that have an internal total reflection metal layer (Au, Ag) [33].

The sensitivity of fibre-optic sensors can be increased by interference principles, e.g. by fibre Bragg gratings (FBGs) or long-period fibre gratings (LPFGs) [34–37]. These principles also allow to measure spatially distributed measurands where, e.g. light of a particular wavelength is influenced and reflected at a certain location with a corresponding FBG, reflecting this particular wavelength [38]. The reflection at a FBG is shown in Figure 6.9. The grating is written as pattern with a refractive coefficient $n_1^* > n_1$ into the core of the fibre. Light with a small bandwidth is reflected due to the interference pattern of the Bragg grating. The centre wavelength λ_B of the filter characteristics in single-mode fibres obeys the Bragg condition:

$$\lambda_B = n_{eff} \cdot 2\Lambda \tag{6.10}$$

where Λ is the grating period and n_{eff} the effective refractive index depending on the diameters of core and cladding, respectively; the refractive indices n_1, n_1^*, and n_2; and the wave modes.

6.3.1.4 Interaction Between Light and External Measurand

In general, light guided in an optical fibre can interact with an external measurand in two ways (Figure 6.10):

- *Extrinsic sensors:* The fibre acts only as a transmission line for the light. The light is coupled out into a volume where it can interact with the measurand, e.g. the target gas, and will then be coupled back into the fibre.
- *Intrinsic sensors:* The interaction takes place in the evanescent field of the optical fibre or by diffusion of sample molecules into the optical fibres. In most of the intrinsic sensors, a chemical dye is used to act as an interface between the target gas and the optical fibre.

Figure 6.11 shows widely used techniques for extrinsic (Figure 6.11a) and intrinsic (Figure 6.11b–f) sensors.

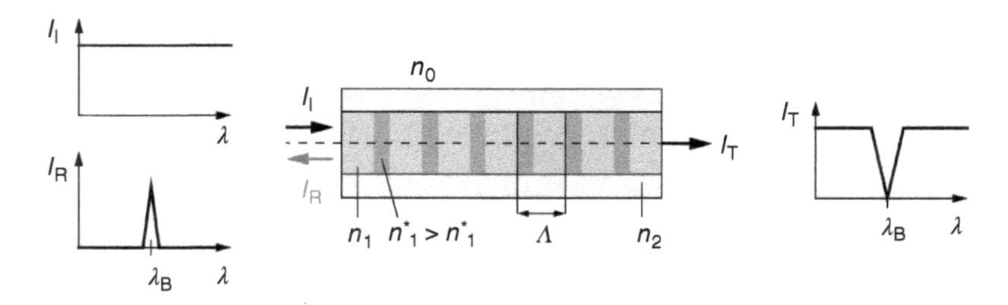

Figure 6.9 Fibre Bragg grating; I_I, I_T, I_R light intensity of incident; transmitted and reflected light; Λ grating period.

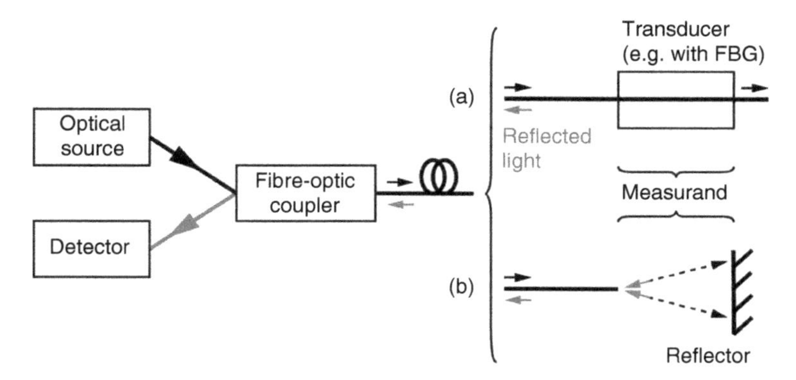

Figure 6.10 (a) Intrinsic and (b) extrinsic fibre-optic sensors.

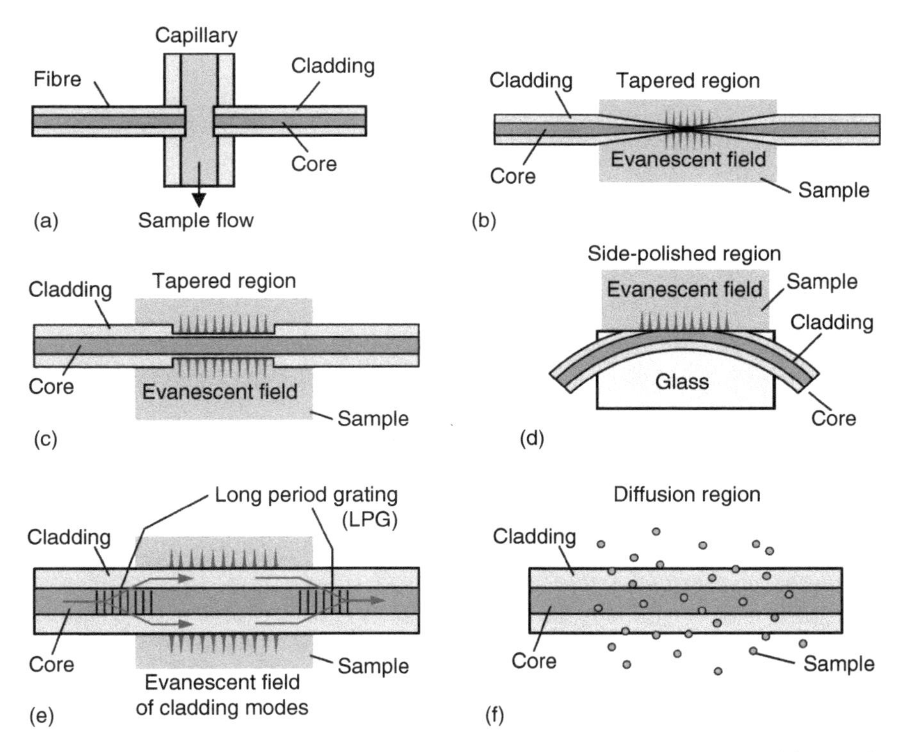

Figure 6.11 Techniques providing interaction between light guided in an optical fibre and an external measurand: (a) light coupled out of the fibre and back in; evanescent fields in (b) a tapered region, (c) an etched region, (d) a side-polished region, or (e) using a LPFG to couple the cladding modes in and out; (f) diffusion of the gas to be measured into the fibre core. Source: From Hodgkinson and Tatam 2013 [39].

6.3.1.5 Advantages of Fibre-Optic Sensors

Fibre-optic sensors offer a large number of advantageous properties:

- They are based on glass, i.e. dielectric material. They are chemically inert and immune to electromagnetic interference. They do not conduct electricity. Therefore, they can be used in high voltage, high temperature, corrosive, and flammable environments or other normally inaccessible areas.
- They are compatible with communications systems and have the capacity to carry out remote sensing where no electrical power is needed at the remote location.
- Both the measurement at many different locations along the length of a fibre and the parallel measurement of several measurands are possible by multiplexing the light wavelength for each sensor or by sensing the time delay as light passes along the fibre through each sensor.
- They enable small sensor sizes.
- They do not contaminate their surroundings.
- They provide high sensitivity, resolution, and dynamic range.

6.3.2 Fibre-Optic Fluorescence Gas Sensors

6.3.2.1 General

The set-up of a fibre-optic chemical sensor follows in general the configuration of Figures 6.7 and 6.11. To make the sensor sensitive to a particular chemical species, the transducer has to ensure the selective recognition of the particular molecule or ion. This could be achieved in different ways:

- By a suitable reagent immobilized at the optical fibre or a cavity with a semipermeable membrane, allowing the target species to pass through (Figures 6.12 and 6.13).
- By chemically doping the fibre with a fluorescent material [41].

A light source is used to stimulate a fluorescence signal at an excitation wavelength λ_{exc}, which then is affected by the concentration of the species to be measured. This leads to a spontaneous emission of light at a usually higher wavelength λ_{em}, i.e. radiation with less energy (Figure 6.14). Many different wavelengths may emerge due to Stokes shift and various electron transitions. This fluorescence signal is now detected as a measure of the level of incident radiation. For high resolution, the wavelength bands of the exciting and of the emitted light should be as wide apart as possible. Similarly, a change in the luminescence wavelength can be transduced in a change of colour as a function of a perturbing environment.

6.3.2.2 Fluorescent Sensor Dyes for CO_2 Detection

The most widely used fluorescence dyes to sense carbon dioxide on the basis of the Severinghaus effect use ion pairs between an pH-sensitive indicator anion and a quaternary ammonium cation [43–45]. The ion pair is then immobilized in various kinds of matrices – on glass or plastic supports, on optical fibres, or in capillaries.

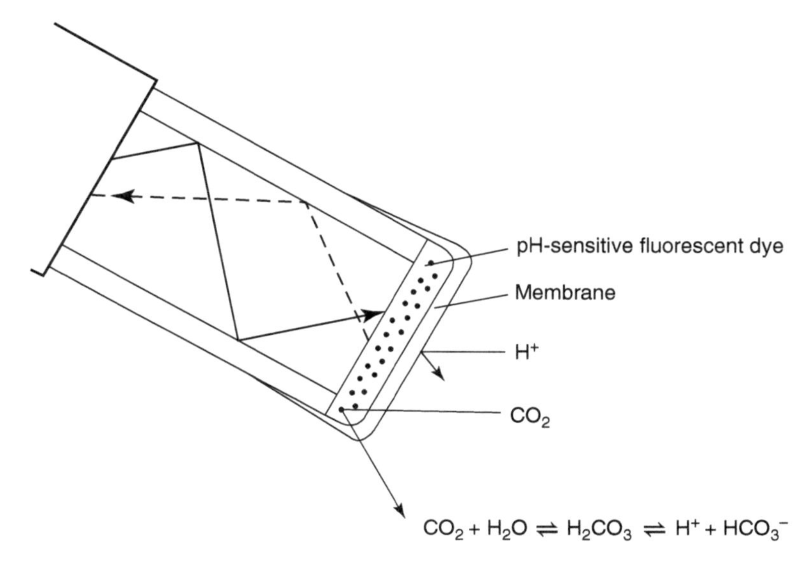

Figure 6.12 Fibre-optic CO_2 sensor based on the Severinghaus effect. Source: From Munkholm et al. 1988 [40].

Figure 6.13 Fibre-optic sensor tip based on a CO_2-permeable Teflon membrane , OF, optical fibre; OFD, optical fibre diameter; CR, capillary reservoir; SA, sensing agent. Source: From Ertekin et al. 2003 [42].

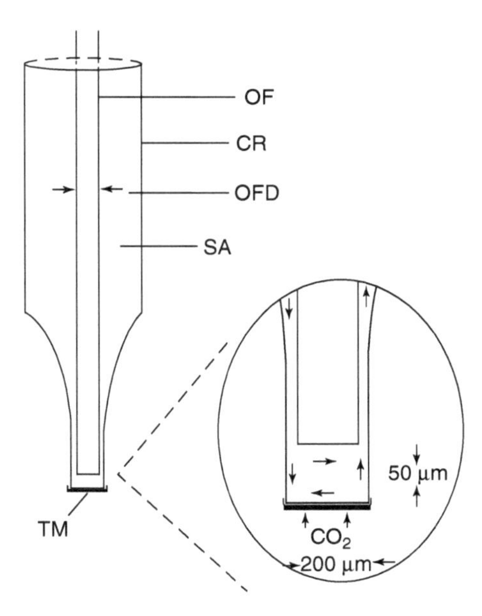

As anionic part (deprotonated dye, i.e. D^-) of the ion pair, the following molecules have been used, for instance [42]:

- 4-Methylumbelliferone.
- 8-Hydroxypyrene-1,3,6-trisulphonate (HPTS) [46].
- 8-Hydroxy-1,3,6-pyrenetrisulphonic acid trisodium salt (HPTS, PTS) [47].
- 1-Hydroxy-3,6,8-pyrenetrisulphonate (also abbreviated with HPTS) [42, 48, 49].

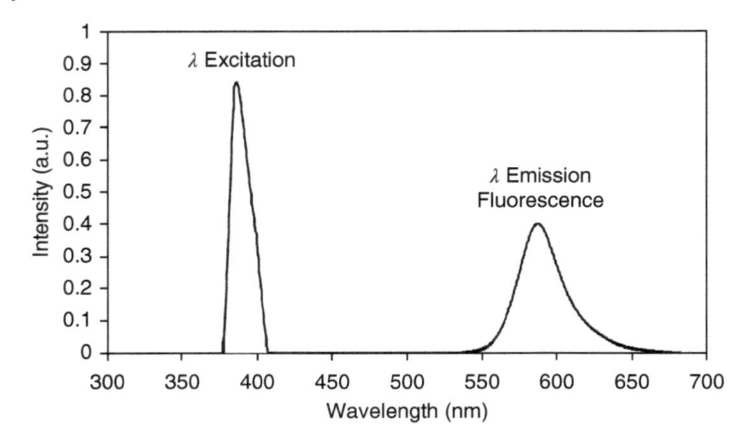

Figure 6.14 Fluorescence with excitation and emission wavelength [34].

Often tetraoctylammonium hydroxide (TOA) is used as the corresponding cation.

The deprotonated dye D^- is usually contained in an aqueous (e.g. sodium bicarbonate) buffer solution where the CO_2 can enter the volume via a gas-permeable membrane. To avoid the need for an aqueous environment, the dye can be incorporated into hydrophobic matrices, e.g. from polymer, sol–gel, or ethyl cellulose [48, 49].

CO_2 dissolves in water to produce carbonic acid and, hence, H^+-ions:

$$CO_2 + H_2O \rightleftharpoons H_2CO_3 \rightleftharpoons H^+ + HCO_3^- \tag{6.11}$$

This protonates the dye, thus changing the fluorescence response:

$$D^- + H^+ \rightleftharpoons D - H \tag{6.12}$$

The ion pair HPTS/TOA can be optically excited at wavelengths of 396 (protonated) and 460 nm (deprotonated), respectively. The wavelengths of the emitted fluorescence light of the protonated and the deprotonated form of HPTS amount then to 430 and 515 nm [48].

6.3.2.3 Fibre-Optic CO_2 Sensors

Chu and Lo presented in [50] a high-sensitive fibre-optic CO_2 sensor that was based on the pH-sensitive dye 1-hydroxy-3,6,8-pyrenetrisulphonic acid trisodium salt (HPTS, PTS) and tetraoctylammonium hydroxide (TOA) as ion pair. Together with silica particles, both were incorporated as dopant in a sol–gel matrix composed of *n*-octyltriethoxysilane (octyl-triEOS)/tetraethylorthosilane (TEOS). The excitation wavelength amounted to 460 nm.

The experimental results show that the relative fluorescence intensity of the ion pair form of HPTS dye decreases as the CO_2 gas phase concentration increases (Figure 6.15).

The sensor exhibits a very low non-linearity for CO_2 concentrations in the range of 0–100%. The sensitivity of the optical fibre CO_2 sensor is quantified in terms of the fluorescence intensity ratio between pure N_2 and pure CO_2. The response time of the sensor amounted to c. 10 and 195 seconds, respectively,

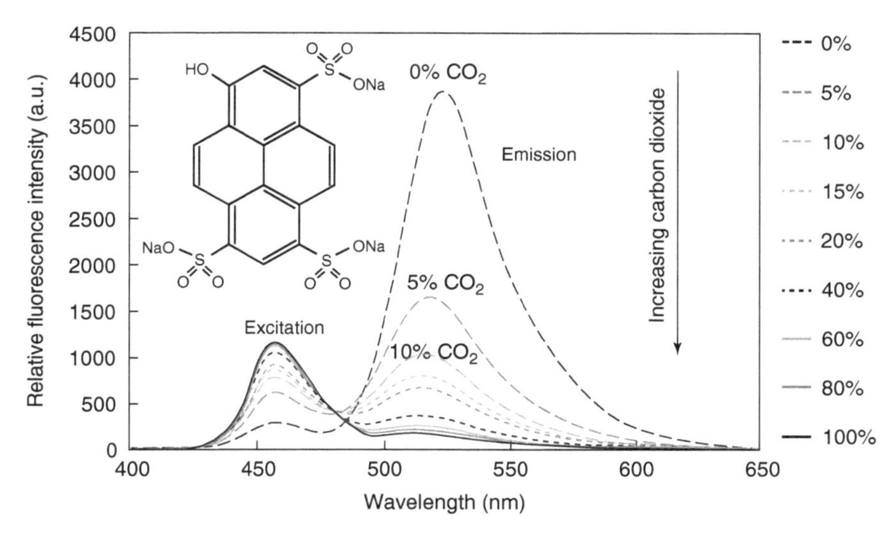

Figure 6.15 Emission spectra versus CO_2 concentration for a fibre-optic fluorescence detector with HPTS-PTS/TOA doped in a TEOS-based sol–gel matrix. Source: From Chu and Lo 2009 [50].

when switching from a pure N_2 atmosphere to a pure CO_2 atmosphere and vice versa.

Davenport et al. describe in [51] a fibre-optic fluorescence sensor for monitoring CO_2 in blood gas. The HPTS dye was bonded to a polymer that was dip-coated onto the distal tip of an optical fibre. To excite the sensing film, light from a blue LED (wavelength 460 nm) was launched into the fibre.

A more advanced set-up was introduced by Wang et al. in [47]. It comprises a 450 nm laser diode as the light source, a CO_2-sensitive capillary array (ion pair HPTS-PTS/TOA; length of 2 cm, outer diameter of c. 3 mm), an auto-suctioning device, a Y-fibre, and a spectrophotometer (Figure 6.16). All joints are sealed by a black thermosetting tube to avoid external light interference. The capillary array itself was composed of 51 capillaries (hole diameter of c. 100 μm, entire

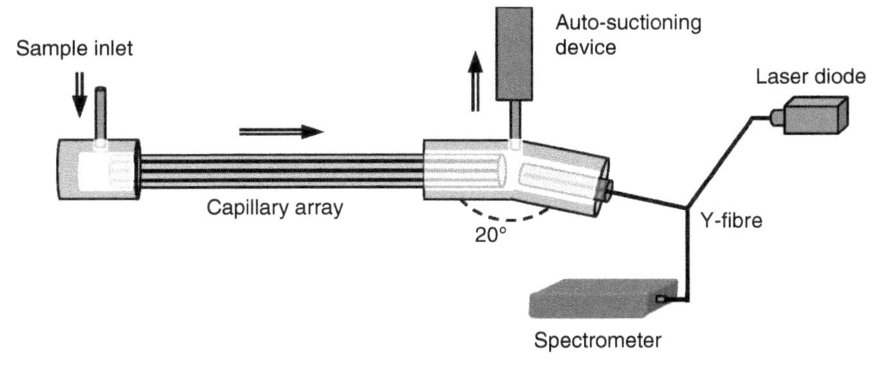

Figure 6.16 Structure of the CO_2-gas sensor with a CO_2-sensitive capillary array. Source: From Wang et al. 2017 [47].

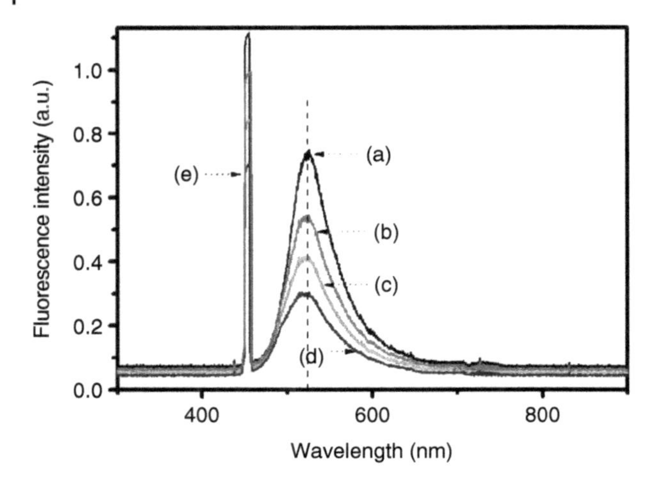

Figure 6.17 Fluorescence spectra of the sensor from Figure 6.16 at different vol% of CO_2 gas, (a) 2%, (b) 6%, (c) 10%, and (d) 60%, and (e) exciting light (450 nm). Source: From Wang et al. 2017 [47].

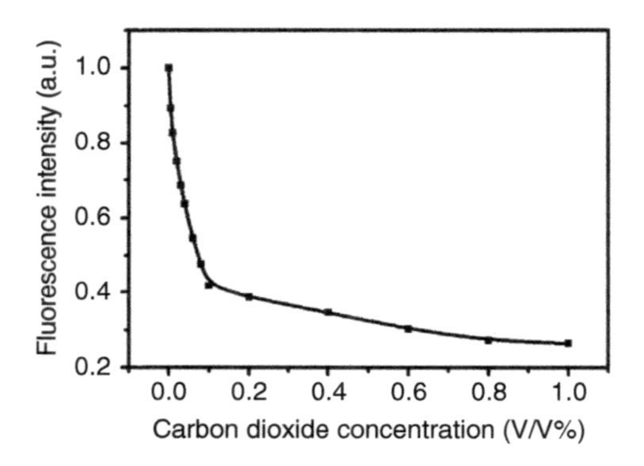

Figure 6.18 CO_2 response curve of the sensor from Figure 6.16 in different CO_2/N_2-gas mixtures. Source: From Wang et al. 2017 [47].

sensing area of c. 13 cm^2) made of glass. The angle of about 20° between the glass capillaries and the Y-fibre avoids effectively the reflected exciting light to enter the optical fibre and the spectrophotometer.

Figure 6.17 shows the fluorescence spectra in the presence of different concentrations of CO_2 in N_2. The sensor has a linear response in the volume range between 0% and 10% (Figure 6.18). The response times amounted to 12 and 93 seconds, respectively, when switching from pure nitrogen to pure carbon dioxide and vice versa.

6.3.2.4 Commercial Fibre-Optic CO_2 Sensor Solutions

The company PreSens Precision Sensing GmbH, Regensburg, Germany, offers commercial solutions for fibre-optic fluorescent sensors to optically monitor oxygen, pH, and CO_2 [52].

The CO_2 sensors measure the partial pressure of dissolved carbon dioxide. The measurement can be performed in three ways:

- Sensor spots with a diameter of a few millimetre that can be fixed to the inner surface of glassware or transparent plastic material. The CO_2 concentration can then be measured from the outside through the vessel wall (Figure 6.19).
- Flow-through cells with the same principle of contactless measurement.
- Dipping probes to dip into liquid samples.

Table 6.3 shows the specification of the sensors and sensor spots, respectively.

Since the application of fibre-optic sensors is non-invasive, this measurement principle also allows to image samples (Figure 6.20). For this, the sample surface area is covered with a fluorescent sensor foil, which translates the spatial distribution of the analyte content into a light signal. This signal is recorded contactless and pixel-wise with a digital camera.

Table 6.4 summarizes main parameters of the VisiSens system combining sensor foils with 2D read-out technology, allowing to visualize oxygen, pH, or carbon dioxide distributions in heterogeneous samples. Up to 300 000 measurement points can be recorded within one image and analysed with the included image processing and evaluation software.

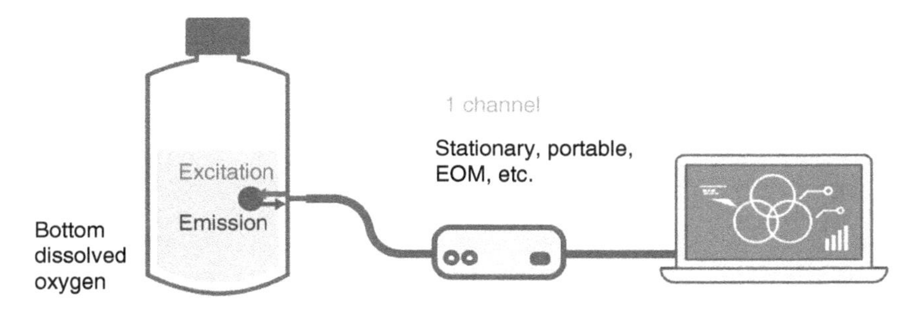

Figure 6.19 Measurement set-up with sensor spots fixed to the inner surface of a flask (from [52]).

Table 6.3 Specification of pCO_2 sensors (PreSens Precision Sensing GmbH, Regensburg, Germany) [52].

Parameter	Value
Measurement range	1–25% CO_2 at atmospheric pressure (1013.15 hPa)
Resolution at 20 °C	±0.06% at 2% CO_2, ±0.15% at 6% CO_2
Accuracy	±5% of reading or 0.2%, whichever is higher
Drift at 37 °C	Typically < 5% of reading per week (at 100% r.H. and 5% CO_2)
Measurement temperature range	+(15–45) °C
Response time t_{90} at 20 °C	<3 min for change from 2 to 5% pCO_2

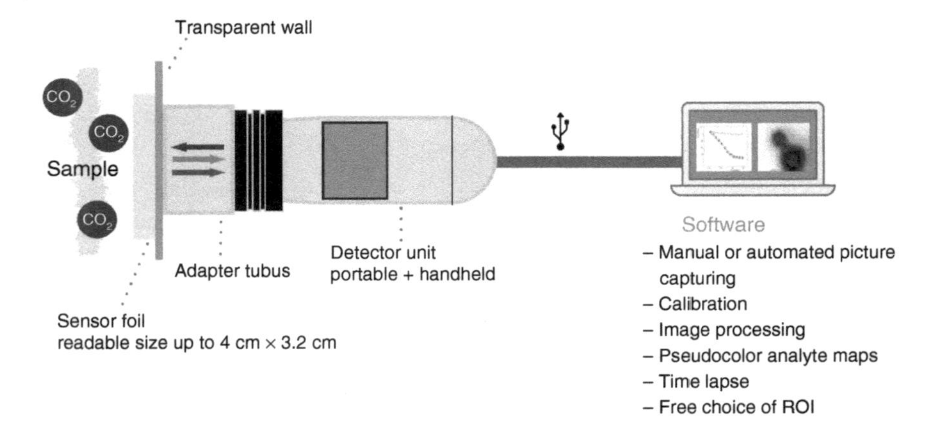

Figure 6.20 VisiSens system to image CO_2 concentration distributions (PreSens Precision Sensing GmbH, Regensburg, Germany) (from [52]).

Table 6.4 Specification of the VisiSens system to image CO_2 concentration distributions (PreSens Precision Sensing GmbH, Regensburg, Germany) [52].

Parameter	Value
Measurement range	0–1%, 1–25%
Size of sensor foil	5×5 mm^2, 40×40 mm^2
Sensing points within one image	300 000
Measurement temperature range	+(15–45) °C
Response time t_{90}	<3 min
Compatibility	Aqueous solutions, pH 4–9

References

1 Wolfbeis, O.S., Weis, L.J., Leiner, M.J.P., and Ziegler, W.E. (1988). Fiber-optic fluorosensor for oxygen and carbon dioxide. *Anal. Chem.* 60 (19): 2028–2030.

2 Orellana, G., Moreno-Bondi, M.C., Segovia, E., and Marazuela, M.D. (1992). Fiber-optic sensing of carbon dioxide based on excited-state proton transfer to a luminescent ruthenium(II) complex. *Anal. Chem.* 64 (19): 2210–2215.

3 Mills, A., Chang, Q., and McMurray, N. (1992). Equilibrium studies on colorimetric plastic film sensors for carbon dioxide. *Anal. Chem.* 64 (13): 1383–1389.

4 Parker, J.W., Laksin, O., Yu, C. et al. (1993). Fiber-optic sensors for pH and carbon dioxide using a self-referencing dye. *Anal. Chem.* 65 (17): 2329–2334.

5 Song, A., Parus, S., and Kopelman, R. (1997). High-performance fiber-optic pH microsensors for practical physiological measurements using a dual-emission sensitive dye. *Anal. Chem.* 69 (5): 863–867.

6 Spichiger-Keller, U.E. (1998). *Chemical Sensors and Biosensors for Medical and Biological Applications.* Weinheim: Wiley-VCH.
7 Mills, A. (2009). Optical sensors for carbon dioxide and their applications. In: *Sensors for Environment, Health and Security* (ed. M.-I. Baraton), 347–370. Dordrecht: Springer Netherlands.
8 Wu, H.P. (1999). System and method for optical chemical sensing. US Patent 6,436,717 B1, filed 12 May 1999 and issued 20 August 2002.
9 https://www.ysi.com/File%20Library/Documents/Manuals%20for%20Discontinued%20Products/YSI-8500-CO2-Monitor-Manual.pdf (Dec. 2002) (retrieved 24 February 2017).
10 von Bültzingslöwen, C.A.J. (2003). Development of optical sensors ("optodes") for carbon dioxide and their application to modified atmosphere packaging (MAP). PhD Thesis. Universität Regensburg, Germany.
11 Pattison, R.N., Swamy, J., Mendenhall, B. et al. (2000). Measurement and control of dissolved carbon dioxide in mammalian cell culture processes using an in situ fiber optic chemical sensor. *Biotechnol. Progr.* 16 (5): 769–774.
12 DeGrandpre, M.D., Baehr, M.M., and Hammar, T.R. (1999). Calibration-free optical chemical sensors. *Anal. Chem.* 71 (6): 1152–1159.
13 DeGrandpre, M.D., Hammar, T.R., Smith, S.P., and Sayles, F.L. (1995). In situ measurements of seawater pCO_2. *Limnol. Oceanogr.* 40 (5): 969–975.
14 DeGrandpre, M.D. (2001). Calibration-free optical chemical sensors. US 7,026 165 B2, filed 31 January 2001 and issued 11 April 2006.
15 Dräger Safety AG & Co KGaA (2011). *Dräger-Tubes & CMS-Handbook, 16th edition Soil, Water, and Air Investigations as well as Technical Gas Analysis.* Lübeck: Dräger Safety AG & Co KGaA.
16 Göpel, W., Hesse, J., and Zemel, M.J. (eds.) (1989). *Sensors, Volume 1, Fundamentals and General Aspects,* 2–3. Weinheim: VCH Verlagsgesellschaft mbH.
17 Lamb, A.B. and Hoover, C.R. (1918). Gas detector. US Patent 1,321,026, filed 28 December 1918 and issued 4 November 1919.
18 Honeywell Analytics (2014). *Honeywell Analytics Gas Detection Tubes and Sampling Handbook - H-010-4003-000,* 120. Honeywell Analytics.
19 Kretschmer, W. (1986). Prüfröhrchen (Test tube). EP Patent 0,225,520, filed 20 November 1986 and issued 16 June 1987 (in German).
20 Manns, A., Wuske, T., Harbaum, J., and Zastrow, D. (1994). Verfahren und Vorrichtung zur quantitativen Analyse von Luftinhaltsstoffen (Analysis of air pollutant present in both aerosol and gaseous form). DE Patent 4,439,433 A1, filed 4 November 1994 and issued 9 May 1996 (in German).
21 May, W. (1989). Vorrichtung zur Messung der Konzentration von gas- und/oder dampfförmigen Komponenten eines Gasgemisches (Arrangement for measuring the concentration of gaseous and/or vaporous components of a gas mixture). DE Patent 3,902,402 C1, filed 27 January 1989 and issued 13 June 1990.
22 Mentrup, C. (1990). Vorrichtung zur Messung der Konzentration von gas- und/oder dampfförmigen Komponenten eines Gasgemisches (Measuring arrangement for gaseous and/or vapour components). DE Patent 4,021,556 A1, filed 6 July 1990 and issued 9 January 1992 (in German).

23 Jayprakash, K.C., Mondal, K.N., Dhar, B.B., and Mahto, M.L. (1997). A composition useful for detecting carbon dioxide. Indian Patent Application Number 1710/DEL/1997, filed 24 June 1997 and issued 17 July 2004.

24 NASA TECH BRIEF (1965). *Test Strips Detect Different CO_2 Concentrations in Closed Compartments*. Brief 65-10390. Springfield, VA: Clearinghouse for Federal Scientific and Technical Information.

25 Mitschke, F. (2016). *Fiber Optics – Physics and Technology*, 2e, 288. Heidelberg, Dordrecht, London, New York: Springer.

26 Castrellon-Uribe, J. (2012). Optical fiber sensors: an overview. In: *Fiber Optic Sensors* (ed. M. Yasin, S.W. Harun and H. Arof). London: Intech Open.

27 Seitz, W.R. and Sepaniak, M.J. (1988). Chemical sensors based on immobilized indicators and fiber optics. *CRC Crit. Rev. Anal. Chem.* 19 (2): 135–173.

28 Shah, R.Y. and Agrawal, Y.K. (2011). Introduction to fiber optics: sensors for biomedical applications. *Indian J. Pharm. Sci.* 73 (1): 17–22.

29 Méndez, A. and Morse, T.F. (2007). *Specialty Optical Fibers Handbook*, 798. Burlington, VT: Academic Press.

30 http://www.fiberguide.com/product/optical-fibers/ (accessed 30 June 2017).

31 Pospíšilová, M., Kuncová, G., and Trögl, J. (2015). Fiber-optic chemical sensors and fiber-optic bio-sensors. *Sensors* 15 (10): 25208–25259.

32 Lezal, D., Petrovska, B., Kuncova, B. et al. (1987). Chalcogenide - halide glasses for optical waveguide. In: *Proc. SPIE 0799*, 44–53. New Materials for Optical Waveguides https://doi.org/10.1117/12.941147.

33 Nemec, M., Jelinkova, H., Fibrich, M. et al. (2007). Mid-infrared radiation spatial profile delivered by COP/Ag hollow glass waveguide. *Laser Phys. Lett.* 4 (10): 761–767.

34 Elosua, C., Matias, I.R., Bariain, C., and Arregui, F.J. (2006). Volatile organic compound optical fiber sensors: a review. *Sensors* 6 (11): 1440–1465.

35 Rao, Y.-J. (1997). In-fibre Bragg grating sensors. *Meas. Sci. Technol.* 8 (4): 355–376.

36 Bhatia, V. and Vengsarkar, A.M. (1996). Optical fiber long-period grating sensors. *Opt. Lett.* 21 (9): 692–694.

37 Wang, Y. (2010). Review of long period fiber gratings written by CO_2 laser. *J. Appl. Phys.* 108 (8): 081101.

38 Wolfbeis, O.S. (2004). Fiber-optic chemical sensors and biosensors. *Anal. Chem.* 76 (12): 3269–3284.

39 Hodgkinson, J. and Tatam, R.P. (2013). Optical gas sensing: a review. *Meas. Sci. Technol.* 24 (1): 012004.

40 Munkholm, C., Walt, D.R., and Milanovich, F.P. (1988). A fiber-optic sensor for CO_2 measurement. *Talanta* 35 (2): 109–112.

41 Starecki, F., Charpentier, F., Doualan, J.-L. et al. (2015). Mid-IR optical sensor for CO_2 detection based on fluorescence absorbance of Dy^{3+}: $Ga_5Ge_{20}Sb_{10}S_{65}$ fibers. *Sens. Actuators, B* 207 (Part A): 518–525.

42 Ertekin, K., Klimant, I., Neurauter, G., and Wolfbeis, O.S. (2003). Characterization of a reservoir-type capillary optical microsensor for pCO_2 measurements. *Talanta* 59 (2): 261–267.

43 Weigl, B.H. and Wolfbeis, O.S. (1994). Capillary optical sensors. *Anal. Chem.* 66 (20): 3323–3327.

44 Weigl, B.H. and Wolfbeis, O.S. (1995). Sensitivity studies on optical carbon dioxide sensors based on ion pairing. *Sens. Actuators, B* 28 (2): 151–156.
45 Weigl, B.H. and Wolfbeis, O.S. (1995). New hydrophobic materials for optical carbon dioxide sensors based on ion pairing. *Anal. Chim. Acta* 302 (2–3): 249–254.
46 Müller, B. and Hauser, P.C. (1996). Fluorescence optical sensor for low concentrations of dissolved carbon dioxide. *Analyst* 121 (4): 339–343.
47 Wang, J., Wen, Z., Yang, B., and Yang, X. (2017). Optical carbon dioxide sensor based on fluorescent capillary array. *Results Phys.* 7: 323–326.
48 Chu, C.-S., Lo, Y.-L., and Sung, T.-W. (2011). Review on recent developments of fluorescent oxygen and carbon dioxide optical fiber sensors. *Photonic Sens.* 1 (3): 234–250.
49 Dansby-Sparks, R.N., Jin, J., Mechery, S.J. et al. (2010). Fluorescent-dye-doped sol–gel sensor for highly sensitive carbon dioxide gas detection below atmospheric concentrations. *Anal. Chem.* 82 (2): 593–600.
50 Chu, C.-S. and Lo, Y.-L. (2009). Highly sensitive and linear optical fiber carbon dioxide sensor based on sol–gel matrix doped with silica particles and HPTS. *Sens. Actuators, B* 143 (1): 205–210.
51 Davenport, J.J., Hickey, M., Phillips, J.P., and Kyriacou, P.A. (2015). A fiberoptic sensor for tissue carbon dioxide monitoring. In: *37th Annual International Conference of the IEEE on Engineering in Medicine and Biology Society (EMBC)*, Milan, Italy (25–29 August 2015), 7942–7945.
52 https://www.presens.de/products/co2.html (accessed 30 June 2017).

7

Non-dispersive Infrared Sensors

Gerald Gerlach

Technische Universität Dresden, Faculty of Electrical and Computer Engineering, Institute of Solid-State Electronics, 01062 Dresden, Germany

7.1 Basic Principle and General Set-Up

7.1.1 General Set-Up

Non-dispersive infrared (NDIR) sensors use the concentration-dependent absorption of electromagnetic radiation in the IR range (see Section 3.1). Unlike spectrometers, often used to identify materials or gas mixtures, NDIR sensors do not comprise any dispersive optical component but colour filters, hence avoiding the most cost-driving elements in spectrometers. Therefore, NDIR analysis can be considered a special operating case of infrared absorption spectrometers where not the entire spectrum but only the transmission for single or several selected wavelengths – corresponding to characteristic absorption bands of the gases to be detected – is recorded. Figure 7.1 compares the functional principle of NDIR gas sensors (see also Table 7.1) with that of IR spectrometers.

The gas is pumped or diffuses into the sample chamber. The IR light is directed through the sample chamber towards the detector. It is absorbed at a gas-specific wavelength in the IR. According to the Lambert–Beer law from Eq. (3.15), absorption depends on both the path length-dependent attenuation of the radiant intensity I during passage through the absorbing gas and its gas concentration c:

$$I_1 = I_0 \mathrm{e}^{-\varepsilon c d} \tag{7.1}$$

where I_0 and I_1 are the radiant intensities of incident and transmitted light, $\varepsilon(\lambda)$ denotes the extinction or absorption coefficient, and d denotes the thickness of the material sample or the length the light is travelling through the gas.

NDIR gas sensors use often a reference channel besides the measurement channel to compensate disturbing effects, e.g. by temperature or by ageing of the IR radiation source. Both the measurement and the reference channel are built up identically but with no gas in the reference channel. This results therein that the difference of the sensor signals between both channels depends only on the gas concentration, whereas all other influences from the both channels cancel each other.

Carbon Dioxide Sensing: Fundamentals, Principles, and Applications,
First Edition. Edited by Gerald Gerlach, Ulrich Guth, and Wolfram Oelßner.

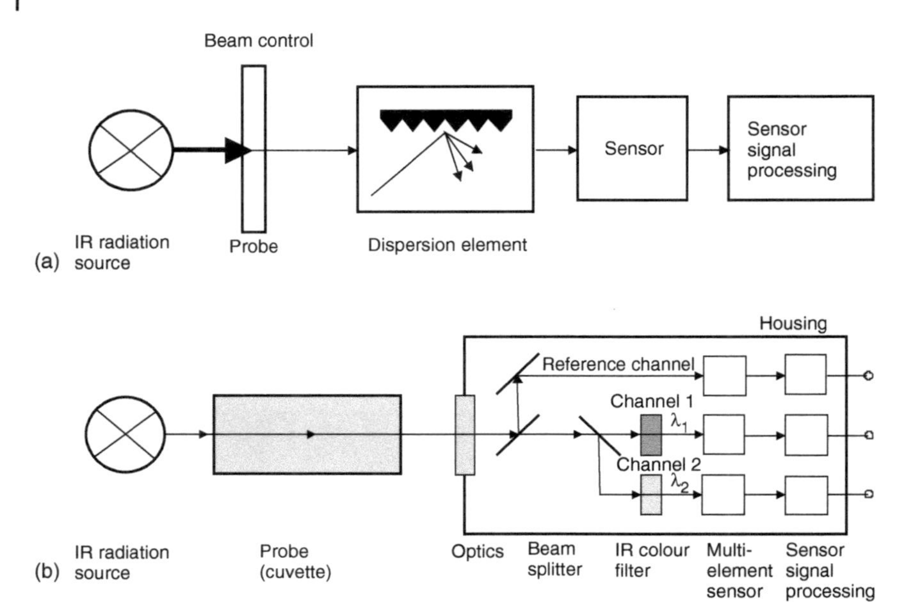

Figure 7.1 Schematic set-up of (a) an absorption spectrometer and (b) a multispectral NDIR gas sensor for two gases. The key components of NDIR sensors are an IR source (lamp), a sample chamber (cuvette), a wavelength filter, and – depending on the number of gases to be detected – one or more infrared detectors (Table 7.1). Source: Budzier and Gerlach (2011) [1]. Reproduced with permission of John Wiley & Sons.

Table 7.1 Structural components of NDIR gas sensors.

Component	Functions
IR radiation source	• Spectral broadband thermal emitter, mainly IR filament lamps • Laser diodes with an emission wavelength that is specifically selected according to the gas to be detected (see Table 7.2)
Probe	• Transmitting atmosphere • Gas in cuvettes
Optics (propagation path)	• Entrance aperture for sensor • Limits the spectral range
Beam splitter (propagation path)	• Splits the radiation as evenly as possible between reference and measuring channels (mainly mirror systems, e.g. mirrored pyramids for four-channel sensors)
IR colour filter (propagation path)	• Filters out the specific wavelength of the respective gas component (mainly interference filters due to its narrowband nature and high edge steepness)
Multi-element sensor	• Converts the incident (projected) radiation according to elements into an electric signal (mainly thermopiles or pyroelectric sensors; compare Section 7.2.1)
Sensor signal processing	• Amplifies with the lowest noise possible • Implements the best signal-to-noise ratio possible or, respectively, maximum detectivity • Compensates interferences from ambient temperature fluctuation • Compensates the effects of the atmosphere in the propagation path by reference measurements in the reference channel

Source: Budzier and Gerlach (2011) [1]. Reproduced with permission of John Wiley & Sons.

Table 7.2 Preferably used wavelengths for gas detection [1].

Gas	Wavelength (µm)	Wave number (cm^{-1})
CH_4	3.33	3003.0
HC	3.40	2941.2
CO_2	**4.24**	**2358.5**
CO	4.66	2145.9
NO_x	5.30	1886.8
SO_2	7.30	1369.8

The bold was used to highlight the wavelength of CO_2 as main subject of the book.

7.1.2 Gas Selectivity

To determine the concentration of a certain gas, a particular wavelength has to be chosen:

- That corresponds to the characteristic absorption bands of the gas components and
- That is sufficiently far away from the characteristic wavelengths (wave numbers) of the other gases.

The molecular transmission of the different gases can be taken from HITRAN (high resolution transmission), the worldwide standard for calculating or simulating atmospheric molecular transmission and radiance from the microwave through the ultraviolet region of the spectrum including also the IR range (Section 3.1.5) [2–5]. Spectra are given for molecular species along with their most significant isotopologues. Due to its abundance of 98.42%, the CO_2 isotopologue 626 is the most decisive [6] in determining the wavelengths to be used (cp. Section 3.1.5). Usually, CO_2 is preferably detected at a wavelength of 4.24 µm corresponding to a wave number of 2358.5 cm^{-1} (Table 7.2). This wavelength is sufficiently separated from that of other gases with which CO_2 often occurs in mixtures.

7.2 NDIR Components

7.2.1 Infrared Detectors

Most common detectors in NDIR sensors are pyroelectric detectors and thermopiles [7]. Both detector types are thermal sensors where the incident radiation "heats" up the sensor (temperature rise in the range of mK). The radiation noise that limits temperature resolution in thermal sensors and, hence, detectivity in NDIR sensors has a $\sqrt{T}$ dependence from temperature, which means that cooling as in photon detectors does not substantially improve detectivity. Therefore,

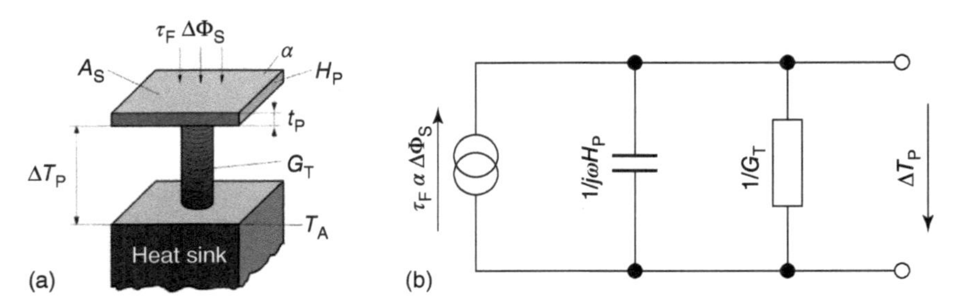

Figure 7.2 (a) Simplified thermal model and (b) equivalent electrical circuit for the temperature change ΔT_p between "heated" sensor element (top) and "cold" surroundings (bottom) of a thermal sensor due to an incident radiation flux $\Delta\Phi_S$. τ_F, transmission rate of the IR window; α, absorption rate of the sensor element; T_A, ambient temperature; G_T, thermal conductance between sensor element and surroundings; H_P, heat capacity; A_S, t_p, sensitive area and thickness of the pyroelectric element; ω, modulation frequency of the incident radiation. Source: From [7, 8] reproduced by permission of InfraTec GmbH, Dresden, Germany.

thermal detectors are operated without any cooling and are particularly suitable for small, light, and portable applications. The use of modern microelectronic and micromechanical manufacturing methods has resulted in a rapid development of miniaturized and inexpensive sensors with high resolution.

Thermal detectors are characterized by the absorbed heat flux, the heat capacity H_P, and the thermal conductance G_T to its surroundings that acts as a heat sink with a given temperature T_A (Figure 7.2). The absorbed radiation flux $\alpha\tau_F\Delta\Phi_S$ is smaller than the incident radiation flux $\Delta\Phi_S$ due to absorption (α) of the IR window and reflection/transmission (τ_F) of the sensor element, in particular of the infrared filter assembled to the sensor. The resulting temperature difference ΔT_p for harmonically modulated incident radiation yields

$$\Delta\widetilde{T}_P = \frac{\alpha\tau_F\widetilde{\Phi}_S}{G_T} \cdot \frac{1}{\sqrt{1+(\omega\tau_T)^2}} \tag{7.2}$$

where the tilde denotes root-mean-square values and $\tau_T = H_P/G_T$ the thermal time constant. To achieve significant temperature differences, the product $\alpha\tau_F$ should be as high as possible, i.e. close to unity. The sensor element has to be thermally isolated to avoid heat conduction (G_T) away from the sensor element. The heat capacity value H_P and, hence, the element thickness t_p should be low, which requires miniaturization of the sensor.

7.2.1.1 Pyroelectric IR Sensors

For a number of materials, particularly crystalline materials, positive and negative charges in the unit cell of a crystal lattice do not coincide, but are spatially separated. This is called polarization. Polarization takes place in dielectric materials. Pyroelectric sensors utilize the temperature dependence of the spontaneous (i.e. without any external excitation) polarization of certain dielectric materials – so-called pyroelectrics – for the conversion of heat from IR radiation into an electrical output signal. The pyroelectric properties disappear above the Curie temperature T_C, which therefore limits the temperature range for applications.

The temperature increase ΔT_p in the pyroelectric sensor element causes a pyroelectric current:

$$I_P = pA_S \frac{\Delta T_P}{dt} \tag{7.3}$$

where p is the pyroelectric coefficient of the pyroelectric sensor material. Together with Eq. (7.2) this leads to a pyroelectric current:

$$\widetilde{I}_P = \omega p A_S \frac{\alpha \tau_F \widetilde{\Phi}_S}{G_T} \cdot \frac{1}{\sqrt{1 + (\omega \tau_T)^2}} \tag{7.4}$$

The sensor element itself can be described by this current source, the capacity C_p of the detector element, its DC resistance R_p, and the resistance $R_{\tan\delta}$ due to dielectric losses (Figure 7.3) [1].

Figure 7.4 presents the basic circuits of a pyroelectric sensor in current and voltage mode. Due to its simplicity the voltage mode is the most commonly

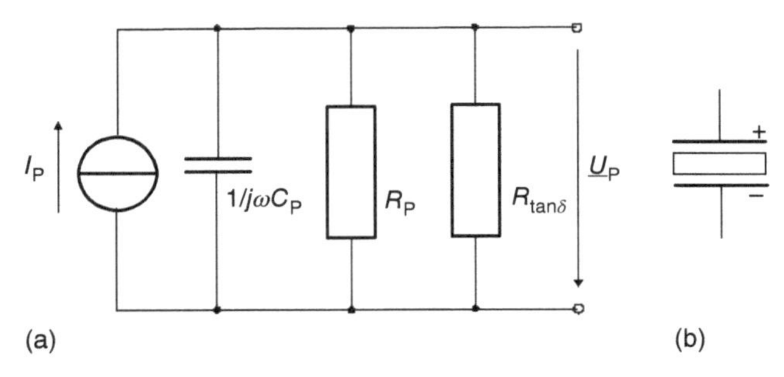

Figure 7.3 (a) Electric small-signal model of the responsive element of a pyroelectric sensor and (b) electric circuit symbol. I_p, pyroelectric current due to temperature change ΔT; C_p, DC resistance; $R_{\tan\delta}$, dielectric loss resistance. Source: Budzier and Gerlach (2011) [1]. Reproduced with permission of John Wiley & Sons.

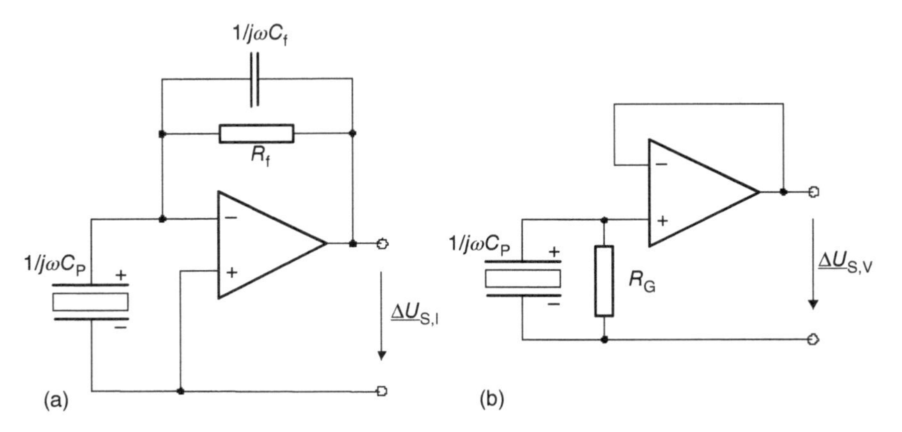

Figure 7.4 Basic circuit of a pyroelectric sensor element (a) in current mode and (b) in voltage mode. Source: Budzier and Gerlach (2011) [1]. Reproduced with permission of John Wiley & Sons.

used operating mode for pyroelectric detectors. The signal voltage U_S and the responsivity $R_V = \widetilde{U}_S / \widetilde{\Phi}_S$ for the voltage mode amount to

$$\widetilde{U}_S = \frac{\omega \alpha \tau_F A_S p R \widetilde{\Phi}_S}{G_T} \cdot \frac{1}{\sqrt{1+(\omega\tau_T)^2}} \cdot \frac{1}{\sqrt{1+(\omega\tau_E)^2}} \tag{7.5}$$

$$R_V = \frac{\omega \alpha \tau_F A_S p R}{G_T} \cdot \frac{1}{\sqrt{1+(\omega\tau_T)^2}} \cdot \frac{1}{\sqrt{1+(\omega\tau_E)^2}} \tag{7.6}$$

with $R = R_G$ and $\tau_E = R_G C_P$. For current mode, Eqs. (7.5) and (7.6) also hold, but with $R = R_f$ and $\tau_E = R_f C_f$.

As Eqs. (7.3) and (7.5) as well as Figure 7.5 show, pyroelectric sensors exclusively detect signals that change over time. Since pulsed radiation sources are used in spectroscopy as well as in gas analysis, this is not a drawback. The signal-to-noise ratio of IR detectors is determined by their specific detectivity:

$$D^* = \frac{\sqrt{A_S \cdot B}}{\text{NEP}} = \frac{\sqrt{A_S} R_V}{\widetilde{U}_n} \tag{7.7}$$

where NEP is the noise-equivalent power, A_S the sensitive sensor area, B the frequency bandwidth, R_V the responsivity, and $\widetilde{U}_n$ the effective noise voltage related to a noise bandwidth of 1 Hz. The unit of D^* is cm Hz$^{1/2}$ W^{-1}. NEP of pyroelectric sensors depends on the chopper frequency and the operation mode (current or voltage output mode) and is influenced by many effects [1].

The radiation noise constitutes the smallest detectable radiation flux and, hence, marks the ultimate physical limit of D^* for (theoretically noiseless) thermal sensors:

$$D^*_{\max} = 1.8 \cdot 10^{10} \text{cm Hz}^{1/2}\,\text{W}^{-1} \tag{7.8}$$

Noise of the loss resistance due to the dielectric loss of the pyroelectric sensor material is the dominating and, hence, limiting noise contribution for chopper frequencies larger than 100 Hz. The Johnson noise of the feedback resistor of the transimpedance amplifier in current mode and that of the series resistance of

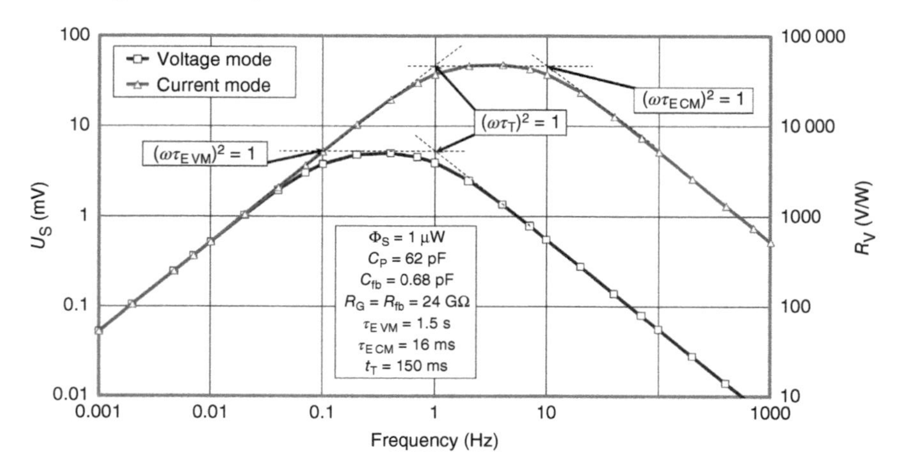

Figure 7.5 Frequency dependence of signal voltage U_S and responsivity R_V of pyroelectric sensors. Source: From [7, 8] reproduced by permission of InfraTec GmbH, Dresden, Germany.

the impedance amplifier in voltage mode are the limiting components for lower chopper frequencies.

The advantages of pyroelectric sensors are their simple and inexpensive structures together with high detectivity and excellent long-term stability.

Recent sensors based on crystalline lead zirconate titanate ($(PbZr_xTi_{1-x})O_3$ with $0 < x < 1$; also called PZT) with a drastically reduced element thickness of less than 1 μm and ultra-thin absorption layers show a value of the specific detectivity D^*(3 Hz, 500 K) of 5.35×10^9 cm $Hz^{1/2}$ W^{-1} that is quite close to the fundamental noise limit [9–11]. PZT thin-film detectors with its high potential for miniaturization reach D^* values of currently not more than 5×10^8 cm $Hz^{1/2}$ W^{-1} [9].

In order to compensate interferences of the ambient temperature, there is the option to calculate the difference in a series or parallel connection of two pyroelectric sensor elements with only one element being exposed to infrared radiation (Figure 7.6).

Examples of different types of pyroelectric sensors are shown in Figure 7.7.

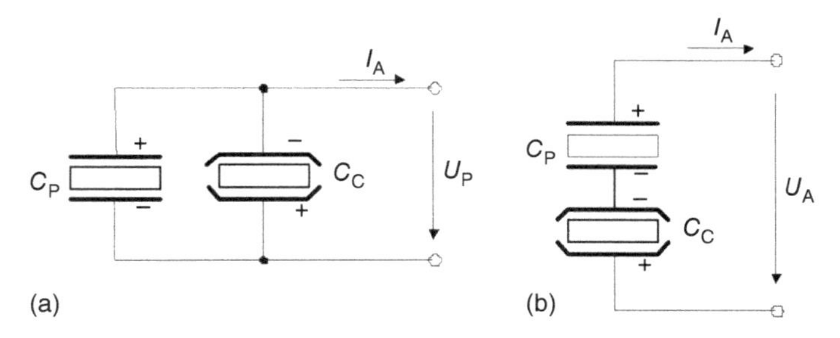

Figure 7.6 Compensated pyroelectric sensors in (a) parallel circuit and (b) series connection. C_P, active pyroelectric sensor element; C_C, inactive pyroelectric sensor element for temperature compensation. Source: Budzier and Gerlach (2011) [1]. Reproduced with permission of John Wiley & Sons.

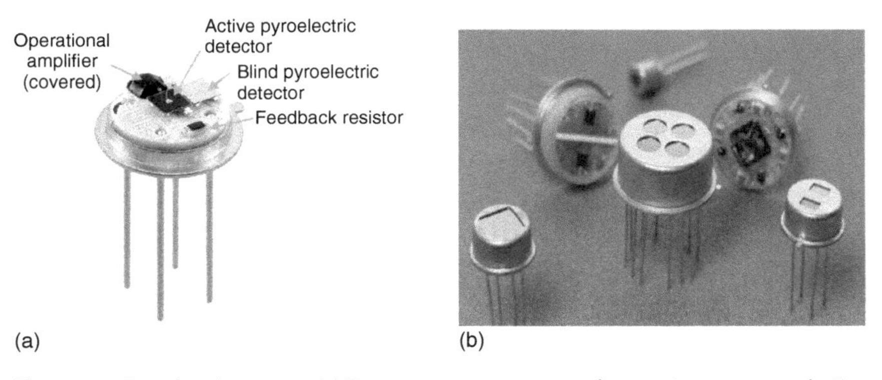

Figure 7.7 Pyroelectric sensors. (a) Temperature-compensated sensor in current mode, Type LME-335; detector element area $A_S = 2 \times 2$ mm^2 with black layer; $R_f = 100$ GΩ; $C_f = 0.2$ pF; housing TO39 (socket diameter 9.2 mm). (b) Types of pyroelectric sensors: single-channel, four-channel, and two-channel detectors (from left to right) (by courtesy of InfraTec GmbH, Dresden, Germany).

7.2.1.2 Thermopiles

Due to the large progress of micromechanical manufacturing procedures, thermopiles became inexpensive thermal sensors that are available in large numbers [12]. They do not necessarily require a chopper and can be easily integrated using standard technologies of semiconductor manufacturing – which leads to low-cost mass production.

Figure 7.8 shows the basic principle of thermopile detectors. They are thermoelectric sensors utilizing the Seebeck effect. A thermoelectric voltage V_{th} occurs in a conductor when applying a temperature difference between the measuring point (temperature T_1, hot junction) and a reference temperature (reference point with temperature T_2, cold junction) (Figure 7.8a). This voltage is a thermal diffusion current caused by thermal diffusion of charge carriers. The thermoelectric voltage V_{th} is directly proportional to the temperature difference ΔT_{S}:

$$V_{\text{th}} = \alpha_{\text{S}}(T_1 - T_2) = \alpha_{\text{S}} \cdot \Delta T_{\text{S}} \tag{7.9}$$

The proportionality factor is a material constant, the thermoelectric coefficient α_{S}, also called Seebeck coefficient or thermal power. It has the unit $\text{V}\,\text{K}^{-1}$. For platinum, it is 0 V. A thermocouple uses a pair of two wires of the materials A and B (Figure 7.8b). The corresponding output voltage due to the temperature difference ΔT_{S} amounts then to

$$V_{\text{th}} = V_{\text{th,A}} - V_{\text{th,B}} = (\alpha_{\text{S,A}} - \alpha_{\text{S,B}})(T_1 - T_2) = \alpha_{\text{AB}} \cdot \Delta T_{\text{S}} \tag{7.10}$$

It only depends on the temperature difference ΔT_{S} and not on the spatial course of the temperature.

A good temperature resolution always requires many thermocouples (number N) connected in series (thermopile; Figure 7.8c):

$$V_{\text{th}} = N\alpha_{\text{AB}}\Delta T_{\text{S}} \tag{7.11}$$

The space required for these many thermocouples sets limits to the miniaturization of thermoelectric sensors.

If the ends of the thermocouple are short-circuited, one gets a thermoelectric current I_{th} that is only limited by the resistance R_{th} of the thermocouple and can become very large:

$$I_{\text{th}} = \frac{V_{\text{th}}}{R_{\text{th}}} \tag{7.12}$$

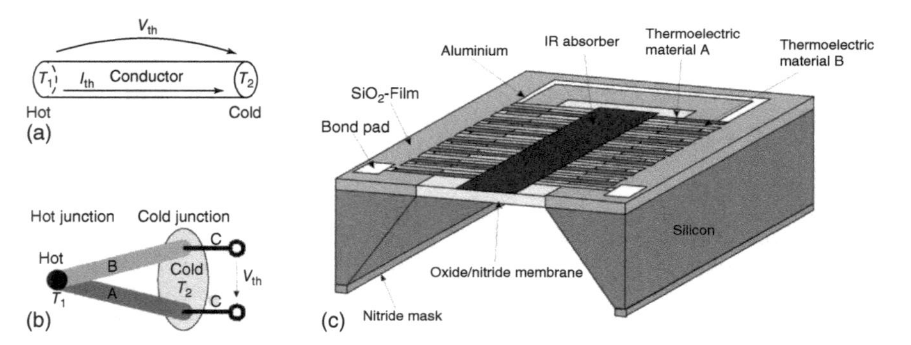

Figure 7.8 (a) Thermoelectric effect, (b) principle of a thermocouple, and (c) thermopile. Source: Budzier and Gerlach (2011) [1]. Reproduced with permission of John Wiley & Sons.

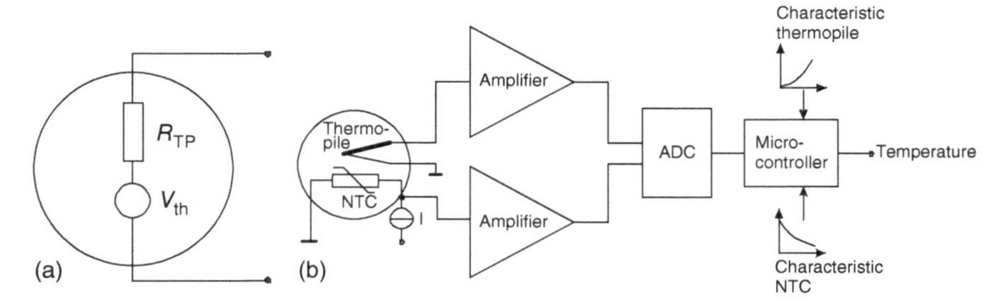

Figure 7.9 (a) Electronic model of a thermopile and (b) typical integrated evaluation circuit of a thermopile with integrated NTC resistor for measuring the reference temperature and with digital output. Source: Budzier and Gerlach (2011) [1]. Reproduced with permission of John Wiley & Sons.

The most important noise source of thermopiles is the thermal resistance noise (Johnson noise) of the thermocouples. This means that the signal-to-noise ratio and, hence, specific detectivity D^* improve with the square root of number N of the series thermocouples [1]. Advanced thermopile detectors with high N numbers show D^* values of about 10^9 cm Hz$^{1/2}$ W^{-1} [13].

Figure 7.9a presents the model of a thermocouple for the electronic simulation of such circuits. In addition to the thermoelectric voltage V_{th}, it also includes the internal resistance R_{TP} of the thermocouples. In order to evaluate the thermoelectric voltage, one needs the temperature at the reference point. Therefore, often a temperature sensor, e.g. an NTC resistor, is integrated into thermopiles to measure the reference point temperature (Figure 7.9b) [1, 14]. In order to achieve a maximum signal-to-noise ratio, the electronics for signal evaluation should also be integrated into the sensor housing (integrated signal processing).

A complete sensor with thermopile, temperature sensor, signal processing, housing, and – if required – optical components is commonly called thermopile module. It can be manufactured – according to application specifications – in a large variety of designs (Figure 7.10).

7.2.1.3 Comparison of Detectors

Table 7.3 gives the main characteristics of uncooled thermal detectors as well as for photon detectors (photodiodes, photoresistors) for the mid-IR radiation.

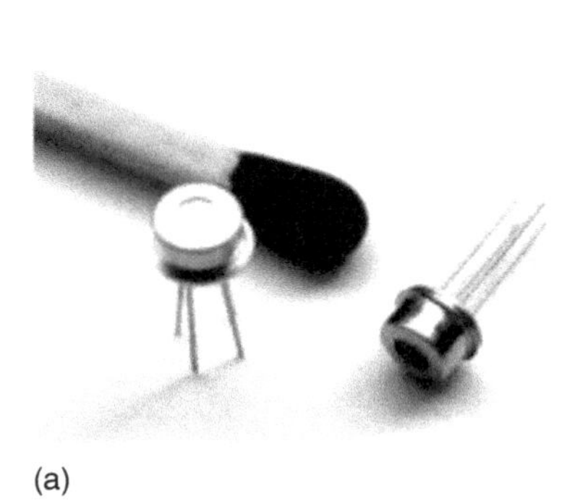

(a)

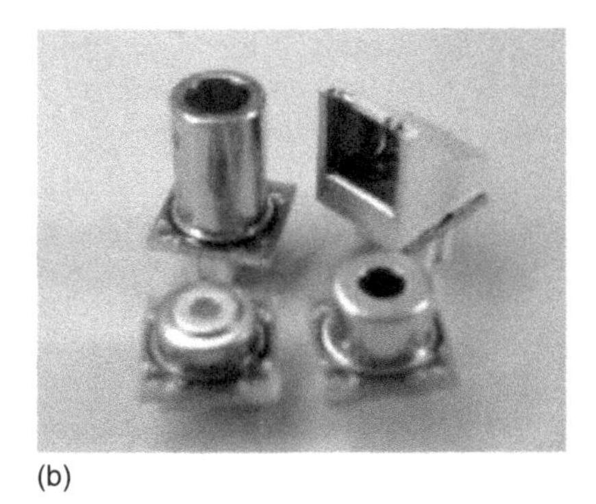

(b)

Figure 7.10 Miniaturized thermopile modules: (a) sensor and (b) sensor with integrated signal processing (by courtesy of Heimann Sensor GmbH, Dresden/Germany).

Table 7.3 Comparison of the sensor properties for NDIR CO_2 detectors. See also [22].

Sensor type	Manufacturer	Type	Response time (ms)	Active/ detection area $\sqrt{A}$ (cm)	Detectivity D^* (cm $Hz^{1/2}$ W^{-1}) (295 K)
Pyroelectric detectors	InfraTec [7, 8]	LM-395	150	0.2	4.5×10^8
	Heimann Sensors [15]	HPS D10E	N/A	0.2	4×10^8
	DIAS Infrared [16]	2LTSI Q1	N/A	0,1	7×10^8
	Micro-Hybrid Electronic [17]	MH_PS2x1C2-A-S1.5	17	0.08	2.8×10^8
Thermopiles	Heimann Sensors [15]	HTS E21	10	0.12	1.4×10^8
	Hamamatsu [18]	T11722-01	20	0.12	1.3×10^8
	Micro-Hybrid Electronic [17]	TS2x200B-A-S1.5	30	0.12	3.6×10^8
Photodiodes	VIGO System [19]	PVI-4TE-5	0.08	0.05	1×10^{11}
	Ioffe LED [20]	PD42Su/Sr WB	0.02	0.32	1.7×10^{10}
Photoresistors	VIGO System [19]	PC-5	0.2	0.0025–0.4	1×10^9
	ICO Ltd. [21]	PR1-38	0.008	0.15	1.5×10^9

Photon detectors have a strong wavelength-dependent sensitivity that rises proportionally with wavelength up to a cut-off wavelength λ_C where sensitivity falls down to zero (Figure 7.11). Here, the particular wavelength to be covered for the targeted gas has to lie in the sensitive range below but close to λ_C. By nature, photodetectors and photoresistors react very fast to changing radiation; however, this is usually not a main criterion for the design of NDIR sensors.

The main advantage of thermal detectors is the uniformity of their spectral characteristics. Therefore, pyroelectric detectors and thermopiles are the preferred sensors for NDIR applications. However, their operation speed is limited due to the thermal time constant. Thermopiles have a considerably smaller response time; but their sensitivity is smaller by one order of magnitude than those of pyroelectric detectors.

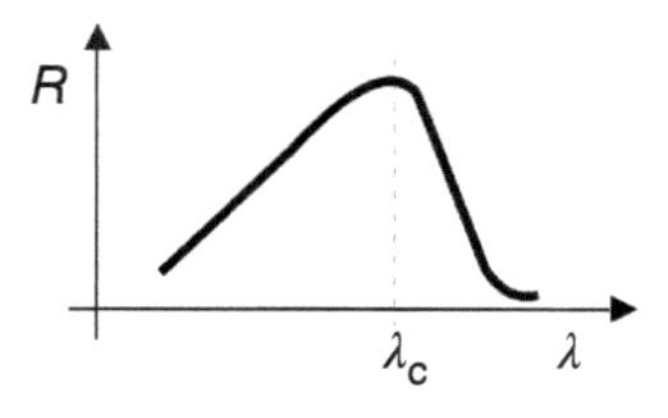

Figure 7.11 Characteristic wavelength-dependent responsivity (sensitivity) of a photon detector. λ_C, cut-off wavelength. Source: Budzier and Gerlach (2011) [1]. Reproduced with permission of John Wiley & Sons.

7.2.2 Wavelength Selection by IR Filters

7.2.2.1 IR Filters

Narrow bandpass (NBP) filters are relatively inexpensive wavelength selectors that allow the transmission of a predetermined wavelength while rejecting or blocking other wavelengths. The filter characteristic is determined by the layer stack. The relatively low cost makes them the preferred wavelength selector for applications such as CO_2 measurement, where the required wavelength is well known. The entire aperture of an interference filter can be illuminated, resulting in high throughput and an excellent signal-to-noise ratio.

Typical NBP filters for the CO_2 measurement have a central wavelength of $CWL = 4.26 \pm 0.04\,\mu m$ with a full width at half maximum of $FWHM = 0.12 \pm 0.03\,\mu m$ [23]. The central wavelength of an interference filter can shift with temperature due to the thermal expansion of the spacer layers and the change in their refractive indices. This wavelength shift is extremely small at room temperature ($0.25\,nm\,K^{-1}$; Figure 7.12).

Typical NDIR sensors use a dual-beam dual-wavelength structure where the second beam at the second wavelength serves as reference (see Figure 7.1b). Other applications such as in combustion control require the measurement of more than one gas and, hence, more than two different wavelengths. Such multi-wavelength sensors are also called multicolour detectors. For these purposes, the similar basic NDIR sensor set-up, consisting of an IR lamp, the cuvette, and the IR detector, can be used if the IR lamp is a broadband thermal emitter and if several sensors with appropriate filters are used depending on the known gases to be measured (cf. Table 7.2). One way to realize such multicolour detectors is by using beam splitters (Figure 7.13a) [24, 25]. Here, the IR radiation enters the detector through the aperture and is divided by the beam splitter in even parts. Each of the partial beams goes through an IR filter and finally hits a detector chip aimed for the particular wavelength of the corresponding

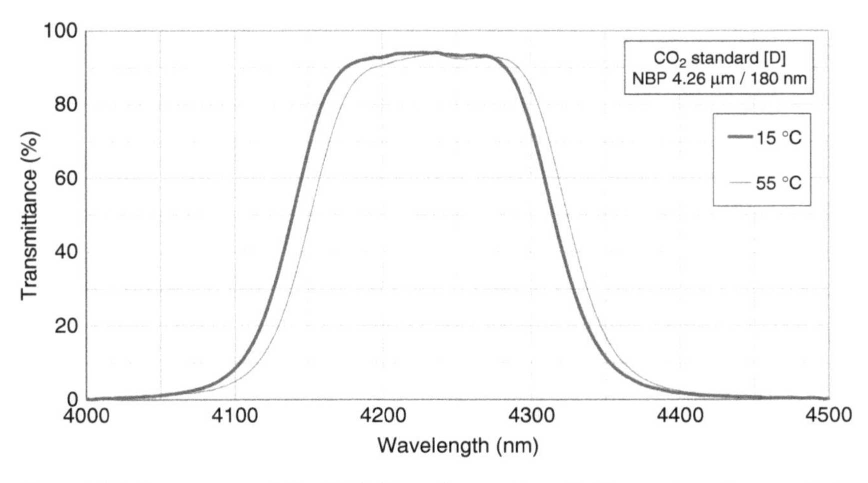

Figure 7.12 Temperature shift of NBP filters. Source: From [7, 8] reproduced by permission of InfraTec GmbH, Dresden, Germany.

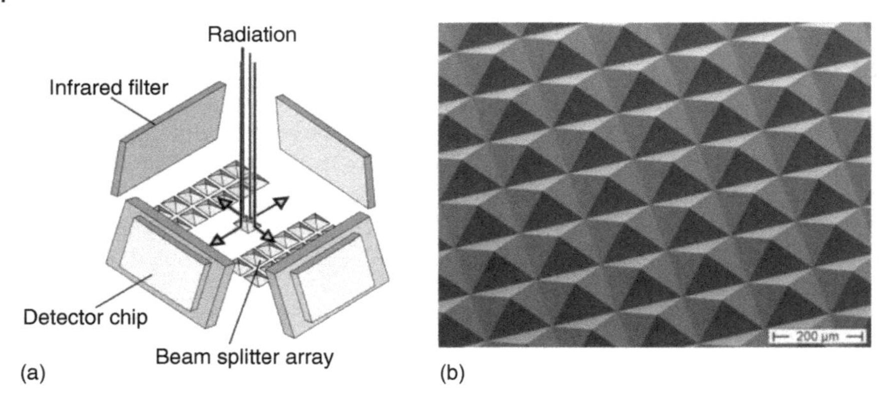

Figure 7.13 Beam splitting of the incident radiation in a 4-channel NDIR module: (a) principle set-up and (b) micro pyramid array. Source: From [7, 8] reproduced by permission of InfraTec GmbH, Dresden, Germany.

gas to be measured. Mostly, four-sided micro pyramids (Figure 7.13b) or micro V-grooves, fabricated by MEMS micromachining [26], are used for four- and two-channel detectors, respectively.

Detector solutions using beam splitters can advantageously be applied when single narrow-beam sources are used or in situations where contaminations (dust, insects) in the light path could disturb the measurement.

7.2.2.2 Fabry–Pérot Filters

Another way to adjust the wavelength to the particular gas in a gas mixture is by replacing the conventional interference filter by a tunable Fabry–Pérot interferometer filter (FPF). Figure 7.14 shows the basic set-up of such a FPF. The FPF is a tunable narrow bandpass (NBP) filter that is assembled to the detector. In the case of the electrostatically driven FPF in Figure 7.14, a control voltage applied to the filter allows one to freely select the wavelength within a certain spectral range or to sequentially measure a continuous spectrum.

The operation is based on the well-known principle of a Fabry–Pérot interferometer (FPI). Two flat and partially transmitting mirrors with reflectance R are arranged in parallel at a distance d, forming an optical resonator. Depending on

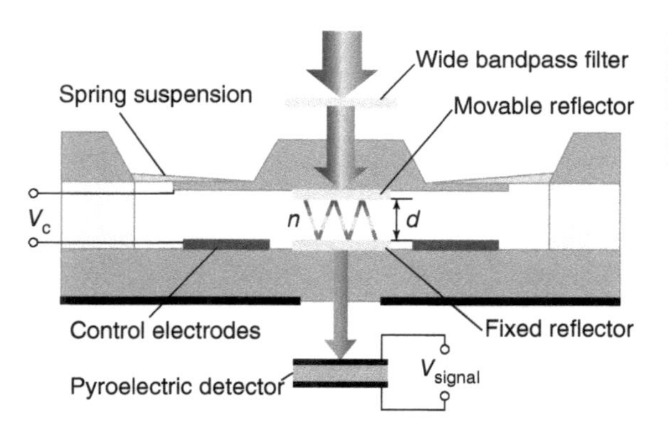

Figure 7.14 Fabry–Pérot interferometer filter. Source: From [7, 8] reproduced by permission of InfraTec GmbH, Dresden, Germany.

the actual distance, only radiation of certain wavelengths can be transmitted that satisfy the resonator condition:

$$\lambda_m = \frac{2nd \cdot \cos\beta}{m} \tag{7.13}$$

where n is the refractive index inside the gap, d the optical gap thickness, β the angle of incidence (mostly 0°), and m the interference order. Usually, the first interference order ($m = 1$) is used, whereas higher orders are blocked by means of an additional optical bandpass filter. The wavelength-dependent transmittance for $n = 1$ and $\beta = 0$ yields from Eq. (7.13)

$$T(\lambda) = \frac{T_{\max}}{1 + F \cdot \sin^2(2\pi d/\lambda)} \tag{7.14}$$

with

$$F = \frac{4R}{(1-R)^2} \tag{7.15}$$

The half-power bandwidth HPBW determines the spectral resolution:

$$\text{HPBW} = 2d\left(\frac{1-R}{\pi\sqrt{R}}\right) \tag{7.16}$$

The spectral distance between two adjacent interference peaks in Figure 7.15 limits the maximum usable tuning range for the particular interference order (FSR – free spectral range).

The first Fabry–Pérot interferometer filter was introduced by Grasdepot et al. who used a pressure sensor-based FPF [27]. Blomberg et al. introduced for the very first time an electrostatically tunable FPF for detecting CO_2 [28]. The wavelength shift was caused by an electrostatically excited displacement of one reflector of a Fabry–Pérot double-reflector structure. In this particular case the wavelength was shifted to either side of the CO_2 wavelength. The ratio of

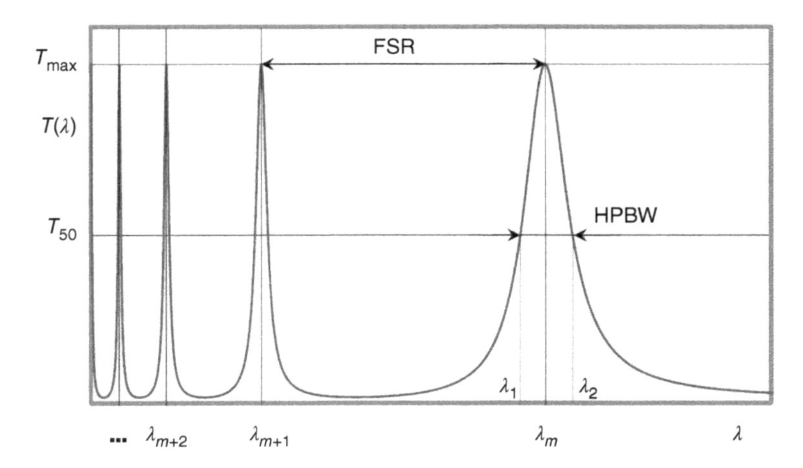

Figure 7.15 Transmittance spectrum of a Fabry–Pérot interferometer filter. Source: From [7, 8], reproduced by permission of InfraTec GmbH, Dresden, Germany.

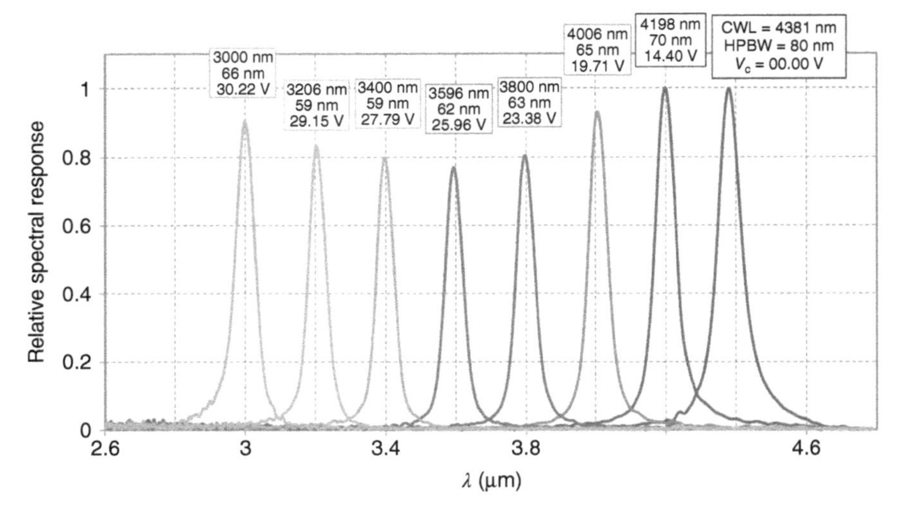

Figure 7.16 Relative spectral response of the FPF detector type LFP-3041L-337 at several tuning voltages. Source: From [7, 8] reproduced by permission of InfraTec GmbH, Dresden, Germany.

these signals then indicates the degree of light absorption and, thus, the gas concentration. The ratio of the measurement to the reference channel changed only 0.6% when the power of the IR source was reduced to 50%. Figure 7.16 shows the relative spectral response of the FPF detector LFP-3041L-337 (InfraTec GmbH, Dresden, Germany).

Noro et al. described an FPF with a wide wavelength range from 2.5 to 4.5 μm that has been developed for a combined NDIR CO_2/H_2O gas sensor. The resolutions defined by standard deviation are about 13 ppm for CO_2 of 2000 ppm and 0.35 g m^{-3} for H_2O of 22.5 g m^{-3} [29].

Recent developments focus on the integration of FPFs with IR detectors (Figure 7.17) and the improvement of the optical characteristics of FPFs, in particular on the increase of spectral resolution and the improvement of the

Figure 7.17 Variable colour detector with integrated Fabry–Pérot interferometer filter. Source: From [8] reproduced by permission of InfraTec GmbH, Dresden, Germany.

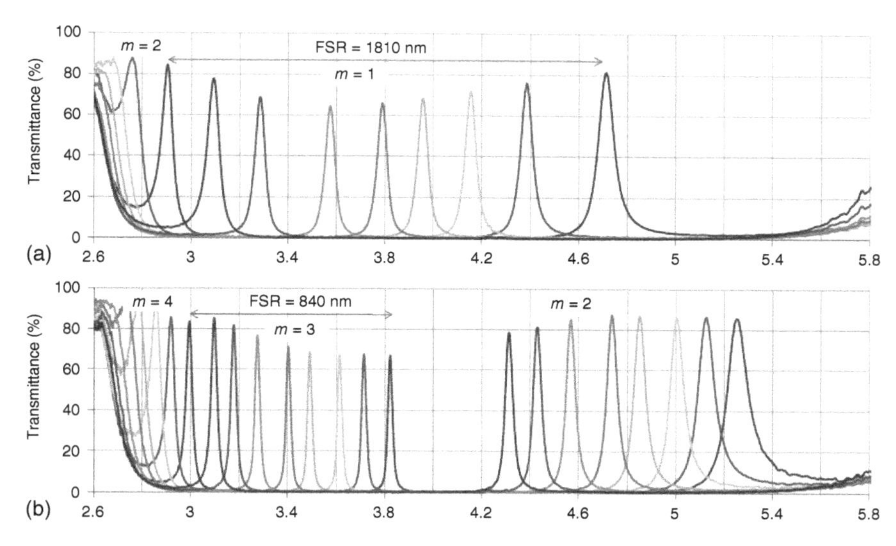

Figure 7.18 Spectral transmittance of a micro-Fabry–Pérot interferometer filter at different tuning voltages: (a) first-order, (b) second-, and third-order interference [34].

signal-to-noise ratio [30, 31]. Main applications are the measurement of gas mixtures of two or many more gases for industrial process control and medical gas analysis by means of infrared absorption spectroscopy [32, 33].

For the measurement of more complex gas mixtures with a large wavelength spectrum, FPFs can be used, which exploit not just the first interference order but two or more, e.g. in both the MWIR and LWIR range (Figure 7.18) [34, 35].

7.2.3 IR Radiation Sources

7.2.3.1 Requirements

NDIR gas sensors need an infrared source for the excitation of the gas molecules in the wavelength range for the particular gas (cf. Table 7.2; 4.24 μm for CO_2). The most desirable lamp characteristics are:

- Rugged construction,
- High emissivity,
- Long lifetime,
- Low cost,
- Small size,
- Low power consumption, and
- High pulse rates and, hence, a low thermal time constant for light modulation.

The latter – light modulation – is strictly required for continuous-wave light-insensitive pyroelectric IR sensors but provides in general the opportunity to offset thermal background signals for all types of sensors.

7.2.3.2 IR Radiation Source Selection

IR lamps can be distinguished between thermal and quantum emitters (Table 7.4) [36]:

Table 7.4 IR radiation sources [22, 36–38].

Emission	Type	Properties
Thermal	Wound filament	High power output with high reliability
	Ribbon filament	Higher pulse rates of up to 200 Hz, modulation depth of 50%, often used with reflectors to direct entire radiation out of package
	Thin-film filament	High volume production, low output power due to small filament size
	MEMS system (hot plate)	Thin-film filaments on thin membranes (of silicon), 1200 °C with 10.7 mW radiation power from 1 mm^2 emission area
Photon	IR LED	Optically pumped LEDs based on III–V semiconductors for CO_2 detection, emitting power of 10–30 μW
	Laser diode	Narrower bandwidth than LEDs

- Thermal emitters generate photons by heating material. For that, they use heated filaments. Such incandescent IR lamps are, by nature, broadband emitters.
- Quantum emitters – e.g. IR LEDs and laser diodes – offer good efficiency and can emit radiation well into the IR region. They are useful in spectroscopic applications where a monochromatic source is preferred.

The choice of an IR radiation emitter to be used in an NDIR sensor is driven by the requirements of the particular application. The emitter must also match the detector to give an optimal "optopair" configuration (Table 7.5). It can be seen that pyroelectric detectors and thermopiles in combination with incandescent IR lamps show the best performance, even though LEDs are useful for fast pulsing.

Table 7.5 Optimal optopairs of radiation source and detector for CO_2 detection (4.2 μm). The relative limit of detection is normalized to the thermal + photodiode combination.

Optopair	Time constant in s	Relative limit of detection
Thermal + pyrodetector	0.1	50–160
Thermal + thermopile	0.1	160
Thermal + photodiode	0.1	1
Thermal + photoresistor	0.1	0.5
LED + pyrodetector	0.1	100
LED + thermopile	0.01	100
LED + photodiode	10^{-8}	0.5×10^{-3}
LED + photoresistor	10^{-6}	4×10^{-3}

Source: Adapted from Sotnikova et al. 2010 [22].

7.2.3.3 Thermal Emitters

Mostly, thermal emitters are employed as thermal radiators because they are very sensitive and cost-effective. As given by Planck's law, their operating temperature should be as high as possible to obtain a sufficiently large output intensity and detector signal (Figure 7.19).

To reduce oxidation of the filament, bulbs are filled with noble gases or evacuated. Filaments of Kanthal, a FeCrAl alloy with a maximum operation temperature of 1350 °C, are used instead of tungsten [39]. However, the transmission of the glass bulb limits the useful spectral range and constrains the types of gas molecules that can be measured by NDIR. As an alternative, TO (transistor single outline) transistor standard housings with an appropriate IR window, e.g. of sapphire (Al_2O_3), ZnSe, or CaF_2, can be used as an envelope to avoid transmission losses of glass or quartz (transmission of only 50% at its peak value at 4.3 μm) bulbs [39].

Modern developments focus on the increase of the radiant power, a better near-black-body emission characteristic, and advanced properties with respect to the electrical modulation of the radiation flux. In [40] a novel type of NiCr-alloy-based thermal IR emitter with nanostructured surface is presented (Figure 7.20). It is fabricated from a thin NiCr film, which leads to a small heat capacity and, hence, to excellent modulation properties. Due to the excellent thermal isolation from the heat sink, the electrical power consumption is very

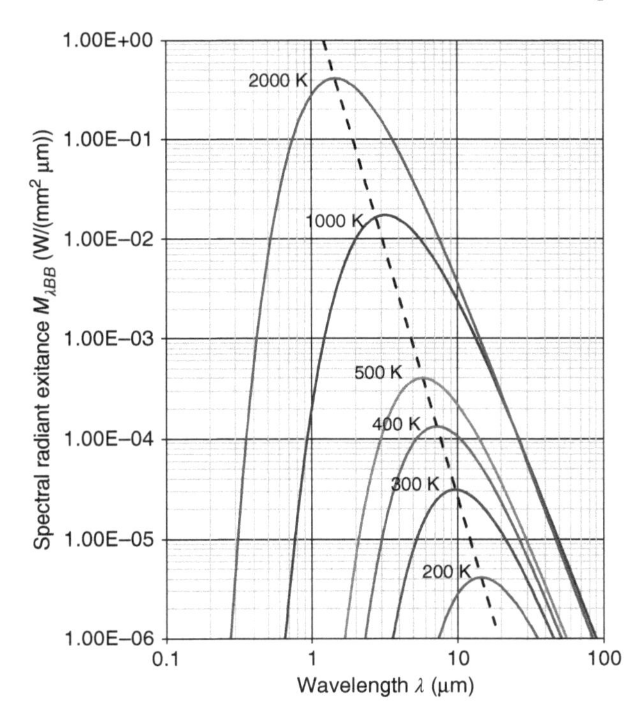

Figure 7.19 Planck's radiation law in double-logarithmic representation. Parameter: temperature. The dashed line marks Wien's displacement law. Source: Budzier and Gerlach (2011) [1]. Reproduced with permission of John Wiley & Sons.

Figure 7.20 Novel thermal IR emitter with nanostructured surface made from NiCr alloy in TO-39 package: (a) prototype operating at about 1170 K, (b) prototype with spiral-shaped filament for homogeneous temperature distribution and suppressed thermal-induced deformation for optimized radiation angle, and (c) nanostructured NiCr emitter surface. Source: Ott et al. 2015 [40]. Reproduced with permission of Copernicus.

low. The used NiCr alloy allows a permanent operation at temperatures of up to 1400 K.

As mentioned above, NDIR sensors often use pulsed radiation. This modulates both the radiation as required for pyroelectric sensors and reduces power consumption by an order of magnitude (pulse duration of several 100 ms, duty cycle time of several seconds). The CO_2 sensitivity at 2000 ppm in pulse mode and in chopper mode amounted to 95 $\mu V\ ppm^{-1}$ at 0.5 s pulse duration and 75 $\mu V\ ppm^{-1}$ at 1.5 Hz modulation frequency, respectively [41]. Nevertheless, the pulse mode does not achieve the same signal-to-noise ratio or accuracy as the common chopping mode. Therefore, a smart measurement algorithm alternates between the two modes and recalibrates the pulse mode from time to time [42]. By this, the daily power consumption was reduced to 8.2 mW (23.5 hours in pulse mode and 0.5 hour in chopping mode) compared with 300 mW in chopping mode.

7.2.4 Gas Sensors for Measuring CO_2 in Gas Mixtures

As already mentioned in Section 7.2.2.1, many applications require the measurement of concentrations of a limited number of other gases in addition to CO_2. For example, exhaust gases of automobiles contain CO_2, CO, NO_x, and water vapour;

air quality sensors measure CO_2 but should also detect C_2H_6 as gas for heating and cooking for safety reasons.

Usually, multicolour NDIR sensors with multiple channels are used as shown in Figure 7.1b. A beam splitter distributes the incident IR radiation equally to the corresponding detector elements. To provide high accuracy, the field of view for each detector element should be the same, which is fulfilled by solutions as shown in Figure 7.13.

Another approach for measuring gas mixtures was presented by Rubio et al. [43]. The detector is a single-module thermopile array that is equipped with an array of broadband IR filters. Here, the filter elements are not designed for a specific absorption band. This increases not only the overall system flexibility but also the complexity of signal processing.

Andrews and King proposed a gas sensor that works without IR filters [44]. Their approach uses the correlation of the detector signal and the non-linear temperature-dependent emission spectrum of thermal radiation sources. Using the output signals at different emission temperatures, simultaneous monitoring of CO_2 and water vapour in the atmosphere could be demonstrated. However, the sensor's performance crucially depends on the stability of the IR source that limits the uncertainty.

Graf et al. investigated another principle lying in between both approaches of Rubio et al. and Andrews and King [45]. The so-called ANDIR concept presents a highly adaptable NDIR sensor using broadband filters instead of interference filters with a specifically designed narrow bandwidth. These broadband filters do not have to be adapted to specific chemical species but can overlap in their spectral transmission range. Using n emitter temperatures and m standard broadband filters enables differentiation of up to $m \cdot n$ different gases. Such a concept can be flexibly applied to many different detection and measuring tasks because no specific transmission filters are needed [46]. Since the detector signals are highly correlated, particular algorithms have to be applied for signal processing. Support vector machines (SVM) have been used as the favourable method for data analysis [47].

7.3 NDIR Sensors

7.3.1 Commercial NDIR Sensors

Currently, there is an abundance of NDIR CO_2 sensors available in the market. Most sensors are intended for measuring air quality, ventilation control in incubators and greenhouse, environmental [48] and combustion control, and automotive applications [49, 50] as well as for measuring dissolved CO_2 in freshwater [51].

For each particular application – as for all kinds of spectrometers – a compromise between spectral resolution and signal-to-noise ratio *SNR* has to be found (Figure 7.21). Beam divergence can be minimized by using a light source with collimated output or by means of an additional prefixed aperture (Figure 7.21a). The optical throughput can be maximized by a focusing optics, but at the expense of larger cone angles (Figure 7.21b).

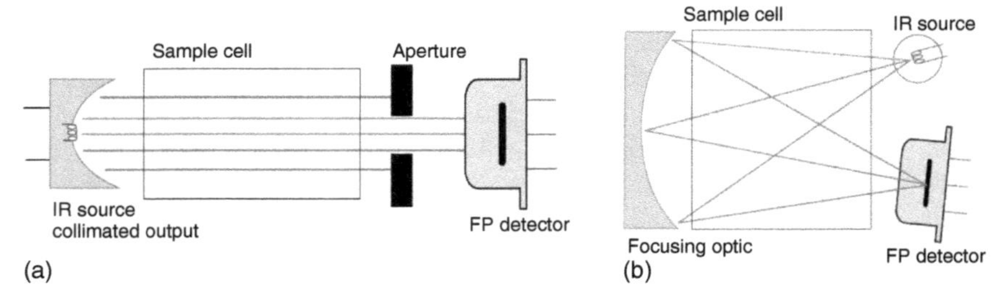

Figure 7.21 Optimization for the optical design of NDIR sensors: (a) illumination with a parallel beam leads to a high resolution and (b) illumination with a large cone angle gives a better signal-to-noise ratio. Source: From [7, 52] reproduced by permission of InfraTec GmbH, Dresden, Germany.

Most commercial NDIR products are single-channel sensors due to lower costs. They are smaller and use less energy, which is most important for handhelds in battery operation. Two-channel sensors exhibit better long-term stability. The second channel (comparison channel) enables the sensor to compensate the drift of the IR lamp, the temperature drift of the IR detector, and changes to the measuring stretch, e.g. due to dirt.

Table 7.6 gives an overview of the characteristic properties of a selected variety of NDIR CO_2 sensors (see also [63, 64]). The values were taken from data sheets

Table 7.6 Characteristic properties of commercial NDIR CO_2 sensors (selection; values as given by the data sheets).

Manufacturer, model	CO_2 range (pm)	Accuracy	Operating temperature range (°C)	References
Alphasense Ltd., IRC-A1	0–5000	±50 ppm	−20 to 50	[53]
Digital Control Systems, 305e	0–2500	±5% of reading or ±100 ppm, whichever is greater	0–50	[54]
ELT SENSOR Corp., B-530	0–2000 to 0–5000	±30 ppm ± 3%	0–50	[55]
Honeywell, C7232A1008/U	0–2000	±5% FS	0–50	[56]
MB-Systemtechnik, CO2 S 200	0–2000 to 0–30,000	±30 ppm ± 5% FS	0–50	[57]
Micro-Hybrid Electronic	0–20 vol%	±0.2 vol% ± 2% of reading	0–190	[58]
Senseair, CO_2 Engine K30 STA	0–6000	±30 ppm ± 3% of reading	0–50	[59]
Vaisala, GMT222	0–2000 to 0–10 000	±1.5% FS ± 2% of reading	−20 to 60	[60]
VTI Valtronics, 2015SPI-1	0–2000 to 0–20 000	±50 ppm ± 5% of reading	0–50	[61]
ZILA GmbH, ZMF-100-IR	0–3000 to 0–50 000	±2%	10–50	[62]

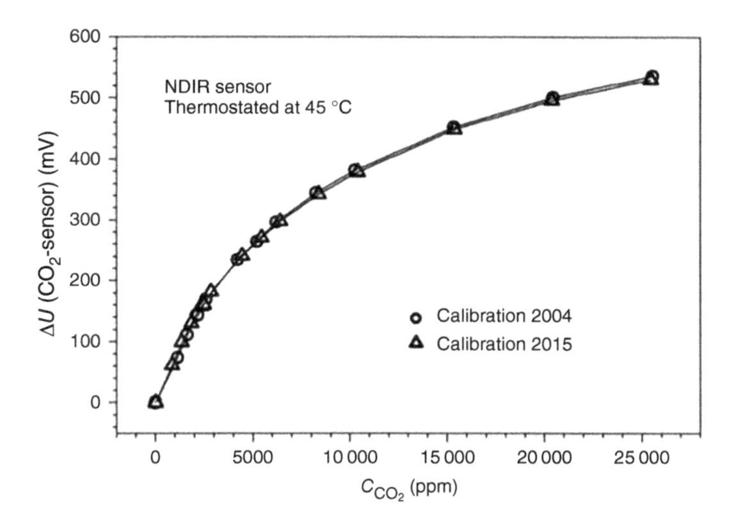

Figure 7.22 Long-term test of a single-beam NDIR sensor. Source: From [65] with courtesy of Gatron GmbH Greifswald, Germany.

in which the lower detection limits of the various sensors were not specified in detail.

Such sensors can work over 10 years with a minimum in maintenance (Figure 7.22) [65]. They should be calibrated using dry and CO_2-free air.

7.3.2 Application for Very Small Concentrations and for Liquid Samples

7.3.2.1 Pre-Concentrators for Low Gas Concentrations

Diffusion of sample gas into the cuvette takes usually between 30 seconds and 2 minutes. Response time is in the same range. Very critical is the direct measurement of gases with very low concentration (trace gases) in samples with complex matrices, due to either analytical sensitivity limitations or matrix interferences. Pre-concentration procedures are generally needed to eliminate matrix interferences and to enrich minute amounts of analytes to a level for reliable measurements. This is usually achieved by pre-concentrator modules [66].

7.3.2.2 Measurement of Dissolved CO_2 in Liquids by Using Permeation Methods

Dissolved CO_2 in liquids can also be measured continuously by means of NDIR detectors or by solid electrolyte gas sensors if the dissolved gas is transferred to a defined gas atmosphere. This is mostly performed by the introduction of a gas-permeable membrane into the solution to be measured and flushing the backside of the membrane with a carrier gas. According to Figure 7.23, the gas sensor is often positioned at the downstream side of the permeation cell.

Carbon dioxide permeates through the membrane and is then carried to the CO_2 sensor. At constant carrier gas flow, a linear relationship exists between the partial pressure of carbon dioxide inside the measuring solution and the carbon dioxide concentration downstream of the diffusion cell. An example for this permeation method, which is already available in the market, is the

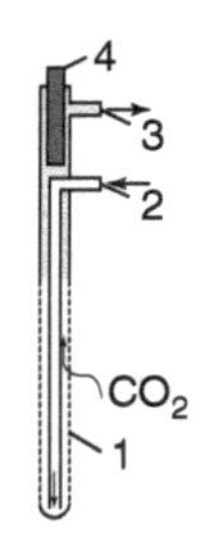

Figure 7.23 Schematic drawing of a membrane-covered gas transfer probe for the measurement of dissolved carbon dioxide: (1) permeation membrane, (2) carrier gas inlet, (3) gas outlet, and (4) CO_2 sensor. Source: Adapted from Canongate 2016 [67].

steam-sterilizable silicone tubing probe Carboline developed by Biotechnologie Kempe GmbH [68]. This probe is equipped with a modular CO_2 sensor operating on the NDIR single-beam dual-wavelength principle. The main advantages of this system are the high long-term stability and the broad measuring range. The system is conceived for the monitoring of biotechnological processes and for quality control in the production of beverages or at breweries. Another commercially available system working on the same principle is the Embra CarboCheck of Rototherm [67]. In contrast to the above-described Carboline probe, this system uses a constantly evacuated chamber behind the membrane.

7.4 IR Spectrometers

7.4.1 Types of IR Spectrometers

Infrared spectrometers are used for the qualitative and quantitative determination of the structure of chemical species, in particular if the composition of, for example, a gas mixture is not known. In such a case, the a priori knowledge, which could be used to select particular wavelength filters for gas mixtures with given components (cp. Figures 7.1b and 7.13), is not available. Instead the entire spectrum has to be taken into account. This regards, for example, the measurement of CO_2-containing samples with unknown composition.

The schematic set-up of a spectrometer is shown in Figure 7.24a. The main difference to non-dispersive devices is the use of a dispersion element that unfolds the wavelength spectrum and, hence, enables the reading of a larger spectrum instead of only a limited number of wavelengths. Commonly used dispersion elements are either gratings or interferometers (e.g. Fabry–Pérot-interferometer; see Section 7.2.2.1 [69]).

The functional principles of the most important types of infrared spectrometers are presented in Figure 7.24 [1]:

- *Absorption spectrometer*: The radiation of the white light source is focussed on the probe, where the respective wave number range is absorbed and directed wavelength-selectively through a grating (analyser) to the sensor. Often, there is a mechanical or electronic radiation modulator inserted between light source and probe, converting the constant light signal into an alternating light signal with a fixed frequency of between 10 and 100 Hz. In comparison with

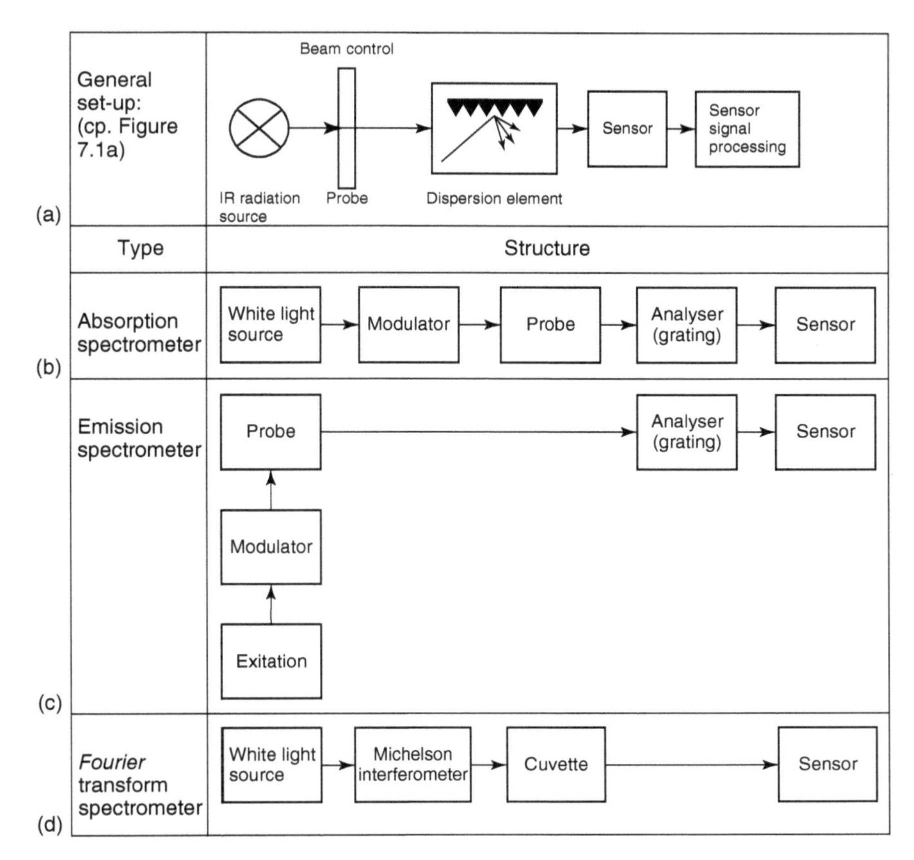

Figure 7.24 Functional principles of infrared spectrometers.

direct voltage signals, the signal evaluation of alternating voltage signals is easier and more reliable. It also has a larger signal-to-noise ratio as interfering signals can be better suppressed with other frequencies.

- *Emission spectrometer*: The probe is externally excited to emit radiation and, thus, becomes a radiation source itself. The excitation can be caused thermally or by discharging. However, usually, it is caused by electromagnetic radiation.
- *FTIR (Fourier transform IR) spectrometer (Figure* 7.25*)*: For the absorption and emission spectrometry, due to the diffraction of the dispersion element, only IR light of a specific wavelength reaches the sensor. In order to record the complete spectrum, one has to cover the entire spectrum, which requires long measuring times. Applying a Michelson interferometer with a movable mirror causes constructive and destructive interference with the reference radiation. The movable mirror M_2 causes a time signal $V(t)$ at the sensor that now can be Fourier-transformed and, after a short measuring time, provides the complete wavelength or wave number absorption spectrum of the probe. The Fourier transform spectrometer does not require any dispersive elements,

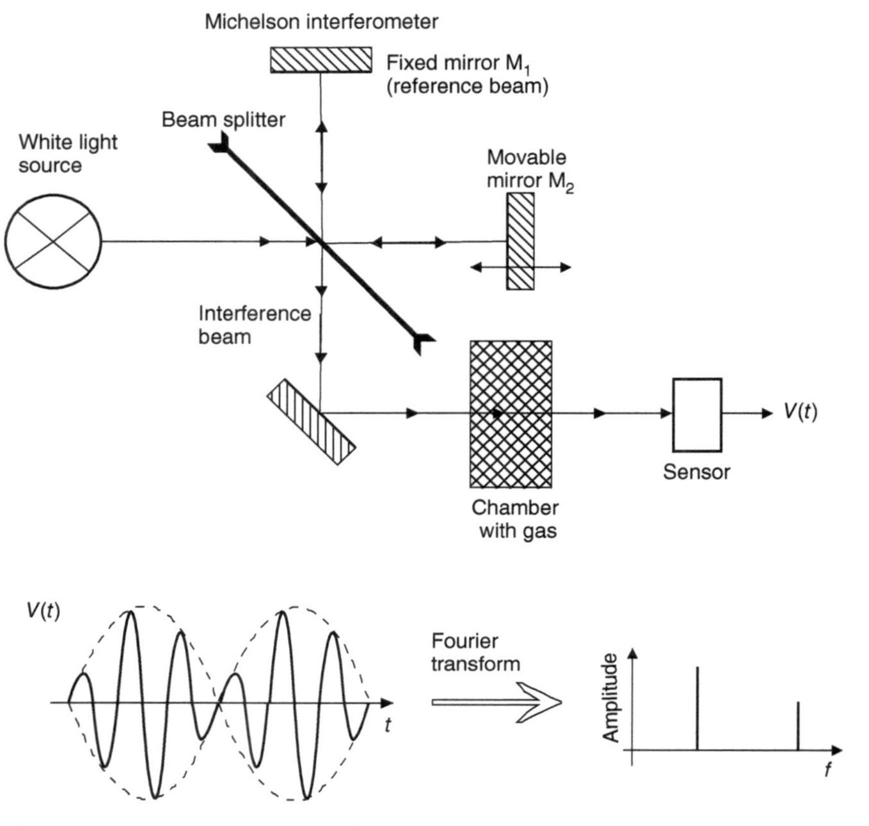

Figure 7.25 Main components and of IR spectrometers and illustration of Fourier transform.

such as prisms or gratings. FTIR spectrometry has the following advantages [70–72]:

- High sensitivity;
- Short measuring times in the range of seconds for the complete spectrum (Fellgett's advantage or multiplex advantage);
- Enables dynamic processes, such as for gas or fluid chromatography;
- Larger signal-to-noise ratio due to the simultaneous absorption of all wave numbers falling within a certain wave number range and to a highly reliable divergence of the radiation in the interferometer (throughput advantage or Jacquinot advantage); and
- Results in a high spectral resolution capability.

Due to its high sensitivity and good spectral resolution capability, it is possible to record interpretable spectra even for highly absorbing probes, diffuse reflectance of the probe, and complicated probe arrangements (e.g. for attenuated total reflectance, ATR [73, 74]).

IR spectrometers usually consist of the structural components represented in Table 7.7 and illustrated in Figure 7.25.

Table 7.7 Structural components of IR spectrometers.

Component	Function
IR radiation source	• Spectral broadband thermal emitter • Typical: "Nernst rod" made of oxides of rare earths or "Globar" consisting of silicon carbide
Beam control (propagation path)	• Guiding and concentrating of the emitter by aluminium- or silver-vaporized mirrors (prevents the absorption occurring in lenses and common mirrors) • Focal point of the emitter in the probe
Window (propagation path)	• Surrounds the probe and protects the detector • Consisting of IR-transparent mineral salts: NaCl or KBr for wave numbers above 650 or 400 cm^{-1} • For aqueous solutions: AgCl cuvettes (>430 cm^{-1}) or CaF_2 cuvettes (<1200 cm^{-1})
Probe	• Fluids: confined between IR-transparent mineral salt plates (similar to the windows), layer thicknesses of c. 10 μm, for diluted solutions 0.1–10 mm, in order to achieve a sufficiently large absorption length • Gases: cuvettes (glass cells) of a length of 5–10 cm, at the ends closed by alkali halide windows. At low pressures, large absorption lengths for gases are achieved by using multiple reflections in the cuvette • Solid substances: mainly in solvents or as suspension to prevent reflection and scattering of the IR radiation at individual particles
Dispersion element (propagation path)	• Mainly gratings due to the smaller absorption in comparison with prisms (grating spectrometers) • Revolving grating for projecting the wavelength-dependent diffracted radiation to single-element sensor, fixed gratings at the detector lines
Sensor	• Mainly pyroelectric sensors or thermopiles (compare Section 7.2.1). Thermal sensors show only a limited wavelength/ wave number dependence over the entire measuring range
Sensor signal processing	• Amplifies with the lowest noise possible • Calculates the spectrum (e.g. for FTIR spectrometers applying Fourier transform based on the time signal)

7.4.2 Applications

Recent advances in FTIR spectroscopy have allowed for convenient and rapid analysis of chemical species like CO_2 to be carried out both in a gaseous and in an aqueous environment. Examples are:

- Determination of the purity of gases [75],
- Trace gas analysis [76],
- Multicomponent analysis of an unknown mixture of gases [77],
- Absorption and desorption of CO_2 and CO at surfaces [78–82],
- Absorption and desorption of CO_2 by chemical absorbents (carbon capture) [83, 84],
- Catalytic processes [85, 86],
- CO_2 interactions with polymers including for polymer processing [87, 88], and
- CO_2 in environmental studies [72].

7.5 IR Imaging for CO_2 Detection

Infrared cameras (or thermal imaging cameras) are devices that capture images using infrared radiation, similar to photo cameras that take images using visible light [1, 89]. Due to the usage of infrared light, such IR cameras can be used for the detection of gases like infrared detectors in NDIR sensors (see Section 7.2.2).

Figure 7.26 shows the principal structure of thermal imaging cameras. The infrared optics projects the measuring scene onto the image sensor. The lenses of the optics are made of germanium, silicon, or special glasses with high transmissivity in the far-infrared range. Automatic focusing and zoom optics are the technological state of the art. The smallest feasible spatial resolutions are produced at a scale of unity, i.e. the object is projected 1 : 1 to the sensor matrix (infrared microscope).

A system-imminent component of the optical channel is the shutter. It is cyclically closed for the recalibration of the system. The focal-plane array (FPA) is an image-sensing device and consists of an array of IR-light-sensing pixels. It is located at the focal plane of the optics. Following the analogue-to-digital conversion of the sensor signal with 14–16 bits, the image is processed using signal processors and is then available to the user. Signal processing includes the pixel-specific correction of the characteristic curve for each pixel, the replacement of dead pixels, and the temperature calibration in order to achieve – for a homogeneous radiation of the sensor array – a uniform output signal.

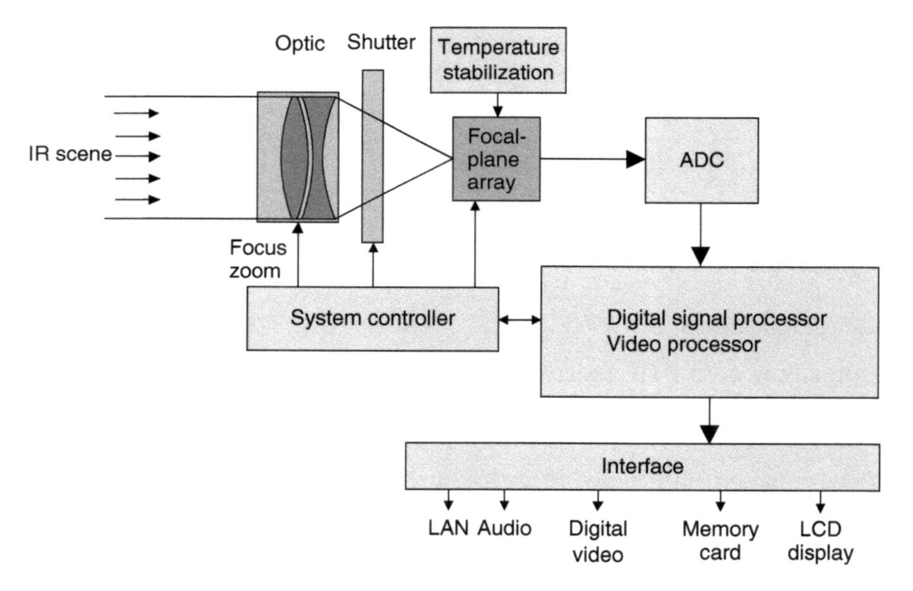

Figure 7.26 Schematic set-up of a thermal imaging camera. Source: Budzier and Gerlach 2011 [1]. Reproduced with permission of John Wiley & Sons.

IR cameras are imaging systems that approximate the temperature distribution of a scene. They represent the radiation emitted by a body or object in a two-dimensional image (thermal image, infrared image). The radiation emitted by the object is composed of the object's own radiation, the reflected ambient radiation, and, possibly, the transmitted background radiation.

For the application as imaging camera, the gas – usually the atmosphere – is the medium through which the IR radiation is transmitted towards the camera. This gas should neither absorb nor emit any IR radiation in order to not change the camera signal as response to the IR scene [89, 90].

However, when the gas to be detected influences this transmission, such IR cameras can be used for gas detection. This was introduced for the first time in 1985 by Strachan et al. [91]. It is a qualitative measurement method but provides information on the two-dimensional distribution of gas concentrations.

In the meanwhile, special IR cameras are available for the detection of various gases, e.g. of volatile organic compounds (VOCs), sulphur hexafluoride (SF_6, a gaseous dielectric medium used in the electrical industry, e.g. for high-voltage circuit breakers), and CO_2 and CO (Table 7.8). For example, the GF-343 camera is an optical gas imaging camera that allows to detect CO_2 leaks easily and from a remote, safe distance.

The wavelength sensitivity of these commercial cameras is defined by the type of the used detector and not by the characteristic of a filter in the transmission path as for NDIR sensors (cp. Section 7.2.2). Here, photon detectors with its distinctive wavelength detection and a matching cut-off wavelength (see Section 7.2.1 and Figure 7.11) are applied.

For applications with lower demands, thermal IR cameras, e.g. less expensive IR cameras with thermal IR detectors, can be used too. In such cases, the wavelength has to be selected by narrowband filters. Figures 7.27 and 7.28 illustrate the potential of IR imaging for the detection of CO_2.

Table 7.8 Infrared cameras for gas detection (GasFindIR cameras; FLIR).

Model	Detector type	Spectral range (μm)	Temperature range (°C)	Application
GF300	Cooled InSb	3.2–3.4	−20 to +350	20 gases in petrochemistry, e.g. hydrocarbons
GF306	Cooled QWIP	10.3–10.7	−40 to +50	SF6 and 25 other harmful gases
GF343	Cooled InSb	4.2–4.4	−20 to +50	CO_2 leak detection
GF346	Cooled InSb	4.52–4.67	−20 to +300	Leak detection of CO and 17 other gases

Frame rate 60 Hz; dynamic range 14 bit; 320 × 240 = 76 800 pixels; InSb, indium antimonide; QWIP, quantum well infrared photon detector.
Source: Adapted from FLIR 2016 [92].

Figure 7.27 CO_2 content in breathing air while exhaling through the nose; IR camera ThermoVision SC6000 (FLIR [92]) with uncooled narrowband filter, imaged in front of a black-body emitter at 50 °C. Source: Vollmer and Möllmann 2010 [89]. Reproduced with permission of John Wiley & Sons.

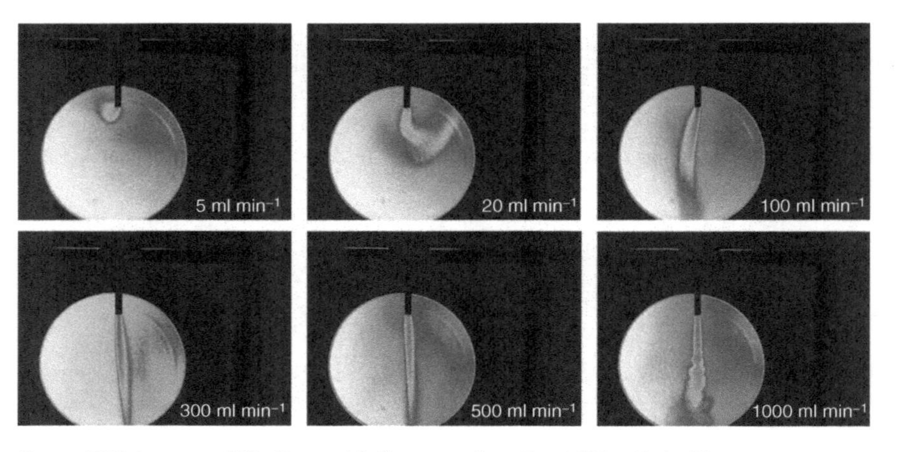

Figure 7.28 Images of CO_2 flows with flow rates from 5 to 1000 ml/min; IR camera ThermoVision SC6000 (FLIR [92]) with uncooled narrowband filter, imaged in front of a black-body emitter at 50 °C. Source: Vollmer and Möllmann 2010 [89]. Reproduced with permission of John Wiley & Sons.

References

1 Budzier, H. and Gerlach, G. (2011). *Thermal Infrared Sensors – Theory, Optimisation and Practice*, 324 pp. Chichester: Wiley.

2 Rothman, L.S., Gordon, I.E., Barbe, A. et al. (2009). The HITRAN 2008 molecular spectroscopic database. *J. Quant. Spectrosc. Radiat. Transfer* 110 (9–10): 533–572.

3 Rothman, L.S., Gordon, I.E., Babikov, Y. et al. (2012). The HITRAN 2012 molecular spectroscopic database. *J. Quant. Spectrosc. Radiat. Transfer* 130 (1): 4–50.

4 http://hitran.org/docs/hitran-papers (retrieved 15 August 2016).

5 www.cfa.harvard.edu/hitran/molecules.html (retrieved 15 August 2016).

6 De Bievre, P., Gallet, M., Holden, N.E., and Barnes, I.L. (1984). Isotopic abundances and atomic weights of the elements. *J. Phys. Chem. Ref. Data* 13 (3): 809–891.

7 Pyroelectric & Multispectral Detectors (2009). *Catalog*. Dresden: InfraTec GmbH.

8 InfraTec. Application Notes: Detector Basics; Advanced Features of InfraTec Pyroelectric Detectors; Temperature Behavior; JFET and Operation Amplifier Characteristics. http://www.infratec-infrared.com/sensor-division/sensor-division-knowledge/application-notes.html (retrieved 27 September 2016).

9 Norkus, V., Chvedov, D., Gerlach, G., and Koehler, R. (2006). Performance improvements for pyroelectric infrared detectors. In: *Proceedings of SPIE 6206*, 62062X-1–62062X-11. https://doi.org/10.1117/12.664389.

10 Schossig, M., Norkus, V., and Gerlach, G. (2009). High-performance pyroelectric infrared detectors. In: *Sensor+Test Conference, Proceedings of OPTO 2009 and IRS² 2009*, Nuremberg, Germany (26–28 May 2009), 191–196.

11 Giebeler, C., Wright, J., Freeborn, S. et al. (2009). High performance PZT based pyro-detectors with D^* of 2×10^9 cm $Hz^{1/2}/W$ for pressure, gas and spectroscopy applications. In: *Sensor+Test Conference 2009, Proceedings of OPTO 2009 and IRS² 2009*, Nuremberg, Germany (26–28 May 2009), 185–189.

12 Graf, A., Arndt, M., Sauer, M., and Gerlach, G. (2007). Review of micromachined thermopiles for infrared detection. *Meas. Sci. Technol.* 18 (7): R59–R75.

13 Foote, M.C., Jones, E.W., and Caillat, T. (1998). Uncooled thermopile infrared detector linear arrays with detectivity greater than 10^9 $cmHz^{1/2}/W$. *IEEE Trans. Electron Devices* 45 (9): 1896–1902.

14 (2001). *Remote Temperature Measurement with PerkinElmer Thermopile Sensors (Pyrometry): A Practical Guide to Quantitative Results*. Corporate Publication, PerkinElmer Optoelectronics GmbH.

15 http://www.heimannsensor.com/products.php (retrieved 13 October 2016).

16 https://www.dias-infrared.de/produkte/infrarotsensoren (retrieved 13 October 2016).

17 http://www.micro-hybrid.de/en/ir-components/product-finder-old/thermopile-detectors.html (retrieved 13 October 2016).

18 http://www.hamamatsu.com/eu/en/product/category/index.html (retrieved 13 October 2016).

19 http://www.vigo.com.pl/products/infrared-detectors (retrieved 13 October 2016).

20 http://www.ioffeled.com (retrieved 04 August 2017).

21 http://www.optico.ru/products_en_detectors_pbseu.html (retrieved 13 October 2016).

22 Sotnikova, G.Y., Gavrilov, G.A., Aleksandrov, S.E. et al. (2010). Low voltage CO_2-gas sensor based on III–V mid-IR immersion lens diode optopairs: where we are and how far we can go? *IEEE Sens. J.* 10 (2): 225–234.

23 *Infrared (IR) Bandpass Filters*. Edmund Optics Inc. http://www.edmundoptics.com/optics/optical-filters/bandpass-filters/infrared-ir-bandpass-filters/3387/ (retrieved 13 October 2016).

24 Norkus, V., Schiewe, C., and Nagel, F. (1994). Multispectral sensor. US Patent 5,300,778, filed 9 October, 1992 and issued 5 April 1994.

25 Norkus, V., Gerlach, G., and Hofmann, G. (1998). Uncooled multispectral detectors. In: *Proceedings of SPIE 3436*, 332–339.

26 Gerlach, G. and Dötzel, W. (2008). *Introduction to Microsystem Technology: A Guide for Students*, 351 pp. Chichester: Wiley.

27 Grasdepot, H., Alause, H., Knap, W. et al. (1996). Domestic gas sensor with micromachined optical tunable filter. *Sens. Actuators, B* 36 (1–3): 377–380.

28 Blomberg, M., Torkkeli, A., Lehto, A. et al. (1997). Electrically tunable micromachined Fabry–Perot interferometer in gas analysis. *Phys. Scr.* T69: 119–121.

29 Noro, M., Suzuki, K., Kishi, N. et al. (2003). CO_2/H_2O gas sensor using a tunable Fabry–Perot filter with wide wavelength range. In: *Proceedings, MEMS-03 Micro Electro Mechanical Systems*, Kyoto, Japan (19–23 January 2003), 319–322.

30 Neumann, N., Ebermann, M., Kurth, S., and Hiller, K. (2008). Tunable infrared detector with integrated micromachined Fabry–Perot filter. *J. Micro/Nanolith. MEM MOEMS* 7 (2): 021004, 1–9.

31 Ebermann, M., Hiller, K., Kurth, S., and Neumann, N. (2009). Design, operation and performance of a Fabry–Perot-based MWIR microspectrometer. In: *Sensor+Test Conference 2009, Proceedings of OPTO 2009 and IRS² 2009*, Nuremberg, Germany, 26–28 May 2009, , 233–238.

32 Ebermann, M. and Neumann, N. (2009). Novel MWIR microspectrometer based on a tunable detector. In: *Proceedings of SPIE 7208*, 72080D1–72080D8.

33 Ebermann, M., Neumann, N., Hiller, K., and Meinig, M. (2012). Spektral abstimmbare IR-Sensoren für die industrielle Prozessmesstechnik und die medizinische Gasanalytik (Spectrally adjustable IR sensors for industrial process control and medical gas analysis). *Tech. Mess.* 79 (10): 440–450 (in German).

34 Ebermann, M., Neumann, N., Hiller, K. et al. (2014). Resolution and speed improvements of mid-infrared Fabry–Perot microspectrometers for the analysis of hydrocarbon gases. In: *Proceedings of SPIE 8977, MOEMS and Miniaturized Systems XIII, 89770T*, 1–9. https://doi.org/10.1117/12.2038235.

35 Ebermann, M., Neumann, N., Hiller, K. et al. (2012). Widely tunable Fabry–Perot filter based MWIR and LWIR microspectrometers. In: *Proceedings of SPIE 8374, 83740X*, 1–9, 9p. https://doi.org/10.1117/12.919169.

36 Elias, B.C. (1992). Match the emitter to the task. *Photonics Spectra* 42 (1): 39–41.

37 Elias, B.C. (2008). Infrared emitters for spectroscopic applications. In: *Sensor+Test 2008, Proceedings of OPTO 2008 and IRS² 2008*, Nuremberg, Germany (6–8 May 2008), 237–242.

38 Johnston, S.F. (1992). Gas monitors employing infrared LEDs. *Meas. Sci. Technol.* 3 (2): 191–195.

39 Schick, D., Norkus, V., Sokoll, T., and Gerlach, G. (2002). Infrared radiation sources in TO packages. In: *IRS² 2002, 7th International Conference on Infrared Sensors & Systems, Proceedings*, Erfurt (14–16 May 2002), 243–247.

40 Ott, T., Schossig, M., Norkus, V., and Gerlach, G. (2015). Efficient thermal infrared emitter with high radiant power. *J. Sens. Sens. Syst.* 4 (2): 313–319.

41 Neumann, N., Stegbauer, H.-J., Sänze, H. et al. (2004). Application of fast response dual colour pyroelectric detectors with integrated op amp in a low

power NDIR gas monitor. In: IRS^2 2004, Proceedings, 8th International Conference on Infrared Sensors and Systems, Nuremberg, Germany (25–27 May 2004), 183–188.
42 Gürtner, M., Heinze, M., Neumann, N., and Schneider, F. (2003). Verfahren zur Bestimmung der Konzentration von Gasen und Dämpfen und nichtdispersiver Infrarot-Gasanalysator zur Durchführung des Verfahrens (Method for determining the concentration of gases and vapors and non-dispersive infrared gas analyzer for carrying out the method). Patent DE 10,221,708 B4, filed 16 May 2002 and issued 30 September 2004 (in German).
43 Rubio, R., Santander, J., Fonseca, L. et al. (2007). Non-selective NDIR array for gas detection. *Sens. Actuators, B* 127 (1): 69–73.
44 Andrews, D.A. and King, T.A. (2001). Gas analysis using an infrared source with temporally varying temperature. *Meas. Sci. Technol.* 12 (8): 1263–1269.
45 Graf, A., Sauer, M., Arndt, M., and Gerlach, G. (2007). ANDIR—a highly adaptable NDIR sensor with broad band filters for gas analysis. In: Sensor 2007, The 13th International Conference, vol. 1, Nuremberg, Germany (22–24 May 2007), 127–132.
46 Graf, A. (2009). *Software-Tailored Non-Dispersive Infrared Sensors*, 146 pp. Dresden: TUD Press.
47 Graf, A., Mielenz, H., Gerlach, G., and Rosenstiel, W. (2008). Software based adaptation of a non dispersive infrared sensor for surface classification. In: *Proceedings of Eurosensors XXII*, Dresden, Germany (7–10 September 2008), Düsseldorf, VDI, 1380-3 (CD-ROM).
48 Wang, Y., Nakayama, M., Yagi, M. et al. (2005). The NDIR CO_2 monitor with smart interface for global networking. *IEEE Trans. Instrum. Meas.* 54 (4): 1634–1639.
49 Frodl, R. and Tille, T. (2006). A high-precision NDIR CO_2 gas sensor for automotive applications. *IEEE Sens. J.* 6 (6): 1697–1705.
50 Arndt, M., Graf, A., and Sauer, M. (2006). Low cost infrared carbon dioxide sensor for automotive applications. In: *Sensor+Test 2006, Proceedings of OPTO 2006 and IRS^2 2006*, Nuremberg, Germany (30 May-1 June 2006), 267–271.
51 Johnson, M.S., Billett, M.F., Dinsmore, K.J. et al. (2010). Direct and continuous measurement of dissolved carbon dioxide in freshwater aquatic systems-method and applications. *Ecohydrology* 3 (1): 68–78.
52 InfraTec. Application Notes from InfraTec: Detector Basics; Advanced Features of InfraTec Pyroelectric Detectors; Temperature Behavior; JFET and Operation Amplifier Characteristics. http://www.infratec-infrared.com/sensor-division/sensor-division-knowledge/application-notes.html (retrieved 27 September 2016).
53 http://www.alphasense.com/index.php/products/ndir-air/ (retrieved 14 October 2016).
54 http://www.dcs-inc.net/m305.htm (retrieved 14 October 2016).
55 http://eltsensor.co.kr/2012/eng/product/co-co2-detector.html (retrieved 14 October 2016).
56 https://customer.honeywell.com/en-US/Pages/Department.aspx?cat=HonECC+Catalog&category=CO2+Sensors&catpath=1.4.31.4 (retrieved 04 August 2017).

57 http://www.mb-systemtechnik.de/de/CO2-Messung/ (retrieved 04 August 2017).

58 https://www.micro-hybrid.de/en/products/gas-sensors/ndir-co2-sensors.html (retrieved 04 August 2017).

59 http://senseair.se/products/oem-modules/k30/ (retrieved 14 October 2016).

60 http://www.vaisala.com/en/products/carbondioxide (retrieved 04 August 2017).

61 http://www.val-tronics.com/product-data-spec-sheets.html (retrieved 04 August 2017).

62 http://www.zila.de/en/products/co2-detection (retrieved 14 October 2016).

63 Calvet, S., Campelo, J.C., Estelles, F. et al. (2014). Suitable evaluation of multipoint simultaneous CO_2 sampling wireless sensors for livestock buildings. *Sensors* 14 (6): 10479–10496.

64 Yasuda, T., Yonemura, S., and Tani, A. (2012). Comparison of the characteristics of small commercial NDIR CO_2 sensor models and development of a portable CO_2 measurement device. *Sensors* 12 (3): 3641–3655.

65 Sasum, U. and Guth, U. (2015). Wasserstoffmessung in Prozessgasen (Determination of hydrogen in process gases). In: *12. Dresdner Sensor-Symposium*, Dresden (07–09 December 2015), 141–146. http://www.ama-science.org/proceedings/details/2186 (retrieved 22 December 2016) (in German).

66 Hildenbrand, J., Eberhardt, A., Peter, C. et al. (2009). Preconcentrator module for the implementation in optical gas measurement systems. In: *Sensor+Test Conference 2009, Proc. OPTO 2009 and IRS^2 2009*, Nuremberg, Germany (26–28 May 2009), 269–274.

67 Canongate Technology Limited Brochure CarboCK211004, Edinburgh. http://www.ct-uk.co.uk/co2.html (retrieved 14 October 2016).

68 Kempe, E. (1989). Probe means for sampling volatile components from liquids or gases. US Patent 4,821,585, filed 12 December 1984 and issued 18 April 1989.

69 Vargas-Rodriguez, E. and Rutt, H.N. (2009). Design of CO, CO_2 and CH_4 gas sensors based on correlation spectroscopy using a Fabry–Perot interferometer. *Sens. Actuators, B* 137 (2): 410–419.

70 Herrmann, K., Walther, L., and Behrend, R. (1990). *Wissensspeicher Infrarottechnik* (Store of Knowledge in Infrared Technology), 436 pp. Leipzig: Fachbuchverlag (in German).

71 Banwell, C.N. and McCash, E.M. (2008). *Fundamentals of Molecular Spectroscopy*, 320 pp. Maidenhead, Berkshire: McGraw-Hill Higher Education.

72 Simonescu, C.M. (2012). Application of FTIR spectroscopy in environmental studies. In: *Advanced Aspects of Spectroscopy* (ed. M.A. Farrukh), 49–84. InTech.

73 Greener, J., Abbasi, B., and Kumacheva, E. (2010). Attenuated total reflection Fourier transform infrared spectroscopy for on-chip monitoring of solute concentrations. *Lab Chip* 10 (12): 1561–1566.

74 Mayer, S.G., Boyd, J.E., and Heser, J.D. (2010). A high-pressure attenuated total reflectance cell for collecting infrared spectra of carbon dioxide mixtures. *Vib. Spectrosc.* 53 (2): 311–313.

75 Valková, M., Pätoprstý, V., and Lawson, M. (2011). The use of FTIR analysis in determining the purity of gases. *Acta Chim. Slovaca* 4 (2): 72–77.

76 Esler, M.B., Griffith, D.W.T., Wilson, S.R., and Steele, L.P. (2000). Precision trace gas analysis by FT-IR spectroscopy. 1. Simultaneous analysis of CO_2, CH_4, N_2O, and CO in air. *Anal. Chem.* 72 (1): 206–215; 2. The $^{13}C/^{12}C$ isotope ratio of CO_2. *Anal. Chem.* 72 (1): 216–221.

77 Saarinen, P. and Kauppinen, J. (1991). Multicomponent analysis of FT-IR spectra. *Appl. Spectrosc.* 45 (6): 953–963.

78 Shafeeyan, M.S., Daud, W.M.A.W., Houshmand, A., and Shamiri, A. (2010). A review on surface modification of activated carbon for carbon dioxide adsorption. *J. Anal. Appl. Pyrolysis* 89 (2): 143–151.

79 Kunimatsu, K., Golden, W.G., Seki, H., and Philpott, M.R. (1985). Carbon monoxide adsorption on a platinum electrode studied by polarization modulated FT-IRRAS. 1. Carbon monoxide adsorbed in the double-layer potential region and its oxidation in acids. *Langmuir* 1 (2): 245–250.

80 Kunimatsu, K., Seki, H., Golden, W.G. et al. (1986). Carbon monoxide adsorption on a platinum electrode studied by polarization-modulated FT-IR reflection-absorption spectroscopy: II. Carbon monoxide adsorbed at a potential in the hydrogen region and its oxidation in acid. *Langmuir* 2 (4): 464–468.

81 Llabrés i Xamena, F.X. and Zecchina, Z. (2002). FTIR spectroscopy of carbon dioxide adsorbed on sodium- and magnesium-exchanged ETS-10 molecular sieves. *Phys. Chem. Chem. Phys.* 4 (10): 1978–1982.

82 Köck, E.-M., Kogler, M., Bielz, T. et al. (2013). In situ FT-IR spectroscopic study of CO_2 and CO adsorption on Y_2O_3, ZrO_2, and yttria-stabilized ZrO_2. *J. Phys. Chem. C* 117 (34): 17666–17673.

83 Jackson, P., Robinson, K., Puxty, G., and Attalla, M. (2009). In situ Fourier Transform-Infrared (FT-IR) analysis of carbon dioxide absorption and desorption in amine solutions. *Energy Procedia* 1 (1): 985–994.

84 Hicks, J.C., Drese, J.H., Fauth, D.J. et al. (2008). Designing adsorbents for CO_2 capture from flue gas-hyperbranched aminosilicas capable of capturing CO_2 reversibly. *J. Am. Chem. Soc.* 130 (10): 2902–2903.

85 Ryczkowski, J. (2001). IR spectroscopy in catalysis. *Catal. Today* 68 (4): 263–381.

86 Mills, A. and Wang, J. (2006). Simultaneous monitoring of the destruction of stearic acid and generation of carbon dioxide by self-cleaning semiconductor photocatalytic films. *J. Photochem. Photobiol., A* 182 (2): 181–186.

87 Nalawade, S.P., Picchioni, F., Marsman, J.H., and Janssen, L.P.B.M. (2006). The FT-IR studies of the interactions of CO_2 and polymers having different chain groups. *J. Supercrit. Fluid* 36 (3): 236–244.

88 Nalawade, S.P., Picchioni, F., and Janssen, L.P.B.M. (2006). Supercritical carbon dioxide as a green solvent for processing polymer melts: Processing aspects and applications. *Prog. Polym. Sci.* 31 (1): 19–43.

89 Vollmer, M. and Möllmann, K.-P. (2010). *Infrared Thermal Imaging – Fundamentals, Research and Applications*, 2e, 612 pp. Weinheim: Wiley-VCH.

90 Vollmer, M. and Möllmann, K.-P. (2012). CO_2 Detektion mit IR Kameras: Grundlagen, Experimente und Anwendungen (IR imaging of CO_2: basics, experiments, and applications). *Tech. Mess.* 79 (1): 65–72 (in German).

91 Strachan, D.C., Heard, N.A., Hossack, W.J. et al. (1985). Imaging of hydrocarbon vapours and gases by infrared thermography. *J. Phys. E: Sci. Instrum.* 18 (6): 492–498.

92 www.flir.com/ogi/ (retrieved 18 October 2016).

8

Photoacoustic Detection of CO_2

Frank Kühnemann

Fraunhofer Institute for Physical Measurement Techniques IPM, Department Gas and Process Technology, Heidenhofstraße 8, 79110 Freiburg, Germany

8.1 Photoacoustic Effect and Photoacoustic Gas Detection

In 1880 Alexander G. Bell reported work on the development of a photophone, an apparatus aimed at transmitting voice with the help of sunlight (Figure 8.1) [2]. A light beam was reflected by a mirror-coated membrane onto a receiver, where the light was absorbed by a selenium element, which was placed in a circuit with the moving coil of a speaker. The sound wave of the speech modulated the position of the membrane mirror and, hence, the intensity received by the selenium element. This led to a modulation of the temperature and the resistance of the selenium element, and sound was audible from the speaker. Using the photophone, Bell and his co-worker Tainter managed to transmit sound over more than 200 m distance.

Shortly after, Bell [3], Röntgen [4], and Tyndall [5] realized in independent studies that a battery-operated speaker was not necessary: the absorption of the modulated radiation (in particular its infrared portion) caused a modulated thermal expansion of the material (solid, liquid, or gas), which then in turn emitted a sound wave itself. This sound wave could then be made audible with a simple hearing tube. The *photoacoustic effect* was discovered. It did take, however, until the 1930s that the photoacoustic effect was exploited for gas analysis [6].

Reviewing the literature on photoacoustic (PA) gas analysis, one should be aware that the term *photoacoustic gas analysis* is used for different spectroscopic detection schemes. In this section, they are only introduced shortly, starting from the most basic scheme for a spectroscopic gas analyser consisting of a light source, the gas sample (in a chamber), and a radiation detector (Figure 8.2). Detailed descriptions are given in the following sections.

The light beam with initial intensity I_0 passes through a sample. The intensity $I(z)$ after a distance z through the sample is then measured with a detector appropriate for the wavelength and power. The absorption in the sample and, hence, the gas concentration are determined using Lambert–Beer law (cp. Eq. (8.1)):

$$I(z) = I_0 e^{-\alpha \cdot z} \tag{8.1}$$

Carbon Dioxide Sensing: Fundamentals, Principles, and Applications,
First Edition. Edited by Gerald Gerlach, Ulrich Guth, and Wolfram Oelßner.

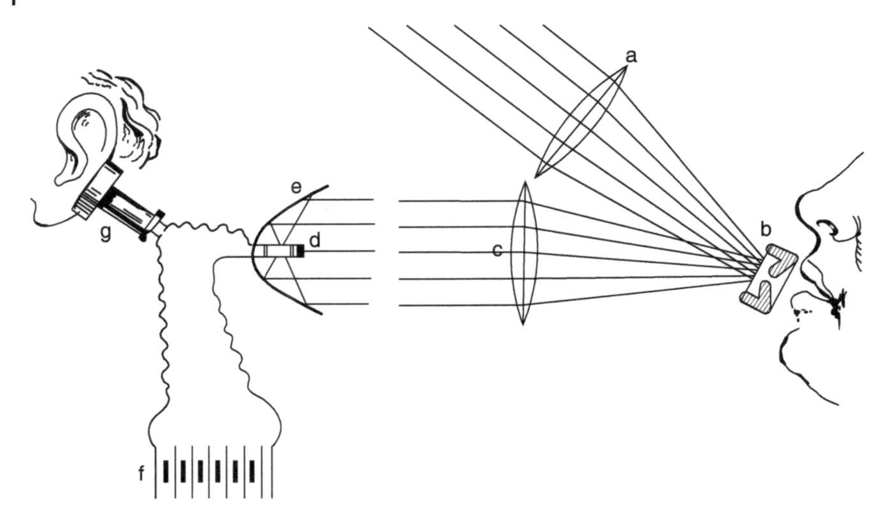

Figure 8.1 Alexander Graham Bell's photophone – technical drawing by Bell. (a, c) Lenses, (b) membrane-mounted mirror, (d) selenium element, (e) parabolic mirror, (f) battery, and (g) speaker. Source: From [1].

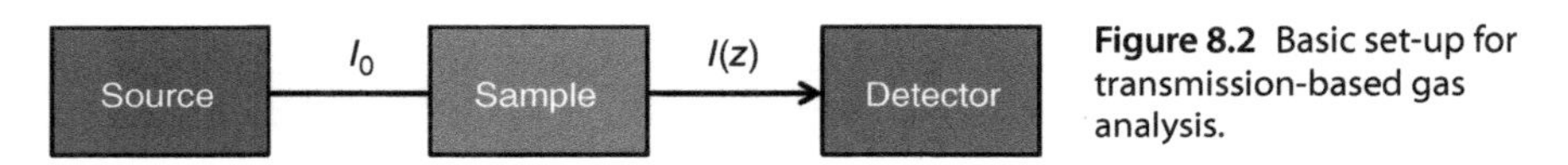

Figure 8.2 Basic set-up for transmission-based gas analysis.

Important parameters of any technique are the smallest measurable change in intensity $I_0 - I(z)$ and the smallest detectable gas concentration.

The photoacoustic effect can now be utilized in two different ways.

8.1.1 Photoacoustic Cell as Gas-Specific Radiation Detector

The radiation detector is replaced by a chamber filled with a high concentration of the gas to be measured in the sample, and a sensitive microphone or another pressure sensor is used to determine the amount of radiation absorbed by the gas in the detector chamber (Figure 8.3). Without target gas in the sample, the light beam reaches the detector without attenuation. With target molecules present

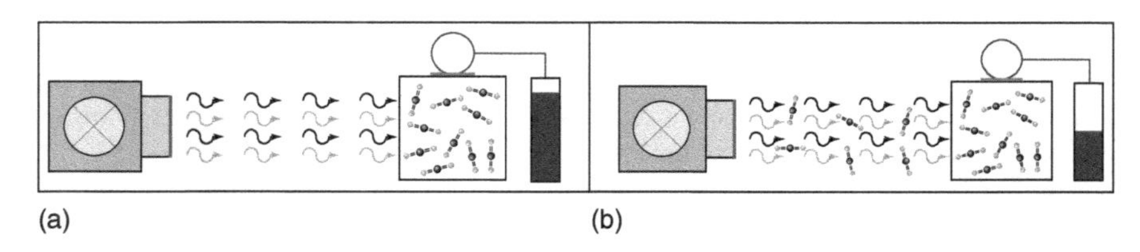

Figure 8.3 Radiation detector based on the photoacoustic effect. (a) No target molecules are present in the analytic path. Radiation reaches the detector without attenuation, and the full detector signal is recorded. Source: From [7]. (b) The presence of target molecules in the light path reduces the amount of radiation arriving in the detector cell, leading to a reduced microphone signal. Source: From [7].

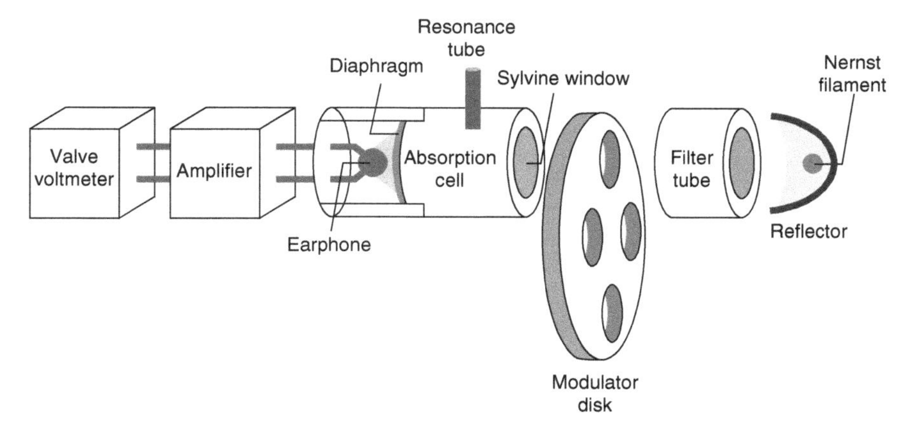

Figure 8.4 Single-channel photoacoustic set-up developed by Veingerov. Source: Adapted with permission from Manohar and Razansky 2016 [8].

in the sample, a part of the radiation is absorbed in the sample. The detector chamber receives less radiation and produces a smaller signal. Since the gas in the detector chamber absorbs only radiation at the characteristic wavelengths of the considered species, it represents a detector with high spectral and molecular selectivity. That opens the door for the use of broadband thermal light sources without sophisticated spectral filters and is especially helpful to reduce cross sensitivities in gas mixtures.

Such a scheme, as shown in Figure 8.4, was first presented by Veingerov who performed CO_2 detection with concentrations down to 2000 ppm [6]. This experiment can be considered the first experimental demonstration of quantitative gas detection using a photoacoustic scheme.

The basic advantage of this scheme is the gas selectivity of the detector chamber. This is particularly useful if one wants to analyse samples of gas mixtures with a (thermal) broadband emitter. They share, however, the drawback of all transmission techniques in that they require a certain minimum absorption – and hence optical path length through the gas sample – to obtain a significant attenuation of the light beam reaching the detector chamber.

8.1.2 Photoacoustic Detection in the Sample Cell

The second basic photoacoustic concept removes that limitation. Instead of detecting the amount of radiation still *present behind the gas sample*, it uses the photoacoustic effect to determine the amount of *absorbed radiation directly in the sample cell*. This approach (Figure 8.5) is primarily used for the detection and analysis of low gas concentrations, when the absorption is so small that the attenuation of the beam cannot be used any longer for a reliable concentration measurement.

This configuration has a major advantage over the above-mentioned one for the analysis of small concentrations: if no absorber is present in the gas sample, the microphone does not record any signal. The measurement is basically performed

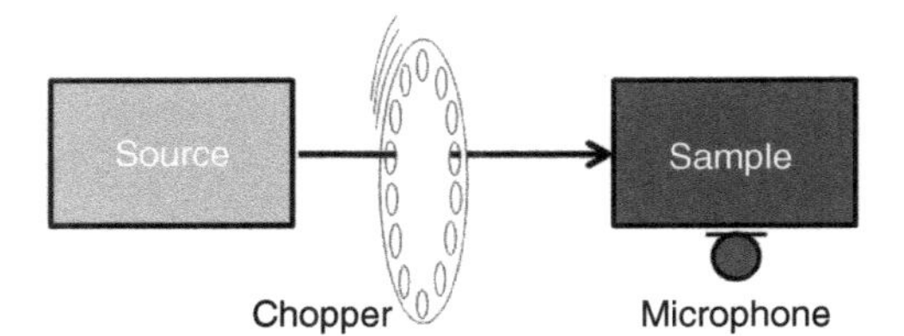

Figure 8.5 Basic photoacoustic sensor set-up with detection in the sample chamber. Modulation can be achieved by intensity modulation (chopper or source) or, in the case of laser-based systems, by wavelength modulation of the laser across the absorption line.

against a zero background and the useful signal scales with the power of the light source. Consequently, the development of this scheme was linked to the availability of powerful (and tunable) lasers, resulting in ever lower detection limits. The microphone signal can be enhanced by shaping the gas cell as an acoustic resonator and modulating the light source at one of its resonance frequencies [9].

Since the acoustic signal is now generated in the gas sample itself, the selectivity of the measurement has to be reached through a spectrally narrow excitation (thermal source with filter or laser). Compared with thermal light sources, lasers offer higher brightness and better selectivity thanks to their narrow spectral emission. Nevertheless, thermal sources have the advantages of a better reliability and a lower price, and the detection limits may be sufficient for a large range of applications.

8.2 Photoacoustic Signal Generation

The theory of photoacoustic signal generation has been described by several authors [10, 11] and shall be summarized here only briefly. The process consists of three major steps:

- The absorption of the radiation by the gas,
- The subsequent relaxation, and
- The formation of the sound wave.

The absorption of the radiation is controlled by the spectrum of the gas, which has been discussed in detail in Section 3.1. After excitation, there are several relaxation channels for an individual excited molecule (Figure 8.6):

- Radiative relaxation (emission of a photon, i.e. fluorescence),
- Non-radiant vibrational–translational relaxation by collision with another molecule and conversion of the energy into translational energy of the gas molecules, and
- Non-radiant collisional vibrational–vibrational energy transfer to another molecule.

The probabilities of the three processes depend on the type of excitation in the molecule (electronic, vibrational, rotational) and the composition and total pressure of the gas (determining the availability of collision partners and the conditions for a resonant collisional energy transfer between molecules). Radiative relaxation has the highest probability in the UV range (electronic excitation). But even for such high transition energies, it will be the dominating process only at very low gas pressures ($\ll$1 mbar).

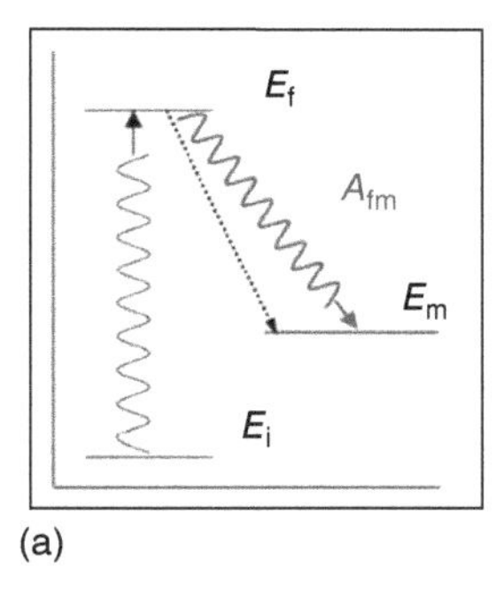

(a)

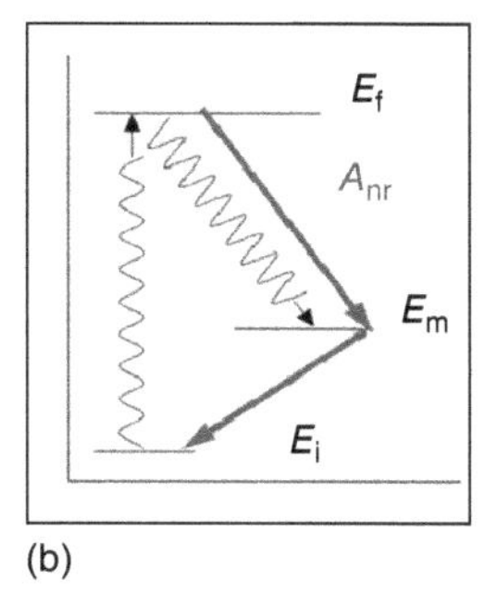

(b)

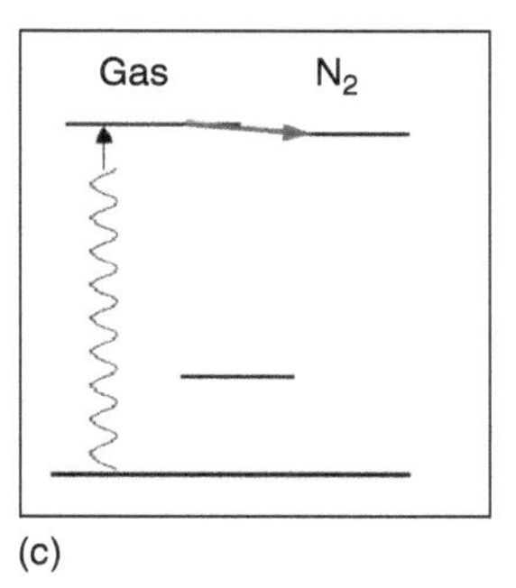

(c)

Figure 8.6 Relaxation processes after excitation: (a) radiative relaxation, (b) non-radiative vibrational–translational thermal relaxation, and (c) collisional vibrational–vibrational transfer.

At gas pressures above that, the collisional relaxation is the dominating transfer process after both electronic and rotational–vibrational excitations. In the latter, the relaxation times ($\tau \approx 10^{-9}$ to 10^{-6} s) dominate over the radiative relaxation times (10^{-2} s) by far. As a result, the gas is heated by the increase of translational energy of the molecules and, subsequently, cools down due to heat flow to the walls of the gas chamber. If the energy flow into the gas is periodically modulated, the gas undergoes periodic temperature and pressure cycles. The resulting acoustic wave is detected by a microphone or another sensor (see Section 8.4).

The third process (vibrational energy transfer) may be in competition with the second one. If the excited state of the trace gas molecule under study is in (near) resonance to an excited state of a major component of the gas (e.g. N_2 or O_2 in air), then the vibrational energy of the excited molecule may be transferred to the N_2 molecule in a collision. Here it gets trapped, since the much more likely collisions with other N_2 molecules result mostly just in a redistribution of vibrational energy. As a consequence, the vibrational–translational relaxation times can reach 1 ms or more even at atmospheric pressure. This can be longer than the modulation period in laser-based photoacoustics, making the PA signal undetectable. The relaxation can be considerably accelerated, and the photoacoustic signal increased by adding other molecules (like H_2O or SF_6) or noble-gas atoms (like argon) to the gas mixture, which provide a dense ladder of rotational–vibrational states for thermal relaxation (Figure 8.7) or which yield an efficient collisional energy transfer. For laser-based CO_2 detection around 2 μm [12], it has been demonstrated that the vibrational–translational relaxation time at a pressure of 8 kPa can be reduced by a factor of 100 (from 139 to 1.24 μs) by adding 2.5% water vapour to the CO_2/N_2 mixture. At a detection frequency of 33 kHz (see Section 8.4.3.1), this results in a strong increase of the PA signal.

Assuming the periodic release of thermal energy through relaxation, one has then to consider the formation of the acoustic wave in the gas chamber. The wave is treated as the temporal change in pressure relative to the static background pressure:

$$\frac{\partial p}{\partial t} = \rho_0 C_p \frac{\partial T}{\partial t} - \kappa \nabla^2 T - W(t) \tag{8.2}$$

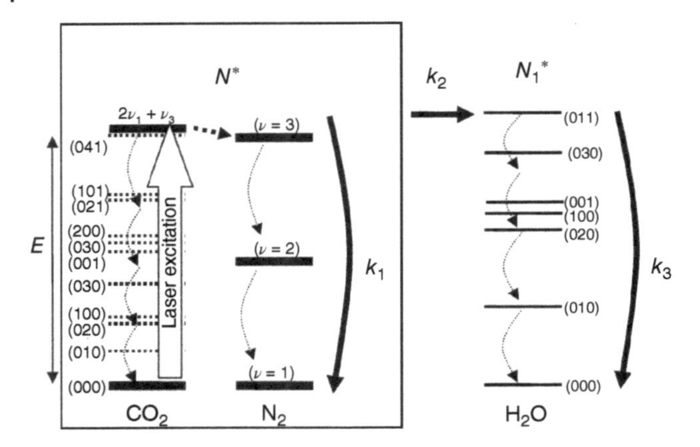

Figure 8.7 Vibrational relaxation channels for CO_2/N_2 after excitation of the $2\nu_1 + \nu_3$ band around 2 μm. Vibrational energy is transferred from CO_2 to "metastable" N_2 states, resulting in a slow relaxation (139 μs at a pressure of 8 kPa). Addition of water vapour opens additional relaxation channels, reducing the relaxation time to 1.24 μs. Source: Wysocki et al. 2006 [12]. Reproduced with permission of Springer Nature.

with κ, ρ_0, and C_p the thermal conductivity, mass density, and heat capacity, respectively, T the gas temperature variation, and p the acoustic pressure signal.

The solution of Eq. (8.2) will depend on the actual shape of the PA cell and on the modulation frequency ω. In the case of PA detection with thermal radiation sources, typical modulation frequencies are in the range of 5–20 Hz. The cells are operated in a non-resonant mode, and the pressure amplitude can be considered to be the same everywhere across the PA cell, since the wavelength of sound in air amounts to several metres at these low frequencies.

In the case of laser-based photoacoustics, the light is modulated at several kilohertz, and the cells are formed as resonators. In the case of quartz-enhanced photoacoustics (see Section 8.4), the laser is modulated at half the resonance frequency of the quartz tuning fork.

In addition to the PA signal generated by the gas, contributions may also come from the windows and the walls of the PA cell. The wall signal is minimized by using a material with low absorption and high thermal capacity and conductivity to minimize the temperature increase in the wall material and the subsequent heat transfer into the gas.

A window signal is generated by the residual absorption of the infrared radiation by the window material which may be between 10^{-2} and 10^{-5} cm^{-1} depending on the material and the wavelength range. In the case of non-resonant cells with thermal sources, it is just considered a fixed background signal like the wall signal [13]. Due to the additional heat-transfer step from the wall or window into the gas, these signals have a phase that is different from the gas signal. Amplitude and phase are determined from a zero-calibration measurement. In subsequent measurements, the actual gas signal is determined by subtracting the background from the total signal.

8.3 Photoacoustic Gas Analysis with Thermal Sources

8.3.1 Photoacoustic Cell as Gas-Specific Radiation Detector

In the 1930s, the work of Veingerov [6], Pfund [14], and Lehrer and Luft [15] led to the first gas analysers based on the photoacoustic effect. The system by Lehrer and Luft (described in detail in [16]) was developed in search for a sensitive and selective gas analyser for the detection of butadiene in air (for the monitoring of potentially explosive mixtures) and for CO in air (to monitor potential health hazards in industry and to ensure workplace safety in mining industry). Their system, the *U*ltrar*o*t *A*bsorptions*s*chreiber (infrared absorption recorder) (URAS), consisted of two channels (measurement and reference; see Figure 8.8), which were periodically (6 Hz) illuminated from two Cr–Ni filaments. Measurement and reference chambers were gold-coated and had typical lengths between 1 and 300 mm, depending on the targeted concentration range. The two detection chambers were filled with the gas to be detected (possibly with some other infrared-inactive gas added), and a membrane capacitor was used to determine the pressure difference between measurement and reference channel. Any imbalance in absorption between measurement and reference cell leads to a modulated signal at the capacitor, allowing one to register pressure fluctuations as small as 0.1 μbar [17] and detecting gas concentrations in the low ppm range.

As one can see from the infrared spectra of CO, H_2O, and CO_2 in Figure 8.9, there is only a small overlap between the CO absorption band and the bands of the two possibly interfering gases. The selective detection of the infrared radiation at the typical CO wavelengths in the detection chambers results in a very

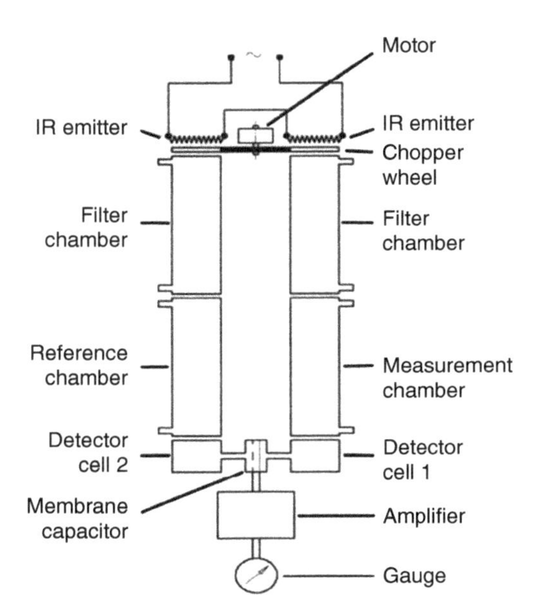

Figure 8.8 Basic set-up of the URAS concept of Lehrer and Luft [15]. A capacitor between the two detector cells measures the differential signal between the measurement channel and the reference channels. Source: Wiegleb 2016 [17]. Modified with permission of Springer Nature.

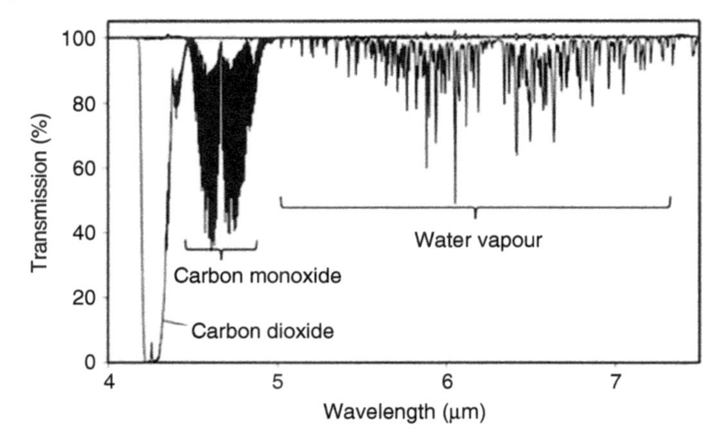

Figure 8.9 Typical transmission spectrum of a process gas containing CO, CO_2, and H_2O. A photoacoustic chamber filled with CO acts as a gas-selective detector, thus effectively suppressing cross sensitivities in the CO detection due to H_2O or CO_2. Source: Wiegleb 2016 [17]. Reproduced with permission of Springer Nature.

small cross sensitivity. Assuming equal concentrations of CO and CO_2, the latter would modify the CO signal only by about 2×10^{-4} [18]. Of course, stronger cross sensitivities are to be expected with a stronger overlap between the spectra. This is the case, for instance, in the analysis of mixtures with several hydrocarbons. If the interfering species are known, filter chambers can be added in front of the measurement and reference cells (Figure 8.8). They would be filled with the interfering gases at high concentration. In this way all radiation at ambiguous wavelengths is removed from the light path.

As mentioned in Section 8.2, the relaxation process in the detection chamber depends on the gas mixture. However, two important aspects have to be considered: the absorption of radiation should take place along the whole available way in the chamber, not just over the first few mm. Otherwise, the heat transfer to the chamber window would be a relevant loss channel and reduce the useful signal. In addition to the target gas, some non-absorbing buffer gas shall be added to ensure an effective thermal relaxation for maximum useful signal (see Section 8.2).

In the context of the topic of this book, it may be added here that the potential application of the URAS for the measurement of CO_2 in breath and assimilation studies was already suggested by Luft [18].

In a later stage, Luft designed a single-channel system placing two detector chambers behind each other (UNOR [19]), thus avoiding background drifts in the two-channel system, which could arise from imbalances in the illumination or the gas pressure in the two channels. This configuration is currently still being used for gas analysers for a wide range of applications from industrial process control to the monitoring of emissions of stationary and marine engines. Comparing the original designs with the current products, one notices the progress in electronic data processing, software control of the operation, and the integration of the gas analyser into comprehensive process analytical solutions for industry [20].

8.3.2 Miniaturized PA Detection Systems

Systems like URAS and UNOR have been developed for process gas monitoring aiming at detection limits in the low ppm range. An emerging new application is the measurement of carbon dioxide concentrations for ambient air quality monitoring in residential buildings and in offices or cars. Elevated CO_2 concentrations above the regular ambient ones (approximately 400 ppm) are an indicator for poor air quality. Continuous monitoring in connection with automated ventilation systems can help to keep concentrations below a threshold of 1000 ppm. Such sensors have to be cheap, robust, and compact and should cover a concentration range between 100 and (application dependent) up to 20 000 ppm (2%). Photoacoustic sensors have been developed here implementing the basic PA configuration of Figure 8.3 and making use of the current developments of MEMS (microelectromechanical system) technology. As emitters, current-modulated hot plates are used (see Section 7.2.3). The detector chamber (Figure 8.10) contains a MEMS-based microphone (Figure 8.11), initially developed for smartphones and featuring a sophisticated on-chip signal processing. The detector chamber is filled with CO_2 and a buffer gas ensuring efficient vibrational relaxation of the excited CO_2 molecules and is sealed with an IR-transparent window. The air to be analysed simply diffuses into the space between emitter and detector chamber. The overall size of the sensor (Figure 8.12) is approximately $3 \times 3 \times 3\,cm^3$. A limit to miniaturization is set by the required minimum absorption path length depending on the targeted CO_2 concentration range, typically a few mm. Figure 8.13 shows a series of

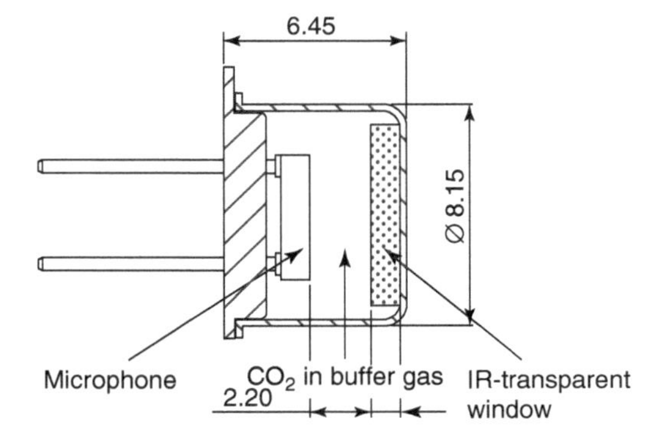

Figure 8.10 CO_2 PA detection chamber with MEMS microphone. Source: Huber et al. 2014 [21]. Reproduced with permission of Elsevier.

Figure 8.11 MEMS microphone SMM310 (Infineon) on a TO socket without cap and window. Source: Huber et al. 2014 [21]. Reproduced with permission of Elsevier.

Figure 8.12 Compact single-path PA sensor for CO_2 measurements. Left side: MEMS microphone in TO39 socket. Right side: IR emitter in TO5 case. Source: Huber et al. 2015 [22]. Reproduced with permission of Elsevier.

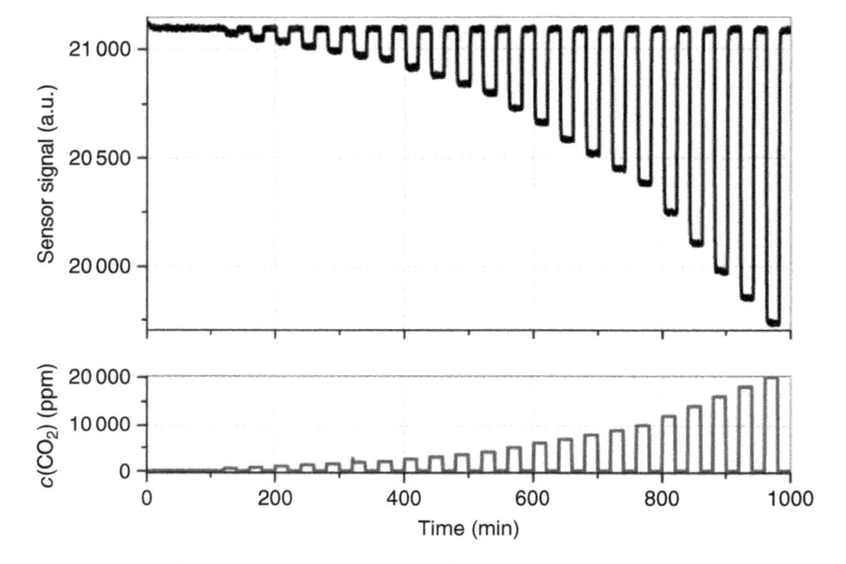

Figure 8.13 Laboratory measurement of a CO_2 concentration series (250–20 000 ppm CO_2 in N_2) at room temperature. Source: Huber et al. 2016 [23]. Reproduced with permission of Elsevier.

CO_2 measurements between 250 and 20 000 ppm. Since the detector records the radiation behind the sample gas, increased concentrations result in a lower detector signal. The state of the MEMS microphone development allows one to detect relative changes in the absorbed radiation at a level below 10^{-3} through phase-sensitive detection and averaging despite the low power level from the compact thermal light sources. This is equivalent to resolve concentration changes at the 100 ppm level, fully sufficient for the targeted application.

8.3.3 Photoacoustic Detection in the Gas Sample

The basic schemes presented in Sections 8.3.1 and 8.3.2 are all transmission-measurement set-ups where the PA cell is used as a gas-selective radiation detector. When the fraction of the radiation absorbed in the sample chamber drops below 0.1%, it becomes challenging to perform a reliable measurement using the

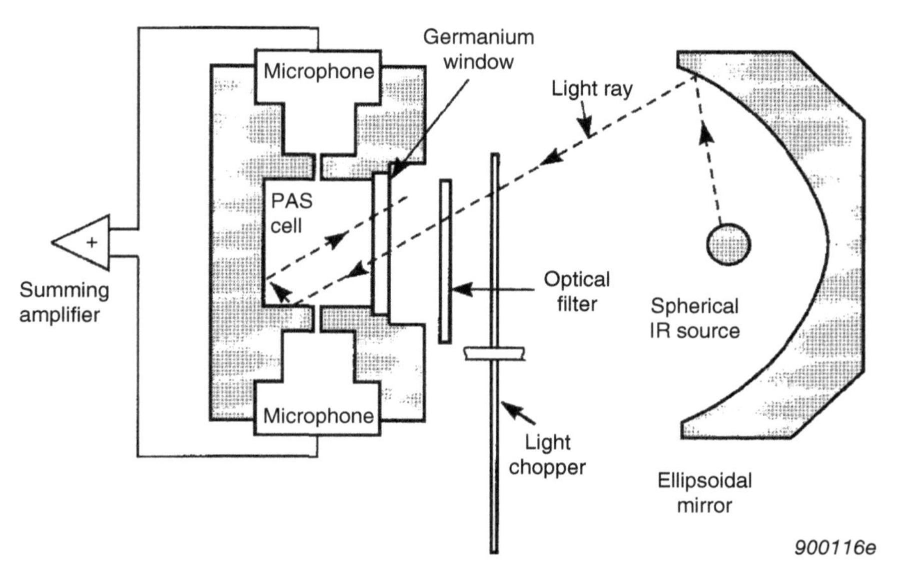

Figure 8.14 Basic scheme of the Brüel & Kjaer PA system with thermal IR source. Source: Christensen 1990 [13]. Reproduced with permission of Brüel & Kjaer.

transmission scheme. As an alternative, the radiation absorbed by the sample gas can be directly determined in the sample chamber using the PA effect. This can be done using either thermal sources or lasers (see Section 8.4).

A commercial system with thermal source is offered by LumaSense, Ballerup, Denmark (formerly Brüel & Kjaer (B&K); Figure 8.14). Key components for a sensitive detection are the two condenser microphones. As radiation source, a black-body emitter at about 800°C is used. To generate the periodic PA signal, radiation is modulated with a mechanical chopper at 20 Hz. As one can see from Figure 8.14, two microphones are put in opposite orientation into the PAS cell, and the signals are added up. At the low detection frequencies used here, external noise or background contributions will lead to signals with opposite phase at the two microphones. Carefully selected microphones with matching characteristics, allow one to suppress effectively such external contributions.

To determine the sensitivity limit for the system, several contributions have to be considered like noise of the detector/preamplifier system, background signals originating from the walls and windows of the cell, acoustic noise, and mechanical vibrations of the cell.

Typical data for the B&K gas analyser 1306 as given from the manufacturer are summarized in Table 8.1. As can be seen, only the wall and window signal exceed the noise level of the microphone-amplifier system. These background signals are, however, stable in amplitude and phase at constant operating conditions. They are determined in the zero-point calibration process using a non-absorbing gas and later subtracted from the analytical total signal.

Using a broadband source in connection with the PA detection in the gas sample itself requires measures to ensure the gas selectivity. In the B&K scheme, high-quality narrowband bandpass filters are used for that purpose. As an

Table 8.1 Noise and background characteristics for the B&K 1306 gas analyser.

Contribution	Value (Pa)
Microphone-amplifier system, RMS	2.5×10^{-6}
Wall signal	1.5×10^{-4}
Window signal	9.5×10^{-6}
Ambient noise (office environment)	6.3×10^{-8}
Mechanical vibration (unbalanced)	6.3×10^{-6}
Mechanical noise (balanced)	1.3×10^{-7}

Modulation frequency 20 Hz, bandwidth 1/8 Hz.
Source: Adapted from Brüel & Kjaer 1990 [13].

example, for the detection of residual CO_2 traces in pure gases, a filter is used with a centre at 2347 cm^{-1} and a 50 cm^{-1} pass width, allowing the detection of CO_2 down to 7 ppb [24].

To analyse gas mixtures with several (known) infrared-active components, the signals are measured using up to five different filters consecutively, and the concentrations are determined through a chemometric analysis of the measured data.

8.4 Laser-Based Photoacoustic Trace Gas Detection

8.4.1 General Overview

The narrow spectral linewidth and high beam quality make lasers an advantageous alternative for PA detection. The first experiments for the utilization of lasers for photoacoustic (trace) gas detection were already started in the 1960s [25], soon demonstrating trace gas detection at the 10 ppm level [26]. The first widely used lasers for mid-infrared photoacoustic trace gas detection were CO_2 and CO gas lasers. Their (line) tunability between 9.6 and 10.6 μm (CO_2 laser), 4.8 and 8.4 μm (CO laser), and 2.8 and 4.0 μm (CO overtone laser) offered possibilities for a sensitive and selective trace gas detection. By placing the PA cells into the respective laser cavity, circulating powers between 1 (CO overtone laser) and 100 W (CO_2 laser) could be used for the PA detection, yielding detection limits at the sub-ppb level for selected gases. The complexity of these laser systems was, however, detrimental to their use as trace gas detectors outside the laboratory.

Considerable new momentum was brought into the development of PA trace gas detection schemes and applications through the appearance of the quantum cascade lasers [27]. Owing to their design principle, their emission wavelength can be tailored for gas-specific absorption regions in the 4–12 μm range, reaching a tunability of more than 100 cm^{-1} and output powers exceeding 100 mW.

As of today, laser-based systems dominate the development of systems for *trace* gas detection, i.e. for concentrations in the sub-ppm range, even in complex mixtures. The possibility to modulate the lasers internally or externally with frequencies of more than 30 kHz has led to the dominating role of resonant detection schemes. New momentum was brought to the PA trace gas analysis with the introduction with quartz tuning forks in 2002 [28] and optically read-out cantilevers [29] as replacement for the classical electret microphones as detectors.

8.4.2 Resonant Photoacoustic Cell Design

Laser-based photoacoustic trace gas detection is performed in a resonant mode, as this allows one to enhance the acoustic signal for a given amount of deposited energy. This approach was first demonstrated in 1973 [30]. Resonant PA cells typically have a cylindrical design with the laser beam propagating along the cell axis. The microphone is attached to the sidewall. Among the possible acoustic modes (azimuthal, longitudinal, radial; see Figure 8.15), one would favour the latter for a perfect coupling between laser beam, resonance, and microphone. However, for the range of modulation frequencies accessible with mechanical choppers and detectable with standard electret microphones (i.e. up to 5 kHz), this would require cell diameters of several centimetres, leading to large cell volumes and long turnover times for the gas exchange in the case of flow-through measurements.

Different cell designs were developed to facilitate a sensitive background-free detection [9].

The PA signal S generated in a gas sample as a function of wavelength is generally described as

$$S(\lambda) = F \cdot \alpha(\lambda) \cdot P(\lambda) \tag{8.3}$$

depending on the cell constant F, the absorption coefficient α of the gas, and the input power P. The cell constant F describes the generated pressure amplitude upon modulation in relation to the absorbed laser power per unit length, which is given by $\alpha \cdot P$, assuming weak absorption. In the non-resonant case the cell

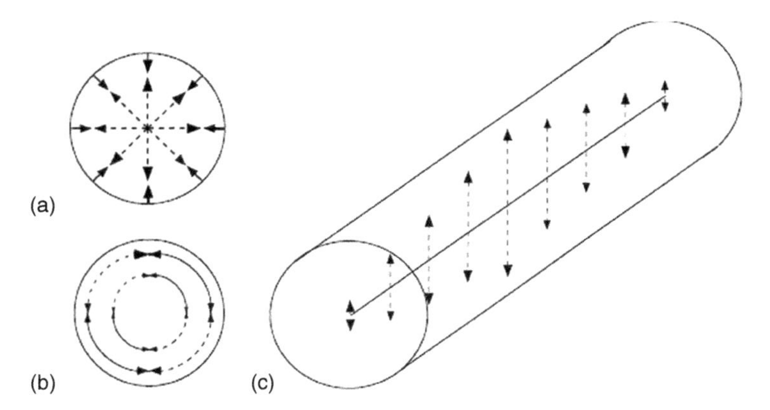

Figure 8.15 Modes of an acoustic resonator: (a) radial, (b) azimuthal, and (c) longitudinal. Note that in (c) the arrows show the spatial distribution of the amplitudes, not the direction of the particles' motion.

constant F for a cylindrical cell with length L and volume V can be described by

$$F_{nr} = \frac{G(\gamma - 1)L}{\omega V} \tag{8.4}$$

with the angular modulation frequency ω and the specific heat ratio γ. G is a geometrical correction factor, which includes, among others, the influence of the thermal losses from a gas layer to the wall of the PA cell. The thickness of this boundary diffusion layer is given by the thermal diffusion length

$$\mu = \sqrt{\frac{2\kappa}{\omega \rho_0 C_p}} \tag{8.5}$$

which itself depends on the thermal properties of the gas (κ, ρ_0, and C_p being the thermal conductivity, mass density, and heat capacity, respectively) and the modulation frequency ω. The volume of the layer compared with the total volume of the chamber indicates the order of magnitude of the wall losses. Higher modulation frequencies result in thinner effective boundary layers. For an air sample at atmospheric pressure, the diffusion length at 20 Hz (typical modulation with thermal sources) is about 0.56 mm; at 2 kHz (as in laser-based PA systems), it reduces to about 56 μm.

The amplitude of the sound wave can be enhanced by designing the PA cell as an acoustic resonator and modulating the laser beam at a resonance frequency of the cell. The enhancement of the cell constant F_{res} in relation to the non-resonant case (F_{nr}) is then given by the acoustic Q-factor of the cell:

$$F_{res} = F_{nr} \cdot Q \tag{8.6}$$

While it is possible to design cells with Q-factors of more than 1000, this would imply challenges to the practical use of the cells. High Q-factors are equivalent to narrow resonance curves. At 2 kHz resonance frequency and a $Q = 1000$, a deviation of the modulation frequency from the actual resonance by just 1 Hz would result in a 50% drop of the signal. Preventing this requires a stabilization of the cell conditions or a tracking of the resonance and adjustment of the modulation frequency.

As in the case of the non-resonant cell (see Section 8.3.3), unwanted background signals may arise from laser light absorbed by the walls of the PA cell and by the windows. Because wall signals can be reduced by careful alignment of the collimated laser beam through the PA cell, window signals are more relevant. They are effectively suppressed by placing buffer volumes between the windows and the actual acoustic resonator, as can be seen from the example shown in Figure 8.16. Residual background signals are determined through a "zero" measurement and subtracted from the gas signal. As a result, only the variation in the background signal limits the detection.

8.4.3 Acoustic Detectors

For decades, the standard PA detectors were condenser microphones or electret microphones as they are typically used in hearing aids. Their high sensitivity, in connection with phase-sensitive lock-in detection, was an essential contribution

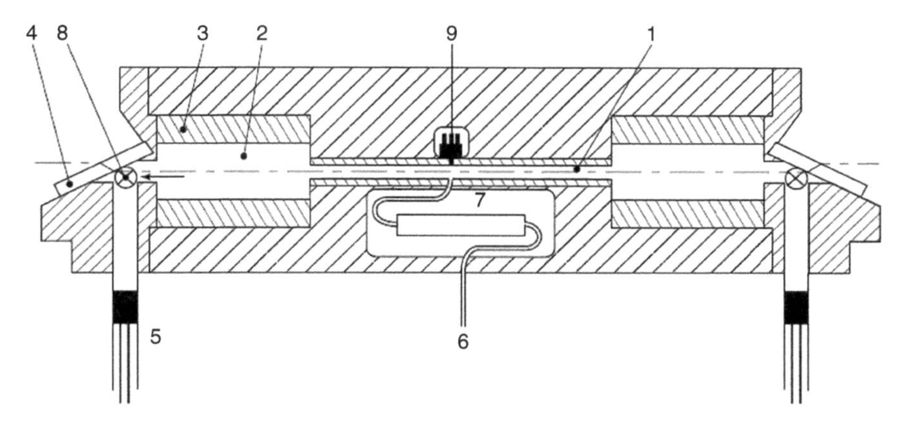

Figure 8.16 Cross section of a resonant PA cell with buffer volumes for the suppression of window signals. 1, resonator; 2, buffer volumes; 3, buffer ring to adapt buffer radius; 4, ZnSe Brewster windows; 5, adjustable notch filter to suppress window signals; 6, inlet gas flow; 7, notch filter to suppress external noise through the flow; 8, gas flow outlet; and 9, microphone. Source: Harren et al. 2006 [31]. Reproduced with permission of John Wiley & Sons.

to the very low detection limits observed in laser-based photoacoustics. In addition to processing signals at an SNR level of −50 dB, the phase-sensitive detection allows one to distinguish the useful gas signal from wall and window signals, which may have a different phase depending on the construction details of the cell.

8.4.3.1 Quartz-Enhanced Photoacoustic Spectroscopy

The development of photoacoustic trace gas analysers gained new momentum with the implementation of alternative detectors for the sound waves. In 2002, Kosterev et al. published the first work on quartz-enhanced photoacoustic spectroscopy (QEPAS) [28]. They used a standard quartz tuning fork commonly found as oscillators in quartz watches (Figure 8.17) and available at ultra-low cost. The prongs are 3.2 mm long and have a cross section of 0.33 mm × 0.33 mm and a gap of 0.3 mm. For the symmetric vibration of the prongs, the resonance frequency is 2^{15} Hz ≈ 33 kHz with very high Q-factors, reaching 10^6 in vacuum

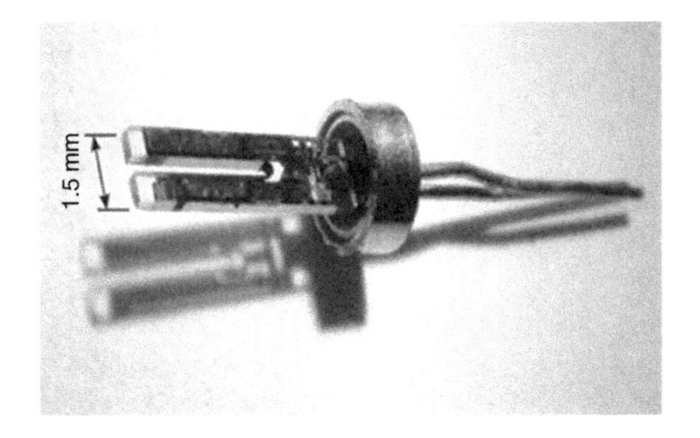

Figure 8.17 Shape and dimensions of a standard quartz tuning fork. Source: Reproduced with permission of Tittel 2014 [32].

and 10^5 in air. For the PA measurements, the laser beam is aimed into the gap between the prongs, and the generated pressure wave excites the symmetric vibration. External background signal or noise in the frequency range of the detection would excite the asymmetric vibration of the prongs, since the sound wavelength is larger than 1 cm. The high Q-factors require a careful tracking of the actual resonance frequency, which depends on the composition and temperature of the gas sample.

The standard technique to generate the periodic acoustic wave in the sample is wavelength modulation. The laser is directly current-modulated at half the resonance frequency to perform $2f$ detection, as described in Section 3.1.7. As a result, (spectrally flat or smooth) background signals as well as small intensity modulations associated with the wavelength modulation are effectively suppressed. An example is given in Figure 8.18, showing the $1f$ and $2f$ signal when scanning across a CO_2 absorption line for CO_2 in N_2. Here, the total gas pressure was set to 6.7 kPa.

When reviewing the literature on QEPAS (see, e.g. [33]), it is found that the pressure in the PA gas cell ranges from several kilopascals to atmospheric pressure. For the determination of the optimum pressure, several aspects are relevant:

- The collisional relaxation rates are lower at reduced pressure.
- Reduced linewidths at lower pressure help to prevent the overlap of individual lines of an individual gas, hence improving specificity, and to avoid cross-interferences from faraway strong lines, like water vapour.
- In the case of a quartz tuning fork as detector, lower pressure reduces the energy losses and helps in maintaining a high Q-factor.

As a result of such considerations, the optimum pressure has to be determined experimentally for each specific gas-analytical situation.

In the basic QEPAS set-up, the absorption in the volume between the fork prongs is the major source for the photoacoustic spectroscopy (PAS) signal. An enhancement for the QEPAS detection can be achieved when an acoustic micro-resonator tube is added (Figure 8.19), similar to the resonant PA cell in the classical laser-PAS scheme. The length of the tube is matched to the resonance frequency of the fork but has a smaller Q-factor. Small holes in the wall of the

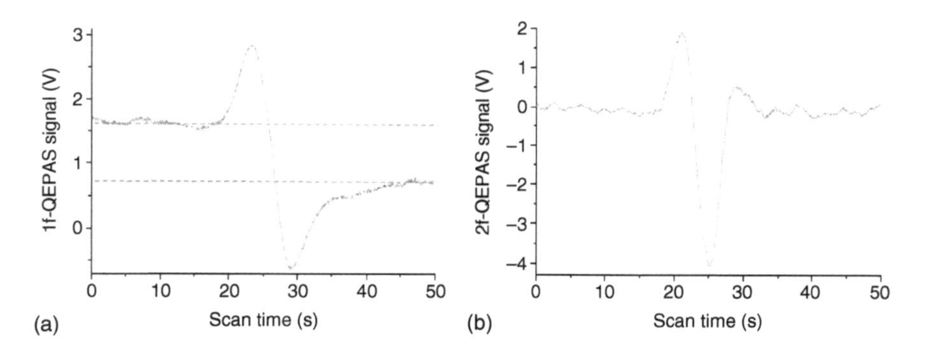

Figure 8.18 QEPAS spectral scan across a CO_2 absorption line at 2311.515 cm^{-1} showing (a) the $1f$ signal and (b) the $2f$ signal. 0.26% CO_2 in N_2, total pressure 6.7 kPa. Source: From [33].

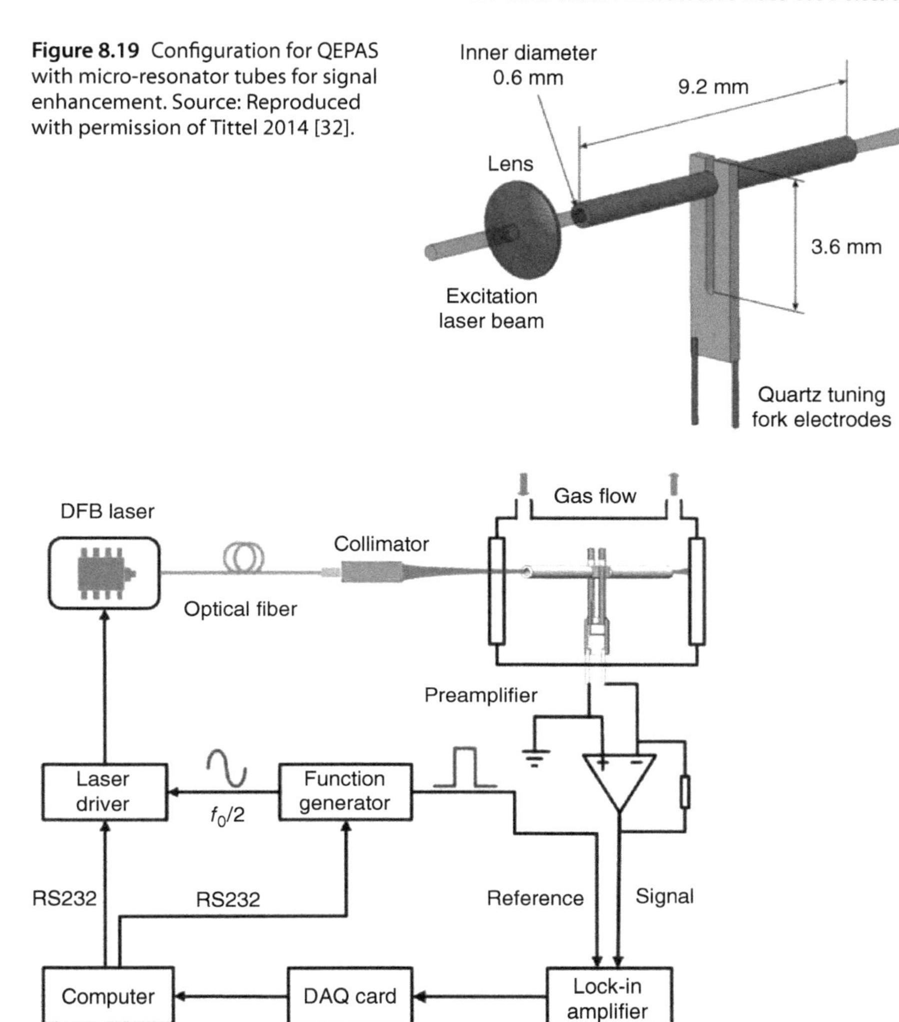

Figure 8.19 Configuration for QEPAS with micro-resonator tubes for signal enhancement. Source: Reproduced with permission of Tittel 2014 [32].

Figure 8.20 Set-up for a QEPAS system with acoustic micro-resonator tubes. Source: Zheng et al. 2016 [34]. Reproduced with permission of The Optical Society.

tube couple the acoustic wave inside the tube to the quartz tuning fork. Using this approach, a signal enhancement of 30–40 can be reached compared with a bare tuning fork. However, special care is required to provide a precise alignment of the laser along the whole length of the micro-resonator tube.

In Figure 8.20, the principal set-up is shown for such a tube-enhanced QEPAS set-up using a fibre-coupled DFB laser as light source. Thanks to the compact dimensions of the tuning fork and resonator tubes compared with the classical laser-PAS set-ups, much smaller detector volumes can be realized requiring less sample gas.

The standard tuning forks have two drawbacks: The resonance frequency of 32 kHz may be too high with respect to the relaxation times found in certain gas mixtures. In addition, the narrow gap between the two prongs of ca. 0.3 mm requires very careful alignment. Consequently, custom tuning forks have been developed with increased gap widths and lower resonance frequency [35].

8.4.3.2 Cantilever-Enhanced Laser-PAS

A second alternative to the classical microphones for the detection of the acoustic wave is cantilever-enhanced photoacoustic spectroscopy (cePAS). Here, the sound wave in the photoacoustic cell excites oscillations of a small cantilever made from silicon (Figure 8.21) [37]. The motion of the cantilever is then detected with an optical read-out based on a Michelson interferometer.

The dynamic range of the detection is given by the maximum detectable cantilever displacement (typically half the interferometer laser wavelength) on the one hand and the minimum detectable displacement on the other hand, which goes down to 1 pm. The high sensitivity of the cantilever requires extra measures to protect it from external acoustic noise (acceleration noise). This noise is transferred via the mechanical frame of the cantilever and via the air masses surrounding it. With a proper geometrical design of the photoacoustic cell, this technical acoustic noise can be effectively eliminated [38], and the sensitivity limit is finally given by the Brownian motion of the gas molecules. Figure 8.22 shows

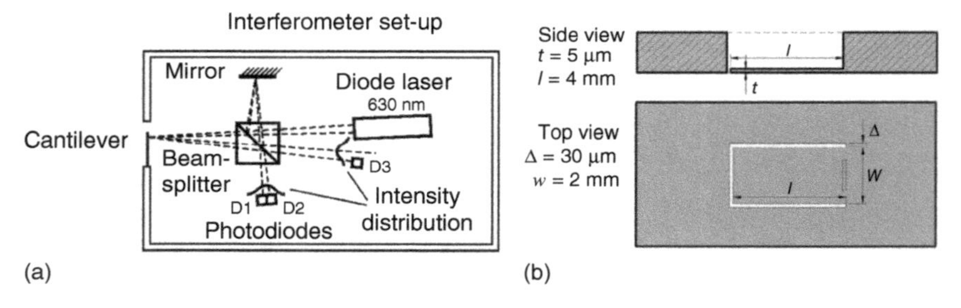

Figure 8.21 PAS with interferometer for the cantilever read-out (a) schematic set-up and (b) typical dimensions of the silicon cantilever. Source: Uotila 2007 [36]. Reproduced with permission of Elsevier.

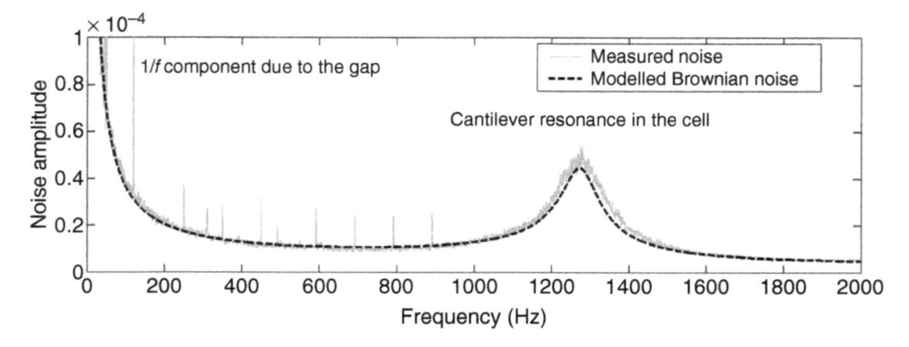

Figure 8.22 Noise spectrum of the cantilever with dominating Brownian noise. The broad peak around the resonance indicates the low *Q*-factor of the oscillating cantilever. Source: From [39].

the measured noise spectrum of a cantilever in comparison with a simulation assuming dominating Brownian noise. Clearly visible is the broad peak around the mechanical resonance of the cantilever, indicating the low Q-factor of the cantilever due to the mechanical damping. At low frequencies, the noise level is dominated by the $1/f$ noise due to the air passing through the gap between cantilever and frame.

cePAS is operated in a non-resonant mode, at a frequency well below the mechanical resonance. Since the detection is limited by the Brownian noise of the gas particles [38], neither an acoustic resonance of the gas cell nor a mechanical resonance of the cantilever would improve the signal-to-noise ratio. To perform gas measurements in laser-based cePAS, the laser is wavelength-modulated across the absorption line at the half detection frequency as in QEPAS. This helps to suppress wall and window signals that do not show a wavelength dependence upon modulation.

It is worth mentioning that cePAS detector modules are not only used for laser-based trace gas detection. They are also utilized as detection units for Fourier transform infrared (FTIR) spectrometers, combining broad spectral coverage and low detection limits for multi-gas analysis [40].

8.4.4 Detection Limits of CO_2 Gas Analysis with Laser-Based PAS

In photoacoustic trace gas detection, the established quantity for system evaluation is the normalized noise-equivalent absorption (*NNEA*). In this quantity, expressed in cm^{-1} $W/Hz^{-1/2}$, the noise-equivalent minimum absorption coefficient (α_{min}) is combined with the available laser power P and expressed for a measurement bandwidth of 1 Hz. Since laser-based PA gas detection is in first approximation a background-free technique, higher laser powers lead to lower detection limits. The *NNEA* is used as a figure of merit to compare different analysers independently from the individual target gas. System and performance data of different laser-photoacoustic CO_2 gas analysers are summarized in Table 8.2. For practical applications, however, one is more interested in the smallest detectable concentration of a particular gas, here expressed as (system- and gas-specific) noise-equivalent concentration (*NEC*). How a given *NNEA*

Table 8.2 System and performance data of different laser-photoacoustic gas analysers for CO_2 detection.

Gas	System	Wave number (cm^{-1})	Pressure (torr)	Laser power (mW)	NNEA (cm^{-1} W /$Hz^{-1/2}$)	NEC (ppm)	References
CO_2	QEPAS	6514.25	90	7	1×10^{-8}	890	[41]
CO_2	QEPAS	4991.26	50	4.4	1.4×10^{-8}	18	[12]
CO_2	QEPAS	6321.20	770	38	4.0×10^{-9}	123	[42]
CO_2	QEPAS	6361.25	150	45	8.2×10^{-9}	40	[43]
CO_2	I-QEPAS	2311.105	38	3	3.2×10^{-10}	3×10^{-4}	[44]
CO_2	cePAS	6361.25	760	30	3.2×10^{-10}	1.9	[45]

translates into an *NEC* will depend on the absorption coefficient of the respective gas at the operation wavelength and on the available laser power. As has been shown in Chapter 3, the CO_2 absorption coefficient varies strongly between the fundamental and overtone bands. As a result, similar *NNEA*s for systems operating in different wavelength regions may translate into large differences in the *NEC*, as can be seen for the first two systems in Table 8.2.

Comparison of PA detectors using different detector types shows that the cePAS detectors reach NNEAs, which are a factor of 100 lower than for classical condenser microphones (see, e.g. [39]).

References

1 Bell, A.G. (1988). Alexander Graham Bell's photophone. http://de.wikipedia.org/wiki/Datei:Bells_Photophon_Schema.jpg (accessed 15 December 2017).

2 Bell, A.G. (1880). On the production and reproduction of sound by light. *Am. J. Sci. Ser.* 3 (20): 305–324.

3 Bell, A.G. (1881). The production of sound by radiant energy. *Science* 2 (48): 242–253.

4 Röntgen, W.C. (1881). Ueber Töne, welche durch intermittirende Bestrahlung eines Gases entstehen (About sounds which are generated by intermitted radiation of a gas). *Ann. Phys.* 248 (1): 155–159. (in German).

5 Tyndall, J. (1881). Action of an intermittent beam of radiant heat upon gaseous matter. *Proc. R. Soc. London* 31: 307–317.

6 Veingerov, M.L. (1938). A method of gas analysis based on Tyndall–Roentgen optoacoustic effect. *Dokl. Akad. Nauk SSSR* 19: 687–688. (in Russian).

7 Huber, J. and Wöllenstein, J. (2015). Photoacoustic CO_2 sensor system – design and potential for miniaturization and integration in silicon. In: *Proceedings of the SPIE 9517 Smart Sensors, Actuators, and MEMS VII; and Cyber Physical Systems* (ed. J.L. Sanchez-Rojas and R. Brama), 951715-1–951715-6.

8 Manohar, S. and Razansky, D. (2016). Photoacoustics: a historical review. *Adv. Opt. Photonics* 8 (4): 586–617.

9 Miklos, A., Hess, P., and Bozoki, Z. (2001). Application of acoustic resonators in photoacoustic trace gas analysis and metrology. *Rev. Sci. Instrum.* 72 (4): 1937–1955.

10 Kreuzer, L. (1977). The physics of signal generation and detection. In: *Optoacoustic Spectroscopy and Detection* (ed. Y.-H. Pao), 1–26. New York, NY: Academic Press.

11 Sigrist, M.W. (1994). Air monitoring by laser photoacoustic spectroscopy. In: *Air Monitoring by Spectroscopic Techniques* (ed. M.W. Sigrist, J.W. Winefordner and I.M. Kolthoff), 163–238. New York, NY: Wiley.

12 Wysocki, G., Kosterev, A.A., and Tittel, F.K. (2006). Influence of molecular relaxation dynamics on quartz-enhanced photoacoustic detection of CO_2 at $\lambda = 2\,\mu m$. *Appl. Phys. B* 85 (2–3): 301–306.

13 Christensen, J. (1990). The Brüel & Kjaer photoacoustic transducer system and its physical properties. In: *Brüel & Kjaer Technical Review*, vol. 1-1990, no. 1, 1–39. Brüel & Kjaer.

14 Pfund, A.H. (1939). Atmospheric contamination. *Science* 90 (2336): 326–327.

15 Lehrer, E. and Luft, K.F. (1938). Verfahren zur Bestimmung von Bestandteilen in Stoffgemischen mittels Strahlenabsorption (Method for the determination of compounds in mixtures of substances by absorption of radiation). DRP 730478, 09 March 1938 (in German).

16 Luft, K.F. (1943). Über eine neue Methode der registrierenden Gasanalyse mit Hilfe der Absorption ultraroter Strahlen ohne spektrale Zerlegung (About a new method for registering gas analysis by the absorption of ultra-red radiation without spectral decomposition). *Z. Tech. Phys.* 24: 97–104. (in German).

17 Wiegleb, G. (2016). IR-Absorptionsfotometer (IR-absorption photometer). In: *Gasmesstechnik in Theorie und Praxis*, 363–477. Wiesbaden: Springer Vieweg (in German).

18 Luft, K.F. (1947). Anwendung des ultraroten Spektrums in der chemischen Industrie (Application of the ultrared spectrum in the chemical industry). *Angew. Chem. B* 19 (1): 2–12. (in German).

19 Luft, K.F., Kesseler, G., and Zörner, K.H. (1967). Nichtdispersive Ultrarot-Gasanalyse mit dem UNOR (Non-dispersive ultrared gas analysis with the UNOR). *Chem. Ing. Tech.* 39 (16): 937–945. (in German).

20 ABB (2017). Advance Optima Integriertes Analysensystem (Advance Optima integrated analysis system) (in German). http://new.abb.com/products/measurement-products/de/analytische-messtechnik/kontinuierliche-gasanalysatoren/advance-optima-und-easyline-serie/ao2000 (accessed 11 December 2017).

21 Huber, J., Ambs, A., Rademacher, S., and Wöllenstein, J. (2014). A selective, miniaturized, low-cost detection element for a photoacoustic CO_2 sensor for room climate monitoring. *Procedia Eng.* 87: 1168–1171.

22 Huber, J., Ambs, A., and Wöllenstein, J. (2015). Miniaturized photoacoustic carbon dioxide sensor with integrated temperature compensation for room climate monitoring. *Procedia Eng.* 120: 283–288.

23 Huber, J., Weber, C., Eberhardt, A., and Wöllenstein, J. (2016). Photoacoustic CO_2-sensor for automotive applications. *Procedia Eng.* 168: 3–6.

24 LumaSense (2017). Photoacoustic gas monitor – INNOVA 1412i – detection limits for various gases. https://www.lumasenseinc.com/EN/products/gas-sensing/innova-gas-monitoring/photoacoustic-spectroscopy-pas/field-monitor-1412i/photoacoustic-gas-monitor-innova-1412i.html (accessed 12 December 2017).

25 Kerr, E.L. and Atwood, J.G. (1968). The laser illuminated absorptivity spectrophone: a method for measurements of weak absorptivity in gases at laser wavelengths. *Appl. Opt.* 7 (5): 915–921.

26 Kreuzer, L.B. (1971). Ultralow gas concentration infrared absorption spectroscopy. *J. Appl. Phys.* 42 (7): 2934–2943.

27 Tittel, F.K., Curl, R.F., Dong, L., and Lewicki, R. (2011). Infrared semiconductor laser based trace gas sensor technologies: recent advances and applications. In: *Proceedings of the SPIE 8073 Optical Sensors 2011 and Photonic Crystal Fibers V*, vol. 8073 (ed. F. Baldini), 807310-1–807310-10.
28 Kosterev, A.A., Bakhirkin, Y.A., Curl, R.F., and Tittel, F.K. (2002). Quartz-enhanced photoacoustic spectroscopy. *Opt. Lett.* 27 (21): 1902–1904.
29 Kauppinen, J., Wilcken, K., Kauppinen, I., and Koskinen, V. (2004). High sensitivity in gas analysis with photoacoustic detection. *Microchem. J.* 76 (1-2): 151–159.
30 Dewey, C.F., Kamm, R.D., and Hackett, C.E. (1973). Acoustic amplifier for detection of atmospheric pollutants. *Appl. Phys. Lett.* 23 (11): 633–635.
31 Harren, F.J.M., Cotti, G., Oomens, J., and te Lintel Hekkert, S. (2006). Photoacoustic spectroscopy in trace gas monitoring. In: *Encyclopedia of Analytical Chemistry* (ed. R.A. Meyers), 2203–2226. Chichester: Wiley.
32 Tittel, F.K. (2014). Mid-IR semiconductor lasers enable sensors for trace-gas-sensing applications. *Photonics Spectra* https://www.photonics.com/Article.aspx?AID=56289 (accessed 13 November 2018).
33 Patimisco, P., Scamarcio, G., Tittel, F.K., and Spagnolo, V. (2014). Quartz-enhanced photoacoustic spectroscopy: a review. *Sensors* 14 (4): 6165–6206.
34 Zheng, H., Dong, L., Sampaolo, A. et al. (2016). Single-tube on-beam quartz-enhanced photoacoustic spectroscopy. *Opt. Lett.* 41 (5): 978–981.
35 Wu, H., Sampaolo, A., Dong, L. et al. (2015). Quartz enhanced photoacoustic H_2S gas sensor based on a fiber-amplifier source and a custom tuning fork with large prong spacing. *Appl. Phys. Lett.* 107 (11): 5. https://doi.org/10.1063/1.4930995.
36 Uotila, J. (2007). Comparison of infrared sources for a differential photoacoustic gas detection system. *Infrared Phys.Technol.* 51 (2): 122–130.
37 Wilcken, K. and Kauppinen, J. (2003). Optimization of a microphone for photoacoustic spectroscopy. *Appl. Spectrosc.* 57 (9): 1087–1092.
38 Kuusela, T. and Kauppinen, J. (2007). Photoacoustic gas analysis using interferometric cantilever microphone. *Appl. Spectrosc. Rev.* 42 (5): 443–474.
39 Uotila, J. (2009). Use of the optical cantilever microphone in photoacoustic spectroscopy. Thesis. Department of Physics and Astronomy, Turku University, Turku.
40 Hirschmann, C.B., Uotila, J., Ojala, S. et al. (2010). Fourier transform infrared photoacoustic multicomponent gas spectroscopy with optical cantilever detection. *Appl. Spectrosc.* 64 (3): 293–297.
41 Weidmann, D., Kosterev, A.A., Tittel, F.K. et al. (2004). Application of a widely electrically tunable diode laser to chemical gas sensing with quartz-enhanced photoacoustic spectroscopy. *Opt. Lett.* 29 (16): 1837–1839.
42 Kosterev, A.A., Dong, L., Thomazy, D. et al. (2010). QEPAS for chemical analysis of multi-component gas mixtures. *Appl. Phys. B* 101 (3): 649–659.

43 Lewicki, R., Jahjah, M., Ma, Y. et al. (2013). Mid-infrared semiconductor laser based trace gas sensor technologies for environmental monitoring and industrial process control. In: *Proceedings of the SPIE 8631 Quantum Sensing and Nanophotonic Devices*, vol. 8631 (ed. M. Razeghi), 86320W-1–86320W-10.

44 Borri, S., Patimisco, P., Galli, I. et al. (2014). Intracavity quartz-enhanced photoacoustic sensor. *Appl. Phys. Lett.* 104 (9): 4. http://dx.doi.org/10.1063/1.4867268.

45 Koskinen, V., Fonsen, J., Roth, K., and Kauppinen, J. (2007). Cantilever enhanced photoacoustic detection of carbon dioxide using a tunable diode laser source. *Appl. Phys. B* 86 (3): 451–454.

9

Acoustic CO_2 Sensors

Gerald Gerlach

Technische Universität Dresden, Faculty of Electrical and Computer Engineering, Institute of Solid-State Electronics, 01062 Dresden, Germany

Acoustics is the science that deals with the generation, propagation, and reception of mechanical waves in gases, liquids, and solids. Each of the steps in an acoustic process shown in Figure 9.1 is influenced by the involved gases and liquids. This allows to build up gas or liquid sensors by detecting this influence on the generation, the propagation, or the reception of acoustic waves.

The frequency spectrum in acoustics can be divided into three sections: audio (20 Hz to 20 kHz), ultrasonic (US) (>20 kHz), and infrasonic (<20 Hz). The propagation of mechanical waves is always connected to the occurrence of vibrations, i.e. mechanical oscillations of particles (e.g. atoms in a lattice or molecules in a gas) that occur around an equilibrium point.

Acoustic sensors for the measurement of the concentration of chemical or biological species in gases or liquids utilize different effects:

- Change of the resonance frequency of the vibration of a resonator due to the mass of the adsorbed or absorbed chemical species. Technically important are resonators made from quartz crystals, which are covered with a particular adsorption or absorption layer for these species. Such devices are called **quartz crystal microbalance (QCM)**. Since they mostly utilize thickness shear mode (TSM) oscillations, they are sometimes also named **TSM sensors**. Such QCM sensors will be explained in detail in Section 9.2.
- Change of the propagation velocity of waves at the surface of solids (SAWs – surface acoustic waves). This change can be caused by specially adsorbing or absorbing layers on top of that surface where the wave is travelling. Such **SAW sensors** will be discussed in Section 9.3.
- Change of the sound velocity in gases and liquids depending on their composition. For that, **ultrasonic sensors** are used, which usually measure the 'time of flight', which means the travel time of an ultrasound pulse, between a sending and a receiving US transducer (Section 9.4).

Since acoustic sensors, in particular QCM and SAW sensors, are operated in resonance, the basic principles of **resonant sensors** are introduced in Section 9.1.

Carbon Dioxide Sensing: Fundamentals, Principles, and Applications,
First Edition. Edited by Gerald Gerlach, Ulrich Guth, and Wolfram Oelßner.

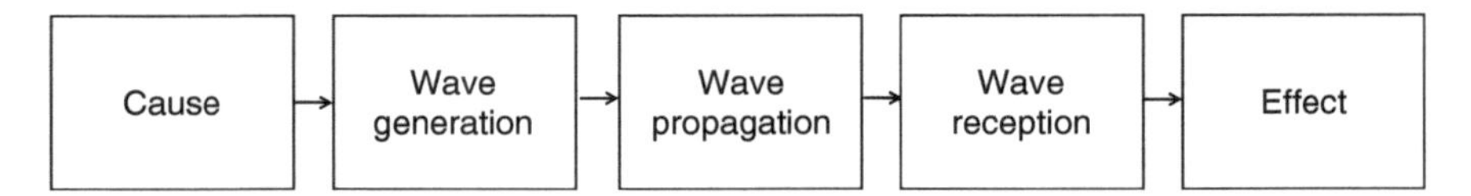

Figure 9.1 Steps in an acoustic process. Source: After [1].

9.1 Basic Principles of Resonant Sensors

Resonant sensors that are operated in its mechanical resonance frequency are a subject of particular practical interest because their output – the resonance frequency – is a quasi-digital signal. This is a great advantage compared with conventional analogue sensors by several reasons [2, 3]:

- Signal processing does not need an analogue-to-digital converter (ADC).
- Time and, hence, frequency are the physical quantities that can be measured with lowest uncertainty of measurement [4]. Digital or quasi-digital signals are much less prone to noise and interference [5].

9.1.1 General Set-Up

Figure 9.2 shows a block diagram of a typical resonant sensor. The core of the resonant sensor is a mechanical mass–spring–damper structure vibrating at its resonant frequency. The vibrations are excited by a suitable harmonic driving force, e.g. a piezoelectric, electrostatic, or magnetoelectric force. The resonance frequency is measured by some form of detection technique. This can be by the same type as the excitation (e.g. piezoelectric or electrostatic) or by a totally different detection principle (e.g. piezoresistive). Feeding back the detected signal to the excitation source ensures that the resonator is maintained vibrating continuously at its resonant frequency. Usually, this feedback is provided either by a simple amplifier or a phase-locked loop [6].

In general, the resonator comprises a mass m; an elastic spring with spring constant k and compliance $n = 1/k$, respectively; and damping influences (energy losses) by the resonator itself or by the interaction of the resonator with the surroundings, e.g. the air (Figure 9.3). The latter can be described in the simplest

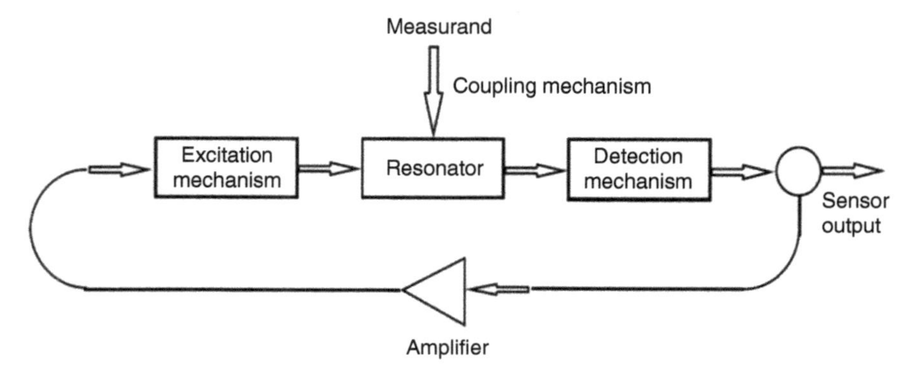

Figure 9.2 General set-up of a resonance sensor. Source: From [3].

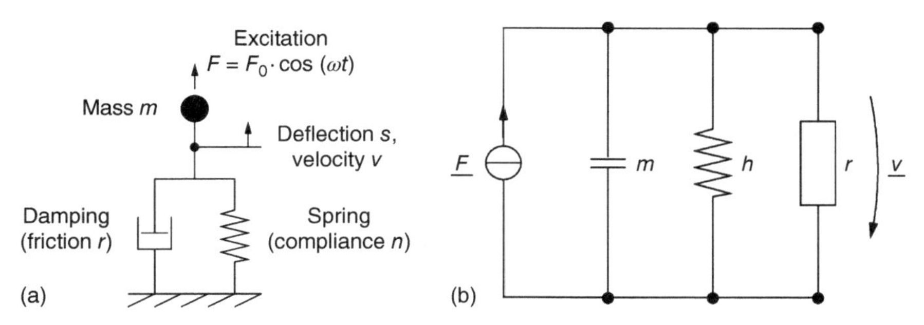

Figure 9.3 Simplified model of a mass–spring–damper resonator of resonance sensors. (a) Mechanical structure. (b) Equivalent mechanical network.

way by a friction r that is proportional to the excitation force F ($r = F/v$). The resonance effect is similar to the vibration of molecules due to excitation by light in spectroscopy (Section 3.1.2, Eqs. (3.1)–(3.4)).

The mass m is deflected sinusoidally due to the harmonic excitation force F. For reasons of better understanding, in the following, the deviation of the resonance behaviour of the resonator will be done in the frequency domain. The excitation force $\underline{F}$ will both deflect the mass m and deform the spring n as well as the damper r. All three elements m, n, and r are subjected to the same velocity $\underline{v}$. The ratio between velocity $\underline{v}$ and force $\underline{F}$ yields the mechanical impedance $\underline{Z}$:

$$\frac{\underline{v}}{\underline{F}} = \underline{Z} = \frac{1}{\frac{1}{j\omega n} + j\omega m + r} = \frac{j\omega n}{1 - \omega^2 mn + j\omega nr} = \frac{j\omega n}{1 - \left(\frac{\omega}{\omega_0}\right)^2 + j\frac{\omega}{\omega_0}\frac{1}{Q}} \tag{9.1}$$

with the resonance frequency

$$\omega_0 = \frac{1}{\sqrt{mn}} \tag{9.2}$$

and the quality factor Q (resonance exaggeration) at resonance frequency

$$Q = \frac{1}{\omega_0 nr} \tag{9.3}$$

The modulus $Z = |\underline{Z}|$ of the impedance yields

$$Z = \frac{\omega n}{\sqrt{\left[1 - \left(\frac{\omega}{\omega_0}\right)^2\right]^2 + \left(\frac{1}{Q}\cdot\frac{\omega}{\omega_0}\right)^2}} = \begin{cases} \omega n & \text{for } \omega \ll \omega_0 \\ Q\omega_0 n = \frac{Q}{\omega_0 m} & \text{for } \omega = \omega_0 \\ \frac{1}{\omega m} & \text{for } \omega \gg \omega_0 \end{cases} \tag{9.4}$$

Figure 9.4 shows the typical resonance peak at $\omega = \omega_0$. Assuming that an initial mass m_0 is increased by a mass increase Δm, the resonance frequency $\omega_0(m_0 + \Delta m)$ becomes for $\Delta m \ll m_0$:

$$\omega_0(m) = \omega_0(m_0 + \Delta m) = \frac{1}{\sqrt{(m_0 + \Delta m)n}} = \frac{1}{\sqrt{m_0 n}} \cdot \frac{1}{\sqrt{1 + \frac{\Delta m}{m_0}}}$$

$$\approx \omega_0(m_0)\left(1 - \frac{\Delta m}{2m_0}\right) \tag{9.5}$$

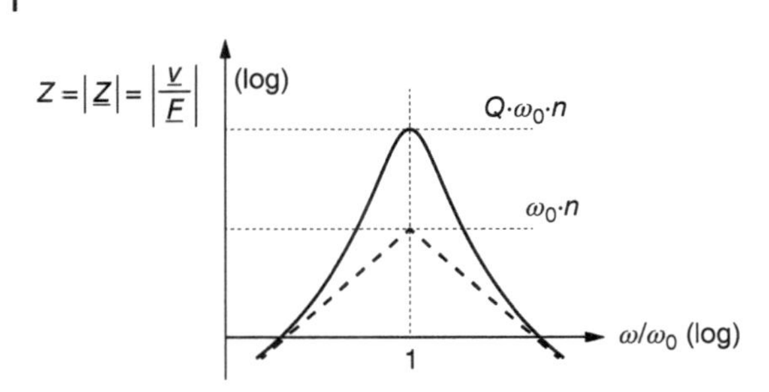

Figure 9.4 Resonance peak of a mass–spring–damper resonator at resonance frequency ω_0.

and

$$\frac{\Delta\omega_0(\Delta m)}{\omega_0(m_0)} = \frac{\omega_0(m_0 + \Delta m) - \omega_0(m_0)}{\omega_0(m_0)} = -\frac{1}{2} \cdot \frac{\Delta m}{m_0} \tag{9.6}$$

Equation (9.6) leads to several conclusions with respect to resonant sensors:

- Frequency change is strictly linearly connected to mass change.
- Increase of mass yields in a decrease of the resonance frequency.
- The higher the base resonance frequency ω_0 is, the larger is the mass-induced frequency change $\Delta\omega$. By nature, the resonance frequency is inversely proportional to the size of the resonator. Therefore, the resonator should be small, i.e. miniaturized.

9.1.2 Piezoelectric Resonators

9.1.2.1 Circuit Model

Often, piezoelectric transducers are used as electromechanical resonators. Piezoelectricity means that an electric charge accumulates in certain solid materials in response to applied mechanical stress (direct effect) or vice versa (indirect or reverse effect). It allows to provoke a mechanical deformation by applying a voltage to the piezoelectric material. Vibrations can easily be generated by using sinusoidal electrical excitations.

According to [7] the behaviour of a piezoelectric element is characterized by its dielectric properties (capacitance C_0), by its elastic properties (spring with mechanical compliance n), and by the piezoelectric interactions between voltage U and force F and vice versa between current I and displacement velocity v, respectively (Figure 9.5a). These latter interactions describe the reversible piezoelectric effect. The force F_T carries the index T because the force F to be measured from outside is smaller than the force F_T generated by the piezoelectric transduction (to deform the spring n_K elastically). The similar applies to the current I_T where a part of the input current I has to load the capacitance C_0. The transducer constant Y depends on geometry parameters (thickness l; cross-sectional area A) and on piezoelectric material constants.

Figure 9.5a also includes the effective mass m of the resonator and losses r due to internal loss mechanisms within the piezoelectric element and due to damping effects as result of lossy interactions of the resonator with the environment (e.g. air). When a load $\underline{y}_L$, e.g. an adsorbed mass, is applied to the piezoelectric element, it can be described by an additional complex mechanical impedance $\underline{Z}_L$.

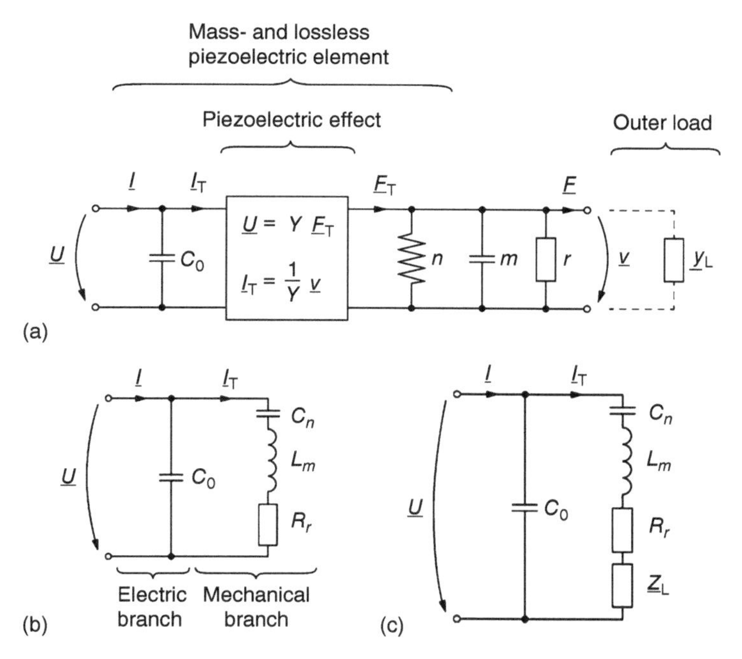

Figure 9.5 Electromechanical model of a piezoelectric resonator. (a) With electrical and mechanical components connected by the piezoelectric coupling. (b, c) With mechanical elements transformed to the electrical side (Butterworth–Van Dyke model [8, 9]), (b) without and (c) with a load $\underline{Z}_L$ applied to the resonator.

For reasons of simplicity, the mechanical elements in the network model of Figure 9.5 can be transformed to the electrical side. Since the piezoelectric transducer is a so-called gyrator [7, 10], connecting flow quantities with potential quantities ($F_T \leftrightarrow U$; $I_T \leftrightarrow v$), the mechanical parallel circuit of n, m, r and the load $\underline{y}_L$ becomes a series circuit of a capacitance C_n, an inductance L_m, a resistance R_r, and a complex impedance $\underline{Z}_L$ representing the load:

$$C_n = \frac{n}{Y^2}, \quad L_m = mY^2, \quad R_r = rY^2, \quad \underline{Z}_L = \underline{y}_L Y^2 \tag{9.7}$$

If the load impedance $\underline{Z}_L$ is determined by a mass increase Δm, as it is used for quartz crystal microbalances, then the model of Figure 9.5c can be reduced to the simpler one of Figure 9.5b with

$$L_m = (m_0 + \Delta m)Y^2 \tag{9.8}$$

where $m = m_0 + \Delta m$.

The complex input impedance $\underline{Z}$ between the input terminals of the circuit from Figure 9.5 yields

$$\underline{Z} = \frac{1}{j\omega C_0 + \frac{1}{R_r + j\omega L_m + \frac{1}{j\omega C_n}}} = \frac{1}{j\omega(C_0 + C_n)} \cdot \frac{1 + j\omega C_n R_r - \omega^2 C_n L_m}{1 - \omega^2 \frac{C_0 C_n L_m}{(C_0 + C_n)} + j\omega R_r \frac{C_0 C_n}{C_0 + C_n}} \tag{9.9}$$

with the absolute value

$$Z = |\underline{Z}| = \frac{1}{\omega(C_0 + C_n)} \cdot \sqrt{\frac{(1 - \omega^2 C_n L_m)^2 + (\omega C_n R_r)^2}{\left(1 - \omega^2 \frac{C_0 C_n L_m}{(C_0+C_n)}\right)^2 + \left(\omega R_r \frac{C_0 C_n}{C_0+C_n}\right)^2}} \tag{9.10}$$

9.1.2.2 Resonance Frequencies

Due to the numerator and the denominator terms in the root of Eq. (9.10), the impedance Z possesses a zero at an angular frequency of

$$\omega_Z = \frac{1}{\sqrt{C_n L_m}} \sim \frac{1}{\sqrt{m}} = \frac{1}{\sqrt{m_0 + \Delta m}} \tag{9.11}$$

and a pole at

$$\omega_P = \sqrt{\frac{C_0 + C_n}{C_0 C_n L_m}} = \sqrt{\frac{C_0 + C_n}{C_0}} \frac{1}{\sqrt{C_n L_m}} \sim \frac{1}{\sqrt{m}} = \frac{1}{\sqrt{m_0 + \Delta m}} \tag{9.12}$$

with $\omega_Z < \omega_P$. Here it is assumed that losses (imaginary terms in Eq. (9.8)) are neglectable ($R_r \to 0$), i.e. that the quality factor at resonance is much larger than unity.

The resulting frequency response curve of impedance Z is depicted in Figure 9.6, showing the general $1/\omega$ proportionality. ω_Z and ω_P are the angular resonance frequencies for the local minimum (zero) and the local maximum (pole, anti-resonance), respectively. Both are proportional to $1/\sqrt{m}$ that corresponds to Eq. (9.5). For this reason, Eq. (9.6) can also be applied to piezoelectric resonators.

All previous considerations have been made by the assumption that frequencies are so small that the wavelength is much larger than the size of the resonator, i.e. that the resonator can be described by lumped elements and wave phenomena can be neglected. Figure 9.7 shows the typical frequency dependencies when larger frequencies are taken into account too. Besides the fundamental wave also higher harmonics and side modes have to be considered [11, 12].

In the case of simple mass loading – as it is used for measuring adsorption or absorption of CO_2 – the simple equivalent circuit from Figure 9.5b can be used where L_m represents the mass $m = m_0 + \Delta m$. For other applications with more complex load cases, corresponding models for the impedance of the load according to Figure 9.5c are available [11–14].

Considering Figure 9.7c, the characteristic frequency of the fundamental (first) mode can be measured practically in different ways:

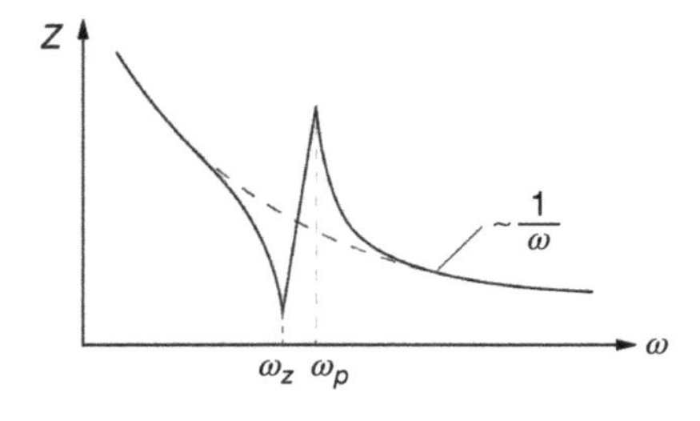

Figure 9.6 General frequency dependence of the absolute value of impedance Z for piezoelectric resonators (fundamental wave).

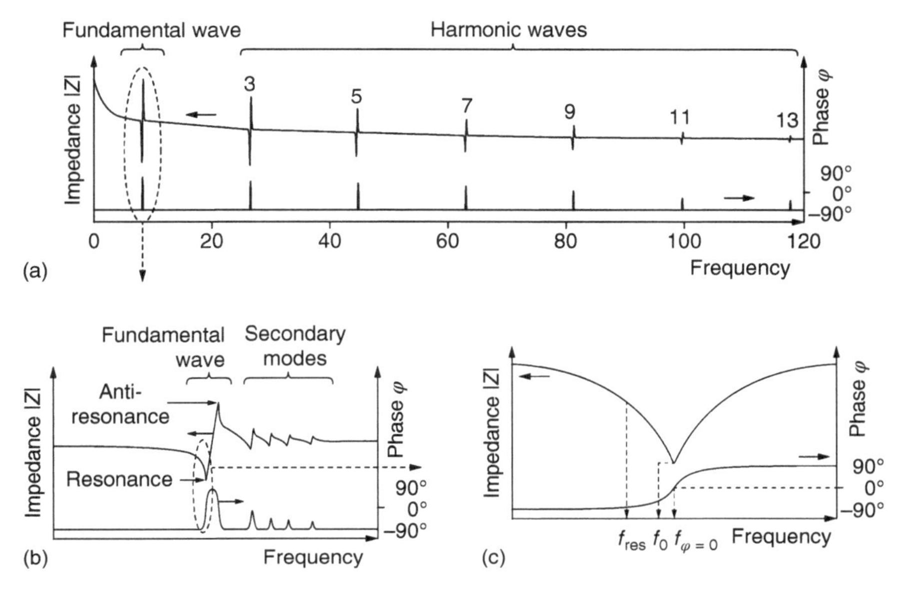

Figure 9.7 Frequency dependence of absolute value Z and phase ϕ of impedance $\underline{Z}$ of a thickness shear mode resonator. (a) Fundamental and harmonic waves up to the 13th overtone. (b) Detailed curse in the region of the fundamental mode with side modes. (c) More detailed curse around the impedance minimum at f_0. Source: From [11] reproduced by courtesy.

- *Eigenfrequency or natural frequency* f_{Eigen}: eigenwert as solution of the eigenvalue problem by considering damping (quality factor Q) besides mass m and compliance n:

$$f_{\text{Eigen}} = f_0\sqrt{1 - 1/Q^2}. \tag{9.13}$$

- The eigenfrequency is that frequency at which a resonating system continues to oscillate after switching off the excitation.
- Frequency where the impedance Z has a minimum:

$$f_0 = f(Z \to \min) \tag{9.14}$$

- Frequency $f_{\phi=0}$ where the phase angle ϕ of the impedance Z has a zero. This is the method by which usually PLL amplifiers are operated [15].

9.1.2.3 Types of Piezoelectric Resonant Sensors

Piezoelectric resonant sensors utilize the propagation of elastic waves in a bulk or at a surface of a bulk. Therefore, they can be divided into three categories [2, 16]:

- *Bulk resonators or bulk acoustic wave (BAW) sensors*: Here, mostly quartz-based sensors are used, in particular for quartz microbalance sensors [12]. BAW-type gas sensors operating by binding molecules to the device surface were first introduced by King [17].
- *SAW sensors*: In such sensors the SAW travels along the surface of a solid medium. The wave propagation is strongly affected by changes occurring at the

surface, e.g. in or on thin films deposited on top of the surface. By that, SAW sensors represent a sensitive probe for characterizing thin films and, hence, for investigating interaction mechanisms between chemical species and the device surface, e.g. the coating materials.
- *Lamb wave (LW) sensors*: Lamb waves are elastic waves that propagate in plates of finite thickness [18]. This allows the operation both in the low-MHz frequency range and while immersed in a liquid. However – compared with SAW sensors – they could not reach wide acceptance.

9.2 Quartz Crystal Microbalance Sensors

Resonators made from quartz crystals own the highest technical importance. If they are covered with a suitable adsorption or absorption layer, corresponding chemical or biological species can be measured. Since such devices increase their mass due to bound species, they are called QCM. They are mostly utilizing TSM oscillations that is why they are sometimes also named TSM sensors.

9.2.1 Quartz as Resonator Material

Quartz is a single crystal. Its chemical formula is SiO_2. The crystal structure is a continuous framework of SiO_4 silicon–oxygen tetrahedra.

Quartz resonator sensors are well known from their use as time bases for frequency control. Compared to other principles and materials – for example, LC circuits, tuning forks, and resonators from other single-crystal materials – quartz crystals show the best properties of all with respect to both short-term and long-term stabilities. These superior properties originate from its single-crystal structure and the high purity:

- The material properties of single-crystal quartz are extremely stable with time and temperature and, hence, very reproducible from one specimen to another. This long-term stability leads to the need for less frequent recalibration.
- The acoustic loss or internal friction of quartz is very low, resulting in an extremely high Q-factor (c. 10^7 at 1 MHz). Even mounted devices have still typical Q-factors in the range from a few 10^4 to several 10^5.
- Quartz resonator is very stable with respect to temperature variations. Selected crystal orientations – in particular the so-called AT-cut (Figure 9.8) – show an almost temperature-independent behaviour [19]. In the AT-cut the quartz blank has the form of a thin plate containing the crystal's X-axis. It is cut in such a way that it is inclined by $35°15'$ from the optical Z-axis.
- Quartz resonators show extremely stable mechanical properties. The frequency drift of commercially available devices amounts to only a few 10^{-6} a^{-1}.

The working temperature range of quartz resonator devices is often limited to c. 100 °C or a few 100 °C. If higher operation temperatures are required, then langasite ($La_3Ga_5SiO_{14}$) or related compounds are promising alternatives. In [20] it was shown that langasite works stable up to temperatures of 900 °C.

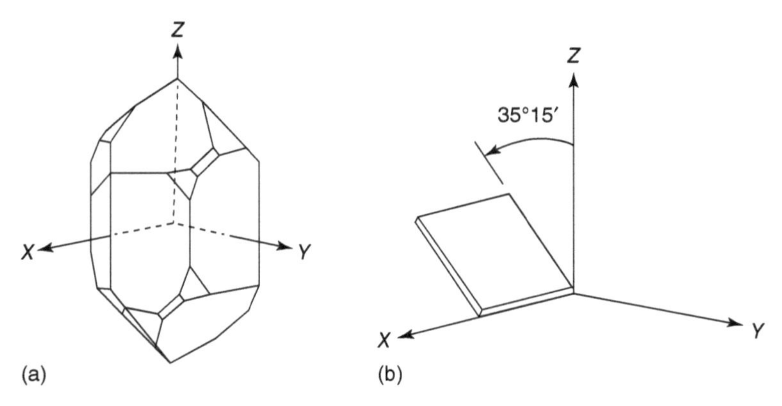

Figure 9.8 Quartz crystal. (a) Assignment of axes to the quartz crystal. (b) AT-cut. Source: O'Sullivan and Guilbault 1999 [19]. Reproduced with permission of Elsevier.

9.2.2 Thickness Shear Mode Sensors

9.2.2.1 Vibration Modes

Piezoelectric materials exhibit different kinds of vibration modes depending on the piezoelectric coupling coefficients (Figure 9.9) [21]:

- Flexural vibrations, i.e. vibrations in bending direction;
- Length (lateral extension, LE) vibrations that expand or contract a thin plate;
- Area expansion vibrations over a surface of a thin plate or disc;
- Thickness shear (TS) vibrations over a surface of a thin plate, where the electric field is perpendicular to the direction of polarization;
- Thickness expansion (TE) vibrations in the thickness direction of a thin plate; and
- Surface acoustic waves (Rayleigh waves; see Section 9.3).

Equation (9.6) has shown that the larger the base resonance frequency ω_0 is the higher the mass-induced frequency change $\Delta\omega$ of a resonator sensor is. As can be seen from Figure 9.9, TSM and thickness expansion mode (TEM) sensors show the highest resonance frequencies. This originates from the small thickness of the resonating plate and the accordingly small compliance n (see Eq. (9.2)).

The corresponding piezoelectric constants describing the mechanical strain component produced by an applied electric field strength (index i) are the so-called strain constants or d_{ij} coefficients. Indices $j = 1$ and $j = 4$ denote normal strain and shear strain, respectively. In quartz the relevant constants are $d_{14} = 0.727 \times 10^{-12}$ C/N for the TSM and $d_{11} = 2.31 \times 10^{-12}$ C/N for the TEM [22].

Table 9.1 compares parameters of resonant sensors in different modes and of different materials with the parameters of QCMs using the TSM. As can be seen, TSM-driven QCMs have the highest values for the quality factor Q, both in air and in water, and for the mass detection resolution R:

$$R = \frac{L}{S} \tag{9.15}$$

which relates to mass sensitivity S:

$$S = \frac{\Delta f / f_0}{\Delta m / A} \tag{9.16}$$

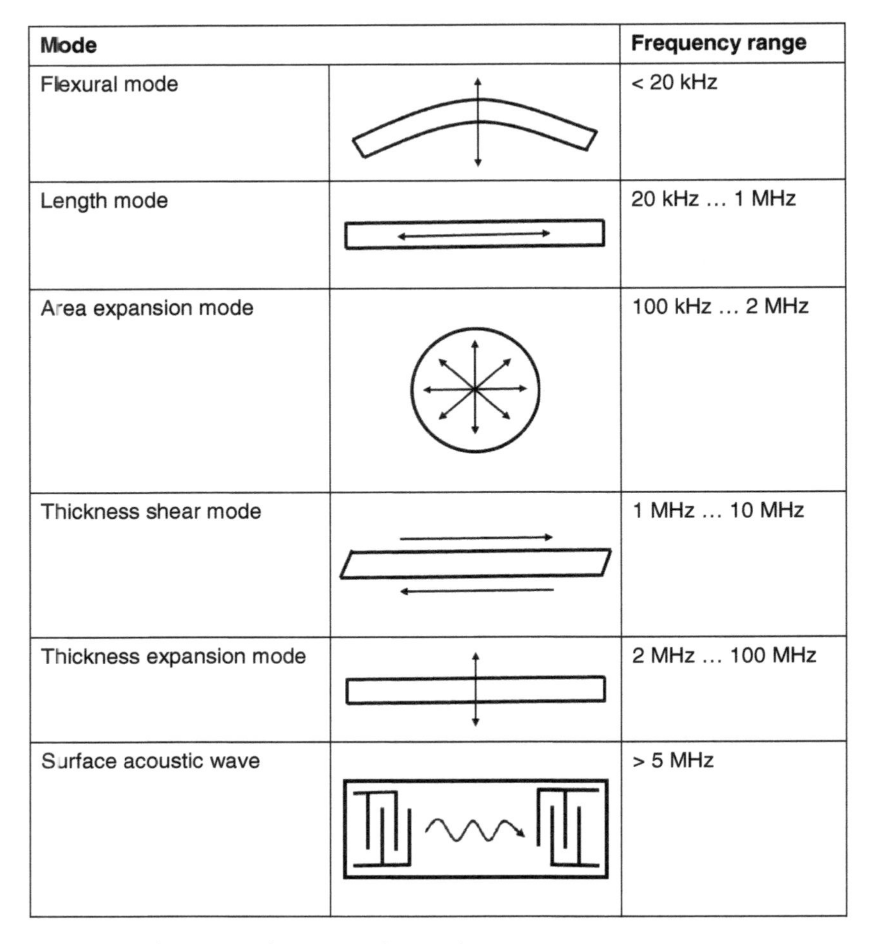

Figure 9.9 Vibration modes in piezoelectric plates.

Table 9.1 Comparison of parameters of piezoelectric microelectromechanical resonant sensors for chemical and biological detection with quartz crystal microbalances [23].

Mode		TE	TS	LE	Flexure	QCM (TSM)
f_0 (MHz)		650–8 000	230–3 100	0.065–160	0.03–3.3	5–60
S ($cm^2\ g^{-1}$)		726–9 000	365–1 150	100–700	165–3 000	14–54
Q	Air	330–2 000	285–408	300–1 400	65–2 500	20 000–50 000
	Water	12–189	150–230	64–189	25	2 000–10 000
R ($ng\ cm^{-2}$)	Air	1		0.05–1	~0.4	0.5–5
	Water	0.37–10	0.3–2.3	1.78 to ~4	~5	0.7–5

TE, thickness expansion; TS, thickness shear; and LE, lateral extension.

L describes the minimum detectable resonance frequency change:

$$L = \frac{\Delta f_{\min}}{f_0} \tag{9.17}$$

$\Delta m / A$ is the mass Δm added to the sensor per unit area A.

9.2.2.2 Sensitivity

The thickness shear vibration of a quartz plate is a stationary transversal wave (Figure 9.10). Parallel layers in the plate displace against each other without being deformed. The shear amplitude depends only on the y-coordinate. The maximum amplitude occurs at the surfaces of the plate. For the fundamental wave the plate thickness d equals half the wavelength. Therefore, the resonance frequency becomes

$$f = \frac{v_{\mathrm{tr}}}{2d} \tag{9.18}$$

where v_{tr} is the propagation velocity of the elastic transversal wave in direction of the plate thickness ($v_{\mathrm{tr}} = 3340\,\mathrm{kHz\,mm}$ for the AT-cut in quartz [24]).

The resonance frequency is determined only by the mass of the surface-near layers of the plate, not by its elasticity. Therefore – for a sufficiently thin layer on top of the plate surface with an additional mass Δm – the relative change of the resonance frequency yields

$$\frac{\Delta f}{f_0} = \frac{\Delta \omega}{\omega_0} = -\frac{\Delta d}{d} = -\frac{\Delta m}{m_0} = -\frac{\Delta m}{\rho A d} \tag{9.19}$$

where Δd, d, Δm, and m_0 are the thicknesses and the masses of the thin layer and of the plate and ρ is the density of the quartz ($2.65\,\mathrm{g\,cm^{-3}}$). This equation is called Sauerbrey equation.

For the mass sensitivity S regarding Eq. (9.16), one gets

$$S = \frac{\Delta f / f_0}{\Delta m / A} = -\frac{1}{\rho \cdot d} \tag{9.20}$$

In [24] Eq. (9.19) is written in a different form:

$$\Delta f = -C_{\mathrm{f}} \frac{f_0^2}{A} \Delta m \tag{9.21}$$

with C_{f} a mass sensitivity constant. For an AT-cut quartz crystal vibrating in the TSM, C_{f} amounts to $2.26 \times 10^{-10}\,\mathrm{m^2\,(g\,s)^{-1}}$.

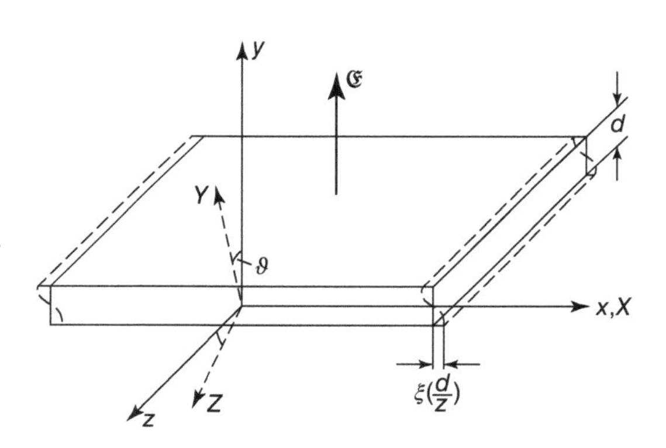

Figure 9.10 Ideal thickness shear vibration of a quartz plate; x, y, z coordinate system, X polar axis, Z optical axis of the quartz crystal, d plate thickness, ξ shear amplitude; AT-cut, $\vartheta = 35°$, electrical field in y-direction. Source: Sauerbrey 1959 [24]. Reproduced with permission of Springer Nature.

Table 9.2 Properties of QCM quartz crystals.

Manufacturer	Type	Basic frequency (MHz)	Resolution	Remarks
Inficon [25]	RQCM	5 or 9	n.a.	Ø 25.4 mm
Laptech Precision [25]	n.a.	1–30	n.a.	Ø 3.8 mm–25.4 mm
QSense [26]	E4	1–70	$-0.5\ ng\ cm^{-2}$	n.a.
ICM Co. [27]	35360	5 or 10	n.a.	13.7 mm

9.2.2.3 Commercial QCM Sensors

Table 9.2 summarizes properties of commercially available QCM quartz crystals (see also [19]). High-frequency stability of quartz crystals can be achieved by an optimized choice of electrodes, electrical contacts, coating of the sensor material, and stable electronics for sensor signal evaluation.

9.2.3 CO_2-Sensitive Coating

If QCM sensors are coated with a suitable, chemically active coating, then it is possible to detect polar and non-polar organic and inorganic compounds in the gas phase [28, 29].

In [28] the interaction of gases like carbon monoxide (CO), carbon dioxide (CO_2), nitrogen dioxide (NO_2), trichloroethylene ($Cl_2C{=}CHCl$), perchloroethylene ($Cl_2C{=}CCl_2$), and octane (C_8H_{18}) with coatings used as absorbents in gas chromatography (triethanolamine, triethylenepentamine, siloxane) and with macrocyclic supramolecular compounds (hexalactam, α- and β-cyclodextrin, crown ether, polyphenolate) have been studied. The amines showed a good sensitivity for CO_2 and NO_2, whereas almost no sensitivity against CO was observed. The supramolecular compounds showed a neglectable sensitivity with respect to CO_2 and CO. The detection limit amounted to c. 1 ng for measuring times in the order of minutes. The time dependences of the frequency shift showed two different characteristic time constants, indicating different mechanisms due to physical sorption and chemical reactions.

The suitability of amides as sensitive coatings for QCMs was confirmed in [30]. In general, the sensitivity increases with the number of amino groups in the coating molecule. For instance, triethylenetetramine shows a larger frequency decrease in response to CO_2 than diethylenediamine. The most sensitive coating was *N,N,N′N′*-tetrakis (-2-hydroxyethyl) ethylenediamine (THEED), a tertiary amine (Figure 9.11).

THEED-coated quartz crystals have been used by the same authors to measure the CO_2 content in wines [31]. Carbon dioxide in wine, arising from fermentation processes, is responsible for the freshness of taste in wine and enhances its fragrance.

9.2.4 Other Applications of CO_2-Sensitive QCMs

In the same manner as different coatings on QCM sensors show different sensitivity to gases, e.g. CO_2, QCM devices can also be used to measure the solubility of

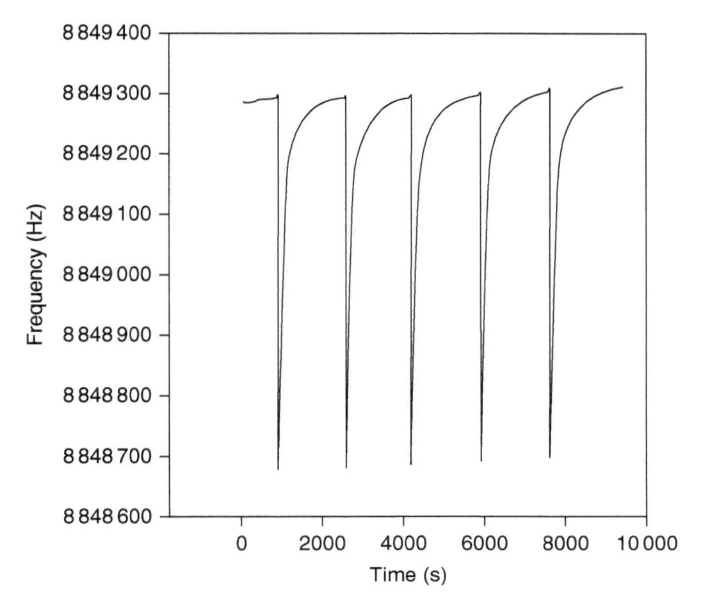

Figure 9.11 Frequency change of a QCM coated with THEED during a series of five 5 ml CO_2 injections. AT-cut quartz crystal, frequency 9 MHz. Source: Gomes et al. 1995 [30]. Reproduced with permission of Elsevier.

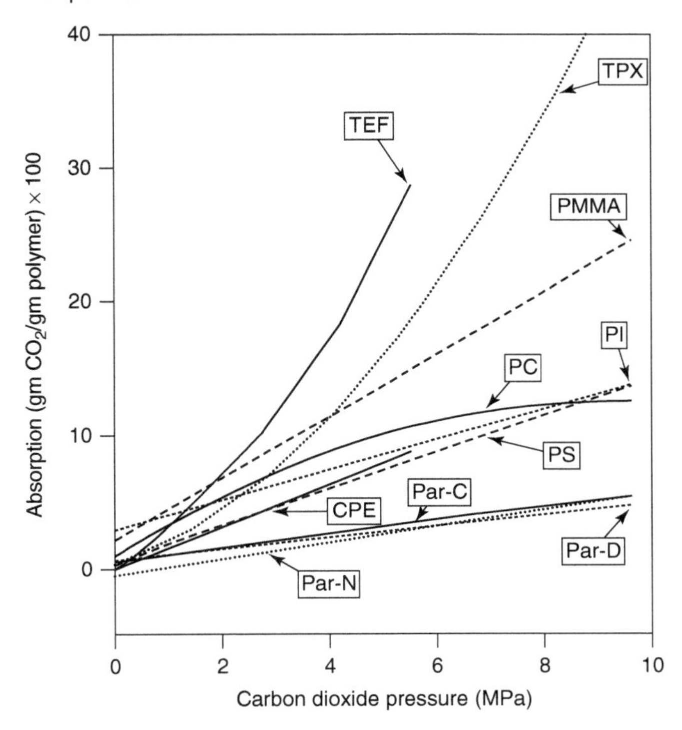

Figure 9.12 Absorption of CO_2 in polymers: TEF, Teflon AF-1600; TPX, poly 4-methyl-1-pentene; PMMA, polymethylmethacrylate; PC, polycarbonate; PI Ultem 1000 polyimide; CPE, chlorinated polyethylene; PS, polystyrene; Par-D, poly(2,3-dichloroxylylene); Par-C, poly(2-chloroxylylene); and Par-N, poly(xylylene). Source: Webb and Teja 1999 [33]). Reproduced with permission of Elsevier.

CO_2 in these coatings. In [32, 33] the solubility of CO_2 in polymers was studied (Figure 9.12). The measurement showed that the QCM technique is reversible and can be used for real-time monitoring of solution processes. It might be able to extend it to measure the partitioning of a component present in CO_2 into the polymer.

Another application of QCM sensors is the measurement of molecular flow and condensation of CO_2 in vacuum [34]. For that purpose, the quartz resonator has to be cooled. Because of the fast measurement and the spatial resolution due to the small size of the sensor, measurements can be performed with both very high temporal and spatial resolution.

The QCM technique can also be applied to study solid–fluid interfaces, for instance, in supercritical high-pressure CO_2 systems where the adsorption of CO_2 on metal surfaces is of interest [35].

9.3 Surface Acoustic Wave Sensors

SAW sensors, such as quartz crystal microbalance sensors, are devices that are based on high-frequency mechanical oscillators. However, the energy of SAWs is constrained to the surface. SAWs usually comprise different waves, but the most important are the Rayleigh waves. They show both longitudinal and transverse motions that decrease exponentially in amplitude with increasing distance from the surface.

Similar to QCM sensors, SAW sensors are sensitive to interactions between chemical vapours and the coating material of the SAW. In this sense, they also can be used as chemical sensors [2].

9.3.1 Operation Principle

9.3.1.1 Excitation of Surface Acoustic Waves

SAWs can be excited using interdigital transducers (IDTs; Figure 9.13) [36, 37]. IDTs consist of a conducting interdigitated finger pattern deposited on a piezoelectric substrate. Because of the piezoelectric effect, an applied AC voltage causes a spatially periodic field between the IDT fingers. The corresponding

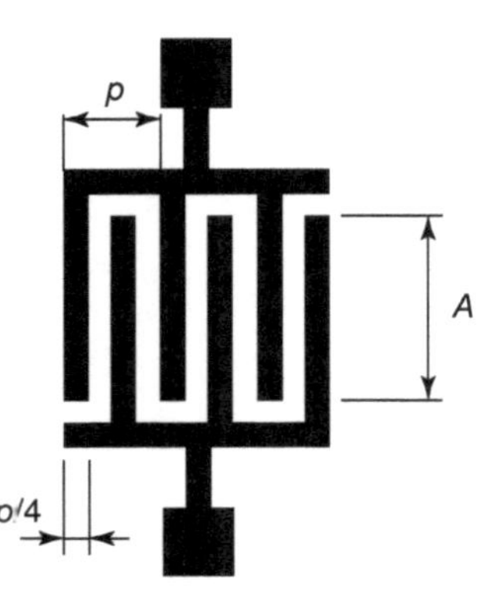

Figure 9.13 Uniform interdigital transducer with period p and constant electrode overlap A. Here, the electrode width is equal to the space between electrodes. Source: Vellekoop 1998 [36]. Reproduced with permission of Elsevier.

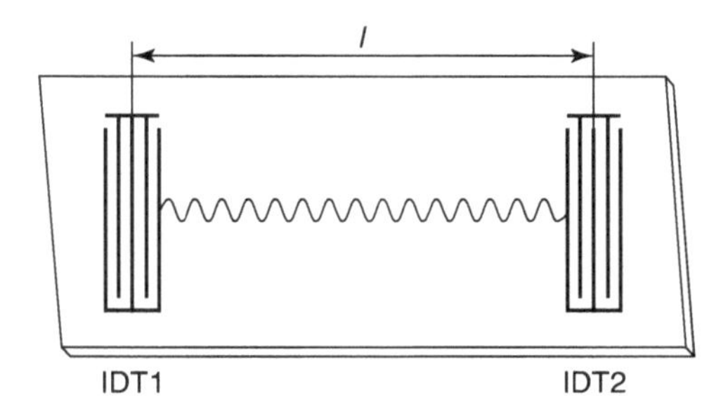

Figure 9.14 Surface acoustic wave generated by an interdigital transducer. Source: Benes et al. 1995 [16]. Reproduced with permission of Elsevier.

periodically alternating mechanical strain pattern causes acoustic waves propagating away from the IDT, in directions perpendicular to the electrodes (Figure 9.14). For an applied sinusoidal voltage with frequency f and period $1/T$, respectively, vibrations interfere constructively if the distance p between adjacent electrodes is equal to the wavelength λ [36, 37]:

$$p = \lambda \tag{9.22}$$

This resonance frequency depends on the phase velocity v of the acoustic wave and the period p:

$$f_0 = v/p \tag{9.23}$$

9.3.1.2 Operation Modes of SAW Sensors

An SAW sensor consists of two IDTs, namely, a transmitting IDT and a receiving IDT (Figure 9.14). Acoustic waves that are excited at the transmitting IDT1 will reach the receiving IDT2 after a certain time, the so-called time delay. This time delay depends on the distance l between the two IDTs. Such a delay line is placed in the feedback loop of an electronic amplifier (Figure 9.15). If the oscillation

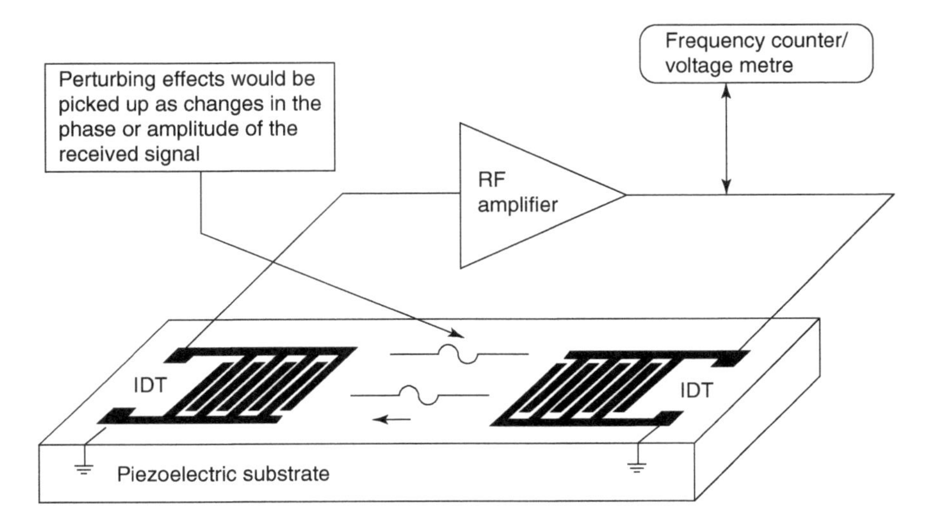

Figure 9.15 Delay line-based SAW sensor as part of an oscillator circuit. Source: Gardner et al. 2002 [37]. Reproduced with permission of CCC Republication.

conditions are fulfilled and correspond to Eq. (9.23), then oscillations occur. The oscillation frequency is related to the phase of the wave [36]. It can be determined using a frequency counter or a voltage metre.

If the surface of a SAW sensor is covered by a chemically active thin film that is capable of selectively adsorbing molecules from the environment (gas, liquid), then the mass change Δm of the film affects the propagation velocity v (see Eq. (9.23)) and the damping of the SAW [38, 39]:

$$\frac{\Delta v}{v} = \frac{\Delta f}{f_0} = \frac{1}{f_0} \cdot \left(\frac{\delta f}{\delta m} \Delta m \right) \tag{9.24}$$

Also, electric fields or physical properties of liquids, such as density and viscosity, can affect the propagation of acoustic waves.

In principle, several types of surface waves can be used in SAW sensors (Figure 9.16) [36–38, 40]:

- Rayleigh waves are two-dimensional waves that propagate along the surface of a solid bulk (Figure 9.16a). They consist of a longitudinal and a shear component (Figure 9.17). The penetration depth is in the order of only one wavelength [36].
- Acoustic plate mode (APM) waves (Figure 9.16b), also called shear horizontal waves, deform the entire volume between the upper and the lower surface. The particle displacement is predominantly parallel to the substrate surface and normal of the direction of propagation. The smaller the thickness of the substrate (which acts as an acoustic waveguide) is, the more acoustic energy is located near the surfaces and the higher is the sensitivity. Such as thickness shear mode sensors (see Section 9.2.2), APM devices are mostly made of quartz.
- Lamb waves in thin membranes or plates can be considered a special case of Rayleigh waves (Figure 9.16c). If the membrane is thicker than two wavelengths, then two separate Rayleigh waves will propagate at the upper and the lower surfaces. If the membrane thickness is in the order of one wavelength or less, then one wave is confined in the membrane, comprising a symmetrical and an asymmetrical wave. Since only the asymmetric zero-order mode decreases monotonously to zero with decreasing membrane thickness, this wave does not excite compressional waves in a loading liquid if its phase velocity is lower than the compressional wave velocity in the liquid [36]. This makes Lamb wave sensors attractive for measurements in liquids [18].
- Surface transverse waves (STWs, Figure 9.16d) are horizontally polarized shear waves. Differently to APMs, they are trapped at the surface by a periodic surface perturbation, e.g. a metallic grating. Due to the surface confinement, they show a high mass sensitivity.
- Love waves (Figure 9.16e) are guided in a thin layer deposited on a substrate. The acoustic energy is concentrated in this guiding layer, leading to a high mass sensitivity of this wave type. Since the thin guiding layer is mechanically backed by the substrate, it is mechanically very robust.

The mass sensitivity for SAW sensors depends strongly on the concentration of acoustic energy near the surface of the substrate where the waves propagate.

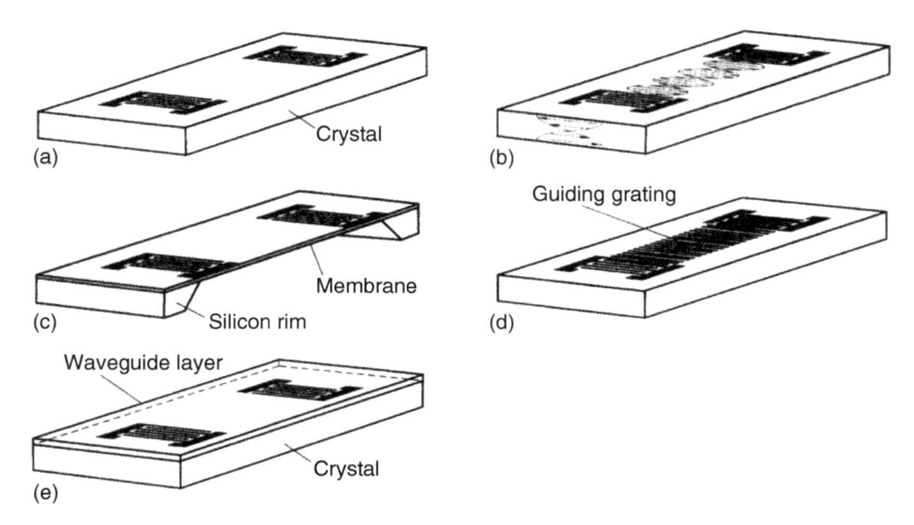

Figure 9.16 Wave propagation in SAW devices. (a) Rayleigh wave, (b) acoustic plate mode (APM) wave, (c) Lamb wave, (d) surface transverse wave (STW), and (e) Love wave. Source: Vellekoop 1998 [36]. Reproduced with permission of Elsevier.

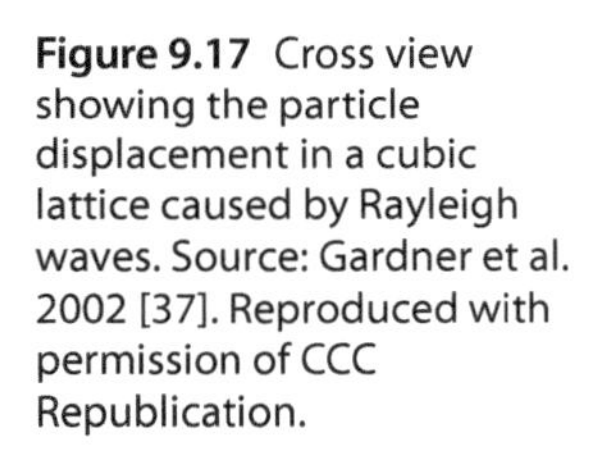

Figure 9.17 Cross view showing the particle displacement in a cubic lattice caused by Rayleigh waves. Source: Gardner et al. 2002 [37]. Reproduced with permission of CCC Republication.

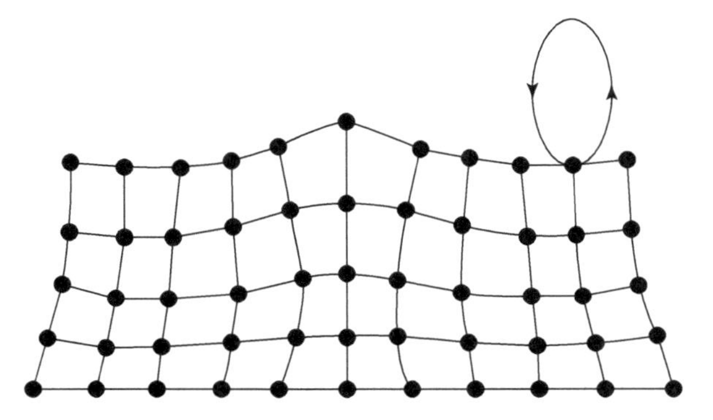

Therefore, the sensitivities of Rayleigh wave, Lamb wave, and Love wave devices are the highest. However, Rayleigh wave sensors have the advantage of lowest fabrication costs and highest mechanical robustness. For that reason, they are the most widely used SAW sensor type for gas detection. Rayleigh waves can only be used in gases but not in liquids because acoustic radiation into the liquid causes high propagation losses.

9.3.2 SAW Sensor Materials

SAW sensors use piezoelectric substrates where the SAW is excited by an IDT and the phase shift due to the quantity to be measured is detected by a second IDT. Single-crystal piezoelectrics are preferred materials due to their excellent long-term stability. Mostly, lithium niobate ($LiNbO_3$), lithium tantalate ($LiTaO_3$), or ST-cut quartz is used (Table 9.3). Other materials, such as zinc oxide (ZnO), are only of little importance.

Table 9.3 Some widespread SAW sensor materials.

Piezoelectric substrate	Cut	Propagation direction	Phase velocity v (m s^{-1})
$LiNbO_3$	Y	Z	3487.7
$LiTaO_3$	Y	Z	3229.9
Quartz	Y	X	3159.3
ZnO	Z	$X + 45°$	2639.4

Source: From [37, 41].

9.3.3 SAW Devices

SAW sensors for gas detection use the mass-loading effect described by Eq. (9.24). Here, the SAW velocity is affected by a gas-absorbing thin film covering the propagation path, i.e. the delay line, at the surface of the piezoelectric substrate.

In practice, often dual-delay-line oscillator systems are used to compensate for small variations of temperature, pressure, and other unwanted interfering effects (Figure 9.18). Here two identical acoustic delay lines are formed where the second path serves as a reference. The output signal is the difference between the two oscillator frequencies f of the gas-sensitive SAW element and f_0 of the reference SAW element, respectively. Usually, the remaining frequency uncertainty amounts to only 10–30 Hz, whereas the frequency change Δf due to the gas to be measured is in the order of several 10–100 kHz [40]. The operating frequency f_0 ranges from 10 MHz to a few gigahertz [37].

SAW sensors can be excited remotely by electromagnetic waves if the transmitting IDT is connected to an antenna (Figure 9.19) [42]. Then the transmitting ITD serves also as receiving ITD. The operating principle of such systems is shown in Figure 9.20 [43]. A high-frequency electromagnetic (EM) request signal is picked up by the antenna of the passive SAW device and conducted to the IDT that converts the received signal into a SAW. The SAW propagates towards reflectors

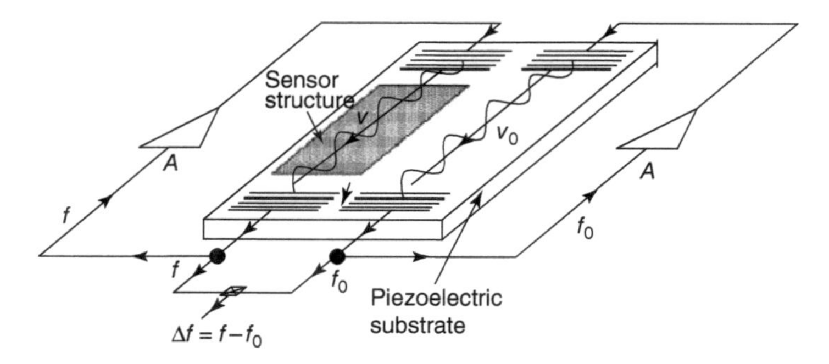

Figure 9.18 Dual-delay-line configuration of the SAW-based gas sensor. Source: Jakubik 2011 [40]. Reproduced with permission of Elsevier.

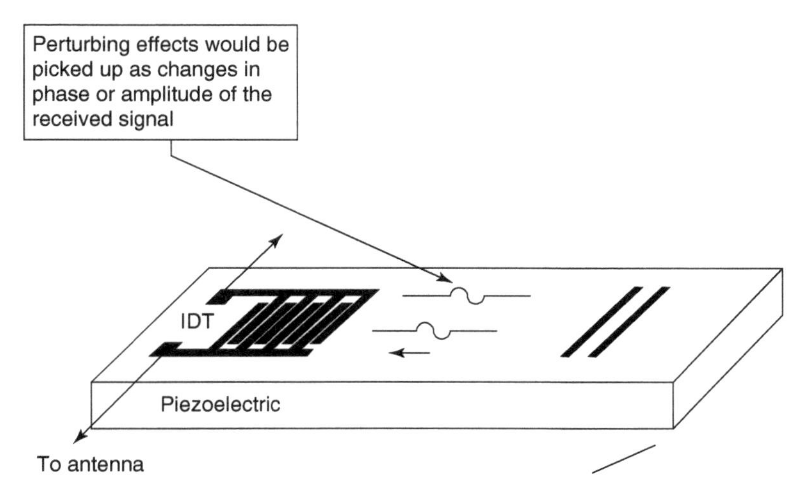

Figure 9.19 Schematic diagram of a passive, i.e. remotely operated, SAW sensor. Source: Gardner et al. 2002 [37]. Reproduced with permission of CCC Republication.

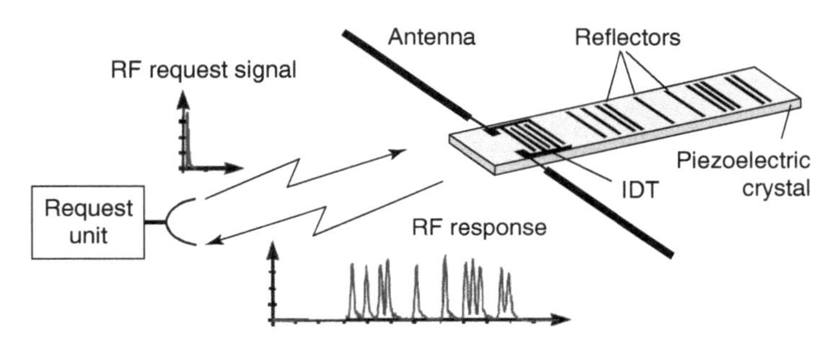

Figure 9.20 Operation principle of a passive SAW sensor where the output signal has been coded through multiple reflectors in the acoustic path. Source: From [43].

distributed in a characteristic barcode-like pattern and is partially reflected at each reflector. The acoustic wave packets returning to the IDT are reconverted into electrical signals by the IDT and sent back to the reading unit by the antenna. This response depends on the properties of the SAW delay line as well as of the number and location of reflectors and the propagation and reflection properties of the SAW. The sensor signal can now be extracted from the coded response signal. Advantages of such devices are:

- They can be read out remotely.
- They do not need a power supply.
- They can be placed in hazardous environments such as contaminated and high-voltage areas.

However, the reading range is limited to a few metres up to c. 10 m and depends on the frequency [44, 45].

Table 9.4 SAW sensors for the detection of CO_2.

CO_2-sensitive layer	Number of delay lines	Piezoelectric substrate	Frequency f_0 (MHz)	CO_2 concentration range	Remarks	References
Poly(ethylene-imine) (PEI)	1	n.a.	40	10 vol%	Degradation phenomena; strong interference from relative humidity; sensitive above 30 °C	[46]
Homo-[a] and copolyimide[b] (for CO_2); poly(*N*-vinylpyrrol-idone) (PNVP, for humidity)	2	ST-quartz	97	95 vol%	Simultaneous determination of the CO_2 concentration and humidity; uncertainty of CO_2 measurement c. 30% due to humidity influence	[47]
N,*N*-bis-(*p*-Methoxybenzylidene)-α, α'-bi-*p*-toluidine (BMBT); poly(ethylene-imine) (PEI); Versamid 900	1	ST-quartz	250	240 ppm	Higher response to CO_2 at the third harmonic than at the fundamental frequency (factor of 3) PEI with highest sensitivity to CO_2 (1 kHz for 240 ppm CO_2), but also to humidity (5.7 kHz per % r.H.)	[48]
Teflon AF 2400[c]	1	41°-YX LiNbO3	440	450 ppm	Wireless SAW sensor with reflective delay line; good linearity and repeatability	[49]
Polyethylene-imine- (PEI-) starch-functionalized single-walled carbon nanotubes (SWNTs)	1	128°-XY $LiNbO_3$	286	40 vol%	Sensitivity due to conductivity change in the coating; cross sensitivity to humidity; resolution 5 000 ppm	[50]
Polyallylamine (PAA), filled with amino-carbon nanotubes; Polyethylene-imine (PEI), filled with amino-carbon nanotubes	1	Quartz	77	5 000 ppm	Improved sensitivity due to amino-carbon nanotubes; long-term drift effects after CO_2 changes	[51]
Teflon AF 2400[c]	2	$LiNbO_3$	300	2 000 ppm	Resolution 500 ppm	[52]

a) Homopolyimide prepared from 6FDA and 2,3,5,6-tetramethyl-1,4-phenylenediamine (4MPD).
b) Copolyimide prepared from FDA and 3,3′,4,4′-benzophenonetetracarboxylic dianhydride (BTDA) and 2,4,6-trimethyl-1,3-phenylenediamine (3MPD).
c) 13 mol% tetrafluoroethylene and 87 mol% 2,2-bis(trifluoromethyl)-4,5-difluoro-1,3-dioxide.

9.3.4 CO_2-Sensitive SAW Sensors

In recent years a large variety of SAW gas sensors to measure CO_2 concentration have been introduced (Table 9.4), but so far no commercial solutions have entered the market.

In general, all presented sensors use CO_2-sensitive coatings on the delay line of the SAW (see Figure 9.15). In principle, the same coatings as introduced for quartz crystal microbalance sensors in Section 9.2.3 can be used. Often polymers are the material of choice for the sensitive coating because they can easily be deposited and patterned on the sensor elements, e.g. by spin coating and photolithography. This enables the usage of advantageous and cost-efficient MEMS fabrication techniques.

Favourable solutions are realized as dual-delay-line set-up (as shown in Figure 9.18) to compensate unwanted interfering effects like temperature and humidity and, hence, to enhance accuracy and resolution. Unfortunately, relative humidity has a substantial influence on measurement causing a significant cross sensitivity (see Table 9.4).

9.4 Ultrasonic CO_2 Sensors

9.4.1 Operation Principle

9.4.1.1 Velocity of Sound in Gases

The velocity of sound in gases (and liquids) depends on their composition. Depending on its particular molar weight for each gas, a certain velocity of sound exists (Table 9.5). This property can be used to measure the time of flight or travel time of an ultrasound pulse in a gas mixture with known gas components and to calculate from that the concentration of the constituents.

Sound waves propagate in gases by alternating compression or expansion of the gas causing a periodic pressure wave. Since the duration of a single compression is too short to release energy from the wave to the surrounding gas, wave propagation is an adiabatic (isentropic) process. The isentropic velocity v of sound in a

Table 9.5 Velocity of sound in selected gases.

Gas	Velocity of sound v (m s^{-1})	Molar mass M (g mol^{-1})
Nitrogen	349	28.0135
Oxygen	326	31.9988
Argon	319	39.9480
Carbon dioxide	267	44.0095
Methane	446	16.0425
Hydrogen	1270	2.0159

Source: Zuckerwar 2002 [53]. Reproduced with permission of Elsevier.

pure ideal gas of molecular mass M is given by the relationship

$$v = \sqrt{\frac{\gamma RT}{M}} \quad (9.25)$$

where γ is the adiabatic constant, characteristic of the specific gas, $R = 8.314\,\mathrm{J\,mol^{-1}\,K^{-1}}$ the universal gas constant, and T the absolute temperature. For dry air, the adiabatic constant and the average molar mass amount to $\gamma_{\mathrm{Air}} = 1.4$ and $M_{\mathrm{Air}} = 28.95\,\mathrm{g\,mol^{-1}}$, respectively. This leads to

$$v_{\mathrm{Air}} = 20.05 \cdot \sqrt{\frac{T}{\mathrm{K}}} \cdot \frac{\mathrm{m}}{\mathrm{s}} \quad (9.26)$$

The adiabatic constant γ is the ratio between the specific heat capacities c_{p} and c_{V} at constant pressure and constant volume, respectively, and depends on the effective degree of freedom f_{eff} of the participating atoms in the molecule due to translation, rotation, and vibration:

$$\gamma = \frac{c_{\mathrm{p}}}{c_{\mathrm{V}}} = \frac{f_{\mathrm{eff}} + 2}{f_{\mathrm{eff}}} \quad (9.27)$$

Regarding to the classical model of Eq. (9.25), the frequency of a sound wave should not affect the velocity of sound at all. However, sound waves at higher frequencies have higher values of v [54]. This effect is caused by intermolecular interactions of polyatomic gases during sound transmission and is called sound dispersion. The velocity of sound shows a sigmoidal course corresponding to the dispersion equation of Kneser [55] (Figure 9.21):

$$\gamma(f) = 1 + R\frac{c_{\mathrm{V}} + (2\pi f)^2\tau^2 c_{\mathrm{Va}}}{c_{\mathrm{V}}^2 + (2\pi f)^2\tau^2 c_{\mathrm{Va}}^2} \quad (9.28)$$

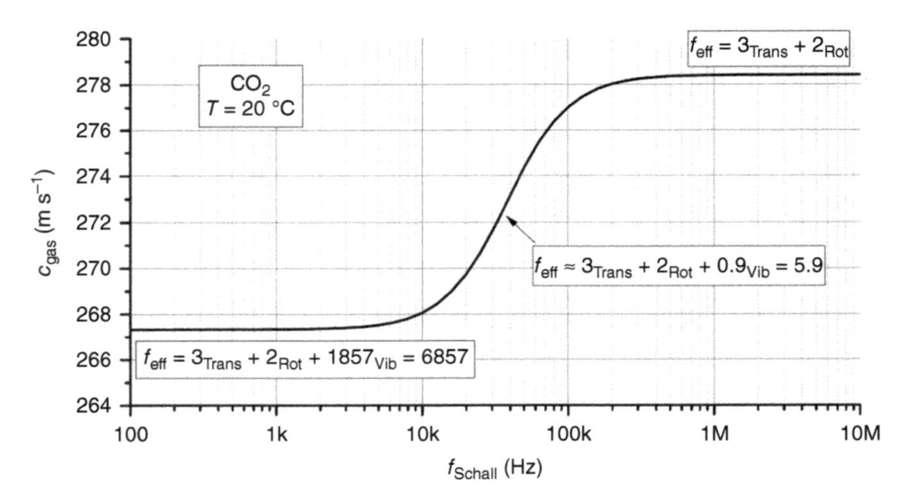

Figure 9.21 Velocity of sound v in CO_2 versus frequency f of the sound wave at 20 °C. f_{eff}, effective degree of freedom of participating atoms in the molecule; Trans, translational; Rot, rotational; and Vib, vibrational. Source: From [56].

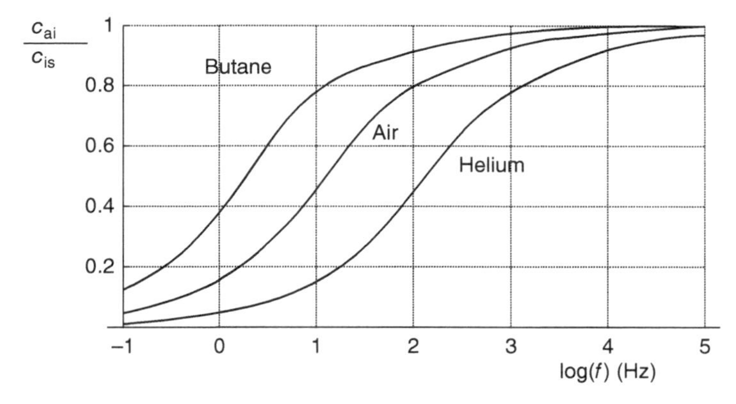

Figure 9.22 Ratio $v_{anisentropic}/v_{isentropic}$ of the velocity of sound for anisentropic and isentropic conditions. Source: Zipser et al. 2000 [58]. Reproduced with permission of Elsevier.

here, f is the frequency of the acoustic wave, τ is a characteristic relaxation time ($\tau = 5.7\ \mu s$ for CO_2 [57]), and c_{Va} is the specific heat capacity when only translational and rotational degrees of freedom are effective at high frequencies.

The velocity of sound in a gas mixture can be calculated from Eq. (9.25) by summing over the substance amount fractions x_i of all constituents i:

$$v_{mix} = \frac{R \cdot T \cdot \sum_i (x_i \cdot \gamma_i)}{\sum_i (x_i \cdot M_i)} \tag{9.29}$$

As can be seen from this equation, the concentration of a gas constituent, i.e. the molar fraction of the amount of substances x_i, can clearly be determined if it is a binary gas mixture with $x_1 + x_2 = 1$. A sufficiently high sensitivity can be achieved when the molar mass differences of both components are large enough.

The concentration of gas mixtures with more than two gases can be determined if multiple measurements take place at different temperatures and the temperature dependence of the adiabatic constant $\gamma(T)$ is exploited. However, the achievable accuracy will be worse than for binary gas mixtures. This is the reason why in most applications ultrasonic CO_2 sensors are only applied to binary gas mixtures.

Another way to measure ternary gas mixtures is by using the propagation of sound in narrow channels at low frequencies. Under such circumstances the acoustic wave has sufficient time and contact area to interact with the channel walls, leading to an essential loss of energy and a rise of entropy. The corresponding velocity of sound $v_{anisentropic}$ for the anisentropic case is smaller than the isentropic velocity of sound (Figure 9.22):

$$v_{anisentropic} < v_{isentropic}$$

This makes it possible to analyse ternary gas mixtures by simultaneously using two measuring chambers with narrow (for anisentropic conditions) and wide channel (for isentropic conditions) walls [58, 59].

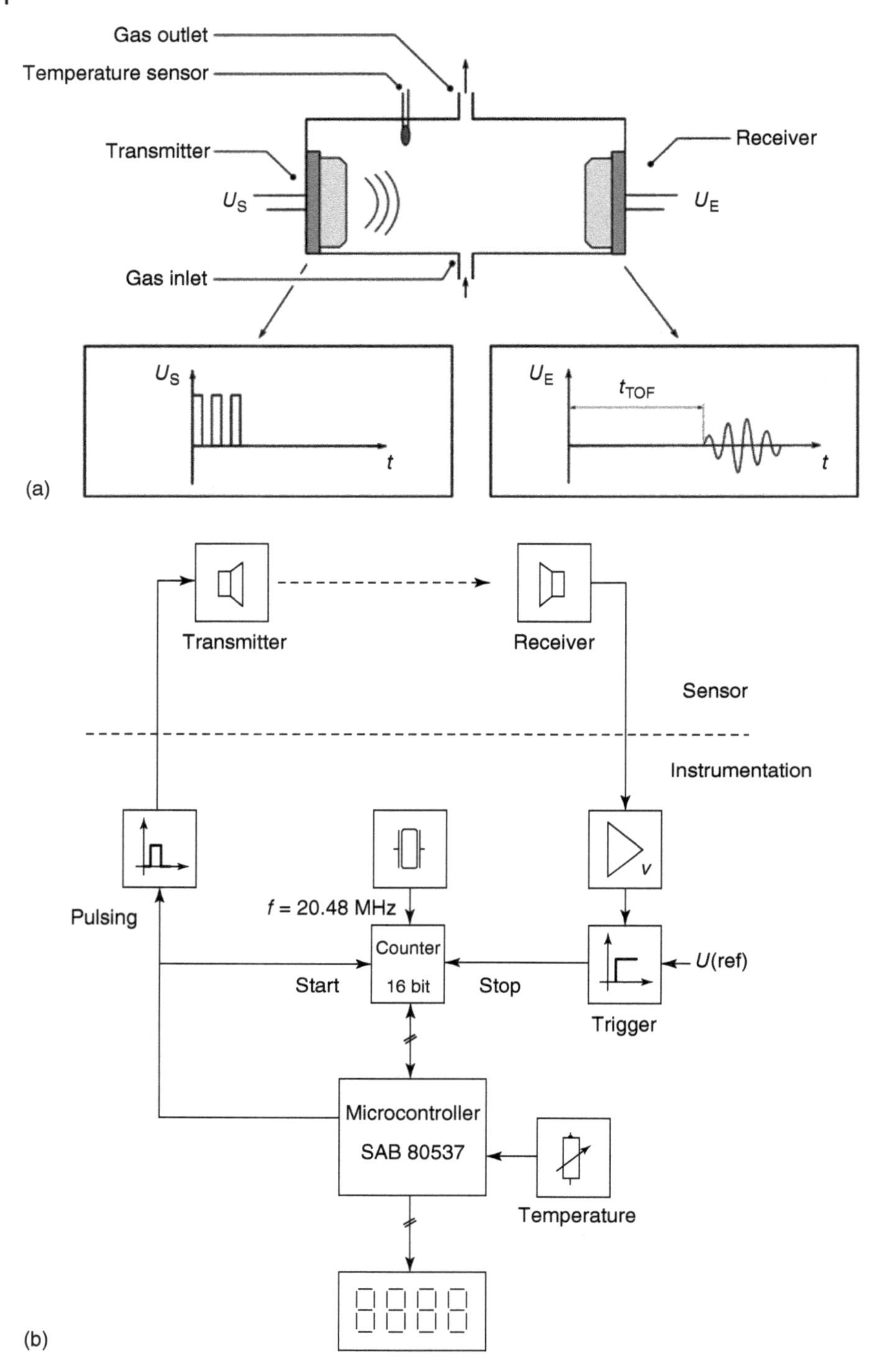

Figure 9.23 (a) Operational principle. Source: From [56]. (b) Instrumentation of an ultrasonic gas sensor to determine the concentration in binary gas mixtures. Source: Joos et al. 1993 [60]. Reproduced with permission of Elsevier.

9.4.1.2 Basic Set-Up

Figure 9.23 shows the basic set-up of an ultrasonic gas sensor utilizing the different velocities of speed of different gases to determine the concentration of a particular constituent in binary gas mixtures. An ultrasound transmitter emits an ultrasound wave – usually as a pulse or a series of pulses – that propagates through the gas to be analysed to the receiver. The time of flight *TOF* – i.e. the time the sound wave takes to reach the receiver – is proportional to the velocity of sound *v*. Mostly, piezoelectric elements are used as transmitter and receiver [61].

Figure 9.24 underlines that this type of ultrasonic gas sensors is well suited to the measurement of the concentration of gas constituents in binary gas mixtures. The curve shows a very linear and precise relationship between the time of flight *TOF* and CO_2 gas concentration in air within the concentration range 0–1% relevant for monitoring of CO_2 in air.

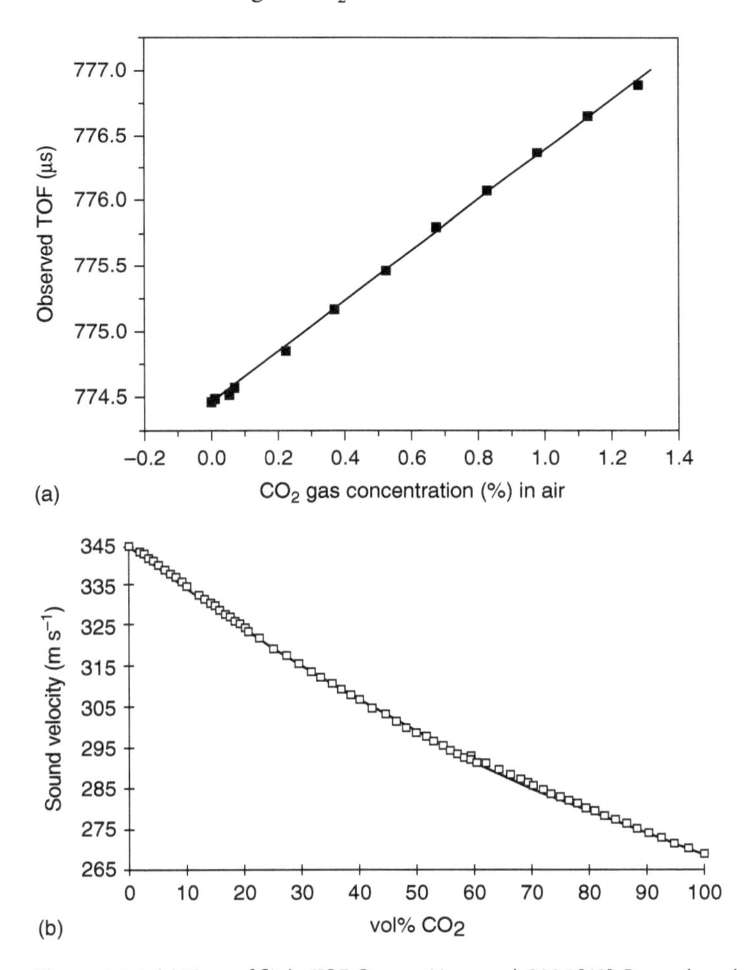

Figure 9.24 (a) Time of flight TOF. Source: Vyas et al. 2006 [62]. Reproduced with permission of Elsevier. (b) Velocity of sound at 22 °C versus CO_2 gas concentrations in air. □ ■ experimental data – calculated. Source: Joos et al. 1993 [60].Reproduced with permission of Elsevier.

9.4.2 Ultrasonic Sensors for CO_2 Detection

Ultrasonic gas sensors for binary gas mixtures exhibit numerous advantages [63, 64]:

- High-precision measurement of sound velocity.
- Fast sensor response.
- Non-invasive measurement.

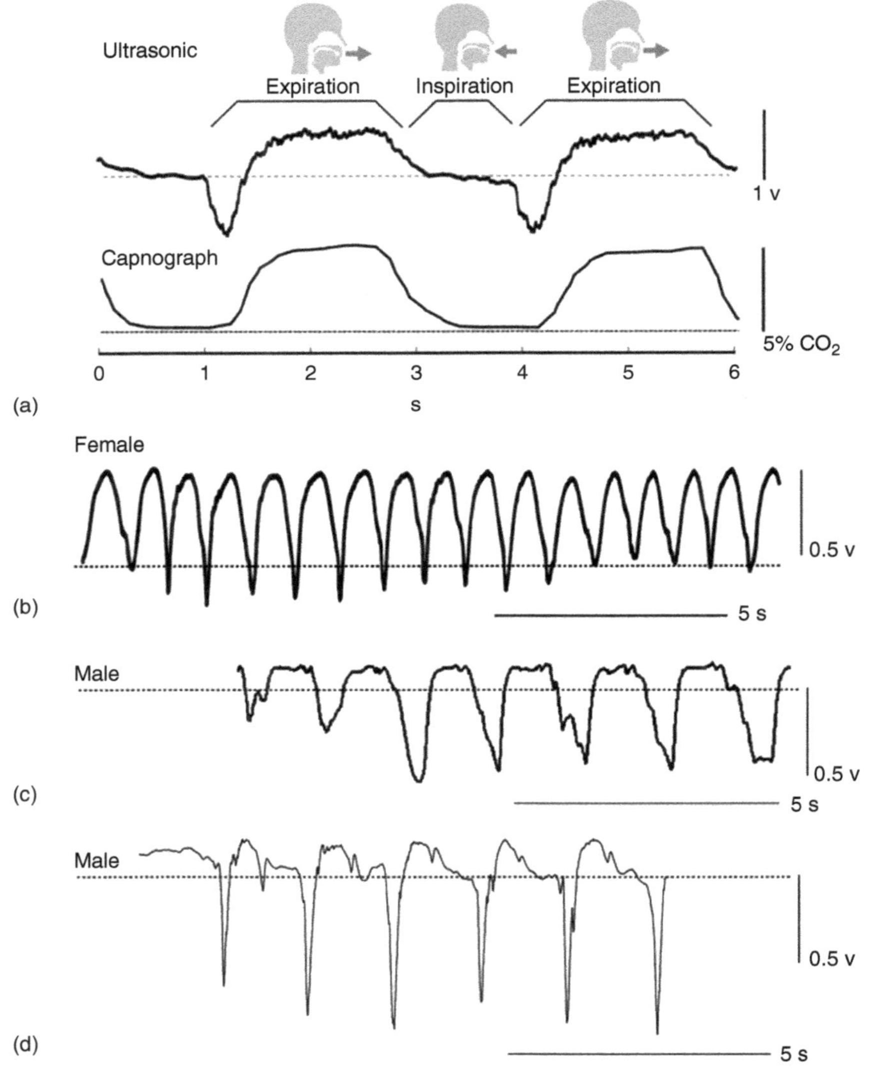

Figure 9.25 Real-time human respiration process analysis based on CO_2 output patterns using an ultrasound gas concentration sensor. (a) Comparison with NDIR-based capnography. (b–d) Characteristic courses of (b) hyperventilation, (c) asthma, and (d) bronchial asthma patients. Sampling speed 1 kHz. Source: Toda 2012 [68]. Reproduced with permission of SAGE Journals.

- Robustness and long-term stability.
- Suitability for use in hazardous atmospheres.

Limitations are given by the sometimes non-monotonic velocity behaviour (depending on concentration and temperature) and the strong influence of temperature.

Several commercial manufacturers offer measurement solutions for the determination of the velocity of sound in liquids [65, 66]. Related applications are, for instance, the determination of the content of solved CO_2 in beverages and the monitoring of liquids in pipelines or tanks. However, no ready-made CO_2 gas sensors are available yet.

Ultrasonic CO_2 sensors have been used successfully for a number of particular research tasks where other types of CO_2 sensors could not fulfil the requirements. This regards, for instance, applications with high demands for accuracy, fast response time, or small size.

Toda used an ultrasonic sensor [67] to monitor the human respiration in real time (Figure 9.25) [68]. The comparison with traditional capnography, which uses non-dispersive infrared sensors, gives an impression that the fast response time allows for a much more detailed behaviour that is impossible with non-dispersive infrared gas sensors, gas chromatography, and solid-state gas sensors. This fast response is mainly determined by the gas exchange in the measuring chamber and can be decreased by reducing the volume.

References

1 https://en.wikipedia.org/wiki/Acoustics#Fundamental_concepts_of_acoustics (retrieved 21 October 2016).

2 Hauptmann, P. (1991). Resonant sensors and applications. *Sens. Actuators, A* 26 (1–3): 371–377.

3 Tudor, M.J. and Beeby, S.P. (1997). Resonant sensors: fundamentals and state of the art. *Sens. Mater.* 9 (3): 1–15.

4 Parthier, R. (2010). *Messtechnik* (Measurement Technology), 5e ed., 210 pp. Wiesbaden: Vieweg+Teubner. (in German).

5 Huijsing, J.H. (1986). Signal conditioning on the sensor chip. *Sens. Actuators* 10 (3–4): 219–237.

6 Sell, J.K., Niedermayer, A.O., and Jakoby, B. (2010). Digital phase-locked loop circuit for driving resonant sensors. *Procedia Eng.* 5: 204–207.

7 Lenk, A., Ballas, R.G., Werthschützky, R., and Pfeifer, G. (2011). *Electromechanical Systems in Microtechnology and Mechatronics*, 472 pp. Berlin: Springer.

8 Butterworth, S. (1915). On electrically-maintained vibrations. *Proc. Phys. Soc. London* 27: 410–424.

9 Van Dyke, K.S. (1925). The electric network equivalent of a piezo-electric resonator. *Phys. Rev.* 25 (6): 895.

10 Tellegen, B.D.H. (1948). The gyrator, a new electric network element. *Philips Res. Rep.* 3 (2): 81–101.

11 Guhr, G. (2010). *Mikroakustische und elektrische Charakterisierung komplexer biologischer Fluide. (Microacoustic and Electrical Characterization of Complex Biological Fluids)*, 180 pp. München: Verlag Dr. Hut (in German).

12 Brünig, R. (2012). *Modellierung von akustischen Dickenscherschwingern im Frequenzbereich. (Modeling of Acoustic Thickness Shear-Mode Resonators in the Frequency Range)*, 151 pp. München: Verlag Dr. Hut (in German).

13 Weihnacht, M., Bruenig, R., and Schmidt, H. (2007). More accurate simulation of quartz crystal microbalance (QCM) response to viscoelastic loading. In: *2007 IEEE Ultrasonics Symposium Proceedings*, 28–31 October 2007, 377–380. New York, NY.

14 Bruenig, R., Schmidt, H., Guhr, G., and Weihnacht, M. (2010). Complex loadings on thickness shear mode resonators. In: *2010 IEEE International Ultrasonics Symposium Proceedings*, 11–14 October 2010, 1886–1889. San Diego, CA.

15 Rocha, D., Ferrari, V., and Jakoby, B. (2004). Improved electronic readout circuit for resonant acoustic sensors. In: *IEEE Sensors Conference 2004*, 24–27 October 2004, 32–35. Vienna.

16 Benes, E., Gröschl, M., Burger, W., and Schmid, M. (1995). Sensors based on piezoelectric resonators. *Sens. Actuators, A* 48 (1): 1–21.

17 King, W.H. (1964). Piezoelectric sorption detector. *Anal. Chem.* 36 (9): 1735–1739.

18 Wenzel, S.W. and White, R.M. (1988). A multisensor employing an ultrasonic lamb-wave oscillator. *IEEE T. Electron Dev.* 35 (6): 735–743.

19 O'Sullivan, C.K. and Guilbault, G.G. (1999). Commercial quartz crystal microbalances – theory and applications. *Biosens. Bioelectron.* 14 (8–9): 663–670.

20 Fritze, H. and Tuller, H.L. (2001). Langasite for high-temperature bulk acoustic wave applications. *Appl. Phys. Lett.* 78 (7): 976–977.

21 http://www.murata.com/products/timingdevice/ceralock/basic/vibration (retrieved 28 October 2016).

22 Bechmann, R. (1958). Elastic and piezoelectric constants of alpha-quartz. *Phys. Rev.* 110 (5): 1060–1061.

23 Pang, W., Zhao, H., Kim, E.S. et al. (2012). Piezoelectric microelectromechanical resonant sensors for chemical and biological detection. *Lab Chip* 12 (1): 29–44.

24 Sauerbrey, G. (1959). Verwendung von Schwingquarzen zur Wägung dünner Schichten und zur Mikrowägung (Use of quartz oscillators for the weighing of thin layers and for micro weighing). *Z. Phys.* 155 (2): 206–222.

25 http://products.inficon.com/en-us/nav-products (retrieved 16 December 2016).

26 https://www.mrl.ucsb.edu/spectroscopy-facility/instruments/q-sense-e4-quartz-crystal-microbalance-qcm-d (retrieved 16 December 2016).

27 https://www.icmfg.com/quartzmicrobalance.html#types (retrieved 16 December 2016).

28 Lucklum, R., Henning, B., Hauptmann, P. et al. (1991). Quartz microbalance sensors for gas detection. *Sens. Actuators, A* 27 (1–3): 705–710.

29 Hauptmann, P., Lucklum, R., Hartmann, J. et al. (1993). Using the quartz microbalance principle for sensing mass changes and damping properties. *Sens. Actuators, A* 37–38: 309–316.
30 Gomes, M.T., Duarte, A.C., and Oliveira, J.P. (1995). Detection of CO_2 using a quartz crystal microbalance. *Sens. Actuators, B* 26 (1–3): 191–194.
31 Gomes, M.T., Duarte, A.C., and Oliveira, J.P. (1996). The utilisation of a piezoelectric quartz crystal for measuring carbon dioxide in wine. *Anal. Chim. Acta* 327 (2): 95–100.
32 Aubert, J.H. (1998). Solubility of carbon dioxide in polymers by the quartz crystal microbalance technique. *J. Supercrit. Fluids* 11 (3): 163–172.
33 Webb, K.F. and Teja, A.S. (1999). Solubility and diffusion of carbon dioxide in polymers. *Fluid Phase Equilib.* 158–160: 1029–1034.
34 Baker, M.A. and Holland, L. (1969). Use of a cooled quartz crystal microbalance to study the molecular flow and condensation of CO_2 in vacuum. *J. Vac. Sci. Technol.* 6 (6): 951–954.
35 Wu, Y.-T., Akoto-Ampaw, P.-J., Elbaccouch, M. et al. (2004). Quartz crystal microbalance (QCM) in high-pressure carbon dioxide (CO_2): experimental aspects of QCM-theory and CO_2 adsorption. *Langmuir* 20 (9): 3665–3673.
36 Vellekoop, M.J. (1998). Acoustic wave sensors and their technology. *Ultrasonics* 36 (1–5): 7–14.
37 Gardner, J.W., Varadan, V.K., and Awadelkarim, O.O. (2002). *Microsystems, MEMS, and Smart Devices*, 522 pp. Chichester: Wiley.
38 Hoummady, M., Campitelli, A., and Wlodarski, W. (1997). Acoustic wave sensors: design, sensing mechanisms and applications. *Smart Mater. Struct.* 6 (6): 647–657.
39 Ricco, A.J., Martin, S.J., and Zipperian, T.E. (1985). Surface acoustic wave gas sensor based on film conductivity changes. *Sens. Actuators* 8 (4): 319–333.
40 Jakubik, W.P. (2011). Surface acoustic wave-based gas sensors. *Thin Solid Films* 520 (3): 986–993.
41 Khlebarov, Z.P., Stoyanova, A.I., and Topalova, D.I. (1992). Surface acoustic wave gas sensors. *Sens. Actuators, B* 8 (1): 33–40.
42 Bao, X.Q., Burkhard, W., Varadan, V.V., and Varadan, V.K. (1987). SAW temperature sensor and remote reading system. In: *IEEE Ultrasonics Symposium 2*, 14–16 October 1987, Denver, CO, 583–585.
43 Reindl, L., Scholl, G., Ostertag, T. et al. (1998). Wireless remote identification and sensing with SAW devices. In: *Proceedings of the IEEE 1998 MMT/AP International Workshop on Commercial Radio Sensor and Communication Techniques*, 83–96.
44 Hartmann, C.S. and Claiborne, L.T. (2007). Fundamental limitations on reading range of passive IC-based RFID and SAW-based RFID. In: *2007 IEEE International Conference on RFID*, 26–28 March 2007, Grapevine, TX, 41–48.
45 http://www.sawcomponents.de/en/products/saw-sensors-and-rfid/saw-temp (retrieved 03 January 2017).
46 Nieuwenhuizen, M.S. and Nederlof, A.J. (1990). A SAW gas sensor for carbon dioxide and water. Preliminary experiments. *Sens. Actuators, B* 2 (2): 97–101.

47 Hoyt, A.E., Ricco, A.J., Bartholomew, J.W., and Osbourn, G.C. (1998). SAW sensors for the room-temperature measurement of CO_2 and relative humidity. *Anal. Chem.* 70 (10): 2137–2145.
48 Korsah, K., Ma, C.L., and Dress, B. (1998). Harmonic frequency analysis of SAW resonator chemical sensors: application to the detection of carbon dioxide and humidity. *Sens. Actuators, B* 50 (2): 110–116.
49 Wang, W., Lee, K., Kim, T. et al. (2007). A novel wireless, passive CO_2 sensor incorporating a surface acoustic wave reflective delay line. *Smart Mater. Struct.* 16 (4): 1382–1389.
50 Sivaramakrishnan, S., Rajamani, R., Smith, C.S. et al. (2008). Carbon nanotube-coated surface acoustic wave sensor for carbon dioxide sensing. *Sens. Actuators, B* 132 (1): 296–304.
51 Serban, B., Sarin Kumar, A.K., Costea, S. et al. (2009). Polymer-amino carbon nanotube nanocomposites for surface acoustic wave CO_2 detection. *Rom. J. Inf. Sci. Technol.* 12 (3): 376–384.
52 Choi, J.H., Kim, S.J., Jung, M.S., and Kim, S.J. (2013). Development of a polymer-coated SAW sensor for detection of CO_2 gas. In: *Third International Conference on Electrical, Electronics and Civil Engineering (ICEECE'2013)*, 4–5 January 2013, Bali, 38–41.
53 Zuckerwar, A.J. (2002). *Handbook of the Speed of Sound in Real Gases*, vol. 3. Amsterdam: Academic Press.
54 Pierce, G.W. (1925). Piezoelectric crystal oscillators applied to the precision measurement of the velocity of sound in air and CO_2 at high frequencies. *Proc. Am. Acad. Arts Sci.* 60 (5): 271–302.
55 Kneser, H.O. and Zühlke, J. (1932). Einstelldauer der Schwingungsenergien bei CO_2 und N_2O (Response time of the oscillation energies for CO_2 und N_2O). *Z. Phys.* 77 (9–10): 649–652. (in German).
56 van der Weerd, B. (2016). Entwicklung und Charakterisierung von CO_2-Sensoren für die Bestimmung der CO_2-Eliminierung während extrakorporaler Membranoxygenierung (Development and characterization of CO_2 sensors for the determination of CO_2 elimination during extracorporeal membrane oxygenation). PhD Thesis, Universität Regensburg, Fakultät für Chemie und Pharmazie (in German).
57 Eucken, A. and Becker, R. (1934). Die Schalldispersion bei verschiedenen Temperaturen in Chlor und Kohlendioxyd (The dispersion of sound at different temperatures in chlorine and carbon dioxide). *Z. Phys. Chem.* 27 (3–4): 235–262. (in German).
58 Zipser, L., Wächter, F., and Franke, H. (2000). Acoustic gas sensors using airborne sound properties. *Sens. Actuators, B* 68 (1–3): 162–167.
59 Zipser, L. and Wächter, F. (1995). Acoustic sensor for ternary gas analysis. *Sens. Actuators, B* 26 (1-3): 195–198.
60 Joos, M., Müller, H., and Lindner, G. (1993). An ultrasonic sensor for the analysis of binary gas mixtures. *Sens. Actuators, B* 16 (1–3): 413–419.
61 Manthey, W., Kroemer, N., and Mágori, V. (1992). Ultrasonic transducers and transducer arrays for applications in air. *Meas. Sci. Technol.* 3 (3): 249–261.

62 Vyas, J.C., Katti, V.R., Gupta, S.K., and Yakhmi, J.V. (2006). A non-invasive ultrasonic gas sensor for binary gas mixtures. *Sens. Actuators, B* 115 (1): 28–32.

63 Hauptmann, P., Hoppe, N., and Püttmer, A. (2002). Application of ultrasonic sensors in the process industry. *Meas. Sci. Technol.* 13 (8): R73–R83.

64 Henning, B. and Rautenberg, J. (2006). Process monitoring using ultrasonic sensor systems. *Ultrasonics* 44: e1395–e1399.

65 http://www.anton-paar.com/corp-de/produkte/gruppe/co2-und-o2-messgeraet (retrieved 10 August 2017).

66 http://www.rhosonics.nl/applications (retrieved 04 January 2017).

67 Toda, H. and Kobayakawa, T. (2008). High-speed gas concentration measurement using ultrasound. *Sens. Actuators, A* 144 (1): 1–6.

68 Toda, H. (2012). The precise mechanisms of a high-speed ultrasound gas sensor and detecting human-specific lung gas exchange. *Int. J. Adv. Rob. Syst.* 9 (6): 9.

10

Miscellaneous Approaches

Wolfram Oelßner[1], Manfred Decker[1], and Gerald Gerlach[2]

[1] *Kurt-Schwabe-Institut für Mess- und Sensortechnik e.V. Meinsberg, Kurt-Schwabe-Straße 4, 04736 Waldheim, Germany*
[2] *Technische Universität Dresden, Faculty of Electrical and Computer Engineering, Institute of Solid-State Electronics, 01062 Dresden, Germany*

10.1 Hydrogel-Based CO_2 Sensors with Pressure Transducer

In a broader sense hydrogel-based carbon dioxide sensors operate on the Severinghaus concept (see Section 4.1) with a membrane-covered pH glass electrode. But instead of measuring the pH value of a liquid hydrogen carbonate electrolyte by a glass electrode, the hydrogel-based CO_2 sensor exploits a pH-sensitive hydrogel as sensing material and a micro pressure sensor as transducer for the detection of carbon dioxide [1–6].

Hydrogels are cross-linked hydrophilic polymers. By incorporating functional groups they can undergo volume changes in response to changes in stimuli like pH value, temperature, light, electric field, or ion concentration [7–9]. For the hydrogel-based CO_2 sensor – instead of a pH glass electrode – a particularly pH-sensitive hydrogel is used that swells and shrinks reversibly, depending on the pH value of the bicarbonate electrolyte. The changes in the resulting mechanical pressure are measured and indicate the partial pressure of CO_2. Compared with the conventional Severinghaus sensor, the main advantage of this hydrogel-based CO_2 sensor principle is that an electrochemical reference electrode is not necessary.

Figure 10.1 illustrates the working principle and the main components of the hydrogel-based CO_2 sensor. CO_2 from the gaseous or liquid environment permeates through a gas-permeable membrane into the hydrogen carbonate-containing electrolyte and changes its pH value in dependency on the CO_2 partial pressure according to the Severinghaus principle. The hydrogen carbonate electrolyte is in contact with the pH-sensitive hydrogel via a porous layer. The hydrogel is enclosed between a mechanically stiff porous cover layer and the pressure sensor. Since the pH-sensitive hydrogel cannot extend, it will generate a pressure P in response to the pH change. This pressure is measured by the pressure sensor and converted into an electrical signal S, which depends on the partial pressure of carbon dioxide in the measuring medium. The described process is fully reversible.

Carbon Dioxide Sensing: Fundamentals, Principles, and Applications,
First Edition. Edited by Gerald Gerlach, Ulrich Guth, and Wolfram Oelßner.

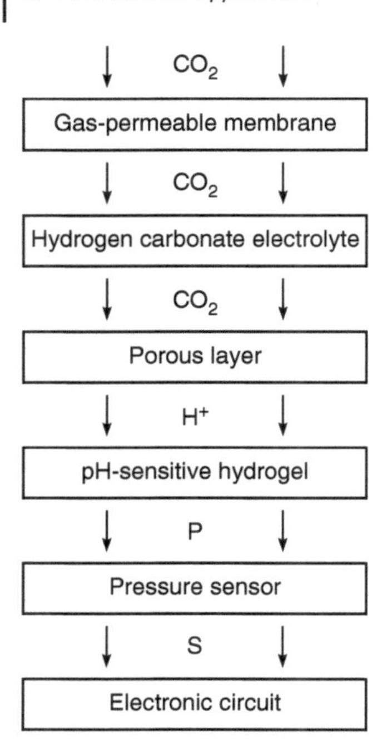

Figure 10.1 Block diagram of the main components and the working principle of hydrogel-based CO_2 sensors according to [3].

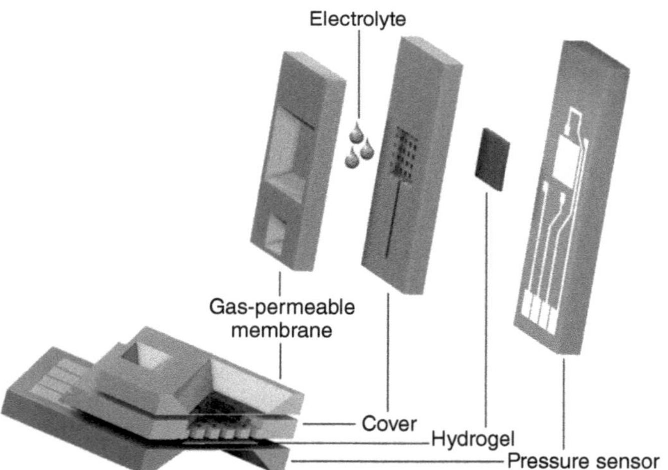

Figure 10.2 Set-up of a hydrogel-based CO_2 sensor [3].

Figure 10.2 shows exemplarily an exploded view of the main components of such a hydrogel-based CO_2 sensor [3]. The gas-permeable membrane is mounted on top of the cover to retain the electrolyte in the so-formed reservoir. It consists of a polydimethylsiloxane layer mounted to a silicon substrate, which functions as a carrier. The pH-sensitive hydrogel is synthesized of 2-hydroxyethylmethacrylate and dimethylaminoethyl methacrylate. Its thickness amounts to c. 5 μm. As pressure sensor a modified Honeywell pressure sensor (26PC Series), consisting of a silicon pressure sensor chip and a plastic housing, was used. For maximum sensor resolution, the optimal electrolyte consists of 17 mM sodium bicarbonate and 8 mM sodium chloride and has an ionic strength of 25 mM. The dimensions of this sensor are $2.92 \times 0.95 \times 0.70\,mm^3$.

Since at the intended application the operating temperature of the sensor will be the same as the human body temperature, CO_2 measurement cycles were performed at 37 °C. Figure 10.3 shows typical measurement results. The sensor responds well to carbon dioxide changes with a response time between two and four minutes per CO_2 step, showing little hysteresis. It enables the detection of steps smaller than 0.5 kPa CO_2. The maximum pressure generation occurred at 20 kPa CO_2 with a pressure of 0.28×10^5 Pa.

With the result of Figure 10.3, the calibration curve of the sensor can be determined. The partial CO_2 pressure pCO_2 is plotted versus the pressure, and a calibration curve is fitted, as shown in Figure 10.4. Because of the non-linear characteristic, it is necessary to calibrate the sensor.

The intended main application of the sensor described here is measuring the partial pressure of CO_2 in the stomach to diagnose gastrointestinal ischemia, which is indicated by high pCO_2 levels [10]. Since this CO_2 measurement is performed *in situ*, miniaturization is required to be able to insert the sensor, applied on a catheter, in the stomach through the nose. This sensor is

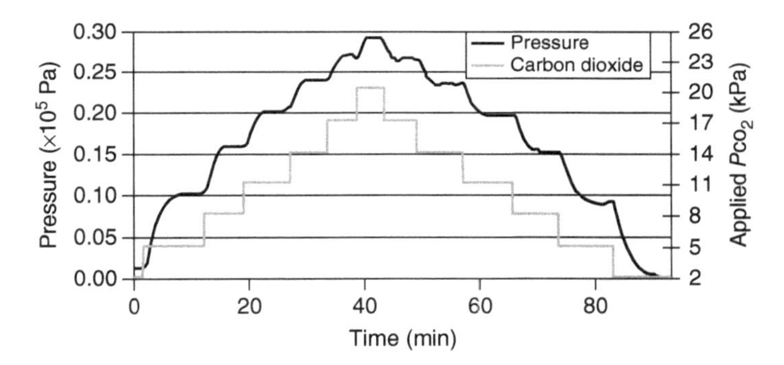

Figure 10.3 Plot of a CO_2 measurement cycle with a hydrogel-based Severinghaus-type CO_2 sensor with gas-permeable membrane; temperature 37 °C.

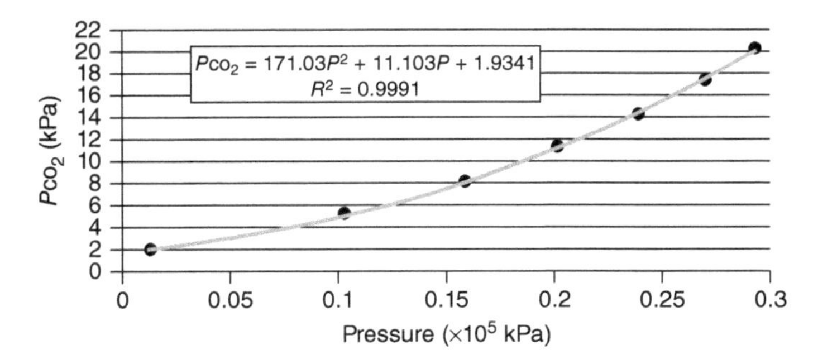

Figure 10.4 Partial CO_2 pressure pCO_2 versus pressure P for the measured equilibrium pressures and a fitted quadratic trend line ($pCO_2 = 171.03P^2 + 11.103P + 1.9341$), representing the calibration curve of the sensor.

considered to fulfil already most of the medical requirements. Further possible applications are expected in the automobile industry, in horticulture, and for environment-monitoring instruments [3].

Reversibly swelling polymer hydrogels have also been used in conductometric CO_2 sensors [11]. In a preliminary study [12], it has been tried to replace the thin water film of a conductometric CO_2 sensor with a hydrogel. However, this was found to be unsuitable because of the instability of the sensor signal and the insufficient reproducibility.

10.2 Miniaturized and ISFET-Based CO_2 Sensors

The 'classical' electrochemical Severinghaus-type CO_2 sensor as described in Section 4.1 is appropriate for many fields of application, but it is not suitable for measurements of very small sample volumes in the microlitre range. The main element of the CO_2 sensor is the pH electrode, which has to be placed near the gas-permeable membrane on the tip of the sensor. This requires special pH electrodes with a small diameter. The dimensions of the surface area of the pH glass electrode can be significantly reduced, but the bicarbonate volume in front of the pH electrode needs a certain amount of permeating carbon dioxide from the analyte sample volume to establish the equilibrium at the pH-sensitive surface. Furthermore, pH glass microelectrodes are difficult to fabricate and have a very high resistance. For this reason, in CO_2 microelectrodes for physiological studies, ion-selective liquid membrane microelectrodes or metal oxide pH electrodes have been used to a large extent instead of pH glass microelectrodes. In Figure 10.5 a schematic drawing of such a CO_2 microelectrode is presented.

Needle-type electrodes for CO_2 detection with tip diameters of less than 20 μm have been developed and applied, e.g. for the investigation of biofilms. Zhao et al. [13] and de Beer et al. [14] used for this purpose an appropriate liquid exchange membrane for the pH detection. This pH-sensitive viscous liquid can be placed in the tip of another needle and – combined together with a reference electrode – allows a reliable pH detection. Beyenal et al. [15], on the contrary, decided to use pH-sensitive iridium oxide for the determination of the pCO_2-dependent pH change. The anodically grown iridium oxide film tips (AIROF tips) were produced by electrochemical oxidation of an Ir wire, melted in a glass tube in 0.5 M sulphuric acid. This kind of sensor has been used successfully for carbon dioxide detection in *Staphylococcus aureus* biofilms.

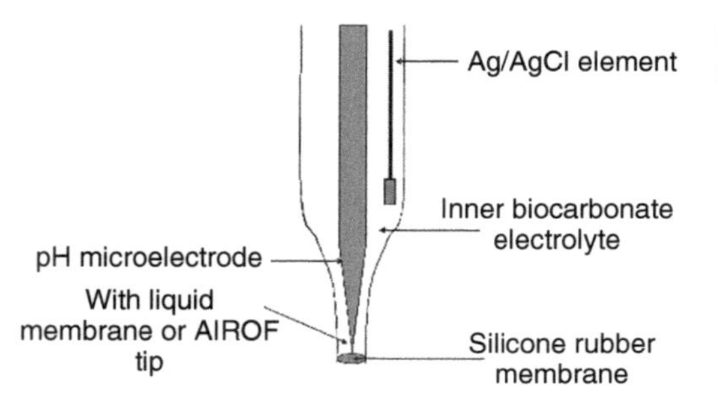

Figure 10.5 Schematic drawing of a needle-type CO_2 microelectrode.

Alternatively, instead of miniaturized pH glass microelectrodes, ion-selective field effect transistors (ISFETs) have been used in CO_2 microelectrodes. ISFETs allow to miniaturize the pH sensing area down to the square millimetre range and are robust and shatterproof, and the low ohmic resistance reduces the requirements for the signal measurement. They show near Nernstian slope and short response times. In [16] ISFETs are compared to the glass electrode concerning their advantages, challenges, and similarities.

The ISFET principle was invented by Bergveld around 1970 [17, 18] and stimulated a boom of research and development activities in this field [19]. The sensitivity of pH-ISFETs is induced by pH-sensitive layers on the gate, mostly based on Si_3N_4, Al_2O_3, or Ta_2O_5 [20–22]. Recent research on alternative material systems has brought forth the group III nitrides (GaN and its ternary alloys) as promising candidates for sensing applications. A GaN-based ISFET was applied in a sensor system for biochemical analysis, which is characterized above all by optical transparency and high chemical and biological inertness and does not require any additional coating for determining the pH value [23].

The focus in the development of ISFET-based CO_2 sensors has to be directed at the integration of the reference electrode and on the covering of the sensing area by the inner electrolyte and by the gas-permeable membrane. Already for a long time, pH-ISFETs have been studied for the pCO_2 determination in biology and medicine. Some examples will be presented more in detail:

- Tsukada et al. [24] combined the pCO_2 sensor with additional ISFET sensors (K^+, Na^+, Cl^-, pH, and amperometric pO_2) to form a multisensor array. However, the service life of the gas sensors proved to be unsatisfactory.
- Shoji and Esashi [25] reported on a micro flow cell for blood gas analysis based on ISFET modules (Figure 10.6). To realize a very small sample volume, a micro flow cell for blood gas monitoring is fabricated on a silicon wafer. It has a Clark-type micro-pO_2 sensor, a Severinghaus-type micro-pCO_2 sensor, and an ISFET-type pH sensor. The cell size is $1.0 \times 0.6 \times 0.8\,cm^3$, and the necessary sample volume amounts to c. 0.34 μl. The Si_3N_4 pH-ISFET was placed on the front side of the wafer, facing the measuring solution. The electrolytic contact to the Ag/AgCl, Cl^- reference electrode positioned on the back of the transducer was established by a channel in the wafer. The gas-permeable membrane consisted of a negative photoresist. To ensure a small gap between ISFET and gas-permeable membrane for the bicarbonate

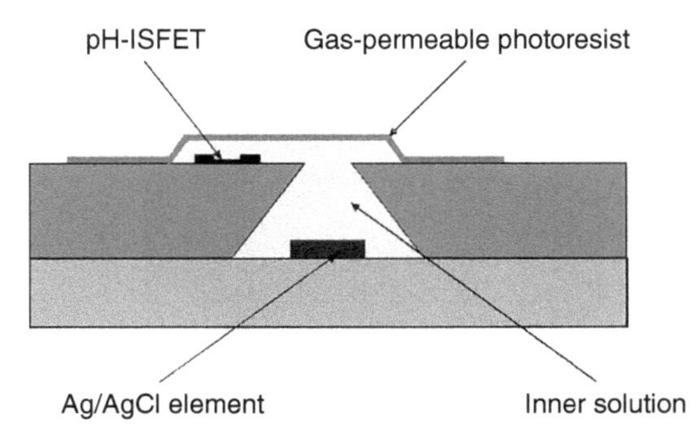

Figure 10.6 Cross section of an ISFET with an Ag/AgCl, Cl^- electrode placed on the back. Source: Shoji and Esashi 1992 [25]. Reproduced with permission of Elsevier.

solution, a positive photoresist had been patterned on the ISFET and removed by washing after printing of the gas-permeable resin layer. The pCO_2 sensor is a miniaturized Severinghaus-type sensor using an ISFET as a pH electrode. It has a gas-permeable membrane, an inner solution chamber, and an Ag/AgCl electrode. The chamber volume was about 5 nl. It was filled with 0.005 M bicarbonate solution, containing 0.02 M sodium chloride to ensure a stable reference potential at the Ag/AgCl, Cl^- electrode. The sensors showed good results in the physiological range of CO_2, but response times of three to five minutes were not acceptable for blood gas analysis. The authors assumed that the poor permeability of carbon dioxide through the negative photoresist was the cause of the slow response.

- A slight improvement of the response time had been achieved by Arquint et al. [26, 27], describing an ISFET-based CO_2 sensor covered with a hydrogel and polysiloxane membrane – both directly patterned on the sensor surface. Beside the Al_2O_3 pH-ISFET on the wafer, a thin-film Ag/AgCl,Cl^- electrode was screen-printed (Figure 10.7). In the range between 50 and 20 000 Pa, the sensitivity was found to be −49 mV/decade pCO_2 and the t_{95} time of two minutes. When used in 37 °C transfusion blood, the response time t_{95} increased to about five minutes and the slope to −52 mV/decade pCO_2. The sensor device was stable for more than two weeks under continuous operation.
- ISFETs covered with carbonate-selective polymer membranes on the surface of the gate, combined with a reference electrode, result in a potentiometric carbonate measuring chain. Abramova et al. [28] showed that a measurement of carbonate with polymer matrix-based carbonate electrodes at constant pH (pH = 8.7) is possible. An influence of penetrating CO_2 and water molecules on the stability of the contact surface between membrane and gate of the ISFET has not been observed in contrast to the assumption of many authors.
- For the use of FETs, only elevated temperatures below 180 °C are acceptable. Therefore, research is focussed on solid electrolytes, which allow lower temperatures or even room temperature. The CO_2-sensitive layers developed by Shimanoe et al. are sensitive to carbon dioxide even at 30 °C [29]. The gate insulated by a Ta_2O_5 layer was covered with the Na^+ form of a cation exchange membrane as ionic conductor. An auxiliary phase for the establishment of the CO_2 sensitivity consisted of $Li_2CO_3BaCO_3$/ITO (indium tin oxide) and was deposited on the surface. The electrochemical reaction for the detection is described by

$$2Li^+ + CO_2 + \tfrac{1}{2}O_2 + 2e^- \rightarrow Li_2CO_3 \tag{10.1}$$

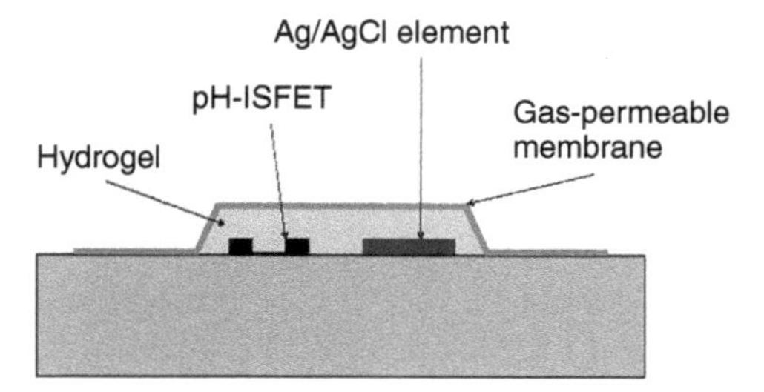

Figure 10.7 Cross section of an ISFET with screen-printed hydrogel and siloxane membrane. Source: Arquint et al. 1993 [26]. Reproduced with permission of Elsevier.

The slope of the sensors was nearly the Nernstian value for an electrochemical reaction involving a two-electron step. The response time was determined as one to two minutes. Especially remarkable is that the sensor was not influenced by humidity between 30% and 70% relative humidity.

Although ISFETs have a lot of advantages, they have not yet reached a breakthrough as pH measuring technique in general and for CO_2 determination in particular. This might be due to the fact that packaging [30] is still a challenge and that adequate reference electrodes are still not available. Nevertheless, ISFETs have proved to be very useful for specific measurement tasks, for example, CO_2 measurements in very small samples of blood or biological fluids [29].

10.3 Thermal Conductivity CO_2 Detectors

Thermal conductivity detectors (TCDs), also called katharometers, are sensors detecting gases on the bases of their thermal conductivity. They are commonly used in gas chromatography (cp. Table 3.7). These detectors make use of the different thermal conductivity values of different gases (Table 10.1) in such a manner that the resistivity of at least two resistors – one in the gas to be analysed and the second one in a reference gas – changes differently due to the different heat losses. In Figure 10.8a two resistors are used in each channel (R_1, R_2 and R_3, R_4, respectively), forming a Wheatstone bridge to generate an electrical output voltage.

In case of gas chromatography, the thermal conductivity λ of the column effluent is measured and compared to a reference flow of the carrier gas (helium, hydrogen). Most gas compounds have a thermal conductivity much less than that of helium or hydrogen, which provides a large difference in the resistivity changes in both channels.

Assuming that all resistors in the Wheatstone bridge have a basic resistance R_0 and that the resistance changes in the resistors are (Figure 10.8b) in the gas channel

$$R_1 = R_2 = R_{gas} = [1 + r_{gas}(\lambda_{gas})] \tag{10.2}$$

Table 10.1 Thermal conductivity λ of selected gases (in mW m^{-1} K^{-1}) [31, 32].

Temperature (K)	100	200	300	400	500	600
Air	9.4	18.4	26.2	33.3	39.7	45.7
Carbon dioxide, CO_2	—	9.6	16.8	25.1	33.5	41.6
Carbon monoxide, CO	—	—	25.0	32.3	39.2	45.7
Helium, He	75.5	119.3	156.7	190.6	222.3	252.4
Hydrogen, H_2	68.6	131.7	186.9	230.4	—	—
Water vapour, H_2O	—	—	18.7	27.1	35.7	47.1
Oxygen, O_2	9.3	18.4	26.3	33.7	41.0	48.1
Argon, Ar	6.2	12.4	17.9	22.6	26.8	30.6
Methane, CH_4	—	22.5	34.1	49.1	66.5	84.1
Ammonia, NH_3	—	—	24.4	37.4	51.6	66.8

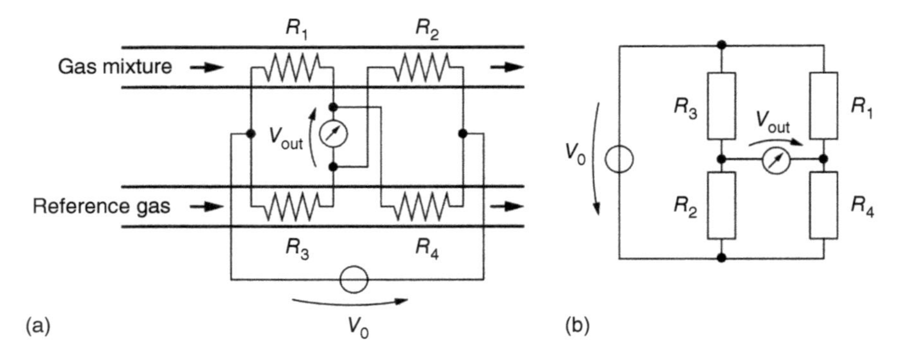

Figure 10.8 Thermal conductivity detector with Wheatstone bridge configuration. (a) Schematic [33]. (b) Electric circuit.

and in the reference channel

$$R_3 = R_4 = R_{\text{ref}} = R_0[1 + r_{\text{ref}}(\lambda_{\text{ref}})] \tag{10.3}$$

where r_{gas} and r_{ref} are the relative resistance changes of the respective sensors due to the heat conduction in the surrounding gas, the measurable voltage change U_{out} of the Wheatstone bridge yields

$$U_{\text{out}} = U_0 \left(\frac{R_2}{R_2 + R_3} - \frac{R_4}{R_1 + R_4} \right) \tag{10.4}$$

Inserting Eqs. (10.2) and (10.3), one gets for $r_{\text{gas}}, r_{\text{ref}} \ll 1$:

$$U_{\text{out}} \approx \frac{U_0}{2}[r_{\text{gas}}(\lambda_{\text{gas}}) - r_{\text{ref}}(\lambda_{\text{ref}})] \sim (\lambda_{\text{gas}} - \lambda_{\text{ref}}) \tag{10.5}$$

Equation (10.5) shows clearly that – for a large sensitivity – the difference in resistance change between gas and reference channel and, hence, between the thermal conductivities between the gas to be detected and the reference gas should be as large as possible. This also means that some gases in air, e.g. oxygen, nitrogen, and carbon monoxide, cannot be measured practically as their thermal conductivities are too close.

In general, the thermal conductivity λ_i of a gas component depends non-linearly on temperature [34]:

$$\lambda_i(T) = A_i + B_i \cdot T + C_i \cdot T^2 + D_i \cdot T^3 \tag{10.6}$$

where A_i, B_i, C_i, and D_i are tabulated gas-specific coefficients. Based on the rule of mixture of Wassiljewa, the thermal conductivity of a gas mixture is [35, 36]

$$\lambda_{\text{mix}}(T) = \sum_i \frac{X_i}{\sum_j \Phi_{ik}(T) X_k} \cdot \lambda_i(T) \tag{10.7}$$

The X denotes the relevant volume fractions and the $\phi_{ik}(T)$ correction terms, depending on the molar mass M and the viscosity η of the corresponding components:

$$\phi_{ik}(T) = \left(1 + \sqrt{\frac{\eta_i(T)}{\eta_k(T)} \cdot \sqrt{\frac{M_k}{M_i}}}\right)^2 \cdot \left(\sqrt{8} \cdot \sqrt{1 + \frac{M_i}{M_k}}\right)^{-1} \tag{10.8}$$

Due to this temperature dependence, the difference $\lambda_{gas} - \lambda_{ref}$ usually varies with temperature. In [37] a working temperature between 400 and 500 °C is suggested, but TCDs can also be operated at lower temperatures.

Thermal conduction detectors are not only used to identify analytes but also to quantify the components of a gas sample. However, they often allow to measure just binary mixtures with sufficient accuracy. As new investigations show – by applying the method of periodic temperature modulation and using the Fourier transform thermal conductivity analysis (FTTCA) – it is possible to evaluate mixtures of up to four components [38].

To achieve reliable measurement results, a calibration is needed for each gas mixture under test. This is done by measuring a certain number of samples with known concentration to determine a mathematical relationship – mostly approximated by a straight line – between sensor voltage and concentration.

Figure 10.9 depicts an example of a TCD. It shows the four heatable sensor elements (resistors) in a thermostated housing, forming the Wheatstone bridge. The sensor elements are deposited on a ceramic or a glass carrier connected with Pt heaters. To avoid corrosion or unwanted interaction with the gases, the sensor elements are covered with an appropriate protective coating or are fully cast in glass.

Table 10.2 summarizes exemplarily technical data of a carbon dioxide analyser for the measurement of CO_2 concentration in brewing tanks.

Summarizing the above, it can be said that TCDs offer a number of advantages:

Figure 10.9 Thermal conductivity detector module (Henze-Hauck Prozessmesstechnik/Analytik GmbH, Dessau, Germany; from [34]).

Table 10.2 Technical data of the carbon dioxide analyser TC1 for the measurement of CO_2 concentration in brewing tanks [39].

Measuring range	0–100 vol%
Resolution	0.1 vol%
Accuracy at normal pressure	<5%
Maximum temperature	80 °C
Maximum gas pressure	10 bar
Size	Diameter 50 mm, height 100 mm

- They are suitable for large gas concentration ranges.
- They allow continuous measurements in industry.
- They are usable in harsh environments and for flammable gases.
- They show high sensitivity for large differences of thermal conductivity between gas under test and reference gas.
- They have good long-term stability.
- They can be used usually for gas mixtures of two (known) gases but are suitable for up to four gases.

10.4 Membrane-Based CO_2 Sensors with Pressure Measurement

The gas-selective properties of polymeric membranes can be used for direct gas analysis with membrane-based gas sensors [40–43].

Figure 10.10 illustrates the principle of the membrane-based gas sensors. All gas components in the gaseous matrix permeate through the gas-selective membrane according to the solution–diffusion model [45, 46] that is most widely used to describe the permeation process of gases through rubbery polymer membranes (used in these technologies). According to this model the gas permeation proceeds in several steps:

(a) The gas is adsorbed from environment at the outer membrane surface.
(b) Then the adsorbed gas molecules are absorbed into the membrane phase, which occurs relatively slowly compared with the faster adsorption process.
(c) Inside the membrane, the gas molecules diffuse according to the concentration gradient across the membrane. This process is much slower than the processes at membrane surface and is therefore the rate-limiting step. Therefore, constant concentrations on both faces of the membrane can be assumed.
(d) After reaching the inner membrane, mass transfer proceeds in the reverse direction where gas desorbs from membrane phase back to gaseous phase inside the tube.

Therefore, a tube coated with such a gas-selective polymeric membrane responds to a certain gas species like CO_2 due to its strong selectivity towards the particular species in the permeating matrix. The sensor response can be

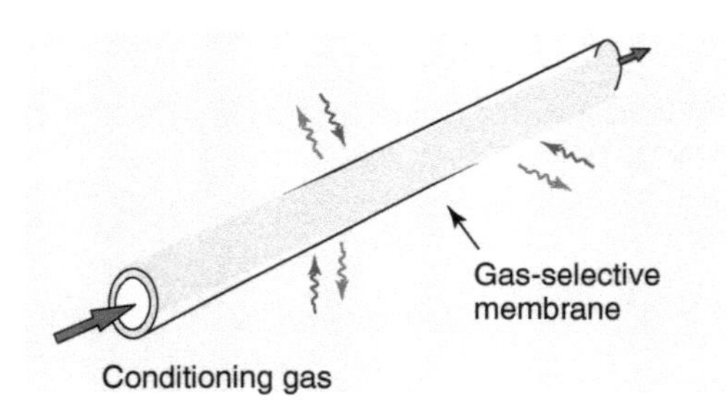

Figure 10.10 Principle of a linear CO_2 sensor, based on the permeation of gases through the walls of a gas-selective membrane into a measurement volume [44].

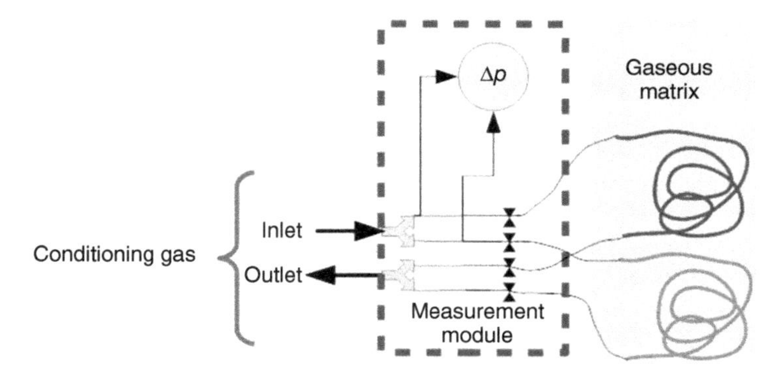

Figure 10.11 Basic set-up for measuring pressure changes Δp as a result of changes in CO_2 concentration in the gaseous matrix. Dark grey colour, gas-permeable membrane; light grey colour, virtually gas-tight reference tube (according to [48]).

expressed either in terms of change of pressure (isochoric operation) or change in volume (isobaric operation) and possibly calibrated as concentration [47].

Figure 10.11 shows the basic physical set-up for measuring changes in CO_2 concentration in a gaseous matrix using such sensors. It consists essentially of a tube coated with a CO_2-permeable polymeric membrane, called measurement tube (dark grey coloured), combined with a tube coated with relatively gas-tight material, called reference tube (light grey coloured) placed together in close vicinity in the gas matrix. The measurement module contains valves to open or close the inlet and outlet of both the tubes and a differential pressure sensor.

The measurement procedure consists of the periodical iteration of two steps [48]: at first is the conditioning/purging step where the inlet and outlet of both the tubes are open for a defined time (e.g. 30 seconds) and the two tubes are purged using a flushing gas of known composition, such as air. Following it is the measurement step where the inlet and outlet of the tubes are closed using valves for a certain defined time (e.g. 5 seconds) to ensure a defined volume. In this step CO_2 from the gaseous matrix permeates into the measurement tube volume, causing a pressure change. The rate of change of pressure difference $\Delta p/\mathrm{d}t$ between both tubes is measured and can be expressed in terms of CO_2 concentrations after necessary calibration.

Compared with electrochemical CO_2 sensors, membrane-based linear gas sensors have some advantages. They need minimal maintenance for practical usage and are easy to handle, low cost, and robust enough to be applied directly in subsurface to measure on relatively larger spatial scale with rapid time response, which can be more meaningful than above surface CO_2 measuring techniques for such application areas. The shape of such sensors is highly adaptable to the measurement problem. Tubular membranes can be used to form linear sensors, which can detect locally fluctuating CO_2 concentrations over large areas with sufficient temporal resolution [48]. Some application examples are presented in Section 2.2.4.

References

1 Herber, S., Olthuis, W., and Bergveld, P. (2003). A swelling hydrogel-based P_{CO_2} sensor. *Sens. Actuators, B* 91 (1-3): 378–382.
2 Herber, S., Olthuis, W., Bergveld, P., and van den Berg, A. (2004). Exploitation of a pH-sensitive hydrogel disk for CO_2 detection. *Sens. Actuators, B* 103 (1-2): 284–289.
3 Herber, S. (2005). Development of a hydrogel-based carbon dioxide sensor. PhD thesis. University of Twente, NL.
4 Herber, S., Bomer, J., Olthuis, W. et al. (2005). A miniaturized carbon dioxide gas sensor based on sensing of pH-sensitive hydrogel swelling with a pressure sensor. *Biomed. Microdevices* 7 (3): 197–204.
5 Herber, S., Bomer, J., Olthuis, W. et al. (2005). A micro CO_2 gas sensor based on sensing of pH-sensitive hydrogel swelling by means of a pressure sensor. In: *Transducers 2005: 13th International Conference on Solid-State Sensors, Actuators and Microsystems*, Seoul, South Korea (5–9 June 2005), 1146–1149.
6 ter Steege, R.W.F., Herber, S., Olthuis, W. et al. (2007). Assessment of a new prototype hydrogel CO_2 sensor; comparison with air tonometry. *J. Clin. Monit. Comput.* 21 (2): 83–90.
7 van den Linden, H., Herber, S., Olthuis, W., and Bergveld, P. (2002). Development of stimulus-sensitive hydrogels suitable for actuators and sensors in microanalytical devices. *Sens. Mater.* 14 (3): 129–139.
8 van der Linden, H.J., Herber, S., Olthuis, W., and Bergveld, P. (2003). Stimulus-sensitive hydrogels and their applications in chemical (micro)analysis. *Analyst* 128 (4): 325–331.
9 Herber, S., Eijkel, J., Olthuis, W. et al. (2004). Study of chemically induced pressure generation of hydrogels under isochoric conditions using a microfabricated device. *J. Chem. Phys.* 121 (6): 2746–2751.
10 Groeneveld, A.B.J. and Kolkman, J.J. (1994). Splanchnic tonometry: a review of physiology, methodology and clinical implications. *J. Crit. Care* 9 (3): 198–210.
11 Sheppard, N.F. Jr. (1991). Design of a conductimetric microsensor based on reversibly swelling polymer hydrogels. In: *International Conference on Solid-State Sensors and Actuators, Digest of Technical papers*, 773–776.
12 Van der Weerd, B. (2016). Entwicklung und Charakterisierung von CO_2-Sensoren für die Bestimmung der CO_2-Eliminierung während extrakorporaler Membranoxygenierung (Development and characterization of CO_2 sensors for the determination of CO_2 elimination during extracorporeal membrane oxygenation). PhD thesis. Universität Regensburg, Germany (in German).
13 Zhao, P. and Cai, W.J. (1997). An improved potentiometric pCO_2 microelectrode. *Anal. Chem.* 69 (24): 5052–5058.

14 de Beer, D., Glud, A., Epping, E., and Kühl, M. (1997). A fast-responding CO_2 microelectrode for profiling sediments, microbial mats, and biofilms. *Limnol. Oceanogr.* 42 (7): 1590–1600.

15 Beyenal, H., Davis, C.C., and Lewandowski, Z. (2004). An improved Severinghaus-type carbon dioxide microelectrode for use in biofilms. *Sens. Actuators, B* 97 (2-3): 202–210.

16 Lübbers, B. and Schober, A. (2009). Comparing the ISFET to the glass electrode: advantages, challenges and similarities. *Chem. Anal. (Warsaw)* 54: 1121–1148.

17 Bergveld, P. (1970). Development of an ion-sensitive solid state device for neurophysiological measurements. *IEEE Trans. Biomed. Eng.* BME-17 (1): 70–71.

18 Bergveld, P. (1972). Development, operation and application of the ion sensitive field effect transistor as a tool for electrophysiology. *IEEE Trans. Biomed. Eng.* BME-19 (5): 342–351.

19 Bergveld, P. (2003). Thirty years of ISFETOLOGY: what happened in the past 30 years and what may happen in the next 30 years. *Sens. Actuators, B* 88 (1): 1–20.

20 Matsuo, T. and Wise, K.D. (1974). An integrated field effect electrode for biopotential recording. *IEEE Trans. Biomed. Eng.* BME-21 (6): 485–487.

21 Abe, H., Esashi, M., and Matsuo, T. (1979). ISFET's using inorganic gate thin films. *IEEE Trans. Electron Devices* 26 (12): 1939–1944.

22 Matsuo, T. and Esashi, M. (1981). Methods of ISFET fabrication. *Sens. Actuators* 1: 77–96.

23 Lübbers, B., Kittler, G., Ort, P. et al. (2008). A novel GaN-based multiparameter sensor system for biochemical analysis. *Phys. Status Solidi C* 5 (6): 2361–2363.

24 Tsukada, K., Miyahara, Y., Shibata, Y., and Miyagi, H. (1990). An integrated chemical sensor with multiple ion and gas sensors. *Sens. Actuators, B* 2 (4): 291–295.

25 Shoji, S. and Esashi, M. (1992). Micro flow cell for blood gas analysis realizing very small sample volume. *Sens. Actuators, B* 8 (2): 205–208.

26 Arquint, P., van den Berg, A., van der Schoot, B.H. et al. (1993). Integrated blood-gas sensor for pO_2, pCO_2 and pH. *Sens. Actuators, B* 13-14 (1–3): 340–344.

27 Arquint, P., Koudelka-Hep, M., de Rooij, N.F. et al. (1994). Organic membranes for miniaturized electrochemical sensors: fabrication of a combined pO_2, pCO_2 and pH sensor. *J. Electroanal. Chem.* 378 (1–2): 177–183.

28 Abramova, N., Levichev, S., and Bratov, A. (2010). The influence of CO_2 on ISFETs with polymer membranes and characterization of a carbonate ion sensor. *Talanta* 81 (4-5): 1750–1754.

29 Shimanoe, K., Goto, K., Obata, K. et al. (2004). Development of FET-type CO_2 sensor operative at room temperature. *Sens. Actuators, B* 102 (1): 14–19.

30 Oelßner, W., Zosel, J., Guth, U. et al. (2005). Encapsulation of ISFET sensor chips. *Sens. Actuators, B* 105 (2005): 104–117.
31 www.engineersedge.com/heat_transfer/thermal-conductivity-gases.htm (retrieved 04 August 2017).
32 www.engineeringtoolbox.com/thermal-conductivity-d_429.html (retrieved 04 August 2017).
33 www.processanalytik.de/index.php/analysetechnik/wld-sensor (retrieved 04 August 2017).
34 Dietrich, S., Kusnezoff, M., Ziesche, S., and Henze, J. (2011) Wärmeleitfähigkeitsdetektoren-Array zur selektiven Gasanalyse (Thermal conductivity detector array for selective gas analysis). 10. Dresdner Sensor-Symposium, Dresden, 5.-7.12.2011, DOI 10.5162/10dss2011/8.3, www.ama-science.org/proceedings/details/611 (retrieved 04 August 2017).
35 Wassiljewa, A. (1904). Wärmeleitung in Gasgemischen. *Physik. Z.* 5 (22): 737–742.
36 Verein Deutscher Ingenieure, VDI-Gesellschaft Verfahrenstechnik und Chemieingenieurwesen (GVC) (2013). *VDI-Wärmeatlas*, 11e. Düsseldorf: VDI-Verlag.
37 www.sgxsensortech.com/products-services/industrial-safety/thermal-conductivity-sensors (retrieved 04 August 2017).
38 Grienauer, H.S. (2012). Temperature modulated thermal conductivity gas analysis sensor properties and applications, 16. GMA/ITG-Fachtagung Sensoren und Messsysteme 2012, 22.-23.05.2012, Nuremberg, DOI 10.5162/sensoren2012/1.2.2; www.ama-science.org/proceedings/details/682 (retrieved 04 August 2017).
39 Kohlendioxidmessung im Bier (Carbon dioxide measurement in beer). Data sheet, Kohlendioxid-Analysator TC1, Henze-Hauck Prozessmesstechnik/Analytik GmbH, Dessau, Germany.
40 Lazik, D. and Geistlinger, H. (2005). A new method for membrane-based gas measurements. *Sens. Actuators, A* 117 (2): 241–251.
41 Kesson, J. (1985). Method of and apparatus for monitoring concentration of gas in a liquid, Patent US 4,550,590, filed 18 August 1983 and issued 5 November 1985.
42 Lazik, D. and Geistlinger, H. (2001). Method for the measurement of the concentration or the partial pressure of gases in fluids in gas sensor. Patent US 6,679,096, filed 2 June 2000 and issued 26 July 2001.
43 Lazik, D., Geistlinger, H., Eichhorst, P., and Kamusewitz, H. (2004). Measurement cell and method for determining the concentration of different gases in a fluid medium. Patent EP 1 359 414 A3, filed 28 April, 2003 and issued 27 October 2004.
44 Lazik, D. and Ebert, S. (2013). Membranbasierte Gassensoren – Ein neues Instrument zur Gas- und Branddetektion (Membrane-based gas sensors – a new instrument for gas and fire detection). *Tech. Sicherheit* 3 (3): 16–19. (in German).
45 Wijmans, J.G. and Baker, R.W. (1995). The solution-diffusion model: a review. *J. Membr. Sci.* 107: 1–21.

46 Ismaila, A.F, Kusworoa, T.D, Mustafaa, A., and Hasbullaha, H. (2005). Understanding the Solution-Diffusion Mechanism in Gas Separation Membrane for Engineering Students. Proceedings of the 2005 Regional Conference on Engineering Education. 12–13 December 2005, Johor, Malaysia.

47 Lazik, D., Ebert, S., Leuthold, M. et al. (2009). Membrane based measurement technology for in situ monitoring of gases in soil. *Sensors* 9 (2): 756–767.

48 Sood, P. (2016). Concept-based improvements of membrane based linear gas sensors for the measurement of CO_2 concentrations in wet soils. Master Thesis. Karlsruhe University of Applied Sciences, Karlsruhe, Germany.

11

Survey and Comparison of Methods

Hans Ulrich Guth[1], *Gerald Gerlach*[2], *and Wolfram Oelßner*[3]

[1] *Technische Universität Dresden, Faculty of Chemistry and Food Chemistry, 01062 Dresden, Germany*
[2] *Technische Universität Dresden, Faculty of Electrical and Computer Engineering, Institute of Solid-State Electronics, 01062 Dresden, Germany*
[3] *Kurt-Schwabe-Institut für Mess- und Sensortechnik e.V. Meinsberg, Kurt-Schwabe-Straße 4, 04736 Waldheim, Germany*

Following general considerations on properties of carbon dioxide and the carbon cycle in Part I of the book, in Part II, Chapters 3–10, the most important principles of carbon dioxide sensors are described in detail. The advantages, disadvantages, and preferred applications of the mentioned CO_2 sensor principles are summarized in Table 11.1.

At present the following methods are commercially available and generally accepted:

- IR (infrared),
- NDIR (non-dispersive infrared),
- GC (gas chromatography),
- MS (mass spectrometric),
- Severinghaus electrode, and
- Solid electrolyte sensor.

Performances and operational conditions of some selected methods, which can be used for CO_2 measurements in gases and liquids in a broader field of applications, are summarized in Table 11.2.

The selection of the measuring principle respective sensor that has to be chosen to solve the analytical task is a question of, e.g. the measuring environment, the requirements on the timescale (delay between sampling and obtained measuring value), the expectation on lifetime and precision of the data, and of course on the price and maintenance of the measuring device. In some cases an *in situ* measurement is necessary to obtain real-time information and/or to avoid errors due to sampling. Therefore, no simple recommendation can be given to the user. Most of them – scientists and technicians – who intent to measure CO_2 are not specialists in measuring technology and rely on the specifications of devices provided by suppliers and producers. Very often, these specifications describe the advantages and possibilities of the particular device, while the limits and drawbacks are not particularly pointed out. One of the aims of this book consists in

Carbon Dioxide Sensing: Fundamentals, Principles, and Applications,
First Edition. Edited by Gerald Gerlach, Ulrich Guth, and Wolfram Oelßner.

Table 11.1 Principles of carbon dioxide sensors.

Measurement principle	Sections	Advantages	Disadvantages	Applications
Spectroscopes	3.1	– Allows determination of unknown gas compositions – High resolution and accuracy – Determination of isotopologues possible – Matured technology – Large number of commercial products	– Complex, expensive – Not inline-capable	– Detection of unknown gas compositions – Trace gas analysis – Determination of the purity of gases
Gas chromatographs (GC)	3.2	– High resolution and accuracy – Matured technology – Large number of commercial products	– Complex, expensive – Not inline-capable – Needs carrier gas – Needs specific detectors for different gases	– Compounds of gases or vapourizable chemical species
Electrochemical (*Severinghaus*) CO_2 sensors	4.1	– Applicable in liquids and gases – Inexpensive production – No operational energy required	– Long response time between 2 and 4 minutes – Needs regular calibrations – Not suited for very low concentrations	– pCO_2 measurements in blood – Biomedical applications – Environment control – Pharmaceutical industry – Biotechnology – Fermentation control
Coulometric and amperometric CO_2 sensors	4.2	– Low drift – High resolution and accuracy	– Long response time – Restricted concentration range – Short service life – Complex preparation – No commercial products available yet	– Monitoring of the fermentation of beer – Monitoring of dissolved carbon dioxide in seawater – Breath-by-breath measurement of CO_2 – CO_2 measurement in greenhouses
Conductometric sensors	4.3	– Cost-efficiency – Fast sensor response – Microfabrication possible	– Low selectivity	– Biomedical applications – Atmospheric carbon dioxide determination – Measuring low ion concentrations

Quinhydrone measurement cell	4.4	– Inexpensive production – Fast sensor response – Miniaturization possible	– Insufficient long-term stability – Poor selectivity – No commercial products available yet	– Biomedical applications – Breathing gas monitoring
Solid electrolyte sensors for indirect measurement of CO_2 in hot water gas	5.1	– CO_2 in equilibrated water gas – In line measurement – Highly selective – Nernst's equation, logarithmic scale – Calibration-free – Maintenance-free – Commercially available	– High-temperature measurement – Complex calculation of CO_2 concentration – Reference gas (air) necessary – Middle expensive	– Complex water gas analysis
Solid electrolyte cells for direct CO_2 measurement	5.2	– CO_2 measurements from 5 ppm to percentage – High temperature measurement (350–750 °C) – Nernst's equation, logarithmic scale – No cross sensitivity against water vapour – Short response time (1 s) – Low calibration efforts – Long-term stability – Low price (few US$) – Commercially available	– Damage by acid gases (SO_2) – Cross sensitivity against hydrocarbons in non-equilibrated gases	– Air monitoring in classrooms and hospitals – Measurements in biogases – Breath-by-breath measurement of CO_2
Solid-state sensors based on changes in capacity and resistivity	5.3	– Simple set-up – Small sensors	– Cross sensitivities – Memory effects – Temperature-dependent – Calibration necessary – Only academic solutions yet	– No applications known yet
Liquid reagent-based opto-chemical CO_2 sensors	6.1	– Very sensitive – Fast response – Can be easily miniaturized	– Expensive – Complex measurement processing	– Medical and biological applications – Environment control – Control of dissolved carbon dioxide in mammalian cell culture processes – *In situ* measurements of seawater pCO_2

(Continued)

Table 11.1 (Continued)

Measurement principle	Sections	Advantages	Disadvantages	Applications
CO_2 detector tubes	6.2	– Low costs – Rapid response – Simplicity of use – No calibration necessary – Very low maintenance effort – Easily portable	– Low accuracy (≤±10%) – Single use only – Individual optical evaluation – No measured value storage	– Control of the concentration at the workplace – Safety control in coal mines, tunnels, caves, and breweries – Monitoring of the breathing air quality – Technical gas analysis
Fibre-optic fluorescence CO_2 sensors	6.3	– Chemically inert – Immune to electromagnetic interference – Capacity to carry out remote sensing without electrical power needed at the remote location – Small sensor size – High sensitivity, resolution, and dynamic range – Spatially resolved sensing with one single fibre – Fast sensor response – Inline-capable – Commercial solutions available		– Allows operation in high-voltage, high-temperature, corrosive, and flammable environments or other normally inaccessible areas – Sensing in remote areas
Non-dispersive infrared (NDIR) sensors	7.1	– Determination of known gases or gas mixtures – High selectivity by choosing preferential wavelengths – Potential for miniaturization – Good long-term stability – Simple set-up – Inline-capable – Matured technology		– Detection of unknown gas compositions
IR spectrometers	7.4	– Like spectroscopy (Section 3.1)	– Expensive – Not inline-capable	– Like spectroscopy (Section 3.1) – Gas or fluid chromatography – [illegible]

Photoacoustic URAS (Ultrarot-Absorptions-schreiber) detectors	8.3	– Mature products – Configuration with reference cell allows one to detect small changes in gas concentration – Gas-specific detection reduces cross sensitivities, “filtering” much more selective then for optical bandpass filters	– Concentrations > 10 ppm	– Analysis of complex mixtures
Single-channel PAS detectors	8.3	– Simple, cheap, robust – No moving parts	– Concentrations > 100 ppm	– Indoor air quality monitoring – Envisioned: safety monitoring for CO_2-based air conditioners
Photoacoustics in general	8.3	– Mature products – High sensitivity – Broad spectral coverage – “Background-free” measurement – Concentrations down to 10–100 ppb	– Gas-specific filters required to avoid cross sensitivities, this may compromise sensitivity	– Analysis of complex mixtures
Laser-based photoacoustics	8.4	– Very high sensitivity (literature data: <1 ppb) – “Background-free” measurement	– High costs, tunable lasers make for a delicate set-up – Limited laser tunability: simultaneous detection of only a few gases – Many “start-up” systems, but no mature product	– Analysis of ultra-traces in nearly pure or few-component matrices
Quartz crystal microbalance (QCM) sensors	9.2	– Extremely sensitive to mass changes of mass from adsorbed/absorbed gas molecules – Long-term stable – Frequency as output signal (quasi-digital) – Inexpensive – Inline-capable	– Tailored absorption layers for gas to be detected needed	– Detection of specific gas components or fluidic species – Measurement of solubility of gases in materials (absorption layer) – Adsorption/desorption of gases at surfaces
Surface acoustic wave (SAW) sensors	9.3	– Like QCM sensors (Section 9.2) – Remote operation without power supply possible (Section 9.3.3)	– Like QCM sensors (Sect. 9.2) – No commercial products available yet	– Like QCM sensors (Sect. 9.2) – Suitable for use in hazardous atmospheres

(Continued)

Table 11.1 (Continued)

Measurement principle	Sections	Advantages	Disadvantages	Applications
Ultrasonic sensors	9.4	– High precision – Fast sensor response – Non-invasive – Robust, long-term stable – Inline-capable	– Tailored absorption layers for gas to be detected needed – Often strong influence of temperature – No commercial products available yet	– Detection of binary gas compositions or liquids – Suitable for use in hazardous atmospheres
Hydrogel-based CO_2 sensor with pressure transducer	10.1	– Electrochemical reference electrode is not necessary – Small dimensions – Manufacturing on wafer scale possible	– Long response time between 2 and 4 minutes – Non-linear calibration curve – No commercial products available yet	– Chemical microanalysis – In medicine: diagnosis of gastrointestinal ischaemia – Various medical applications – Expected applications in the automobile industry and for environment monitoring
Miniaturized CO_2 sensors	10.2	– Extremely small dimensions – Low sample quantities required	– No adequate reference electrode for ion-selective field effect transistors (ISFETs) – Complicated preparation – No commercial products available yet	– Measurements under biofilms – Neurophysiological measurements – Biopotential recording – Blood gas analysis
Heat conduction-based sensors	10.3	– Corrosion resistant – Continuous measurement – Low concentrations down to 1 vol% – Usable in harsh environments and for flammable gases – Excellent long-term stability – Inline-capable – First commercial products available	– Heat conductivity of CO_2 has to be different from carrier gas – Vulnerable to corrosion if not used in glass tube	– Measurement of binary gas mixtures
Membrane-based CO_2 sensors with pressure measurement	10.4	– Robust and long-term stable in the subsurface	– High initial costs and efforts for sensor installation	– Measurements in natural systems like soils, aquifers, gas storage sites, subsurface pipelines etc

Table 11.2 Characteristics of some CO_2 measurement methods.

Method	Severinghaus electrode	Conductivity sensor	Solid electrolyte sensor	NDIR sensor
Application (gas)	Membrane	Membrane	Direct	Direct
Application (liquid)	Membrane	Membrane	Indirect, behind membrane	Indirect, behind membrane
Sensor function	Nernst logarithmic	Linear	Nernst logarithmic	Beer–Lambert linear
Concentration range	ppm to vol%	ppm to 5 vol%	ppm to vol%	0.1 ppm to vol%
Response time	30 s	1–2 s	<1 s	Few seconds
Cross sensitivity	Water, SO_2, NO_x, H_2S	SO_2, NO_x, H_2S	Combustibles SO_2, NO_x, H_2S	Dust, aerosol
Measurement temperature (°C)	<50	20	350–750	<50
Calibration	Yes	Yes	Yes	Yes
Energy consumption	Low	Low	Higher	Higher
Maintenance	High	High	Mean	Low

the complex evaluation of a broad variety of principles and devices for measuring CO_2 to enable the reader to select the method and device most suited for his application.

Selecting a method and device for the measurement of CO_2 always should start with careful reply to the following questions:

- In which medium (gas or liquid) CO_2 is wanted to be measured?
- Should the measurement provide continuous readings or discontinuous values at longer intervals?
- Which response time t_{90} the sensor or device should have to provide reliable real-time values? (Breath gas measurement requires $t_{90} < 300\,\text{ms}$, while monitoring at moored buoys in non-flowing lakes can be carried out at $t_{90} > 10$ minutes.)
- Does the sensor/transducer have to be protected by membranes or filters from other components in the medium causing unwanted interference or damage? (Membranes and filters can prolongate the response time by one or more orders of magnitude.)
- Which interferences the measuring principle might exhibit in a particular application? (This question is often the most difficult one to answer, since interference detection is challenging.)
- What measuring range and precision the device has to be provided? (This question points also to the quality and frequency of calibration, which is often the most expensive maintenance issue.)
- How much technical, organizational, and financial input is reasonable to maintain the point of measurement?

Table 11.3 Decision support for the choice of analysis method.

Question	Decision between
Field of application	Industry Environment Research
Medium to be analysed	Gas Liquid
Mode of measurement	Continuous measurement Intermittent or with sampling
Response time	Short response required (e.g. for *in situ* measurements) Only modest requirements (e.g. for environment control)
Measuring range	Specially very high or low concentrations Wide range or restricted for specific application
Accuracy	High accuracy required (often in medicine) Only modest requirements
Selectivity	Are disturbing substances present or to be expected Well-known medium without interferences
Temperature	Normal temperatures Very high or low temperatures are possible
Service life	Important for industrial or environmental applications In scientific experiments often not critical
Expenditure, costs	Acquisition costs for sensor and installation Costs for maintenance, service, and calibration
Availability	Must be commercially available Custom-made solution for special application

The above-mentioned questions are also of broad scientific interest and guide the development of special solutions for applications, for which appropriate sensors or systems are not commercially available. A decision support is given by a simplified scheme in Table 11.3.

The following examples and trends should provide some answers to the above-mentioned questions by comparing the different types of CO_2 sensors.

Of course it does not make sense to measure the CO_2 concentration in a classroom by means of mass spectroscopy. However, it should be discussed whether a coarse analysis with a solid electrolyte sensor or a low-cost NDIR sensor is suitable for that application. Solid electrolyte sensors can already be purchased for less than US$20 and NDIR sensors, depending on the required precision, for about US$50–550. At present NDIR sensors are the preferable devices for many applications of indoor air quality monitoring, biotechnology, and environmental measurements. They require comparably low operational power (10 W), and their calibration can be carried out easily by dry and CO_2-free gas and a gas with certain amount of CO_2.

Generally, it is obvious that measuring techniques based on physical principles are more preferred than electrochemical-based ones. For stationary high-precision measurements, spectrometric (IR), MS, and chromatographic techniques (GC, HPLC) are preferred. In many cases the problem consists in sampling. Although water vapour does not disturb the determination, water condensation is crucial for nearly every physical method. Compared with optical methods, electrochemical sensors require significantly less manufacturing effort as well as much less complex electronic equipment necessary for operation and for data acquisition. In contrast to these advantages, the effort for maintenance and calibration is considerably higher than with optical sensors if similar precision and selectivity has to be established. Since sensor signals are obtained directly, (*in situ*) real-time information for process control is delivered. Therefore, they are preferred tools for screenings in field applications. On the other hand, electrochemical sensors cannot completely replace the standard methods in laboratories in terms of precision, detection limit, etc.

Thermal conductivity sensors, in prize much higher than NDIR sensors (€1500 compared with €600), can be applied in gas matrices in which only the CO_2 is changing and the influence of the other components on thermal conductivity is less than that of CO_2. Appropriate results are obtained in two-component gas mixtures as well as air and CO_2. These sensors are applied, for instance, to measure CO_2 in the fermentation gas of beer brewing. Due to the relatively high operating temperatures (250 °C), open filaments will degrade in oxygen-containing gas mixtures due to corrosion. Sensors with glass-encapsulated filaments resist these attacks but exhibit higher response times as compared with those having open filaments.

Up to now it is believed that other methods to measure CO_2 (coulometric and conductometric sensors) are niche products. Due to the specificity of some applications, it is necessary to develop special probe systems and to use unusual methods for them, which are not available on the market.

CO_2 solid electrolyte sensors can be applied successfully in all cases of long-term measurements in air, in breath analysis, and in processes measuring especially at higher temperature. As compared with sensors with aqueous electrolytes, the main advantages consist in a short response time and maintenance-free operation with extended calibration-free periods. Water vapour and traces of combustibles can generate cross sensitivities of this kind of sensors as it is described in Section 5.2.2. They can be easily miniaturized and require an electric heating power of 2–3 W due to their elevated operating temperatures ranging between 550 and 700 °C.

As compared with NDIR sensors, Severinghaus and solid electrolyte-based sensors are simpler in their set-up and in most cases cheaper. The main advantage of the Severinghaus sensor consists in energy-free operation of the electrochemical cell. This kind of sensor is the only one for direct application in liquids. However, for its application a basic knowledge of electrochemistry and periodic calibrations as well as maintenance adjusted to the requirements and complexity of the matrix is necessary. Methods using membranes for gas separation from liquids with the subsequent CO_2 analysis by NDIR are frequently excessive in maintenance.

In Part III of the book, Chapters 12–16, CO_2 measurements on a number of different application fields are presented. It can be assumed that in each case the most suitable measuring system was selected. Therefore, Table 11.4 can be helpful in selecting a CO_2 sensor for a similar task.

Table 11.4 Examples for CO_2 measurements on different fields of application.

Field of application	Sections	Examples for CO_2 measurements	Measuring method
Environment	12.2	– Atmospheric CO_2 – Measurements at the Mauna Loa Observatory	– NDIR sensor
	12.3	– Oceanic research – Deep-sea research – CO_2 in freshwater lakes and boreholes	– NDIR sensor – Severinghaus sensor – Colorimetric method
Safety control	13.2, 13.3	– CO_2 at workplaces and laboratories – CO_2 in classrooms – CO_2 in underground mines, caves, cellars, tunnels, and sewers – CO_2 in fruit and vegetable storage depots and wine cellars	– NDIR sensor – Severinghaus sensor – Detector tubes – Solid electrolyte sensor – Gas chromatography
Biotechnology, industry	14	– CO_2 in beverage and food industry – CO_2 in bioreactors – CO_2 in biogas plants	– Severinghaus sensor – NDIR sensor – *pT* sensor – Thermal conductivity sensor – Opto-chemical method – IR sensor
Biology	15.1, 15.2	– CO_2 in fish farming – CO_2 measurements on zebra mussels – CO_2 measurements on butterfly pupae – CO_2 measurements on honeybees	– Severinghaus sensor – Analytical methods
	15.3	– Study of photosynthesis – *In situ* determination of the respiratory quotient (RQ)	– Severinghaus sensor – IR sensor – Solid electrolyte – Cavity-enhanced Raman spectrometry (CERS)
Medicine	16.3, 16.4	– CO_2 measurement in blood	– Severinghaus sensor – ISFET sensor – Optical fluorescence method – Solid electrolyte sensor
	16.5	– Capnography, breathing gas analysis	– Colorimetric sensor – IR sensor – NDIR sensor – Ultrasonic sensor – Solid electrolyte sensor
	16.6	– CO_2 measurements on baby mattresses	– IR sensor – Solid electrolyte sensor

Part III

Applications

12

Environmental CO_2 Monitoring

Detlev Möller[1] *and Wolfram Oelßner*[2]

[1] *Brandenburgische Technische Universität Cottbus und Senftenberg, Fakultät für Umwelt, Verfahrenstechnik, Biotechnologie und Chemie, Institut für Umweltwissenschaften, Platz der Deutschen Einheit 1, 03046 Cottbus, Germany*
[2] *Kurt-Schwabe-Institut für Mess- und Sensortechnik e.V. Meinsberg, Kurt-Schwabe-Straße 4, 04736 Waldheim, Germany*

This chapter will present background knowledge on the CO_2 behaviour and measurements in atmosphere, oceans, and different waters.

12.1 CO_2 and Climate Change

12.1.1 The Carbon Dioxide Environmental Problem

In recent decades, humans have become a very important force in the Earth system, demonstrating that emissions and land-use change are the causes of many of our environmental challenges. These emissions are responsible for the major global reorganizations of biogeochemical cycles. With humans as part of nature and the evolution of a man-made changed Earth system, we also have to accept that we are unable to remove the present system into a pre-industrial or even pre-human state because this would mean disestablishing humans. The key question is which parameters of the climate system allow the existence of humans under which specific conditions. Nevertheless, major regional and global environmental issues, such as acid rain, stratospheric ozone depletion, pollution by persistent organics (POP), and tropospheric ozone pollution, resulting in adverse effects on human health, plant growth, and ecosystem diversity, have been identified and controlled to different extents by various measures. Some key issues remain unsolved, such as the further increase of greenhouse gases (GHG), most importantly that of CO_2.

The biosphere comprises the Earth's crust, atmosphere, oceans, and ice caps and the living organisms that survive within this habitat. Geochemists define the biosphere narrower as being the total sum of living organisms (the "biomass" or "biota" referred to by biologists and ecologists). Climate researchers have introduced the *climate system* being the sphere affecting life, which comprises those parts of the lithosphere, hydrosphere, and atmosphere where life exists (the reader may now see that this definition is close to that of the biosphere in

Carbon Dioxide Sensing: Fundamentals, Principles, and Applications,
First Edition. Edited by Gerald Gerlach, Ulrich Guth, and Wolfram Oelßner.

a wider sense). Today the term anthroposphere is also used. The idea of a close interrelation between humans and the biosphere is topical in understanding the Earth system, i.e. climate change, and is used with the terms global mind [1, 2] and Anthropocene [3] to characterize the present epoch.

In nature, i.e. without humans industrialized world (i.e. before Industrial Revolution), biogeochemical cycles were almost in steady state (i.e. production equals consumption) or, if not (geochemical evolution), they were not influencing actual life on Earth. The today's problem is that humans interrupted such cycles, resulting in the environmental issues known. In case of small fluxes of matter and substances having a short lifetime in the climate system (less than a few years), this not yet results in climate change. Another more economic problem is that the rate of raw material extraction from natural reservoirs is larger than natural recycling. Hence, mankind must search for alternative resources and technologies (such as the "energy change") until end of this century. However, the key problem of anthropogenically released CO_2 is the extremely long residence time in air and the ocean in the order of many hundreds and even thousands of years. That means that – even after CO_2 abatement – the environmental problems, i.e. climate change, will last further for several 100 a.

12.1.2 Rise of Atmospheric CO_2

Since the beginning of the Industrial Revolution, humans have been emitting about 330×10^{15} g CO_2-C from the combustion of fossil fuels and cement production [4] and about 150×10^{15} g CO_2-C from land-use change, mainly deforestation [5]. Since 1950 these sources have been amounting to about 350×10^{15} g CO_2-C, i.e. 73% of the total carbon release. Measurements and constructions of carbon balances, however, reveal that less than half of these emissions remain in the atmosphere [6]. The anthropogenic CO_2 that did not accumulate in the atmosphere must have been taken up by the ocean and by the land biosphere.

Anthropogenic CO_2 emissions have been growing about four times faster since 2000 than during the previous decade despite efforts to curb emissions in a number of Kyoto Protocol signatory countries. Emissions from the combustion of fossil fuel and land-use change reached 10×10^{15} g CO_2-C in 2007. Natural CO_2 sinks are growing but slower than the atmospheric CO_2 growth, which has been increasing at 2 ppm since 2000 or 33% faster than in the previous 20 years. Natural land and ocean CO_2 sinks, which removed 54% (or 4.8 billion tons per year) of all CO_2 emitted from human activities during the period 2000–2007, are now becoming less efficient. Although the size of these sinks continues to grow in response to greater concentrations of CO_2 in the atmosphere, they are losing efficiency as feedbacks between the carbon cycle and climate increase [7].

It is easy to calculate the amount m_c of CO_2 accumulated in the atmosphere. The recent CO_2 mixing ratio (394 ppm) corresponds to 848×10^{15} g CO_2-C. Taking into account the total mass of the atmosphere (5.2×10^{21} g), it follows that

$$m_c = 0.0394 \times 10^{-2} \times \frac{12}{29} \times 5.2 \times 10^{21}\ \text{g} = 848 \times 10^{15}\ \text{g} \tag{12.1}$$

where 12 and 29 are the molar masses of carbon and of air, respectively. The changing air concentrations of trace compounds do not influence the value of

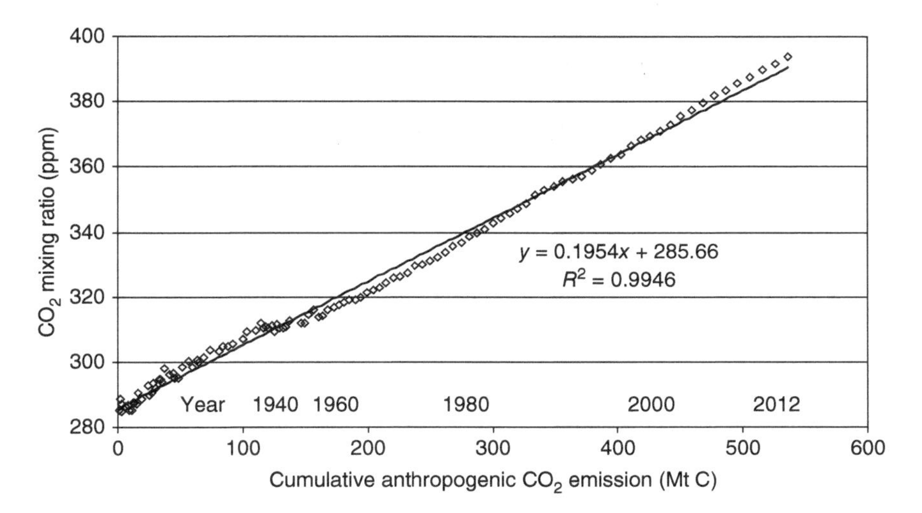

Figure 12.1 Cumulative anthropogenic CO_2 emission (Source: Data from [4].) versus atmospheric CO_2 mixing ratio (data as in Figure 12.2).

the total mass of the atmosphere in any detectable way. Hence, the reference level of 285 ppb (related to about 1850; see Section 12.3.2) is equivalent to 615×10^{15} g CO_2-C. The totally added CO_2 mass since 1900 of 232×10^{18} g CO_2-C is therefore almost half (exactly 43%) of the totally emitted carbon (536×10^{15} g) from fossils fuels, cement production, and land-use change. Due to the linear relationship between accumulated CO_2 in the atmosphere (airborne fraction) and the CO_2 mixing ratio over this period, Figure 12.1 suggests that this uptake fraction will remain constant. This means that the emitted carbon dioxide is directly partitioned among reservoirs. Without ocean and land uptake, the atmospheric CO_2 concentration would have increased to 540 ppm, if it had all stayed there.

The most important reservoir is the backmixed surface layer of the ocean. The further transport of anthropogenic carbon from the surface to the ocean bulk (deep sea) is believed to be extremely slow (hundreds of years). A first quantification of the oceanic sink for anthropogenic CO_2 is based on a huge amount of measured data from two international ocean research programmes [8]. The cumulative oceanic anthropogenic CO_2 sink for 1994 was estimated to be $(118 \pm 19) \times 10^{15}$ g CO_2-C. As the characteristic time for the exchange between surface and deep water is that large that all CO_2 added to the atmosphere so far is stored only within the upper layers. About 30% of the anthropogenic CO_2 is fixed at depths shallower than 200 m and nearly 50% at depths below 400 m. Only 7% of total anthropogenic CO_2 is present deeper than 1500 m [8]. The anthropogenic emissions are added to the atmosphere, continuously increasing the equilibrium carbon content of the surface layer as a result of the increasing partial pressure of the carbon dioxide in the gas phase (see Section 12.3.3).

Table 12.1 shows the total CO_2 emission from fossil fuel use and cement production and the resulting CO_2 increase and uptakes.

For this base year (1994), the cumulative fossil fuel CO_2 emission is 238×10^{15} g CO_2-C [4]. This means that 50% of the CO_2 added to the atmosphere

Table 12.1 Global carbon budgets in 10^{15} g C a^{-1}.

Source	[9]		[8]		
Period	1990s	2000–2005	1800–1994	1980–1999	Total
Industrial CO_2 emission[a)]	6.4 ± 0.4	7.2 ± 0.3	244 ± 20	117 ± 5	329[b)]
Land-use change	1.6[c)]	1.6[c)]	100–180	24 ± 12	156[d)]
Sum (total emission)	8.0	8.9	344–424	141	485
Atmospheric increase	3.2 ± 0.1	4.1 ± 0.1	165 ± 4	65 ± 1	225[e)]
Difference (biospheric uptake)	4.8	4.8	179–259	76	260
Ocean uptake	2.2 ± 0.4	2.2 ± 0.5	118 ± 19	37 ± 8	165[f)]
Terrestrial uptake	2.6[g)]	2.6[g) h)]	61–141	39 ± 18	95

a) From fossil fuel use and cement production.
b) 1751–2005 [4].
c) Range 0.5–2.7.
d) 1850–2005 [10].
e) Calculated from the difference between 384 and 280 ppm CO_2.
f) To be assumed 50% of cumulative industrial CO_2 emission.
g) Range 0.9–4.3.
h) Forest sequestration for the period 1993–2003 estimated to be 0.3×10^{15} g C a^{-1} [7].

is captured by the world's oceans. The total oceanic dissolved carbonate carbon corresponds to 0.028 g l^{-1} as carbon in seawater, taking into account the volume of the world's oceans. The experimentally estimated seawater standard carbonate carbon is 0.0244 g l^{-1} seawater [11]. In the upper 200 m of the ocean, the total deposited anthropogenic CO_2 (Table 2.9 and assuming that 30% is within this layer) only contributes to 3% to dissolved inorganic carbon (DIC). However, it is very difficult to measure trends in the DIC because of man-made changes. Changes between ocean sediment and deep water play no role on a timescale of a few 100 a of anthropogenic perturbation of the surface layer CO_2. But over several 1000 a, nearly all anthropogenic CO_2 will be captured by the ocean.

For calculating the global carbon budget, the problem of the uncertainty of large global values, such as the marine and terrestrial photosynthesis (CO_2 uptake from the atmosphere) and respiration (CO_2 release to the atmosphere), accelerates. The global total value of CO_2 assimilation amounts to c. 200×10^{15} g a^{-1} C. Assuming an optimistic uncertainty of 10%, the bias (20×10^{15} g a^{-1}) is twice as large as the annual anthropogenic carbon release into the atmosphere. Therefore, it is impossible to constrain a budget from the differences of such values.

The total flux F (budget) in terms of net surface CO_2 surface exchange includes fossil fuel burning (including cement production), fires (forest, savanna, and wood fuel), terrestrial biosphere exchange, and exchange with the ocean:

$$F(x, y, t) = F_{ff}(x, y, t) + F_{fire}(x, y, t) + F_{bio}(x, y, t) + F_{ocean}(x, y, t) \tag{12.2}$$

F_{bio} and F_{ocean} can be positive or negative depending on the location (x, y) and time t, thereby showing fluctuation and trends. F_{ff} and F_{fire} are considered positive (this sign is related to a flux into the atmosphere). However, it can be assumed that on average over a few years climatological mean), the natural carbon budget is zero. Small fluctuations in the yearly difference assimilation (gross primary production, GPP) respiration result in fluctuations of the atmospheric CO_2 growth rate.

Assuming that the global net ecosystem production (NEP) is likely to be zero or at least negligible, one has only to consider the budget between atmospheric CO_2 content (its increase based on measurements – the most reliable value in global CO_2 budgeting), the emissions by different sources (fossil fuel burning, cement production, land-use change, and other biomass burning), and the oceanic increase (Table 2.9).

The impact of global climate change on future carbon stocks is particularly complex. These changes might result in both positive and negative feedbacks on carbon stocks. For example, increases in atmospheric CO_2 are known to stimulate plant yields, either directly or via enhanced water-use efficiency, and therefore enhance the amount of carbon added to soils. Higher CO_2 concentrations can also suppress the decomposition of stored carbon because C/N ratios in residues might increase and because more carbon might be allocated below ground. Predicting the long-term influence of elevated CO_2 concentrations on the carbon stocks of forest ecosystems remains a research challenge [6]. The severity of damaging human-induced climate change depends not only on the magnitude of the change but also on the potential for irreversibility. It has been shown [12] that climate change that takes place because of increases in CO_2 concentration is largely irreversible for 1000 a after emissions stop (see Section 12.2.4 for the residence time of anthropogenic CO_2 in the atmosphere). There are strong arguments that the anthropogenic-caused CO_2 increase is largely irreversible. Hence, stopping emissions will not solve (though might smooth) climate change problems. As a consequence, CO_2 capture from the atmosphere remains the challenge for climate sustainability. The oceans seem to be the final sink of anthropogenic CO_2 but after thousands of years. Moreover, the seawater uptake capacity will decrease, and ocean acidification will result in serious ecological consequences (see Section 12.3.3).

12.2 Atmospheric CO_2

12.2.1 Pre-industrial CO_2 Level

The knowledge of the pre-industrial CO_2 level is important from two reasons: (i) such figures are needed to estimate how much CO_2 humans have been added to the atmosphere since beginning of the Industrial Revolution and (ii) this value is the reference for all human-caused impacts that depend from the actual CO_2 level, e.g. climate forcing and ocean acidification. This reference value has been estimated by several authors to be in the range of 243–290 ppm ([13] and citations therein, p. 242). The today's most accepted pre-industrial value is 280 ± 10 ppm, whereas mainly data from bubbles in ice cores have been employed [14–16]. From historical ocean CO_2 data, a smaller value of 268 ± 13 ppm has been estimated [17]. From selected historical CO_2 measurements in the nineteenth century, values of c. 290 ppm [18, 19] and 292 ppm [20] have been determined, respectively.

Joseph Black was the first who found CO_2 in the air of Edinburgh, probably between 1752 and 1754 ([21], p. 74). In 1787 de Saussure (the elder) employed lime water as a test for CO_2 in the air [22]. First attempts to determine CO_2 in atmospheric air were done at the end of the eighteenth century by von Humboldt and Dalton, however only showing that its level is less than 0.1% [23]. Thénard, de Saussure, Boussingault, and von Gilm measured CO_2 levels around 400 ppm (and more) between 1815 and 1850 at different sites in Central Europe. All later researchers agreed that such values were much too high. Letts and Blake published two remarkable papers [24, 25] providing a new accurate CO_2 determining method and the most extensive critical overview on CO_2 observations in the nineteenth century. It is remarkable that many other authors agreed to a value of 0.03% (according to 300 ppm) as natural reference value [24, 26–38].

CO_2 has been analysed wet-chemically by absorbing from air always in alkaline solutions, i.e. aqueous solution of $Ba(OH)_2$. After precipitating of $BaCO_3$ after one to two hours, an extracted part of the clear solution is titrated (mostly by oxalic acid). Many modifications have been made (apparatus such as bottles or tubes with an aspirator, instead of titration gravimetric estimation, other titrating acids such as HCl solution, different indicators for the neutralization point, etc.). It has been found that the Pettenkofer ordinary process (so-called bottle method) and the aspiration method give systematically too high values due to additional absorption of CO_2 by the baryta water, e.g. when handling solutions in laboratory air [24, 25, 27, 39]. *In that time* careful simultaneous sampling and analysis have been made by the Pettenkofer new process [25], showing that the Pettenkofer ordinary process results in values that are c. 25–35% too high. Fresenius [40] estimated for air analysis a precision of ±0.002% absolute CO_2 level (according to ±20 ppm), whereas Haldane [41] gave ±10 ppm as precision (3–7% relatively). Hempel [42] was the first who compared the accuracy of different methods. He stated that accuracy is not better than ±100 ppm (50–100 ppm for ambient air analysis, 100–200 ppm for indoor CO_2 analysis) and ± 30–50 ppm by comparing Pettersson's and Pettenkofer's method. Today it is self-evident that measurements are corrected to standard conditions and related to dry air. In many papers before 1870, this is not clearly expressed.

When discussing historical data, it has to be taken into account that CO_2 in laboratory air can go up to 0.2% [42], and, therefore, any analytical operation without appropriate care might lead to higher analytical figures. Another fact that needs to be considered when comparing different historical CO_2 values is the town influence, which was estimated as 30–200 ppm higher than in the rural surroundings [28, 31], which is very close to today's values of city dome CO_2 (see below). Today, higher town CO_2 is almost a result of traffic, whereas in the past domestic coal combustion (heating, cooking, manufactures) and locomotives were the main CO_2 sources. It was found that in cities (shown for Dresden), the soot pollution around 1900 was likely by a factor of 50 and more higher than today [43]. Of course, CO_2 is not proportional to soot emission but is supported by this finding to be an important local pollutant in that time.

Moreover, seasonal CO_2 variations at remote continental sites amount to 10–20 ppm and at Mauna Loa only about 5 ppm, whereas inter-annual CO_2 variation is comparatively small with 2 ppm Only very few of the historical measurement series were able to reflect a seasonal cycle due to the limited accuracy with 10–20 ppm and a too small number of measurements. However, the seasons influence on CO_2 already was stated by de Saussure, who stated it in summer higher than in winter. This was supported both by Fittbogen and Hässelbarth [44] who found a minimum in December and by Petermann and Graftiou [45] with very small variations of only a few parts per million. Marie-Davy [46] observed a CO_2 maximum in December. Whereas seasons and towns influences on the CO_2 level may be excluded or to be assessed of minor importance in evaluation of historic background levels, there remain two significant factors responsible for large variations among the observers: (i) the sampling height above ground and (ii) the time of sampling. The most serious effect of CO_2 variation is the diurnal cycle, which amounts to 30–80 ppm (as maximum–minimum) and appears as sinusoidal curve. Hence, the difference between continental daytime means and nocturnal means could range between 10 and 30 ppm.

Today, CO_2 is measured continuously, automatically giving a mean daily value, resulting in representative monthly means and, subsequently, annual mean figures (when the time gaps of measurements are not too large). In the past, often only one value (and almost daytime) per day or only few values per month have been determined. According to the time of day, such figure cannot reflect a daily average. Only a large number of measurements, made at different times, can reflect a mean daytime value. The majority of CO_2 measurements have been carried out at business times; hence, a corrected daily mean must be around or higher than 10 ppm The diurnal amplitude depends strongly on the local situation. Principally, this was already known at the end of the nineteenth century, but extensive sampling over longer periods was impossible due to the long time for sampling and single analysis (several hours) and, hence, large manpower needed. More CO_2 is present in the ground air from lower levels than from higher levels. The gradient has a definite connection with the season, rainfall, and the activity of plants (plant transpiration overnight) has a significant effect (see Figure 13.7). Comparable results are only obtained when the influence of plant activities (photosynthesis and plant transpiration) can be excluded. Soil CO_2 emission has been estimated to be $0.175\,\mathrm{l\,m^{-2}\,d^{-1}}$ [47].

In summary, historically published values after 1870 between 270 and 360 ppm can be seen as no principally erroneously but are not representing a true mean CO_2 figure. As found from the few inter-comparisons, the scattering of only single or a few CO_2 measurements can be ±50–100 ppm due to different reasons: methodical errors by different authors and/or methods, timely influences (almost diurnal cycle), and different local influences (but these are likely of minor influence). Consequently, it makes no sense to list and to interpret published historical CO_2 data based on only a few measurements. Therefore, only results from long-term observations and/or more than 50 single measurements will be regarded here in the following.

Also some of the longer times series show large timely (but not periodically as it is known today) variations of the single measurements, which can be interpreted

Table 12.2 Characteristics of two periods of Montsouris Observatory monitoring.

Period	Annual mean	Min.–max.	Number of mean values
1876–1889	283 ± 8	270–294	9
1890–1910	311 ± 6	300–325	20

as the scattering of the analytical figure by errors and likely predominant by the different time of the single measurements and the subsequent average. This is best illustrated by the well-known Montsouris Observatory monitoring from 1876 to 1910. Firstly, the inter-annual (5–15 ppm) and monthly variations (>10 ppm) cannot be explained by natural physical reasons [1], and secondly, the marked rise by 27 ppm in 1890 suggests changes in the sampling and the analytical procedure (Table 12.2).

The obvious problems with the Montsouris analysis (mentioned already in [48]) is illustrated by the data published by Marie-Davy [46] despite his explanation of different air mass influences (without doubt, the Montsouris station north of Paris in a park – today within Paris – is much less suitable than the Dieppe station for background estimates). The statistics of these measurements from April 1876 until December 1879 illustrate the above discussed uncertainties: the standard variations of total monthly means and for annual means amount to 35–56 and 12–41 ppm, respectively (monthly minimum–maximum of 244–359 ppm).

From the cited data one can assess the town contribution to CO_2 (remarkable only in winter) to by as mean 40 ± 20 ppm and a day–night difference of 20–30 ppm Assuming that 285–295 ppm is the range for a "best" daytime average, it follows a daily mean in the range of 295–310 ppm with the most probable mean of 300–310 ppm for continental background. Dumas [49] states 294–310 ppm CO_2 as "normal value" (background in today's sense). According to NOAA, Mauna Loa CO_2 represents the global CO_2 average level within 1 ppm range. Baumann [50] states in an excellent overview that the (mean) CO_2 content of the atmosphere varies only small. However, the absolute minimum and maximum from all these measurements were 260 and 354 ppm, respectively [50]. This is much more than it can be explained by natural timely variations from such relatively remote sites.

The measurement series carried out in Boston by Benedict [32] between April 1909 and January 1912 is probably the longest and most precise of that time. He tried carefully to avoid town influences. By separating the values, he derived a seasonal amplitude of about 9 ppm and a town contribution (adding to the rural background) of about 45 ppm This means that the most likely background mean is 307 ppm.

Krogh [33] states from Benedict's analysis the background averages for CO_2 to be 0.030 ± 0.001% identical with his own measurements (adding 10–70 ppm by town influence from Copenhagen). To give an example, how different values become when averaging, Callendar [51] cites Benedict's analysis as mean of 317.5 ppm, which is much higher for all cases but confirms Krogh's value of 300 ppm.

Callendar [18, 19, 51] and From and Keeling [20] state that the pre-industrial CO_2 levels derived from historical measurements are 290 and 292 ppm, respectively. However, the limited data sets (despite the huge total number of measurements) do not allow any trend construction. We can assume that most data were obtained from daytime (this value would be lower compared with a daily mean) and from the warm season (this value would be lower compared with the annual mean). Hence, a certain value of c. 10 ppm could be added to represent annual mean values. On the other hand, local CO_2 influence from heating can be mostly excluded in the warm season, and the measurement sites have been selected to avoid industrial and town CO_2. The best guess means from measurements between 1870 and 1910 are between 290 and 300 ppm Without going into speculations how large the corrective is, it can be conducted that historical analyses give evidence for a pre-industrial CO_2 (period 1870–1910) mean level to be around 300 ppm (see Table 12.3). This is slightly higher than values derived from ice cores (see Section "The City Dome CO_2") but may be within the uncertainties of chemical air analysis.

12.2.2 Pre-industrial CO_2 Level Derived from Ice Core Data

Ice cores are unique with their entrapped air inclusions, enabling the direct recording of past changes in atmospheric trace gas composition. The CO_2 record presented in Figure 12.2 is derived from DE08 ice cores obtained at Law Dome, East Antarctica, from 1849 to 1978. The Law Dome site satisfies many of the desirable characteristics of an ideal ice core site for atmospheric CO_2 reconstructions including the negligible melting of the ice sheet surface, low concentrations of impurities, regular stratigraphic layering undisturbed at the surface by wind or at depth by ice flow, and a high snow accumulation rate.

Ice cores recovered from the Antarctic ice sheet reveal that the concentration of atmospheric CO_2 at the Last Glacial Maximum (LGM) 21 000 years ago was about one-third lower than during the subsequent interglacial (Holocene) period started 11 700 years ago [70–72]. Longer (up to 800 000) records exhibit similar features, with CO_2 values of c. 180–200 ppm during glacial intervals [73]. Prior to 420 000 years, interglacial CO_2 values were 240–260 ppm rather than 270–290 ppm after that date [74]. The variations in atmospheric CO_2 over the past 11 000 a preceding industrialization is just a fifth of the CO_2 increase observed during the industrial era. During three interglacial periods prior to the Holocene, CO_2 did not increase, and this led to a hypothesis that pre-industrial anthropogenic CO_2 emissions could be associated with early land-use change and forest clearing [75, 76]. The conclusion of the above studies was that cumulative Holocene carbon emissions as a result of pre-industrial anthropogenic land-use and land-cover change were not large enough (50–150 Pg C during the Holocene before 1850) to have had an influence larger than an increase of c. 10 ppm on late Holocene observed CO_2 concentration increase [77].

Air bubbles were extracted using the so-called "cheese grater" technique. Ice core samples weighing 500–1500 g were prepared by selecting crack-free ice and

Table 12.3 Historic CO_2 measurements (mixing ratio in ppm).

Site or station	Year	Mixing ratio	*n*	References
Montsouris	1881–1890	287 ± 8 (277–302)[a]		[52]
		293 ± 2.5 (289–297)[b]		
Gembloux	1889–1891	294 ± 18	525	[45]
Rostock	1863–1864	360 ± 21 (320–405)[c]	41	[53]
	1868–1871	291 ± 12 (272–323)	36	
Innsbruck	1856–1857	415 ± 23 (381–458)	18	[54]
Tabor[d]	1874–1875	342 ± 8 (328–362)	27	[55]
Dorpat[e]	1888	262 (189–336)	556	[56]
	1889	269 (182–375)[f]	601	[57]
Dieppe	1872–1873	294	92	[58] [g]
	1879	298[h]	91	[48] [g]
Vincennes[i]	1881	284 ± 13 (270–317)	37	[59]
Paris	1881/1882	319 ± 26 (301–422)	20	[59]
	1882	293 ± 6 (288–306)	8	[59]
Rural site	1881	300 ± 15 (273–329)	8	[59]
Pic du Midi	1881	286 ± 9 (269–301)	14	[59]
Atlantic Ocean	1866	300		[60]
Halle, Germany	1845	310	150	[61]
Leeds	1879	313		[62]
Belfast	1897	300		[24, 25]
Kew	1898–1901	294 (243–360)	94	[29]
Boston	1909–1912	307	645	[32]
Stockholm	1885–1886	303		[63]
Copenhagen	1917	300	40	[33]
Sweden[i]	1920–1923	304		[64]
United States[j]	1930s	310		[18]

n, number of samples or measurements.

a) Based on annual means.
b) Mean from all monthly means.
c) Doubt by the author himself.
d) Bohemia.
e) Today Tartu, Estonia.
f) Variation among monthly means are 250 to −283 ppm; day–night difference to be on average 28 ppm: daytime average 258 ppm (n = 379) and night-time average 286 ppm (n = 222).
g) All data have been related to standard conditions (dry air, 0 °C, 760 Torr).
h) Daytime mean 289 ppm and night-time mean 308 ppm.
i) Costal site.
j) Rural mean.
k) Suburb of Paris.

trimming away the outer 5–20 mm. Each sample was sealed in a polyethylene bag and cooled to −80 °C before being placed in the extraction flask where it was evacuated and then ground into fine chips. The released air was cryogenically dried at −100 °C and collected in electropolished stainless steel traps, cooled to about −255 °C. The ice cores were dated by counting the annual layers in oxygen isotope ratio ($\delta^{18}O$ in H_2O), ice electroconductivity measurements, and hydrogen peroxide (H_2O_2) concentrations. For these three parameters, each core displayed clear, well-preserved seasonal cycles allowing a dating accuracy of ±2 years in 1805 and ± 10 years in 1350. Further details on the site, drilling, and cores as well as on the extraction technique are provided in [65, 78–80].

These atmospheric CO_2 reconstructions (Figure 12.2) offer a record of atmospheric CO_2 mixing ratios from the years 1006 to 1978, having an excellent overlapping with the Mauna Loa CO_2 record between 1959 and 1978 (Figure 12.3). The uncertainty of the ice core CO_2 mixing ratios has been reported as 1.2 ppm [65]. Pre-industrial CO_2 mixing ratios were in the range of 275–284 ppm, with lower levels between the years 1550 and 1800, probably because of the colder global climate [65]. Law Dome ice core CO_2 records show major growth in atmospheric CO_2 levels over the industrial period, except during 1935–1945 when levels stabilized or decreased slightly.

As stated in Section "Timely Variations", air chemical analysis give evidence for atmospheric CO_2 levels around 300 ppm before 1920. This is in excellent agreement with the value of 300.7 ± 0.9 ppm derived from the Law Dome ice core (1909–1917 average) (Figure 12.3). It is remarkable that between 1939 and

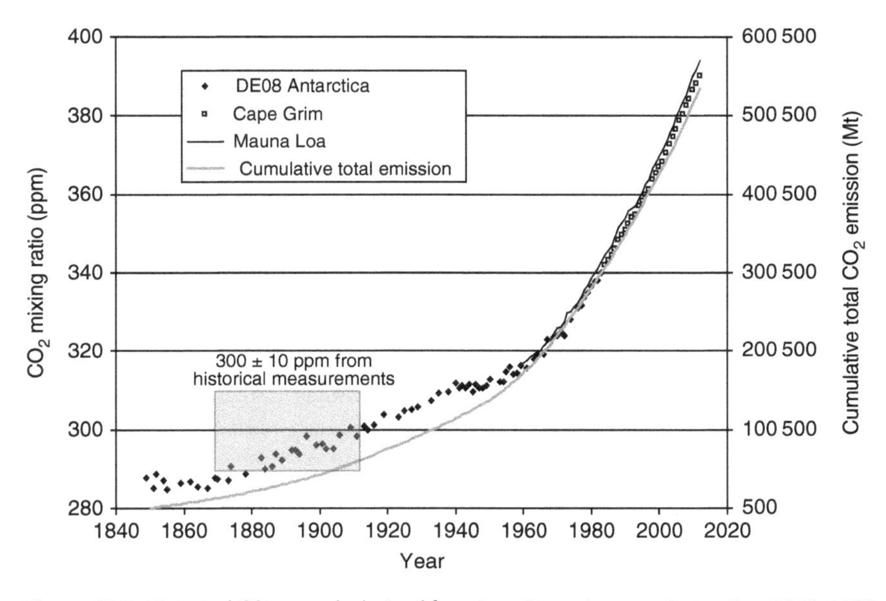

Figure 12.2 Historical CO_2 records derived from Law Dome ice core, Antarctica, 1849–1978 [65, 66], and atmospheric annual CO_2 levels (ppm) derived from *in situ* air samples collected at Cape Grim, 1980–2006 [67], and Mauna Loa, 1959–2012 [68]; cumulative global total anthropogenic CO_2 emission (fossil fuel use, cement production, and land-use change) from [69].

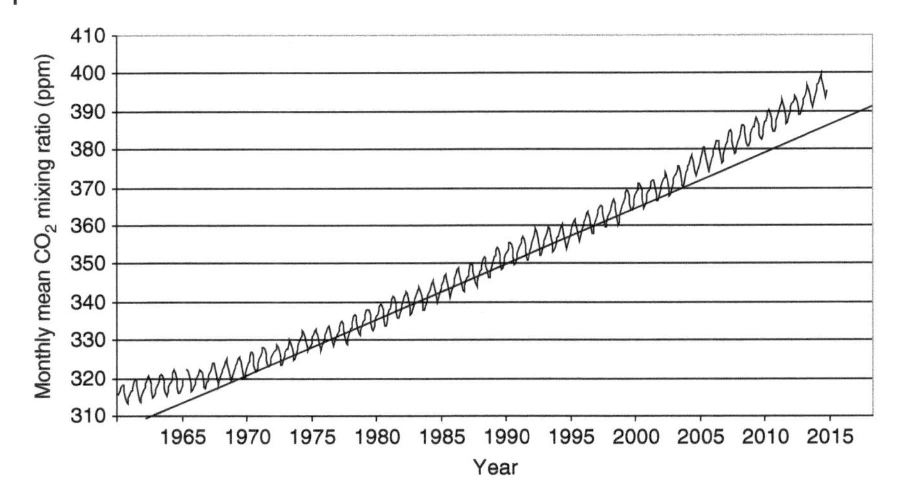

Figure 12.3 Keeling curve: monthly mean of the atmospheric CO_2 content (ppm) between January 1959 and November 2013 derived from *in situ* air samples collected at Mauna Loa, Hawaii, USA [68]. The line represents a linear regression fit between 1970 and 2000.

1946, no CO_2 increase has been observed (310.2 ± 0.1 ppm). From the ice core the average from 1010 to 1850 amounts to 280.9 ± 2.6 ppm, and the 1850–1900 average was 290.6 ± 3.8 ppm Hence, there is evidence to set the pre-industrial CO_2 level to c. 285 ppm.

12.2.3 CO_2 Increase in the Twentieth Century

12.2.3.1 Mauna Loa CO_2 Record

The Mauna Loa record (Figure 12.3), also known as the Keeling curve, is almost certainly the best-known icon illustrating the impact of humanity on the planet as a whole [81]. These measurements have been independently confirmed at many other sites around the world. When Keeling started his CO_2 measurements in the 1950s, he almost always derived the same value of 310 ppm. Previous measurements of CO_2 in the atmosphere did not show such constancy, but these older measurements had been made by wet-chemical methods, which were considerably less accurate than the dry manometric method he deployed. He concluded that (i) the Earth system might behave with surprising regularity and (ii) there was a need to make highly accurate measurements to reveal that regularity. By the early 1970s, this curve was gaining serious attention and played a key role in launching a research programme into the effect of rising CO_2 on the climate. Since then, the rise has been relentless and shows a remarkably constant relationship with fossil fuel burning (Figure 2.7). It can be well accounted for based on the simple premise that 57% of fossil fuel emissions remain airborne. Measurements of the changes in atmospheric molecular oxygen using a new interferometric technique [82] show that the O_2 content of air varies (inversely with CO_2) seasonally in both the northern and southern hemispheres because of plant transpiration and photosynthesis. The seasonal variations provide a new basis for estimating global rates of biological organic carbon production in the ocean, whereupon the inter-annual decrease constrains estimates of the rate of anthropogenic CO_2 uptake by the oceans. In addition, during the process of combustion, O_2 is removed from the atmosphere. Recent work indicates that atmospheric O_2 is decreasing at a faster rate than CO_2 is increasing [83], which demonstrates the importance of the oceanic carbon sink.

For discussing the robustness and precision of CO_2 measurements, it is worth to describe briefly the measurement procedure and kinds of fluctuations. Since 1958, CO_2 concentrations at the Mauna Loa Observatory have been obtained using a non-dispersive, dual-detector infrared (IR) gas analyser. Air samples are obtained from air intakes at the top of four 7 m towers and one 27 m tower. Four samples are collected every hour from air intakes on the taller tower and from one of the 7 m towers. Air is sampled from one tower intake for 10 minutes, followed by a second tower intake for 10 minutes, and then from a reference gas for 10 minutes. The gas analyser is calibrated by standardized CO_2-in-nitrogen reference gases twice daily. Flask samples are taken twice a month for comparison to the data recorded using the IR gas analyser. The uncertainty for measuring the reference gases amounts to c. 0.2 ppm However, agreement differences of less than 0.5 ppm between flask and analysers, or between different analysers on a short-term basis, are difficult to obtain. Monthly averages from May 1964 to January 1969 might have been in error by as much as 1.0 ppm, but the systematic error since 1970 probably has not been exceeding 0.2 ppm The precision of monthly averages is approximately 0.5 ppm In summary, monthly and annual averages of the Mauna Loa data are statistically robust and serve as a precise, long-term record of atmospheric CO_2 concentrations. Daily, monthly, and annual averages are computed for the Mauna Loa data after the deletion of contaminated samples and readjustment of the data.

The steady rise in atmospheric CO_2 concentration shown by this record has been widely interpreted as a global trend. It is remarkable that all CO_2 measurements taken at different remote sites in the world and the ice core data from Antarctica and Greenland show significant overlap in averages, reflecting the global increase of the CO_2 burden (Figure 12.2). Despite differences in seasonal cycles among the stations, which reflect the timely concentration variation (carbon budget concerns sources and sinks; see Section 2.2.3), as well as very small differences in the absolute mixing ratios (Table 12.4), the trends are similar. All variations are smaller than the measurement uncertainty, i.e. are not significant.

The annual mean rate of CO_2 growth in a given year is the difference in concentration between the end of December and the start of January of that year. This represents the sum of all CO_2 added to, and removed from, the atmosphere during the year by human activities and natural processes (Figure 12.4). The estimated uncertainty in the global annual mean growth rate is 0.07 ppm a^{-1}.

The atmospheric CO_2 growth rate averages continuously increase between 1960 and 2008 at 0.026 ppm a^{-1}. As Figure 12.4 clearly shows, the growth rate itself increases, reflecting an exponential growth. The Keeling curve (Figure 12.3) shows three periods with subsequent increasing growth rates but relatively constant rates within the periods (Table 12.5).

Table 12.4 Mean CO_2 mixing ratios (1981–1992 annual average) for different remote measurement stations.

Station	Geographic coordinates	Altitude (m above MSL)	CO_2 mixing ratio (ppm)
Alert, Canada	82°27′ N, 62°31′ W	210	344.7 ± 13.1
Barrow, Alaska	71°19′ N, 156°36′ W	11	349.9 ± 5.9[a)]
Mauna Loa, Hawaii	19°32′ N, 155°35′ W	3397	348.4 ± 5.7
Guam, Mariana Islands, Pacific	13°43′ N, 144°78′ E	2	348.8 ± 5.7
Mahé Island, Seychelles, Indian Ocean	4°67′ S, 55°17′ E	7	348.1 ± 5.9
Samoa, Indian Ocean	14°15′ S, 170°34′ W	30	347.1 ± 5.6
Amsterdam Island, Indian Ocean	37°47′ S, 77°31′ E	70	345.7 ± 5.7
Cape Grim, Tasmania	40°41′ S, 144°41′ E	94	348.3 ± 4.1
Palmer, Antarctica	64°92′ S, 64° W	10	346.7 ± 5.5
South Pole	89°59′ S, 24°48′ W	2810	347.1 ± 5.4
Mean			347.4 ± 1.7[b)]

a) Standard variation based on the annual values.
b) Standard variation based on the stations mean values.
Source: Data from [69].

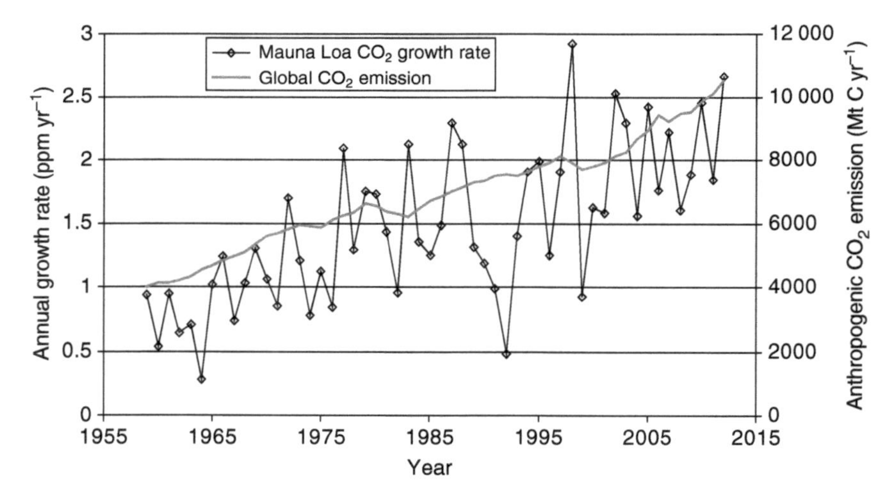

Figure 12.4 Annual mean rate of growth of CO_2 at Mauna Loa [68] and relative global CO_2 emission growth related to the value of 2462 Gt C in 1959 [4].

The overall averaged growth rate amounts to 1.47 ± 0.61 ppm a^{-1}. The 1974–1985 period of continuous atmospheric CO_2 measurements from the NOAA-GMCC (Geophysical Monitoring for Climate Change) programme at the Mauna Loa Observatory in Hawaii shows an average growth rate of CO_2 of 1.42 ± 0.02 ppm a^{-1}, and the fraction of CO_2 remaining in the atmosphere from

Table 12.5 Growth rates of CO_2 increase (ppm CO_2 a^{-1}) at Mauna Loa station.

Period	From growth rate (Figure 12.4)	From linear fit (Figure 12.3)
01/1959 to 12/1969	0.85 ± 0.31	0.76 ($r^2 = 0.62$)
01/1970 to 12/1999	1.46 ± 0.53	1.47 ($r^2 = 0.53$)
01/2000 to 11/2013	2.03 ± 0.41	2.03 ($r^2 = 0.41$)

fossil fuel combustion was 59% [84]. Hence, with constant increases expected, the atmospheric CO_2 mixing ratio is predicted to grow to 500 ppm in 2050 and to 700 ppm in 2100. There is no doubt that CO_2 has been increasing since the Industrial Revolution and has reached a concentration unprecedented for over more than 400 000 years.

The Keeling curve is non-linear but shows a remarkable highly correlated fitted polynomial function (for further discussions see [85]). Purely mathematical the year 1930 can be considered as the starting point for the CO_2 growth, i.e. as a reference level with a value of 300 ppm This does not mean zero emission in 1930 because all CO_2 emitted earlier must be taken up by oceans and terrestrial ecosystems, i.e. the airborne fraction was zero. Excepting the correctness of the ice core data, it has to be concluded that the atmospheric CO_2 level grows in different periods after different models:

- Between 1850 (and assuming that this time corresponds to the pre-industrial reference level of 286.3 ± 1.3 ppm) and 1940 (310.0 ± 1.0 ppm), the ice core record represents a significant ($r^2 = 0.95$) polynomial fit, but with a slower growth rate compared with the Keeling curve (cf. Figure 12.3).
- The period 1937–1949 (mean 310.2 ± 0.1 ppm) shows no CO_2 increase from the ice core data.

Comparing the CO_2 concentration with the global CO_2 emission, it can be seen that in the period 1910–1950, there is no coincidence in contrast to the periods before 1910 and after 1950.

12.2.3.2 Latitudinal Variation

The CO_2 content shows large seasonal cycles, and a major hemispheric gradient – CO_2 – is only well mixed on a multi-year timescale. The seasonal cycle is large in northern (10–20 ppm) latitudes and small in southern latitudes (2–3 ppm).

The natural or unperturbed component is equivalent to that part of the atmospheric CO_2 distribution, which is controlled by non-anthropogenic CO_2 fluxes from the ocean and the terrestrial biosphere. The following key features of the natural latitudinal distribution have been found [86]:

(a) CO_2 concentrations in the northern hemisphere that were lower than those in the southern hemisphere,
(b) CO_2 concentration differences that are higher in the tropics (associated with outgassing of the oceans) than those currently measured, and
(c) CO_2 concentrations over the Southern Ocean that are relatively uniform.

Anthropogenic CO_2 sources are almost located between 30° N and 60° N and negligible in the southern hemisphere. From baseline CO_2 measurement stations, it was noticed that the north–south gradient increases. It amounted to 5 ppm in 1980 (338 ppm/333 ppm) and to 7 ppm in 2009 (385 ppm/378 ppm) for Alert (82° N) and South Pole, respectively [69].

12.2.3.3 Timely Variations

The annual fluctuation in carbon dioxide is caused by seasonal variations in CO_2 uptake by land plants. Since many more forests are concentrated in the northern hemisphere, more carbon dioxide is removed from the atmosphere during northern hemisphere summers than southern hemisphere summers. This annual cycle is shown in Figure 12.5 by taking the average concentration for each month across a year. Interestingly, there is no significant change in the seasonal amplitude over the years. The amplitude amounts to 1.008 ppm, corresponding to 6–7 ppm only. It has been shown [88] that on average the seasonal variation of atmospheric CO_2 at Mauna Loa is influenced mostly by the Siberian CO_2 flux, followed temperately by Asia and North America. The inter-annual variability of less than 2 ppm of the seasonal cycle is caused mainly by the inter-annual variation in the transport of the Siberian signal to Mauna Loa.

Cyclical changes over shorter timescales are harder to spot in the records because they are usually much weaker than seasonal oscillations and can be masked by random variations in the data. Daily cycles have not been observed, but possible ambient error sources at Mauna Loa include volcanic, vegetative, and man-made effects (e.g. vehicular traffic and industry). Daily peaks in measured concentrations occur because of complex wind currents. Downslope winds often transport CO_2 from distant volcanic vents, causing elevations in measured CO_2 concentrations. Upslope winds during afternoon hours are

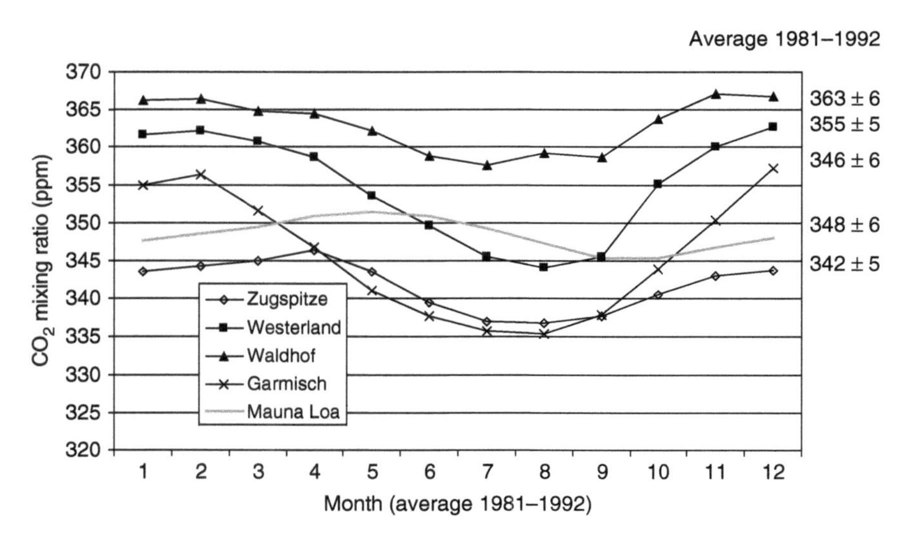

Figure 12.5 Seasonal variation of CO_2 at Mauna Loa (for data source see Figure 12.3) and German stations [69, 87].

often low in CO_2 because of photosynthetic depletion occurring in sugar-cane fields and forests. Recently, a weekly cycle has been found at Mauna Loa [89]. The measurements show that the CO_2 level rises to a peak on Mondays and then declines steadily to a minimum on Saturdays. Crucially, there was not such cycle in carbon dioxide records from the Amundsen–Scott South Pole station in Antarctica, which is far from any sources of pollution. The Antarctic measurements show the same yearly trend and seasonal cycle, but there is no significant difference between average daily values. Such short-term variations have evened out by the time CO_2 pollution reaches Antarctica.

In Germany, several CO_2 monitoring series are available on urban background stations, showing a wider range of mean CO_2 mixing ratios and different seasonal cycles (Figure 12.5). It is remarkable that the growth rate for the 1981–1992 period, which is 1.5–1.7 ppm a^{-1}, is exactly the same as at Mauna Loa. The German average CO_2 level amounts to 352 ± 9 (Mauna Loa average 348.4 ± 5.7 ppm), which is about 4 ppm larger than in the remote background. From the data it can by summarized that:

(a) Seasonal cycles are smoothed with increasing altitude above sea level,
(b) The concentration differences between the sites amounts to up to 20 ppm, and
(c) The inter-annual variations amount to about 2 ppm (similar to that found at remote background stations).

12.2.3.4 The City Dome CO_2

As already mentioned, in the vicinity of large CO_2 sources (big cities, industrial areas), elevated concentrations are measured at a level, strongly depending on meteorological parameters, which influence dispersion. This difference between rural and town sites had already been recognized a century ago [90] within a range of 33–133 ppm. In recent years, an increasing number of studies on CO_2 in urban areas have been published ([76] and citation therein). Concentrations of CO_2 measured in cities reflect a complex response of both the local anthropogenic and biogenic surface–atmosphere exchange of CO_2 and the concentrations accumulated, diluted, or advected over time controlled by meteorological factors. This makes the interpretation of urban CO_2 observations challenging.

The situation is usually different in urban areas where patterns are heavily influenced by anthropogenic emissions, which often cause strong short-term variations, but less visible seasonal patterns over the year.

Measurements of the CO_2 concentration carried out in Rome (Italy) [91] showed a mean yearly value increase from 1995 (367 ± 29 ppm) to 2004 (477 ± 30 ppm). The daily trend had a peak in the early morning when traffic was highest and the atmosphere was more stable. The annual trend showed a peak in winter, 18% greater than the summer one, which also correlated with traffic density. The weekly trend had lowest values (414 ± 19 ppm) during the weekends when traffic density was 72% lower.

Miyaoka et al. [92] *reported* on measurements made in Sapporo (Japan) in 2005 that showed significant diurnal variation (360–400 ppm in summer and 390–400 ppm in winter for lowest daytime values) as well seasonal differences.

Table 12.6 Carbon dioxide offset in cities compared with the rural environment (ppm).

Location	Additional CO_2	References
Phoenix, United States	100–200	[94]
Rome, Italy	80–100	[91]
Essen, Germany	20–40	[93]
Tokyo, Japan	15–60	[96]
Paris, France	20–550	[97]

It amounted for the urban plume to 379 ± 7 ppm in summer and 416 ± 12 in winter for southern air masses and to 374 ± 4 ppm in summer and 396 ± 7 ppm in winter for northern air masses.

In Essen (Germany) a similar pattern has been found [93], showing lower daytime values of 402 ± 20 ppm in winter and 369 ± 16 ppm in summer and higher values at night of 427 ± 23 ppm in winter and 417 ± 20 ppm in summer. Average values amount to 415 ± 21 ppm in winter and 393 ± 18 ppm in summer.

Similar results have been observed [94] in Phoenix (United States) that emphasize that the character of the city's urban CO_2 dome is almost exclusively a product of vehicular emissions and the region's distinctive meteorology. The CO_2 concentration of the air over Phoenix drops off rapidly with altitude, returning to a normal non-urban background value of approximately 378 ppm at an air pressure of 800 hPa [95]. Consequently, Phoenix's urban CO_2 dome did not have much of an impact on its near-surface air temperature, creating a calculated warming of just 0.12 °C at the time of maximum CO_2-induced warming potential. Table 2.11 shows the available data on the city dome CO_2. It is surprising that the supplementary CO_2 to the background value is roughly in the same range as the historical measurement of 100 years ago despite the different CO_2 sources (today traffic and in the past coal combustion).

From all these studies it can be concluded that anthropogenic CO_2 emissions are the primary source of the urban CO_2 dome. The dome is generally stronger in city centres, in winter, on weekdays, at night, under conditions of heavy traffic, close to the ground, with little to no wind, and in the presence of strong temperature inversions (Table 12.6).

12.2.4 Atmospheric CO_2 Residence Time

As already mentioned, the CO_2 cycle in the atmosphere is characterized that there is no direct chemical sink. In nature, CO_2 can only be assimilated by plants as biological sink through conversion into hydrocarbons (Section 2.2) and stored in calcareous organisms, partly buried in sediments, but almost completely turns back into CO_2 by respiration. Hence, CO_2 is distributed between the biosphere and atmosphere. The only definite carbon sink is the transport of DIC to deep ocean – when the ocean–atmosphere system is not in equilibrium, i.e. in case of increasing atmospheric CO_2 levels (due to anthropogenic and volcanic activities). As mentioned in Section 2.2.2, the CO_2 source term by volcanic exhalations is very uncertain but is likely to have a value of much less than 0.1×10^{15} g a^{-1} carbon. Hence, with respect to time periods being of interest for mankind (from decades to hundreds of years), this natural biogeochemical recycling can be regarded to be closed, or, in other words, the net flux is zero. Consequently, all concentrations (pools) in the biosphere and atmosphere remain constant (short-term variations should not be considered because of seasonal and inter-annual fluctuations).

The only driving forces behind removing CO_2 from the atmosphere are dry deposition (including plant uptake) and wet deposition (CO_2 scavenging). As discussed, the marine and terrestrial Earth's surface can be assumed to be carbonate saturated. This equilibrium is only disturbed by the yearly increase of the CO_2 level due to anthropogenic emissions. Hence, the physical (not the biological) surface resistance is zero as is the physical dry deposition flux. Therefore, the only abiogenic removal pathway from the atmosphere is CO_2 scavenging by clouds and finally precipitation. The DIC in precipitation can easily be calculated (assuming equilibrium) using Eq. (12.45b). Thus, one gets 0.21 and 0.28 mg l^{-1} carbon for 280 and 383 ppm CO_2, respectively (pH = 5.6, 10 °C). By using the global precipitation, it results in a very small total of wet removal fluxes:

- *280 ppm CO_2 (pre-industrial)*: 0.08 and 0.02×10^{15} g a^{-1} carbon for marine and terrestrial precipitation, respectively, and
- *400 ppm CO_2 (today)*: 0.10 and 0.03×10^{15} g a^{-1} carbon for marine and terrestrial precipitation, respectively.

The river run-off (Figure 2.4) amounts to c. 0.46×10^{15} g a^{-1} carbon and is much larger than the total wet deposited carbonate (0.13×10^{15} g a^{-1} carbon).

The global volcanic CO_2 emission is uncertain with a value of c. 0.02×10^{15} g a^{-1} carbon. It seems that the global carbon wet removal is significantly larger (by about a factor of 6–7) than the annual volcanic CO_2 release. Hence, it is likely that biogenic CO_2 is precipitated, but this amount is extremely small compared with the assimilation flux (about 0.1×10^{15} g a^{-1} carbon). Moreover, the river run-off is much larger than the total continental wet removal flux (by a factor of ~15). It is likely that it comprises biospheric carbonate from soils, but one cannot exclude anthropogenic CO_2. In summarizing the maximal physical removal fluxes comprise 0.13×10^{15} g a^{-1} carbon wet deposition and 0.46×10^{15} g a^{-1} carbon river run-off.

The residence time τ as the time for the turnover from one reservoir (atmosphere) to another (biosphere) is generally described by

$$\tau = \frac{m}{F_{sink}} \tag{12.3}$$

where m is the mass of the compound in the reservoir and F_{sink} the sink flux. The CO_2 mass in the pre-industrial atmosphere amounts to about 600×10^{15} g a^{-1} carbon with a global assimilation rate (Figure 2.4) of about 200×10^{15} g a^{-1} carbon. The resulting turnover time (a pseudo-residence time) of natural CO_2 amounts to about three years. Some climate sceptics [98] use this time quantity (enlarged due to the rising atmospheric CO_2 mass to about 4 a) to explain that after the cessation of anthropogenic CO_2 emissions, the recovery of the atmospheric CO_2 concentration can be expected soon (within less than 10 years). However, this is misinterpreting the conception of budgets and fluxes [99]. As discussed above, it follows that the natural removal is balanced with the new yearly input by respiration. The residence time defined mathematically by Eq. (12.48) is valid only for removal processes that can be described by a first-order rate equation:

$$F_{\text{sink}} = \frac{\mathrm{d}m}{\mathrm{d}t} = k \cdot m \tag{12.4}$$

where k is the removal rate constant. Since dry deposition (so far as it is driven by physico-chemical sorption) and wet deposition can be mathematically described as pseudo-first-order processes, removal of most atmospheric trace constituents can be modelled by Eq. (12.48). However, CO_2 assimilation must be considered as a zero-order process, i.e. the removal rate is constant and (largely) independent from the atmospheric CO_2 concentration. This becomes reliable when considering the global biosphere as heterogeneous uptake process, only depending from the plant amount. This consideration should not be fully valid but explains that applying Eq. (12.48) is invalid. The estimated pseudo-residence time of CO_2 is the time when all atmospheric CO_2 (assuming no CO_2 source via respiration) is completely consumed and the reaction (photosynthesis) stops abruptly. However, because respiration brings CO_2 back to the atmosphere about at the same rate, the "pseudo-residence time" of CO_2 becomes infinite. Only burial of organic matter (lignin-derived organic matter, almost not biodegradable) in sediments, representing a small excess of photosynthesis over respiration, has been important for control of CO_2 and O_2 over millions of years [100, 101].

The CO_2 measurements clearly show that the accumulative CO_2 increase is due to anthropogenic emission without fully balancing it by a sink. In Section 2.2.3 it was discussed that only about 50% of annual anthropogenic CO_2 is taken up by the biosphere and ocean. The remaining 50% builds up the carbon stock in the atmosphere (airborne fraction) and can only be removed physically.

Taking the present anthropogenic CO_2 mass in the atmosphere (Table 2.9) of about 225×10^{15} g carbon and assuming that the river run-off represents the maximum physical removal rate, it follows from Eq. (12.49) that there is a residence time of about 500 a. Taking into account only the atmospheric wet removal flux (0.13×10^{15} g a^{-1} carbon), the residence time amounts to 1700 a. Moreover, the residence time will increase with increasing airborne CO_2. Having a mixing ratio of 500 ppm (corresponding to about 300×10^{15} g carbon) by 2050, the residence time increases to 650 and 2300 a, respectively. Therefore, it is likely that the removal capacity of the climate system for the recovery of anthropogenic atmospheric CO_2 is in the order of 1000 a. Then, the removal flux amounts to 0.2×10^{15} g a^{-1} carbon, which is only 2% of the present man-made emission flux.

12.2.5 Atmospheric CO_2 Chemistry

Under tropospheric conditions, CO_2 is the final product of oxidation of all other carbon-containing compounds. Its fate is either assimilation by plants or dissolution in water (clouds, rain, and surface water, such as ocean, rivers, lakes, etc.).

Only in the stratosphere, where radiation with wavelength of less than 300 nm is available, CO_2 is photolysed:

$$CO_2 \xrightarrow{h\nu\ (\lambda<220\ \text{nm})} CO + O \tag{12.5}$$

However, carbon monoxide forms back to CO_2 via the reaction

$$CO + OH \rightarrow CO_2 + H \tag{12.6}$$

Compared with the photolytic dissociation of H_2O and CH_4, the CO_2 photolysis is of minor importance for the stratosphere.

In water CO_2 is protolysed (see Section 12.3.3). In the following, the reactions in aqueous solution are described occurring under conditions where radicals are formed, such as in cellular environments, surface water, and cloud droplets. The carbonate radical anion (CO_3^-) is produced from the reaction between the ubiquitous carbon dioxide and peroxonitrite ($ONOO^-$), which is an instable intermediate in NO reduction and firstly forms the CO_2 adduct nitrosoperoxocarboxylate, which then decomposes

$$CO_2 + ONOO^- \rightarrow ONOOCO_2^- \rightarrow CO_3^- + NO_2$$
$$(k_{(12.7)} = 6.5 \times 10^8\ \text{l mol}^{-1}\ \text{s}^{-1}) \tag{12.7}$$

Carbonate radicals react with many organic compounds [102, 103] in the general H-abstraction reaction (competing with OH):

$$CO_3^- + RH \rightarrow HCO_3^- + R \quad (k_{(12.8)} = 10^4\text{–}10^7\ \text{l mol}^{-1}\ \text{s}^{-1}) \tag{12.8}$$

The radical HCO_3 fully dissociates, i.e. it is a strong acid ($pK_a = -4.1$). Moreover, with $E°(CO_3^-/CO_3^{2-}) = 1.23 \pm 0.15$ V, it is a strong oxidizing agent [104–106], which likely exists as the dimer $H(CO_3)_2^-$ [107]. The importance in natural waters is likely limited because it is produced in radical reactions:

$$CO_3^{2-} + OH \rightarrow CO_3^- + OH^- \quad (k_{(12.9a)} = 3.9 \times 10^8\ \text{l mol}^{-1}\ \text{s}^{-1}) \tag{12.9a}$$

$$HCO_3^- + OH \rightarrow CO_3^- + H_2O \quad (k_{(12.9b)} = 1.7 \times 10^7\ \text{l mol}^{-1}\ \text{s}^{-1}) \tag{12.9b}$$

$$HCO_3^- + NO_3 \rightarrow CO_3^- + H^+ + NO_3^- \quad (k_{(12.10)} = 4.1 \times 10^7\ \text{l mol}^{-1}\ \text{s}^{-1}) \tag{12.10}$$

$$CO_3^{2-}\ (HCO_3^-) + Cl_2^- \rightarrow CO_3^{2-} + 2Cl^-\ (+H^+)$$
$$(k_{(12.11)} = 2.6 \times 10^6\ \text{l mol}^{-1}\ \text{s}^{-1}) \tag{12.11}$$

with $k_{(12.9a)}$, $k_{(12.9b)}$, $k_{(12.10)}$, $k_{(12.11)}$ (*C*hemical *A*queous *P*hase *RA*dical *M*echanism (CAPRAM) [108]).

It reacts back to carbonate, which reacts with transition metal ions ($k_{(12.12)}$) as well as with peroxides ($k_{(12.7)}$) representing radical termination in one-electron transfers. Eqs. (12.13a) and (12.14) also represent H-abstraction reactions ($k_{(12.14)}$) [109]:

$$CO_3^- + Fe^{2+}\ (Mn^{2+}, Cu^+) \rightarrow CO_3^{2-} + Fe^{3+}\ (Mn^{3+}, Cu^{2+}),$$
$$(k_{(12.12)} = 2 \times 10^7\ \text{l mol}^{-1}\ \text{s}^{-1}) \tag{12.12}$$

$$CO_3^- + HO_2 \rightarrow HCO_3^- + O_2 \tag{12.13a}$$

$$CO_3^- + O_2^- \rightarrow CO_3^{2-} + O_2 \quad (12.13b)$$

$$CO_3^- + H_2O_2 \rightarrow HCO_3^- + HO_2 \quad (k_{(12.14)} = 4.3 \times 10^5\ l\ mol^{-1}\ s^{-1}) \quad (12.14)$$

The following fast reaction obviously transfers O^- [110]:

$$CO_3^- + NO_2 \rightarrow CO_2 + NO_3^- \quad (k_{(12.15)} = 1 \times 10^9\ l\ mol^{-1}\ s^{-1}) \quad (12.15)$$

Adequate reactions concerning NO to NO_2^- and O_2 to O_3^- are not described in literature. A reaction with ozone is slow and implies the intermediate O_4^- ($\xleftrightarrow{H^+}$ HO_4) [111]:

$$CO_3^- + O_2 \rightarrow CO_2 + O_2 + O_2^- \quad (k_{(12.16)} = 1 \times 10^5\ l\ mol^{-1}\ s^{-1}) \quad (12.16)$$

Another interesting species is given by the carbon dioxide anion radical CO_2^-, produced from aquated electrons [112], *which is an efficient reducing agent with respect to* electron transfer and radical addition ($k_{(12.17)}$; CAPRAM):

$$CO_2 + e^-_{aq} \rightarrow CO_2^- \quad ((k_{(12.17)} = 4 \times 10^9\ l\ mol^{-1}\ s^{-1}) \quad (12.17)$$

$$CO_2^- + H_2O_2 \rightarrow CO_2 + H_2O_2^- \quad (12.18)$$

$$CO_2^- + O_2 \rightarrow CO_2 + O_2^- \quad (12.19)$$

The CO_2^- radical is also given from the oxidation of formate ions [113]:

$$HCOO^{-+} OH \rightarrow CO_2^- + H_2O \quad (12.20)$$

It adds onto organic radicals and double bonds [114]:

$$CO_2^- + RCH(OH) \rightarrow RCH(OH)COO^- \quad (12.21)$$

$$CO_2^- + R{-}CH{=}CH{-}R \rightarrow RCH(COO^-){-}CH{-}R \quad (12.22)$$

It disproportionates and dimerizes to oxalate [114]:

$$CO_2^- + CO_2^- \rightarrow CO_3^{2-} + CO \quad (12.23)$$

$$CO_2^- + CO_2^- \rightarrow {}^-O(O)C{-}C(O)O^- \quad (12.24)$$

Basically, these processes represent a way of sustainable chemistry in the future for CO_2 air capture and CO_2 reduction to fuels [115, 116]. Two-electron steps onto adsorbed CO_2 are favoured compared with Eq. (12.17), whose reduction potential amounts to −1.0 V:

$$CO_2 \xrightarrow{2H^+ + 2e^-} HCOOH \quad (E = -0.61\ V)$$

$$CO_2 \xrightarrow{2H^+ + 2e^-} CO + H_2O \quad (E = -0.53\ V)$$

$$CO_2 \xrightarrow{6H^+ + 6e^-} CH_3OH + H_2O \quad (E = -0.38\ V)$$

In aqueous solutions some metal–ligand complexes form CO_2 adducts, which internally undergo a two-electron step:

$$M^I - L + CO_2 \rightarrow M^I - L(CO_2)\ M^{III} - L(CO_2^{2-}) \xrightarrow{H^+} M^{III} - L + CO, HCOOH, H_2O \quad (12.25)$$

Thus, one of the best routes to remedy the CO_2 problem is to convert it to valuable hydrocarbons using solar energy in photocatalytic farms.

12.3 Oceanic and Water CO_2 and Carbonate Content

12.3.1 CO_2 Water Chemistry

In water, the following chemical carbon(IV) species exist in equilibrium: carbon dioxide (CO_2), carbonic acid (H_2CO_3), hydrogen carbonate (HCO_3^-), and carbonate (CO_3^{2-}). Additionally, the phase equilibriums with gaseous CO_2 and a possible solid body such as $CaCO_3$ and $MgCO_3$ have to be considered. Free carbonic acid is not isolated, but the structure $O{=}C(OH)_2$ in aqueous solution has been confirmed. Often, the expression $CO_2 \cdot H_2O$ is also used for carbonic acid. The sum of the dissolved carbonate species is denoted as total DIC and is equivalent with other terms used in literature:

$$\text{DIC} \equiv \Sigma CO_2 \equiv \text{TCO}_2 \equiv \text{C}_\text{T} = [\text{CO}_2] + [\text{H}_2\text{CO}_3] + [\text{HCO}_3^{-}] + [\text{CO}_3^{2-}] \tag{12.26}$$

The carbon dioxide (physically) dissolved in water – denoted as $CO_2(aq)$ – is in equilibrium with gaseous atmospheric carbon dioxide $CO_2(g)$. There is no way to separate non-ionic dissolved $CO_2(aq)$ and H_2CO_3; therefore, it is often lumped into $CO_2^*(aq)$. Subscripts *s* denote the solid phase, aq the dissolved phase, and g the gaseous phase in all following equations (note that ions are generally dissolved chemical species). Analytically, DIC can be measured by acidifying the water sample, extracting the CO_2 gas produced, and measuring.

The marine carbonate system represents the largest carbon pool in the atmosphere, biosphere, and ocean – meaning it is of primary importance for the partition of atmospheric excess carbon dioxide produced by human activity.

The ocean is saturated with $CaCO_3$, which represents the largest carbon reservoir in sediments in forms of calcite and aragonite [117]. The higher carbonate (in terms of DIC) solubility is because of dissolved CO_2, which converts carbonate (CO_3^{2-}) into higher soluble hydrogen carbonate (HCO_3^-). It follows that the capacity of the ocean for CO_2 uptake is still very large – the system is far away from saturation, but rather in equilibrium. With increasing atmospheric CO_2, the seawater CO_2–carbonate concentration increases, and vice versa. That means that in the case of decreasing atmospheric CO_2 concentrations, the ocean will degas CO_2, thereby leading to a new equilibrium. Therefore, the history of anthropogenic CO_2 is the story of the coupled ocean–atmosphere reservoir.

Seawater is slightly alkaline (pH ≈ 8.2) because of the equilibrium between $CaCO_3$ bottom and dissolved carbonate:

$$(\text{CaCO}_3)_\text{s} \xrightleftharpoons{\text{H}_2\text{O}} \text{Ca}^{2+} + \text{CO}_3^{2-} \tag{12.27}$$

$$\text{CO}_3^{2-} + \text{H}_2\text{O} \rightleftarrows \text{HCO}_3^{-} + \text{OH}^{-} \tag{12.28}$$

According to Eq. (12.28), carbonate acts as a base. From Eq. (12.27) the solubility product K_s follows:

$$K_\text{s} = [\text{Ca}^{2+}][\text{CO}_3^{2-}] \tag{12.29}$$

and from Eq. (12.28) the hydrolysis constant K_1:

$$K_1 = \frac{[HCO_3^-][OH^-]}{[CO_3^{2-}]} \tag{12.30}$$

The solubility of $CaCO_3$ at 20 °C in water is only about 0.007 g l^{-1} as carbon. $CaCO_3$ water solubility decreases linearly with increasing temperature:

$$[CaCO_3] = 80.3 - T\,(\text{in}\,^\circ C)$$

($r^2 = 0.997$) and increases slightly with increasing CO_2 partial pressure:

$$[CaCO_3] = 56.5 + 0.0219 \cdot [CO_2(g)]$$

($r^2 = 0.986$; 20 °C), valid in the range of 20–1000 ppm CO_2 [37]. $CaCO_3$ water solubility is given in mg l^{-1}.

Equations (12.29) and (12.30) describe the solid–liquid equilibrium at the oceanic bottom (sediment–seawater interface) with suspended matter in seawater (calcareous organisms). Moreover, the processes of heterogeneous nucleation and droplet formation from cloud condensation nuclei as well as in general the solid–water interfacial processes are described similarly in Figure 12.6. Hence, seawater is an excellent solvent for acidic gases such as SO_2 (used for flue-gas desulfurization at some coastal site power stations) and atmospheric CO_2.

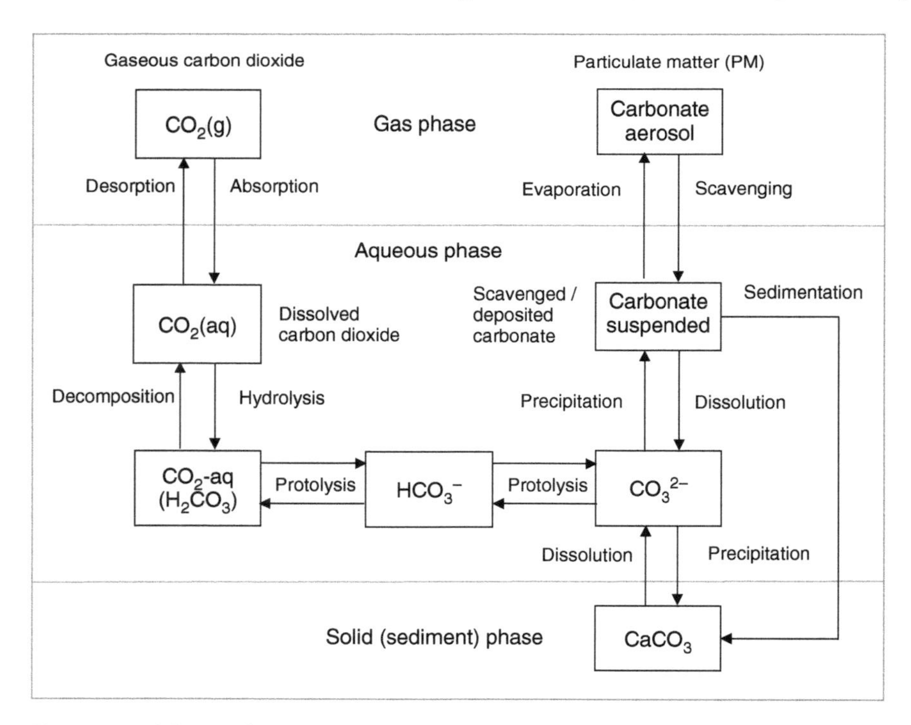

Figure 12.6 Scheme of the multiphase CO_2–carbonate system.

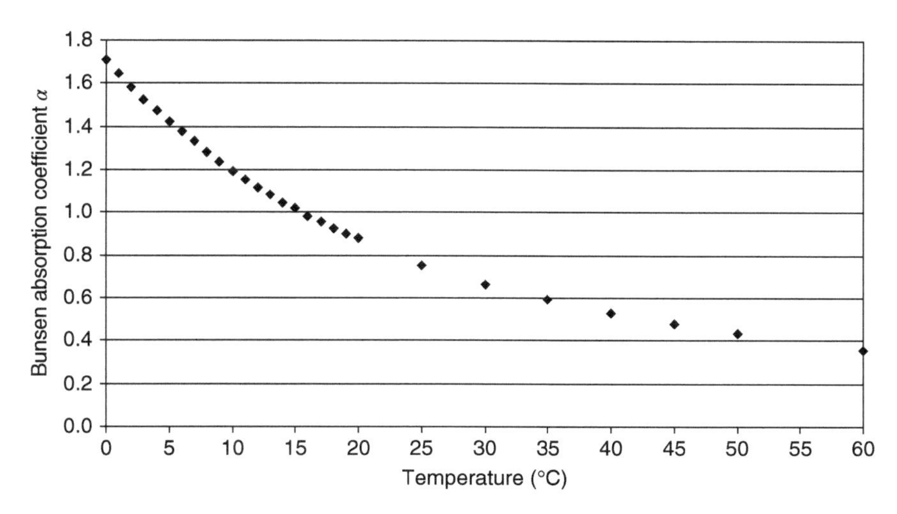

Figure 12.7 Solubility of CO_2 (ratio between volume of dissolved CO_2 and volume of solution) in water at 1 bar CO_2 in terms of the dimensionless Bunsen absorption coefficient α versus temperature. Source: Data from [118].

The ratio between atmospheric and oceanic (water-dissolved) CO_2 is described by the Henry equilibrium (Eq. (12.31)):

$$CO_2(g)\, H_{true} = CO_2(aq) \tag{12.31}$$

where the true Henry constant H_{true} is defined by

$$H_{true} = \frac{[CO_2(aq)]}{[CO_2(g)]} \tag{12.32}$$

depending only on the temperature. As mentioned above, it is not possible to measure $CO_2(aq)$ but only the sum $CO_2(aq) + H_2CO_3$. Hence, all listed equilibrium constants are related to an apparent Henry constant H_p:

$$H_{ap} = \frac{[CO_2(aq)] + [H_2CO_3]}{[CO_2(g)]} \approx \frac{[CO_2(aq)]}{[CO_2(g)]} = H_{true} \tag{12.33}$$

Taking into account the hydrolysis constant of H_2CO_3, the percentage of carbonic acid is very small (~0.3%) so that $H_{ap} \approx H$ is valid within the measurement uncertainty. Figure 12.7 shows the temperature dependency of Bunsen's absorption coefficient α. This coefficient is directly related to the Henry constant via

$$\alpha = H \cdot RT$$

where R is the gas constant and H has the dimension mol l^{-1} bar^{-1}. An empirical fit results in two slightly different ranges of temperature (T in °C):

$$\alpha = 0.0008 \cdot T^2 - 0.0585 \cdot T + 1.6992 \quad (0\text{–}25°C)$$

$$\alpha = 0.0002 \cdot T^2 - 0.0256 \cdot T + 1.283 \quad (25\text{–}60°C)$$

It has been shown that between 0 and 100 °C the dependency of H on T is nearly linear [119]. From the empirical data H yields (T in °C)

$$H = \frac{1}{11.39 + 0.81 \cdot T} \tag{12.34}$$

A true equilibrium is given through subsequent chemical reactions, leading to the higher solubility of CO_2 in water. Dissolved carbon dioxide forms hydrogen carbonate via different steps:

$$CO_2(aq) + H_2O \rightleftarrows HCO_3^- + H^+ \tag{12.35a}$$

$$CO_2(aq) + H_2O \underset{k_{-15.26}}{\overset{k_{15.26}}{\rightleftarrows}} H_2CO_3 \tag{12.35b}$$

$$H_2CO_3 \rightleftarrows H^+ + HCO_3^- \tag{12.36}$$

$$HCO_3^- \rightleftarrows H^+ + CO_3^{2-} \tag{12.37}$$

Reactions are determined by the apparent first dissociation constant K_{ap} in Eq. (12.35a), the hydration constant K_h of Eq. (12.35b), the true first dissociation constant K_1 of Eq. (12.36), and the second dissociation constant K_2 from Eq. (12.32). CO_2 hydration according to Eq. (12.35b) is relatively slow, and the concentration of H_2CO_3 is – in comparison with total dissolved CO_2 – even lower and, hence, negligible. Equation (12.35b) occurs for $pH < 8$, whereas Eq. (12.37) is dominant for $pH > 10$. In the pH region 8–10, both reactions precede in parallel, making it complicated to study their kinetics. Hence, reliable kinetic data are valid only for $pH < 8$ and $pH > 10$, respectively. Equation (12.38) denotes another direct CO_2 protolysis, describing the hydrolysis constant K_2:

$$CO_2(aq) + OH^- \underset{k_{-15.29}}{\overset{k_{15.29}}{\rightleftarrows}} HCO_3^- \tag{12.38}$$

However, this reaction plays basically no role in natural waters with the exception of the initial state of cloud/fog droplet formation from alkaline CCN (e.g. flue ash and soil dust particles). The reaction rate constants have been estimated to be $k_{(12.35b)} = 0.03\ s^{-1}$ (25 °C) and $k_{-(12.35b)} = 20\ s^{-1}$ (25 °C) as pseudo-first-order rates and $k_{(12.38)} = 8400\ l\ mol^{-1}\ s^{-1}$ and $k_{-15.36} = 2 \times 10^{-4}\ s^{-1}$ (minus sign denotes the inverse reaction) [120–122].

The equilibrium constant K_{ap} of the apparent first dissociation regarding Eq. (12.35a) is given by

$$K_{ap} = \frac{[HCO_3^-][H^+]}{[CO_2(aq)]} = K_1 K_h \tag{12.39}$$

K_{ap} can be relatively easily estimated from equilibrium concentration measurements, where the hydration constant K_h is calculated according to Eq. (12.38) and Table 12.7. The adjustment of equilibrium Eqs. (12.36) and (12.37) is fast. However, the direct estimation of the dissociation constants K_1 and K_2 is not possible from concentration measurements, only indirectly through potentiometric and conductometric measurements. The true first dissociation constant K_1 ($pK_1 = 3.8$, similar to the pH value definition, $pK = -\log K$) is 3 orders of magnitude larger than the apparent dissociation constant K_{ap}. Hence, carbonic

Table 12.7 Equilibrium constants in the aqueous CO_2–carbonate system.

T (°C)	0	5	10	15	20	25
H (10^{-2} bar^{-1} mol l^{-1})	7.70	—	5.36	—	3.93	3.45
K_{ap} (10^{-7} mol l^{-1})[a)]	2.64	3.04	3.44	3.81	4.16	4.45
K_1 (10^{-4} mol l^{-1})	—	1.56	—	1.76	1.75	1.72
$K_h \times 10^3$ [b)]	—	1.96	—	2.16	2.52	2.59
K_2 (10^{-11} mol l^{-1})	2.36	2.77	3.24	3.71	4.20	4.29
$K_{ap} = K_s \cdot K_h$ (10^{-7} mol l^{-1})	—	3.06	—	3.80	4.41	4.45

a) Lide and Frederikse [123].
b) $K_h = [H_2CO_3]/[CO_2(aq)]$.
Source: Adapted from Gmelin 1973 [120].

acid is 10 times stronger than acetic acid. However acetic acid can degas CO_2 from carbonic solutions because H_2CO_3 is decomposed to about 99% into CO_2 (which escapes from the water) and H_2O as it follows from K_h. This makes the aqueous carbonic system unique: carbonic acid exists in very low concentrations as H_2CO_3 (in kinetic inhibited equilibriums) and largely as $CO_2 \cdot H_2O$ where CO_2 degassing is also inhibited. The second dissociation constant K_2 characterizes hydrogen carbonate as a very weak acid (pK_2 = 10.4). The aqueous phase concentrations of different C(IV) species can be calculated from the following equilibrium expressions:

$$[CO_2(aq)] = H \cdot [CO_2(g)] \tag{12.40}$$

$$[H_2CO_3] = H \cdot K_h [CO_2(g)] \tag{12.41}$$

$$[HCO_3^-] = H \cdot K_1 K_h [CO_2(g)][H^+]^{-1} \tag{12.42}$$

$$[CO_3^{2-}] = H \cdot K_1 K_2 K_h [CO_2(g)][H^+]^{-2} \tag{12.43}$$

12.3.2 Total Dissolved Carbon (DIC)

At a typical surface seawater pH of 8.2, the speciation between $[CO_2]$, $[HCO_3^-]$, and $[CO_3^{2-}]$ is 0.5%, 89%, and 10.5%, respectively, showing that most of the dissolved CO_2 is existent in the form of HCO_3^- and not CO_2. For the description of the overall gas–aqueous equilibrium, the so-called effective Henry constant is used:

$$H_{eff} = \frac{[CO_2(g)]}{DIC} = \frac{[CO_2(g)]}{[CO_2(aq)] + [H_2CO_3] + [HCO_3^-] + [CO_3^{2-}]} \tag{12.44}$$

Inserting Eq. (12.40) in Eq. (12.43), it follows for the total DIC:

$$DIC = H[CO_2(g)](1 + K_h(1 + K_1[H^+]^{-1} + K_1K_2[H^+]^{-2})) \tag{12.45a}$$

$$DIC \approx H[CO_2(g)](1 + K_{ap}[H^+]^{-1} + K_{ap}K_2[H^+]^{-2}) \tag{12.45b}$$

The true Henry constant H strongly depends on temperature. From data in [118] (up to 50 °C, not listed in Table 12.7), it can be derived:

$$H = 0.061 \cdot \exp(-0.023 \cdot T) \tag{12.46}$$

here T is given in °C. Expressions for K_1 and K_2 depending on temperature T (in K) and salinity S (in ‰) are given for $k^0 = 1\,\mathrm{mol\,kg^{-1}}$ seawater [124]:

$$\log(K_1^*/k^0) = \frac{-3633.86}{T} + 61.2172 - 9.6777 \cdot \ln T + 0.011\,555 \cdot S - 0.000\,115\,2 \cdot S^2 \tag{12.47}$$

$$\log(K_2/k^0) = \frac{-471.78}{T} + 25.929 + 3.169\,67 \cdot \ln T + 0.017\,81 \cdot S - 0.000\,112\,2 \cdot S^2 \tag{12.48}$$

The equilibrium constant K_1^* is expressed as [11, 124]

$$K_1^* = \frac{[H^+][HCO_3^-]}{[CO_2(aq)] + [H_2CO_3]} \tag{12.49}$$

which corresponds to

$$\frac{1}{K_1^*} = \frac{[CO_2(aq)]}{[H^+][HCO_3^-]} + \frac{[H_2CO_3]}{[H^+][HCO_3^-]} = \frac{1}{K_{ap}} + \frac{1}{K_1} \approx \frac{1}{K_{ap}} \tag{12.50}$$

It has to be noted here that constants from different studies in literature are based on different equilibriums and, hence, cannot be easily compared with each other. Moreover, the same name is often used for different definitions.

For S = 35‰ and 25 °C, the constants yield pK_1^* ($\approx pK_{ap}$) = 5.8472 and pK_2 = 8.966 ($pK = -\log(K/k^0)$). From the values listed in Table 12.7, it would follow that pK_{ap} ($\approx pK_1^*$) = 6.3 and pK_2 = 11.63, which is considerably different from the values for seawater. This certainly explains the differently calculated concentration values based on given atmospheric CO_2 partial pressures (see below). At 25 °C and water having pH = 8.1, the equilibrium concentration of total DIC can be calculated from Eq. (12.38):

$$\mathrm{DIC_{water}/mol\,l^{-1}} = 2.3 \cdot [CO_2(g)/10^{-6}\,\mathrm{ppm}] \tag{12.51a}$$

$$\mathrm{DIC_{seawater}/mol\,l^{-1}} = 8.2 \cdot [CO_2(g)/10^{-6}\,\mathrm{ppm}] \tag{12.51b}$$

using K values given for pure water (Table 12.7) and seawater, respectively. With that, it follows for 380 ppm atmospheric CO_2 that DIC in water and in seawater amounts to 11 and 31 mg C l^{-1}, respectively. The measured DIC in surface seawater [11] amounts to 27.6 mg l^{-1} as carbon (0.0023 mol l^{-1}) and lies within the range of the above estimates. This suggests that the K values are either known only with significant uncertainties or, more likely, the seawater DIC is not always in equilibrium. Calculations and laboratory experiments for equilibrium estimates exclude the interaction with solid $CaCO_3$. Assuming that the relationship is valid and taking the measured values (28 mg l^{-1} carbon in surface seawater and 384 ppm atmospheric CO_2), one gets

$$\mathrm{DIC_{seawater}} = 6.1 \cdot [CO_2(g)] \tag{12.51c}$$

or generally

$$\mathrm{DIC_{seawater}} = a(T, \mathrm{pH}) \cdot [CO_2(g)] \tag{12.51d}$$

It is important to note that the factor a depends on the seawater pH and (to a lesser extent) on the temperature T. The value of a in Eqs. (12.51c) and (12.51d) equals 6.1 but is valid only for pH = 8.1. For pH = 7.9 and pH = 8.2, the factor ranges twofold, namely, 3.8 and 7.7, respectively. This means that – with decreasing seawater pH – the uptake capacity for atmospheric CO_2 decreases significantly.

Increasing the sea surface temperature (SST) also leads to decreasing DIC. Taking into account the temperature dependency of H and K_{ap}, a linear curve ($r^2 = 0.99$) with a slope of $-0.13\,mg\,l^{-1}$ carbon in the range of 0–30 °C results from Eq. (12.51d). However, the relationship between atmospheric CO_2 is more complicated because of the buffer capacity of seawater. Here, in a more exact treatment, all buffering chemical species – for example, borate – have to be considered besides carbonate. The buffer capacity of carbonized water (here seawater) is given to complete the acid-based reaction (primarily buffering):

$$CO_2(aq) + CO_3{}^{2-} + H_2O \rightleftarrows 2HCO_3{}^- \qquad (12.52)$$

12.3.3 Changing Seawater Carbonate

Anthropogenic CO_2 dissolves in seawater, produces hydrogen ions, and neutralizes carbonate ions. Hence, the H^+ concentration (and pH) will not change in small ranges depending on the CO_2 partial pressure increase and the available carbonate in seawater. However, when seawater pH declines as a result of rising CO_2 concentrations, the concentration of $CO_3{}^{2-}$ will also fall according to Eq. (12.52), reducing the calcium carbonate saturation state. Marine carbonates also react with dissolved CO_2 through the reaction (secondary buffering):

$$CO_2(aq) + CaCO_3 + H_2O \rightleftarrows 2HCO_3{}^- + Ca^{2+} \qquad (12.53)$$

This reaction depends on the calcium carbonate saturation state Ω:

$$\Omega = \frac{[Ca^{2+}][CO_3^{2-}]}{K_{CaCO_3}} \qquad (12.54)$$

where K_{CaCO_3} is the solubility product. Unless sufficient carbonate is present, $CaCO_3$ will dissolve back into the surrounding seawater. Since the ocean is in contact with carbonate sediments, both on shelves and in the deep sea, the ocean as a first approximation is roughly saturated with respect to calcite. It is assumed that the concentration of Ca^{2+} has remained nearly constant. This is equivalent to a roughly constant carbonate ion concentration. In regions where $\Omega > 1.0$, the formation of shells and skeletons is favoured. Below a value of 1.0, the water is corrosive, and the dissolution of pure aragonite and unprotected aragonite shells will begin to occur. Equation (12.54) is astonishing because it follows from Reaction (12.28):

$$K_s = \frac{[Ca^{2+}][CO_3^{2-}]}{[CaCO_3]}$$
$$K_s[CaCO_3] = [Ca^{2+}][CO_3^{2-}] = K_{CaCO_3} \qquad (12.55)$$

where the $CaCO_3$ activity of the solid body is set by convention to be unity. Therefore, $K_s \equiv K$ and, hence, $\Omega = 1$. Supersaturation and undersaturation remain, hence, in small limits. However, water bodies with insufficient solid $CaCO_3$ are generally undersaturated.

Ocean uptake of anthropogenic CO_2 leads to a shift between carbonate and hydrogen carbonate or, in other words, to a larger solubility of mineral $CaCO_3$. It has been argued that as a consequence of higher CO_2 partial pressures, the surface layers of the ocean will become undersaturated with calcium carbonate, according to Eq. (12.52), with possible catastrophic biological consequences for a variety of marine organisms (for example, coral reefs, shells, and skeletons of other marine calcifying species). However, from general aspects, it seems unlikely that $CaCO_3$ supersaturation is a precondition for carbonate biomineralization for the following three reasons:

(a) Life processes are far out of equilibrium.
(b) Biomineralization depends on the active transport of ions through biomembranes driven by metabolic energy and, thus, do not depend on the free energy of any carbonate reaction.
(c) Calcium carbonate structures of living organisms can be covered by organic tissue, representing a barrier against fast exchange processes [125].

The ecological consequences of a possible change in the calcium carbonate supersaturation, with respect to calcareous organisms, can only be examined experimentally.

Furthermore, CO_2 uptake leads to so-called ocean acidification. With a very good approximation, i.e. neglecting the carbonate concentration compared with that of hydrogen carbonate, Eqs. (12.45a) and (12.45b) can be reduced to

$$[CO_2(g)] = \frac{[H^+][DIC]}{HK_{ap}} \tag{12.56}$$

and further transformed into

$$pH = \log DIC - \log HK_{ap} - \log[CO_2(g)] \tag{12.57}$$

Unfortunately, DIC is a function of pH, and, therefore, no simple relationship between atmospheric CO_2 concentration and seawater pH can be found. Pearson and Palmer [126] held DIC constant to estimate the relationship between changing atmospheric CO_2 concentration and pH, assuming a pre-industrial pH of 8.25 at the pre-industrial CO_2 value of 280 ppm. However, there is no serious argument to hold DIC constant over time. By contrast it has been shown [8] that the oceanic DIC reflects the accumulated atmospheric CO_2. Caldeira and Berner [127] held carbonate constant and used a formula simply derived from Eq. (12.43):

$$[CO_2(g)] = \frac{[H^+]^2[CO_3^{2-}]}{HK_{ap}K_2} \tag{12.58}$$

from which follows

$$\begin{aligned} pH &= 0.5\log[CO_3^{2-}] - 0.5\log HK_{ap}K_2 - 0.5\log[CO_2(g)] \\ &\equiv \text{const} - 0.5\log[CO_2(g)] \end{aligned} \tag{12.59}$$

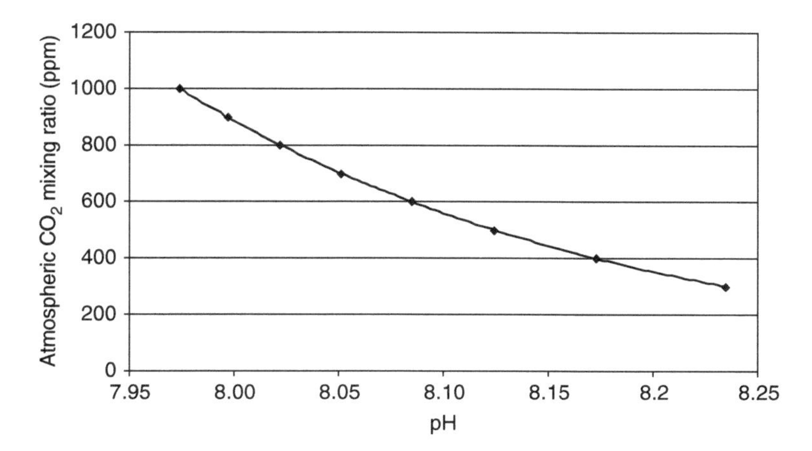

Figure 12.8 Relationship between atmospheric CO_2 mixing ratio and seawater pH assuming a pH of 8.25 at a CO_2 mixing ratio of 280 ppm and constant seawater carbonate concentration.

Assuming a pre-industrial pH of 8.25 at the pre-industrial CO_2 value of 280 ppm (Figure 12.8), the constant in Eq. (12.35b) amounts to 6.474.

It follows that by the year 2100, using a scenario without carbon capture technology (CCS) for about 1000 ppm CO_2, the seawater pH would decrease to 7.96. From Eqs. (12.52) and (12.53), it follows that increasing CO_2 would lead to increasing hydrogen carbonate, whereas carbonate is transformed into hydrogen carbonate. The carbonate concentration is adjusted through hydrogen carbonate dissociation (Eq. (12.1)) and $CaCO_3$ dissolution (Eq. (2.28)), but the relation is strongly non-linear. From Eq. (12.29) it follows that increasing $[H^+]$ would lead to a decreasing carbonate ion concentration. Hence, it is more likely that ocean acidification is faster than predicted from Eq. (12.59) and Figure 12.9, which can be concluded from the measurements of surface water pH [131, 132]. According to the Ocean Station ALOHA (A Long-Term Oligotrophic Habitat Assessment), current pH (2008) amounts to c. 8.08, but from Eq. (12.52) it would follow that pH should be 8.18.

As a conclusion, it does not make any sense to adopt relationships such as Eq. (12.52) or Eq. (12.44) to calculate historical and future seawater pH. There is no scientific evidence to set the pre-industrial seawater pH value to 8.25. When using the reference values of 384 ppm CO_2 and pH 8.08 as measured values for 2008, Eq. (12.52) would deliver a pre-industrial value of pH 8.15 corresponding to 280 ppm CO_2. Exact measurements of pH in natural waters are fully uncertain to the second decimal place.

It has been found in [129, 130] that over two decades of observation (Station ALOHA; see Table 12.8 and Figure 12.9), the surface ocean grew more acidic at exactly the rate expected from the chemical equilibration with the atmosphere. However, that rate of change varied considerably in terms of seasonal and inter-annual timescales and even reversed for a period of nearly 5 a (Figure 12.9). The concentrating/diluting effect of salinity changes on DIC can be removed from the measured data through the normalization to a constant salinity S (=35‰ where normalized dissolved inorganic carbon, nDIC = 35 DIC/S).

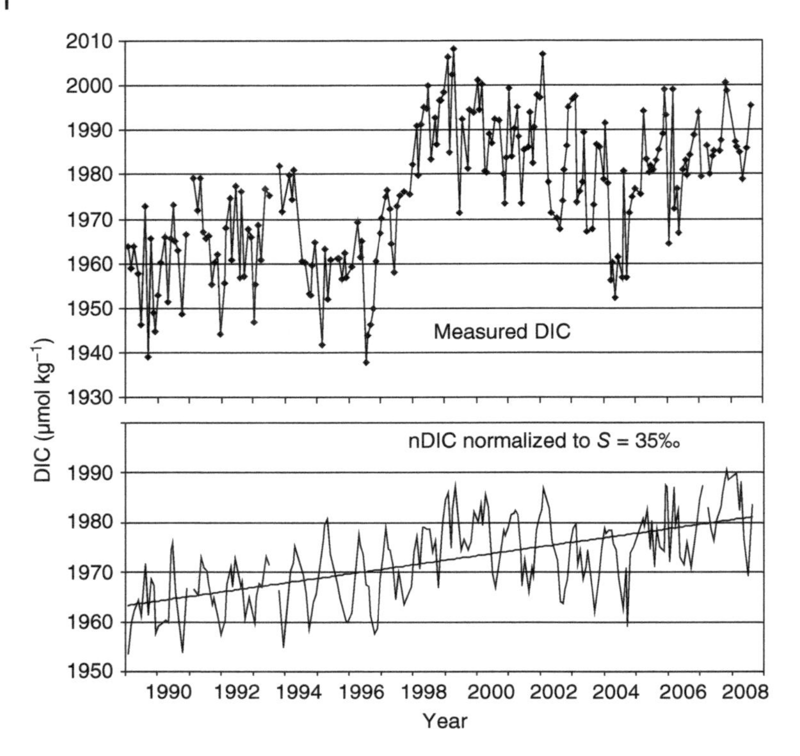

Figure 12.9 Trend of dissolved inorganic carbon (DIC) at Station ALOHA (see Table 12.8 for details). Source: Data from [128]. See also [129, 130].

Figure 12.9 is an interesting example of how to make misinterpretations when not considering collateral factors. Whereas S has no seasonal pattern, DIC and nDIC have distinct annual cycles, largely driven by the input of DIC from below (via winter mixing) and biological drawdown of CO_2 (via photosynthesis).

Year-to-year changes seem to be driven by climate-induced changes in ocean mixing and attendant biological responses to mixing events. Evidence has been found for the upwelling of corrosive 'acidified' water onto the continental shelf [133]. It is not possible to fit the measured data in Table 12.8 with the equilibrium equations derived here, suggesting:

(a) That either different equilibrium constants are valid in seawater compared with pure water and
(b) That the seawater DIC concentration is not fully described by equilibrium conditions. The latter is possible due to the mixing and varying biological activity with respect to assimilation and respiration.

Over the period 1990–2008 from Table 12.8, an increase of DIC by about 15 μmol C l^{-1} or (taking into account the atmospheric CO_2 increase from 354 to 386 ppm) by c. 6 μg C per ppm CO_2 can be derived. In other terms, the oceanic DIC increased yearly by about 10 mg C m^{-3} seawater or about 4 Tg C within the 1 m surface layer of the oceans, corresponding to only 0.3 Gt within the 75 m layer. However, the yearly ocean uptake is assumed to be c. 2 Gt C a^{-1}

Table 12.8 Mean data sets from the deepwater Station ALOHA (A Long-Term Oligotrophic Habitat Assessment; 22°45′ N, 158°00′ W) located 100 km north of Oahu, Hawaii.

Parameter	1988–2008	1992–1998	2003–2008
pH (measured)	—[a]	8.10 ± 0.01	8.09 ± 0.01
T (°C)	24.9 ± 1.1	24.8 ± 1.3	24.9 ± 1.1
n	203	52	57
TA[b] (μeq kg^{-1})	2306 ± 13	2300 ± 14	2308 ± 11
DIC[c] (μmol kg^{-1})	1975 ± 15	1966 ± 12	1980 ± 11
$[CO_2(aq)]$[d] (μmol kg^{-1})	9.8 ± 0.4	9.6 ± 0.2	10.0 ± 0.3
$[CO_3^{2-}]$[e] (μmol kg^{-1})	234 ± 6	235 ± 5	231 ± 5
$[HCO_3^-]$[f] (μmol kg^{-1})	1731	1721	1737
pCO_2 (ppm)[g]	345 ± 16	337 ± 13	356 ± 14
$[CO_2(g)]$ (ppm)[h]	366 ± 10	360 ± 3	380 ± 3

a) Measured pH values only from the selected periods available; density of seawater (S = 35‰, 25 °C): 1.023 343 kg l^{-1} [11].
b) Total alkalinity, measured by open cell titration; TA = $[HCO_3^-] + 2[CO_3^{2-}]$ + others (total alkalinity).
c) Measured by coulometry.
d) Free seawater CO_2 concentration.
e) Seawater carbonate ion concentration.
f) Calculated as difference between DIC and $CO_2(aq) + CO_3^{2-}$.
g) Mean seawater partial pressure, calculated from DIC and TA.
h) Mean atmospheric CO_2 mixing ratio (from Mauna Loa).
Source: Data from [128].

(Table 12.8). Hence, the subtropical station ALOHA cannot reflect the global seawater chemistry. On the other hand, using Eq. (12.51b), a DIC increase would follow for the mentioned period, corresponding to 3.5 Gt within the 75 m layer. This result is much closer to the estimates, but – as discussed in connection with Eq. (12.51c) – changing T and pH have to be taken into account.

Moreover, this calculated inorganic carbon concentration does not incorporate the fact that carbon is continuously supplied into the atmosphere and oceans by degassing from metamorphism and magmatism as well as by the weathering of carbonate minerals and organic carbon and is continuously consumed by the production of carbonate and organic carbon sediments (see Figure 2.3). Hence, the total DIC load of the ocean can be expected to vary over time. Finally, the uptake of atmospheric CO_2 by the sea surface is a dynamic process that depends on oceanic and wind circulation interlinked. Observations suggest that the Southern Ocean sink of CO_2 has weakened between 1981 and 2004 by 0.08×10^{-15} g decade^{-1} relative to the trend expected from the large increase in atmospheric CO_2 and attributed to the observed increase in wind [134].

12.3.4 Oceanic CO_2 Measurements

Oceans represent an enormous carbon dioxide reservoir and are a sink for CO_2. They contain much more CO_2 than the atmosphere and have considerable influence on the content of CO_2 in the atmosphere and, consequently, on the world climate. As is described in detail in Section 12.3.2, the dissolved carbonate species in seawater exist in the forms of hydrogen carbonate HCO_3^- ions, carbonate CO_3^{2-} ions, and a comparatively small portion of dissolved carbon dioxide CO_2. Guideline values are about 90% HCO_3^-, about 10% CO_3^{2-}, and less than 1% CO_2. The relative concentrations of these components depend on pH value, temperature and many other influencing parameters. According to Henry's law in thermodynamic equilibrium, the concentration of a gas in a liquid is proportional to its partial pressure in the gaseous phase above the liquid. However, the solubility coefficient of CO_2 depends on temperature, and the precondition of equilibrium is actually not fulfilled at any time under the prevailing circumstances. The rate of the CO_2 transfer between the air and the ocean and vice versa is comparatively low and depends on the surface roughness of the ocean and the wind speed. Numerous complicated chemical, physical, and biological processes, often referred to as solubility pump and biological pump, result in large seasonal as well as horizontal and vertical changes of the CO_2 concentration in the ocean [135, 136].

Over the last decades numerous oceanic research programmes have been carried out by a number of agencies, commissions, and organizations to obtain data on the complicated carbonate chemistry of the oceans. While ocean acidification is well documented in a few temperate ocean waters, little is known in high latitudes, coastal areas, and the deep sea. At the present time most current CO_2 sensor technologies are quite costly, imprecise, or unstable to allow for sufficient knowledge on the state of ocean acidification. Therefore, further development activities are required in this field. Actually, CO_2 sensors for marine and freshwater applications, based on different measuring principles, are produced and offered worldwide by a variety of manufacturers, only some of which are listed in Table 12.9.

Usually, the $p\text{CO}_2$ of seawater is measured by equilibrating a small volume of gas with a large volume of seawater at a given temperature. Then the mixing ratio of CO_2 in the gas phase is determined either using a gas chromatograph or an IR CO_2 analyser [146, 147]. Figure 12.10 illustrates schematically some of the measuring principles that are applied in the devices of Table 12.9. In most cases, according to Figure 12.10a, dissolved CO_2 molecules diffuse through a thin polymeric membrane into a gaseous or liquid test medium. Special precautions must be taken that the membrane equilibrator withstands the enormous pressure when immersed in high water depths [148]. The equilibrated test sample is then passed to a detector chamber, where the partial pressure of CO_2 is determined optically by means of IR absorption spectrometry (e.g. LI-COR IR gas analysers), an NDIR detector or a colorimetric reagent method [149]. To achieve significantly higher accuracy compared with [149], the indicator medium in the sensor shown in Figure 12.11 is renewed continuously before each measurement point [150, 151]. In Figure 12.10b an air sample is circulated between the CO_2 sensor and the equilibrator, which is the water interface. In the equilibrator, the air sample is bubbled through a column of water, pumping water through the equilibrator and allowing the air sample to reach equilibrium with the dissolved gases in the water. The equilibrated air sample passes through an optical CO_2 sensor (e.g. LI-COR 820 IR sensor) [152].

Table 12.9 Some selected examples of CO_2 sensors for marine and freshwater applications (alphabetical order).

Manufacturer	CO_2 sensor, working principle
AMT Analysenmesstechnik GmbH, Rostock, Germany [137]	Membrane-covered optical CO_2 sensor Principle: optical (single-beam dual-wavelength NDIR), CO_2 permeation through silicone membrane into an indicator solution
Analyticon, Springfield, NJ, USA [138]	CGP-1 portable carbon dioxide metre Principle: Severinghaus-type membrane-covered pH glass electrode
Battelle, Columbus, OH, USA [139]	Seaology Principle: optical (IR absorption), CO_2 extraction from the water by bubbling air through an equilibrator
CONTROS Systems & Solutions GmbH, Kiel, Germany [140]	HydroCTM/CO_2 flow-through sensor Principle: optical (IR absorption), CO_2 permeation through composite membrane into a gas circuit
General Oceanics, Miami, FL, USA [141]	$p$$CO_2$ measuring system Principle: optical (LI-COR IR CO_2 analyser); an equilibrator balances CO_2 in seawater with a headspace gas that is analysed
Kimoto Electric, Osaka, Japan [142]	MOG-701 carbon dioxide monitor Principle: optical (NDIR with flow-through procedure), CO_2 extraction by a gas–liquid equilibrator
Pro-Oceanus Systems, Inc. (PSI), Bridgewater, Canada [143]	CO_2-Pro Principle: optical (NDIR detector), cylindrical membrane interface
SubCtech, Osdorf, Germany [144]	OceanPacks MK2 and MK3 Principle: optical (dual-wavelength NDIR detector LI-COR), silicone flat membrane equilibrator
Sunburst Sensors, Missoula, MT, USA [145]	SAMI-CO2 Principle: optical (colorimetric reagent method), CO_2 permeation through silicone membrane into an indicator solution

The main technical specifications of the manufacturers vary over the values:

- CO_2 measuring ranges: 0–3 mbar (ppm)
- Accuracy, precision: <3 μbar (ppm)
- Maximum immersion depth: 100–6000 m

If it is intended to apply a carbon dioxide sensor in oceanology, then this probe must meet difficult requirements concerning pressure resistance, rapid and correct compensation of pressure and temperature changes, and high speed of response. When heaving or sending down the probe with a speed of 2.5 m s^{-f}, pressure changes of up to 25 kPa s^{-1} and temperature changes of up to 5 K s^{-1} can occur.

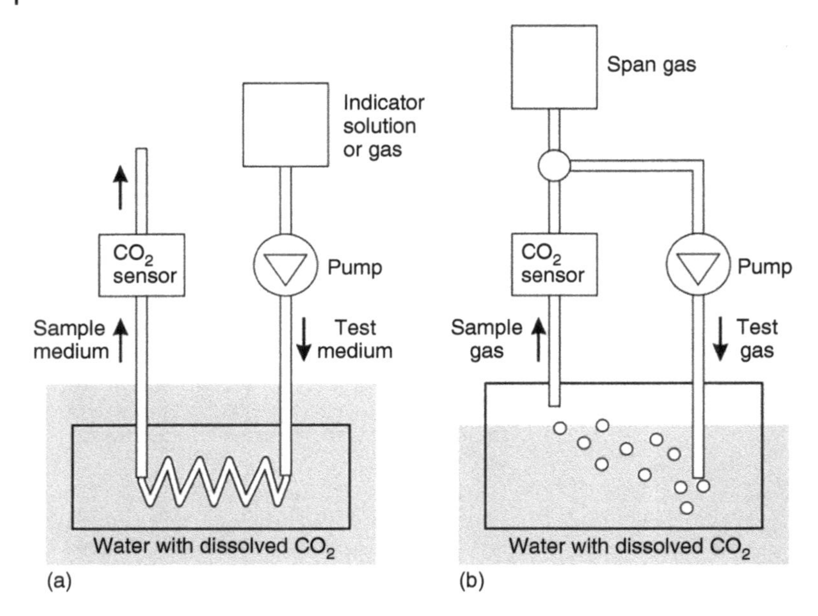

Figure 12.10 Measuring principles for oceanic CO_2 sensors. (a) Membrane equilibrator. (b) Air equilibrator.

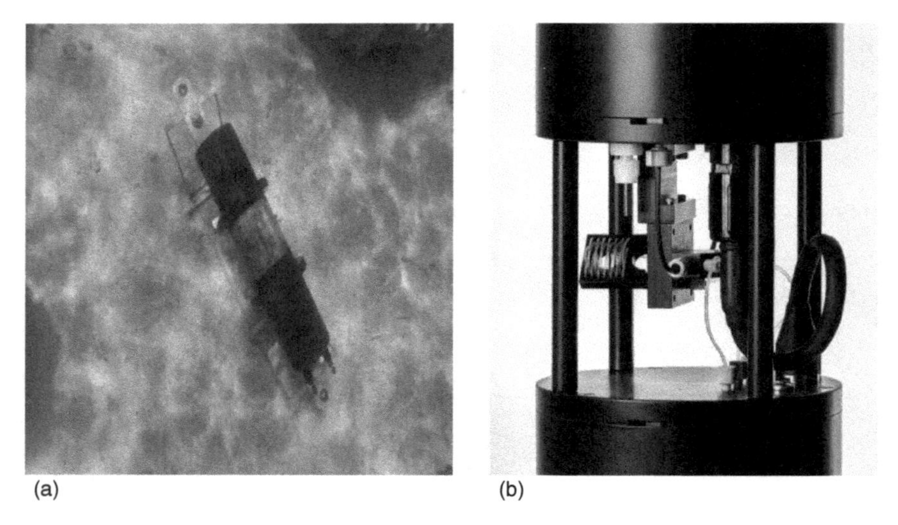

Figure 12.11 SAMI-CO_2 ocean CO_2 sensor (Sunburst Sensors, Missoula, MT, USA). (a) Immersed in the ocean. (b) Detailed view of the measuring unit.

As an outstanding example for an ocean CO_2 sensor, Figure 12.11 shows the SAMI-CO_2 sensor (*S*ubmersible *A*utonomous *M*oored *I*nstrument from Sunburst Sensors, Missoula, MT, USA) [153].

This sensor device for autonomous mooring-based measurements of seawater pCO_2 uses a highly precise and stable colorimetric reagent method [150, 151]. As described in more detail in Chapter 9, it operates by the equilibration of ambient seawater pCO_2 with a colorimetric pH indicator contained in a gas-permeable membrane. To improve stability and sensitivity, the indicator is periodically renewed. By this means, SAMI-CO_2 is capable of measuring seawater pCO_2 with exceptional long-term stability (no detectable drift in one month) and sensitivity comparable with ship-based equilibrator-IR methods. SAMI-CO_2 is designed to operate down to 600 m (or even 3500 m with a titanium housing) and contains sufficient batteries and reagent for deployment up to six months while making 48 measurements per day. The basic performance parameters of the SAMI-CO_2 Ocean CO_2 Sensor are shown in Table 12.10.

The accuracy, weight, size, and cost of SAMI-CO_2 make it feasible for long-term measurements of the seawater pCO_2 dynamics on moorings. In addition, the instrument can be modified (different pressure housing) to operate with ship-based oceanographic measuring devices for the determination of conductivity, temperature, and depth (CTD packages) in the ocean on a wire or towed platform. This approach can be applied substantially easier than performing pCO_2 measurements on discrete samples. The pCO_2 time series can be combined with other physical and biogeochemical time-series measurements to assess short- and long-term controls on pCO_2 in surface waters of the ocean.

A completely different, rather complex construction of an ocean CO_2 sensor with a special electrode holder for pressure compensation has been disclosed in [154]. As illustrated in Figure 12.12, the electrochemical CO_2 sensor comprises a mechanically very sensitive thin polymer membrane, which is arranged in such a way that a tubular sleeve can freely slide. The space volumes above the hydrogen carbonate sensor electrolyte and the inner buffer solution of the pH glass electrode are filled with highly insulating silicone oil by which the external pressure is transmitted through an opening immediately to the inner electrode system of the CO_2 sensor, thus preventing its damage due to enormous and rapid pressure increase when the CO_2 probe in the deep sea is moved down into great water depths.

Table 12.10 Performance parameters of the SAMI-CO_2 ocean CO_2 sensor (Sunburst Sensors, Missoula, MT, USA).

Parameter	Value
pCO_2 measuring range	200–600 μbar (ranges above 600 are available by request)
pCO_2 calibration range	150–700 μbar
Accuracy	±3 μbar
Precision, long-term drift	<1 μbar
Deployable depth	600 m (Delrin housing), 3500 m (titanium housing)
Long-term deployments	~10 000 measurements (a year taking hourly measurements)
Response time	~5 min
Dimensions	55 cm (housing length), 15.2 cm (diameter)
Weight	7.6 kg (in air)/1.1 kg (in seawater)

Source: Adapted from Wendy 2015 [153].

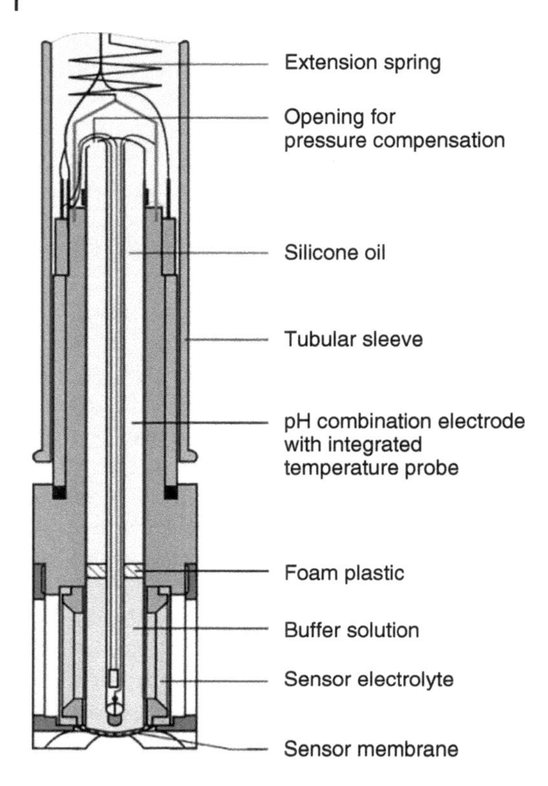

Figure 12.12 Schematic drawing of an electrochemical carbon dioxide sensor with pressure compensation (KSI Meinsberg, Kurt Schwabe Research Institute, Germany).

A prototype of such a pressure-compensated CO_2 probe was tested in a so-called Hydro Bottom Station (HBS) as shown in Figure 12.13. This device was constructed for sampling and monitoring hydrothermal fluids in the deep sea down to 3500 m water depth [155]. Besides the CO_2 probe, the sensor unit of the HBS contained also sensors for pH value, redox potential, O_2 concentration, and temperature.

12.3.5 CO_2 Measurements in Waters and Boreholes

Apart from pH value, oxygen concentration, and conductivity, the concentration of dissolved carbon dioxide is one of the most important parameters for the assessment of the condition of natural waters. It affects substantially the natural equilibrium between the limy substratum and the water, plays a role in the solubility of different inorganic components, and is involved in various biological reactions occurring in the water.

According to Henry's law, the CO_2 concentration in pure water or in sufficiently diluted solutions being in equilibrium with the atmospheric air can be estimated by the equation

$$c(\mathrm{CO_2}) = K_\mathrm{H}(T) \cdot p\mathrm{CO_2} \tag{12.60}$$

where c is the CO_2 concentration in the solution (mostly specified in milligrams per litre or moles per litre), p is the partial pressure of CO_2 in the atmosphere,

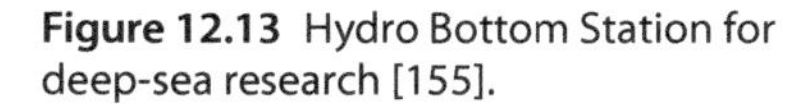
Figure 12.13 Hydro Bottom Station for deep-sea research [155].

and $K_H(T)$ is Henry's law coefficient, which depends exponentially on temperature. Figure 12.14 illustrates the considerable temperature dependence of the CO_2 concentration in water that is at its surface in equilibrium with the atmospheric CO_2 partial pressure of approximately 38 Pa [156]. At 25 °C it amounts to approximately 0.57 mg l^{-1}.

In most cases under natural environmental conditions, equilibrium between freshwater and the atmosphere, which is preconditioned in Henry's law, does not exist. Depending on local conditions and on temperature, the CO_2 content of

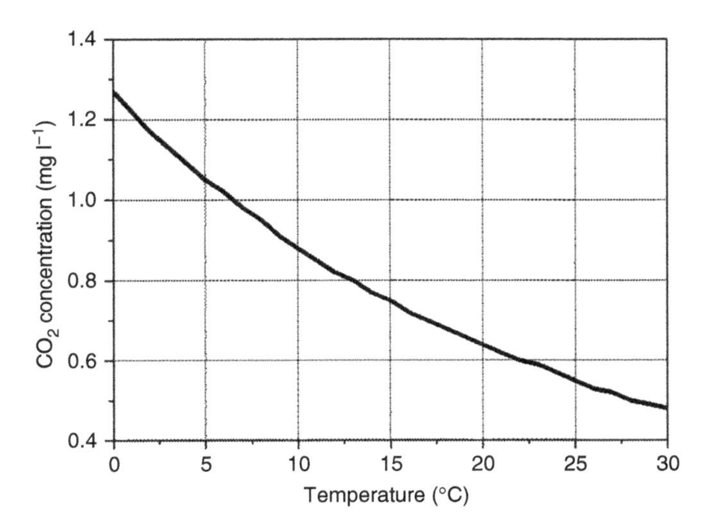

Figure 12.14 Temperature dependence of the CO_2 concentration in water at the total atmospheric pressure of 101 kPa and a CO_2 partial pressure of 38 Pa, calculated from [156].

natural waters can differ considerably and is also subject to seasonal fluctuations. Surface waters contain usually less than 10 mg l^{-1} CO_2 originating from the atmosphere and biological processes. As a result of the dissolution of minerals and the relatively high carbon dioxide content of the pore air of soils, groundwaters can exhibit up to 100 mg l^{-1} of CO_2, mineral waters even more than 1000 mg l^{-1}.

The CO_2 concentration in freshwater lakes is mainly determined by the organic production and decomposition of organic substances such as phytoplankton and algae. Hence, the CO_2 concentration represents a biological production indicator and is an important parameter for characterizing the trophic level of freshwater lakes [157]. For long-term measurements of CO_2 concentrations at different depths of a lake (Willersinnweiher near Ludwigshafen, Germany), an electrochemical CO_2 sensor, together with sensors for measuring other parameters, has been used in a special flow-through system [158, 159]. Furthermore, the CO_2 concentrations were calculated with the general geochemical computer program PHREEQC [160], which is applicable to many hydrogeochemical environments for simulating chemical reactions and transport processes in natural or polluted water. The program is based on equilibrium chemistry of aqueous solutions interacting with minerals, gases, solid solutions, exchangers, and sorption surfaces but also includes the capability to model kinetic reactions. In Figures 12.15 and 12.16, results of measurements with the CO_2 sensor are compared with the calculated values. Generally, the comparison of measured and calculated CO_2 values shows good correlation. In Figure 12.15 some deviations are obvious only in the metalimnion between 9 and 12 m of depth, which can be attributed to the dissolution and precipitation processes of calcite and the oxidation of CH_4.

During the mixing period in springtime, the generally low CO_2 values show a good correlation below 10 m of water depth. Above that depth, the measured values of the sensor show a different development than the calculated ones. This deviation can be explained by the fact that the aqueous carbonate system is not in balance with the changing CO_2 concentrations. The data of the sensor reflects the dissolution of atmospheric CO_2 due to the assimilation by phytoplankton.

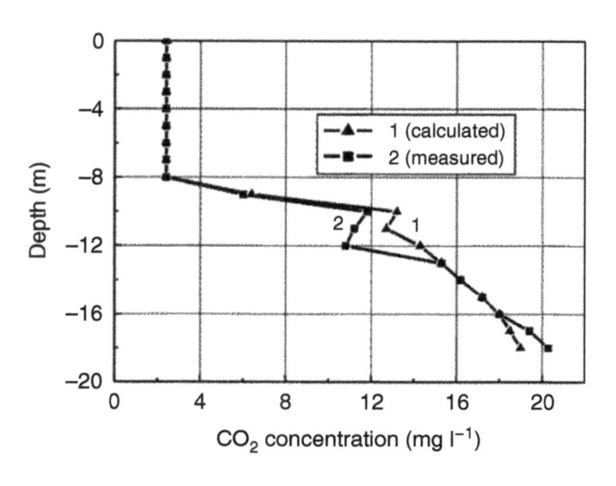

Figure 12.15 Comparison of calculated and measured CO_2 concentrations during the stagnation period in October [158, 159].

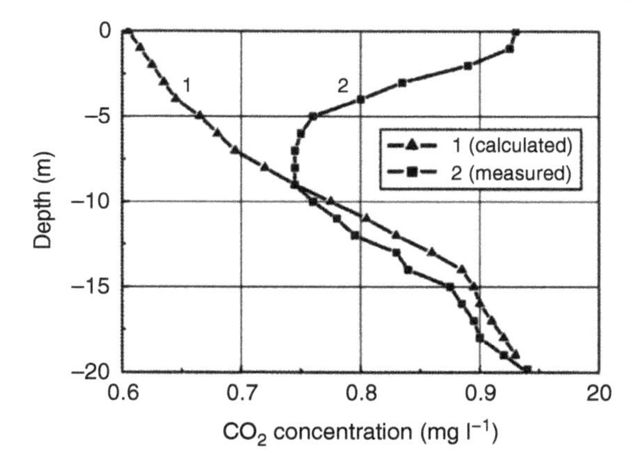

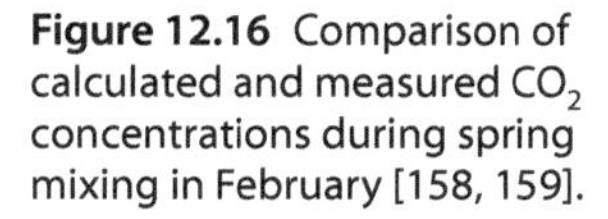
Figure 12.16 Comparison of calculated and measured CO_2 concentrations during spring mixing in February [158, 159].

This represents a general error source for calculated CO_2 concentrations in lake water, since these calculations assume a balanced aqueous carbonate system that does not establish in those dynamic lake water systems.

The measurement results show that the CO_2 dynamic in freshwater lakes is strongly determined by organic processes and the interactions between numerous physical, biological, and chemical influencing parameters. Furthermore, the carbon dioxide fluxes across the air–water interface have an impact on carbon availability in aquatic systems [161, 162]. The gas exchange between the water and the surrounding atmosphere can be studied by directly measuring CO_2 concentrations with the CO_2 sensor immediately below and above the water surface. The comparison of the measured CO_2 concentrations with calculated values on one hand showed a strong correlation but also gave evidence that the *in situ* measurement of CO_2 with a sensor is preferable or even necessary in many cases. Deviations from the CO_2 calculation can be found especially at measurements in disturbed water samples.

For the same reasons, the results of analytical CO_2 determination according to Section 2.1 and the measurement with an electrochemical CO_2 sensor in natural waters can be differ in many cases. Table 12.11 shows exemplarily the comparison of the CO_2 concentration of different water sources measured by an electrochemical CO_2 sensor.

While the results of measurements in tap water and in the river agree very well, the measurements in the pond show noticeable deviations from the calculated value. Such differences have also been observed in other territorially stagnant waters, whereby the relative deviations depend on the special site and the season. This phenomenon may be explained by the fact that the CO_2 sensor measures the partial pressure of CO_2 in the liquid, while the analytical method determines the concentration of carbonate and possibly also some other ingredients in the solution.

Monitoring of CO_2 could also be useful for earthquake prediction. There are internationally many efforts to find more reliable and quantitatively better evaluable methods than the observation and interpretation of strange behaviour patterns of animals (e.g. of ants [163]) before an earthquake eruption. Hydrogeological and hydro-chemical solutions seem to be promising approaches.

Table 12.11 Typical CO_2 concentrations in different water sources measured in September by using an electrochemical sensor (EMCO2, KSI Meinsberg).

Water source	Temperature	pH value	Conductivity	CO_2 sensor	Analytical method
	(°C)	—	(μS cm^{-1})	(mg l^{-1})	(mg l^{-1})
Tap water	22	7.60	690	4.0	4.0
Medium-sized river (Zschopau, Germany)	20	7.34	363	2.7	2.6
Fire protection pond	20	7.26	742	14.3	18.3

It has been found that CO_2-rich springs occur worldwide along major zones of seismicity. Consequently, the presence of such springs may indicate a potentially hazardous region [164].

For already more than two decades, systematic seismo-hydrological studies have been performed in the Vogtland and the neighbouring north-west Bohemian region at the German–Czech border [165]. In this geodynamically active region, CO_2 degassing takes place in numerous mineral springs and the so-called mofettes. The seismicity is characterized by numerous micro earthquakes, which on some occasions occur at high frequencies, then considered as "swarmquakes". First attempts to take water samples from an exploration borehole at a depth of 135 m failed and resulted in a huge water eruption due to the devolatilization of the fluid oversaturated with CO_2 as a consequence of the movement of the scoop [166].

For this reason, the CO_2 content of the water at the bottom of a borehole was determined by means of an electrochemical CO_2 sensor, which was modified especially for this purpose (Figure 12.17) [167]. The diameter of the CO_2 sensor is 10.5 mm. By means of a watertight connecting cable, the CO_2 probe can be submerged up to a depth of 100 m. To transfer the very high ohmic sensor

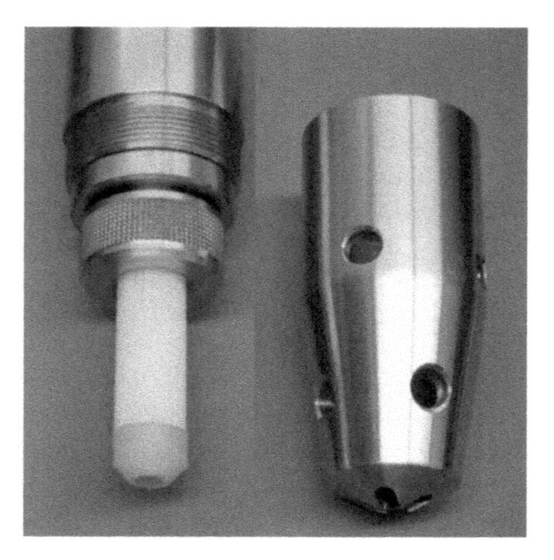

Figure 12.17 CO_2 probe with unscrewed protective cap (KSI Meinsberg) for measuring the CO_2 content of the water at the bottom of a borehole.

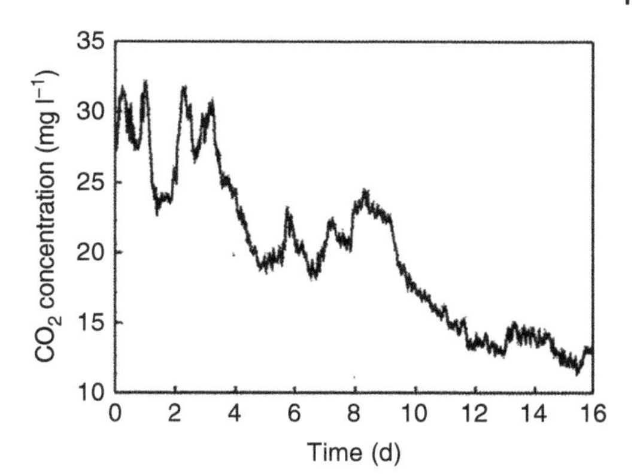

Figure 12.18 Temporal course of the CO_2 concentration in a mineral water spring (Bad Reiboldsgrün, Germany) [167].

signal over this distance, an impedance converter is integrated into the head of the probe.

Figure 12.18 illustrates the temporal course of the CO_2 concentration in a CO_2-rich mineral water spring. During the study period of 16 days, significant changes of the CO_2 concentration have been measured, even though the main influencing parameters like water temperature and atmospheric pressure remained nearly constant during this time. When heaving the sensor after longer immersion time from a water depth of 100 m, not only the hydrostatic water pressure of 981 kPa changes to the normal air pressure. Also the CO_2 concentration from more than 10 mg l^{-1} in the sensor electrolyte decreases to less than 1 mg l^{-1}. The excessive CO_2 tends to escape from the sensor rapidly, which can result in the destruction of the thin polymeric sensor membrane. To prevent this, the sensor must be heaved very slowly from the water depth, thus allowing a gradual adaptation of CO_2 pressure and concentration.

References

1 Schellnhuber, H.-J. (1999). Earth system analysis and the second Copernican revolution. *Nature* 402 (6761): C19–C23.

2 Waterman, L.S. (1983). Comments on 'The Montsouris series of carbon dioxide concentration measurements, 1877–1910', by Stanhill. *Clim. Change* 5 (4): 413–415.

3 Crutzen, P.J. and Stoermer, E.F. (2000). The "Anthropocene". *Global Change Newslett.* 41: 17–18.

4 Boden, T.A., Marland, G., and Andres, R.J. (2009). *Global, Regional, and National Fossil-Fuel CO_2 Emissions.* Oak Ridge, TN: Carbon Dioxide Information Analysis Center, Oak Ridge National Laboratory, U.S. Department of Energy https://doi.org/10.3334/CDIAC/00001.

5 Houghton, R.A. (2005). Tropical deforestation as a source of greenhouse gas emission. In: *Tropical Deforestation and Climate Change* (ed. P. Moutinho and S. Schwartzman), 13–21. Belém, BR: Amazon Institute for Environmental Research.

6 Prentice, I.C., Farquhar, G.D., Fasham, M.J.R. et al. (2001). The carbon cycle and atmospheric carbon dioxide. In: *Climate Change 2001: The Scientific Basis* (ed. J.T. Houghton, Y. Ding, D.J. Griggs, et al.), 183–237. Cambridge: Cambridge University Press.

7 IPCC (2007). Climate change 2007: the physical science basis. In: *Contribution of Working Group I to the Fourth Assessment Report of the Intergovernmental Panel on Climate Change* (ed. S. Solomon, D. Qin, M. Manning, et al.), 996 pp. Cambridge and New York: Cambridge University Press.

8 Sabine, C.L., Feely, R.A., Gruber, N. et al. (2004). The oceanic sink for anthropogenic CO_2. *Science* 305 (5682): 367–371.

9 Denman, K.L., Brasseur, G., Chidthaisong, A. et al. (2007). Couplings between changes in the climate system and biogeochemistry. In: *Climate Change 2007: The Physical Science Basis. Contribution of Working Group I to the Fourth Assessment Report of the Intergovernmental Panel on Climate Change* (ed. S. Solomon, D. Qin, M. Manning, et al.), 499–588. Cambridge: Cambridge University Press.

10 Houghton, R.A. (2008). Carbon flux to the atmosphere from land-use changes: 1850–2005. In: *TRENDS: A Compendium of Data on Global Change*. Oak Ridge, TN: Carbon Dioxide Information Analysis Center, Oak Ridge National Laboratory, U.S. Department of Energy. http://cdiac.ornl.gov/trends/landuse/houghton/houghton.html (retrieved 28 March 2017).

11 Dickson, A.G., Sabine, C.L., and Christian, J.R. (eds.) (2007). *Guide to Best Practices for Ocean CO_2 Measurements*, PICES Special Publication 3. http://cdiac.esd.ornl.gov/oceans/Handbook_2007.html (retrieved 28 March 2017).

12 Solomon, S., Plattner, G.-K., Knutti, R., and Friedlingstein, P. (2009). Irreversible climate change due to carbon dioxide emissions. *Proc. Natl. Acad. Sci. U.S.A.* 106 (6): 1704–1709.

13 Holmén, K. (1992). The global carbon cycle. In: *Global Biogeochemical Cycles* (ed. S.S. Butcher, R.J. Charlson, G.H. Orians and G.V. Wolfe), 239–262. London: Academic Press.

14 Neftel, A., Moor, E., Oeschger, H., and Stauffer, B. (1985). Evidence from polar ice cores for the increase in atmospheric CO_2 in the past two centuries. *Nature* 315 (6014): 45–47.

15 Friedli, H., Lötscher, H., Oeschger, H. et al. (1986). Ice core record of $^{13}C/^{12}C$ ratio of atmospheric CO_2 in the past two centuries. *Nature* 324 (6094): 237–238.

16 Etheridge, D.M., Steele, L.P., Francey, R.J., and Langenfelds, R.L. (1998). Atmospheric methane between 1000 AD and present: evidence of anthropogenic emissions and climatic variability. *J. Geophys. Res.* 103 (D13): 15979–15993.

17 Chen, C.T.A. and Poisson, A. (1984). Excess carbon dioxide in the Weddell Sea. *Antarct. J.* 19: 74–75.

18 Callendar, G.S. (1938). The artificial production of carbon dioxide and its influence on temperature. *Q. J. R. Meteorol. Soc.* 64 (275): 223–240.

19 Callendar, G.S. (1949). Can carbon dioxide influence climate? *Weather* 4 (10): 310–314.

20 From, E. and Keeling, C.D. (1986). Reassessment of late 19th century atmospheric carbon dioxide variations in the air of western Europe and the British Isles based on an unpublished analysis of contemporary air masses by G.S. Callendar. *Tellus B* 38 (2): 87–105.
21 Black, J. (1803). *Lectures on Elements of Chemistry* (ed. J. Robinson). Edinburgh: Mundell & Son.
22 de Saussure, H.-B. (1796). *Voyages dans les Alpes, précédés d'un essai sur l'histoire naturelle des environs de Genève (voyages in the Alps, preceding a history of the environment of Geneve). Tomé IV*, 199–201. Neuchâtel: Fauche-Borel (in French).
23 von Humboldt, A. (1850-1852). *Cosmos: A Sketch of a Physical Description of the Universe* (Translated from German by E.C. Otté), vol. 1 (1850) 275 pp., vol. 2 (1850) 367 pp., vol. IV (1851) 219 pp., vol. IV (1852) 234 pp. New York: Harper & Brothers.
24 Letts, E.A. and Blake, R.F. (1900). The carbonic anhydride of the atmosphere. *Sci. Proc. R. Dublin Soc.* 9: 107–270.
25 Letts, E.A. and Blake, R.F. (1901). On some problems connected with atmospheric carbonic anhydride, and on a new and accurate method for determining its amount suitable for scientific expeditions. *Sci. Proc. R. Dublin Soc.* 9: 435–453.
26 Schloesing, T. (1880). Sur la constance de proportion d'acide carbonique dans l'air (about the constancy of carbonic acid in the air). *C.R. Hebd. Seances Acad. Sci.* 90: 1410–1413 (in French).
27 Blochmann, R. (1887). Ueber den Kohlensäuregehalt der atmosphärischen Luft (about the amount of carbonic acid in the air). *Liebigs Ann. Chem.* 237 (1): 39–90 (in German).
28 Renk, F. (1886). *Die Luft. Handbuch der Hygiene und der Gewerbekrankheiten. Erster Theil. Individuelle Hygiene. (The air. Handbook of hygiene and work-related diseases) 2. Abtheilung. 2. Heft* (ed. M. von Pettenkofer and H. von Ziemssen). Leipzig: F.C.W. Vogel (in German).
29 Brown, H.T. and Escombe, F. (1905). Researches on some of the physiological processes of green leaves, with special reference to the interchange of energy between the leaf and its surroundings. *Proc. R. Soc. London, Ser. B* 76 (507): 29–111.
30 Friedheim, C. (ed.) (1907). *Gmelin-Kraut's Handbuch der anorganischen Chemie*, Band 1, Abteilung 1, 888 pp. Heidelberg: Carl Winters Universitätsbuchhandlung.
31 Lode, A. (1911). Atmosphäre (atmosphere). In: *Handbuch der Hygiene*, I. Band (ed. M. Rubner, M. von Grubner and M. Ficker), 367–518. Leipzig: Verlag Hirzel (in German).
32 Benedict, F.G. (1912). *The Composition of the Atmosphere with Special Reference to its Oxygen Content*. Washington, DC: Carnegie Institution Washington.
33 Krogh, A. (1919). The composition of the atmosphere. *K.D. Vid. Selsk. Math.-fys. Medd.* I (12): 1–19.
34 Lueker, T.J., Dickson, A.G., and Keeling, C.D. (2000). Ocean pCO_2 calculated from dissolved inorganic carbon, alkalinity, and equations for K_1 and K_2:

validation based on laboratory measurements of CO_2 in gas and seawater equilibrium. *Mar. Chem.* 70 (1–3): 105–119.

35 Quinn, E.L. and Jones, C.L. (1936). *Carbon Dioxide,* American Chemical Society Monographs Series. New York: Reinhold Publishing Corporation.

36 Mellor, J.W. (1940). *A Comprehensive Treatise of Inorganic and Theoretical Chemistry,* vol. VI, 1024 pp. London: Longmans.

37 D'Ans, J. and Lax, E. (1943). *Taschenbuch für Chemiker und Physiker (pocket book for chemists and physicists).* Berlin: Springer-Verlag (in German).

38 Remy, H. (1965). *Lehrbuch der anorganischen Chemie (textbook of inorganic chemistry),* Band. I. 12. Auflage. Leipzig: Akad. Verlagsgesell (in German).

39 Walker, J. (1900). Estimation of atmospheric carbon dioxide. *J. Chem. Soc. Trans.* 77: 1110–1114.

40 Fresenius, R. (1875). *Anleitung zur quantitativen chemischen Analyse (manual for the quantitative chemical analysis), In zwei Bänden.* Braunschweig: Vieweg und Sohn (in German).

41 Haldane, J.S. (1918). *Methods of Air Analysis.* London: Charles Griffin & Company.

42 Hempel, W. (1913). *Gasanalytische Methoden (methods of gas analysis),* 4e. Braunschweig: Vieweg & Sohn (in German).

43 Möller, D. (2009). Feinstaubbelastung: Ursachen und Gesundheitsgefährdung. (particulate pollution: reasons and health hazards). *Forum Forsch. (BTU Cottbus)* 22: 117–126.

44 Fittbogen, J. and Hässelbarth, P. (1879). Ueber locale Schwankungen im Kohlensäuregehalt der atmosphärischen Luft (about local fluctuations of the carbonic acid content in the atmosphere). In: *Jahresbericht ueber die Fortschritte auf dem Gesammtgebiete der Agricultur-Chemie,* vol. 22, 67–71. Berlin Verlag Paul Parey (in German).

45 Petermann, A. and Graftiau, J. (1892). Recherches sur la composition de l'atmosphère. Acide carbonique. Combinaisons azotées contenues dans l'air atmosphérique et dans l'eau de pluie. Première partie: Acide carbonique contenu dans l'air atmosphérique (Research on the composition of the atmosphere. Carbonic acid. Combinations of nitrogen contained in atmospheric air and rainwater. Part 1: Carbonic acid contained in atmospheric air). In: *Mémoires couronnés et autres mémoires publiés par l'academie royal de Belgique. Tomé XLVII (1892–1893),* 79 pp. Bruxelles: Académie royale des sciences, des lettres et des beaux-arts de Belgique (in French).

46 Marie-Davy, M. (1880). L'acide carbonique de l'air, dans ses rapports avec les grands mouvements de l'atmosphére (The carbonic acid of the air, in its relations with the great motions of the atmosphere). *C.R. Hebd. Seances Acad. Sci.* 90: 32–35 (in French).

47 von Fodor, J. (1879). Experimentelle Untersuchungen über Boden und Bodengase (experimental investigations concerning ground and ground gases). *Vierteljahresschrift für öffentliche Gesundheitspflege* 7: 205–237 (in German).

48 Reiset, J. (1880). Recherches sur la proportion de l'acide carbonique dans l'air (investigations about the amount of carbonic acid in the air). *C.R. Hebd. Seances Acad. Sci.* 90: 1144–1148 (in French).

49 Dumas, M. (1882). Sur l'acide carbonique normal de l'air atmosphérique (about natural carbon dioxide in the air). *Ann. Chim. Phys. Series 5, Tomé* 26: 254–261 (in French).

50 Baumann, A. (1892). Chemie der Atmosphäre (chemistry of the atmosphere). In: *Jahresberichte über die Fortschritte auf dem Gesamtgebiete der Agrikultur-Chemie. Neue Folge, XV*, 3–18. Berlin: Verlag P. Parey (in German).

51 Callendar, G.S. (1958). On the amount of carbon dioxide in the atmosphere. *Tellus* 10 (2): 243–248.

52 Lévy, A. and Miquel, P. (1891). Annuaire de l'Observatoire Municipal de Montsouris pour l'an 1891 (annual observations of Montsouris in the year 1981); cited from Baumann [39] (in French).

53 Schulze, F. (1871). *Tägliche Beobachtungen über den Kohlen-Säuregehalt der Atmosphäre zu Rostock vom 18. Oktober 1868 bis 31. Juli 1871 (daily observations of the carbonic acid amount in the atmosphere of Rostock from Oct. 18, 1868 until July 31, 1871), Festschrift für die 44.* Rostock: Versammlung Deutscher Naturforscher und Aerzte. Leopoldsche Universitätsbuchhandlung (in German).

54 Gilm, V.H. (1857). Über die Kohlensäure-Bestimmung der atmosphärischen Luft (about the carbonic acid determination in air). *Sitzungsberichte der Kaiserlichen Akademie der Wissenschaften*, Math.-Nat. Classe. Band 24, Wien, 279–284 (in German).

55 Farsky, F. (1877). Bestimmungen der atmosphärischen Kohlensäure in den Jahren 1874–1875 zu Tabor (Böhmen) (determination of atmospheric carbonic acid in the years 1874–1875 in Tabor Bohemia). *Sitzungsberichte der Kaiserlichen Akademie der Wissenschaften, Math.-Nat. Classe.* Band 74, Wien, 67–77 (in German).

56 von Frey, E. (1889). Der Kohlensäuregehalt der Luft in und bei Dorpat bestimmt in den Monaten September 1888 bis Januar 1889 (the amount of carbonic acid in the air in and around Dorpat in the months from September 1888 until January 1889). Doctor thesis. Med. Fakultät der Kaiserl. Universität Dorpat (in German).

57 Heimann, J. (1888). Der Kohlensäuregehalt der Luft in Dorpat bestimmt in den Monaten Juni bis September 1888 (the amount of carbonic acid determined in the air in Dorpat in between June and September 1888). Doctor thesis. Med. Fakultät der Kaiserl. Universität Dorpat (in German).

58 Reiset, J. (1879). Recherches sur la proportion de l'acide carbonique dans l'air (investigations about the amount of carbonic acid in the air). *C.R. Hebd. Seances Acad. Sci.* 88: 1007–1012 (in French).

59 Müntz, A. and Aubin, E. (1882). Sur la proportion d'acid carbonique dans les hautes regions de l'atmosphére (about the amount of carbonic acid in higher regions of the atmosphere). *Ann. Chim. Phys. Series 5, Tomé* 26: 222–254 (in French).

60 Thorpe, T.E. (1867). On the amount of carbonic acid contained in sea air. *J. Chem. Soc.* 20: 189–199.
61 Marchand, R.F. (1850). Ueber die Eudiometrie (about eudiometry). *J. Prakt. Chem.* 49 (1): 449–468 (in German).
62 Armstrong, G.F. (1879). On the diurnal variations in the amount of carbon dioxide in the air. *Proc. R. Soc. London* 30: 343–355.
63 Selander, N.E. (1888). *Luftundersökningar vid Vaxholms fästning okt. 1885 – juli 1886 (investigations of the air at Vaxholm fortress)*, 13, Afd. II, No. 9, 38 pp. Stockholm: Kungliga Svenska Vetenskapsakad (in Swedish).
64 Lüthi, D., Le Floch, M., Bereiter, B. et al. (2008). High-resolution carbon dioxide concentration record 650,000–800,000 years before present. *Nature* 453 (7193): 379–382.
65 Etheridge, D.M., Steele, L.P., Langenfelds, R.L. et al. (1996). Natural and anthropogenic changes in atmospheric CO_2 over the last 1000 years from air in Antarctic ice and firn. *J. Geophys. Res.* 101 (D2): 4115–4128.
66 Malingreau, J.-P. and Zhuang, Y.H. (1998). Biomass burning: an ecosystem process of global significance. In: *Asian Change in the Context of Global Climate Change* (ed. J. Galloway and J. Melillo), 101–127. Cambridge: Cambridge University Press.
67 Steele, L.P., Krummel, P.B., and Langenfelds, R.L. (2007). Atmospheric CO_2 concentrations from sites in the CSIRO Atmospheric Research GASLAB air sampling network (August 2007 version). In: *Trends: A Compendium of Data on Global Change*. Oak Ridge, TN: Carbon Dioxide Information Analysis Center, Oak Ridge National Laboratory, U.S. Department of Energy. http://csiro.au/greenhouse-gases/GreenhouseGas/data/CapeGrim_CO2_data_download.txt.
68 Data were obtained from the Scripps website (scrippsco2.ucsd.edu); Pieter Tans, NOAA/ESRL (www.esrl.noaa.gov/gmd/ccgg/trends/) and Ralph Keeling, Scripps Institution of Oceanography (scrippsco2.ucsd.edu/); data from March 1958 through April 1974 have been obtained by C. David Keeling of the Scripps Institution of Oceanography (SIO). For additional details see also www.esrl.noaa.gov/gmd/ccgg/trends/ (retrieved 25 February 2017).
69 CDIAC (Carbon Dioxide Information Analysis Center) (2013). Fossil-fuel CO_2 emissions. http://cdiac.ornl.gov/ (retrieved 25 February 2017).
70 Delmas, R.J., Ascencio, J.-M., and Legrand, M. (1980). Polar ice evidence that atmospheric CO_2 20,000 a BP was 50% of present. *Nature* 284 (5752): 155–157.
71 Neftel, A., Oeschger, H., Schwander, J. et al. (1982). Ice core sample measurements give atmospheric CO_2 content during the past 40,000 yr. *Nature* 295 (5846): 220–223.
72 Monnin, E., Indermühle, A., Dällenbach, A. et al. (2001). Atmospheric CO_2 concentrations over the last glacial termination. *Science* 291 (5501): 112–114.
73 Petit, J.R., Jouzel, J., Raynaud, D. et al. (1999). Climate and atmospheric history of the past 420,000 years from the Vostok ice core, Antarctica. *Nature* 399 (6735): 429–436.

74 MacFarling Meure, C., Etheridge, D., Trudinger, C. et al. (2006). The law dome CO_2, CH_4 and N_2O ice core records extended to 2000 years BP. *Geophys. Res. Lett.* 33 (14): L14810. 4 pp. https://doi.org/10.1029/2006GL026152.

75 Ruddiman, W.F. (2003). The anthropogenic greenhouse era began thousands of years ago. *Clim. Change* 61 (3): 261–293.

76 Ruddiman, W.F. (2007). The early anthropogenic hypothesis: challenges and responses. *Rev. Geophys.* 45 (4): RG4001, 37 pp. https://doi.org/10.1029/2006RG000207.

77 IPCC (2013). Climate change 2013: the physical science basis. Annex V: Contributors to the IPCC WGI fifth assessment report. In: *Climate Change 2013: The Physical Science Basis. Contribution of Working Group I to the Fifth Assessment Report of the Intergovernmental Panel on Climate Change* (ed. T.F. Stocker, D. Qin, G.-K. Plattner, et al.). Cambridge and New York: Cambridge University Press.

78 Etheridge, D.M., Pearman, G.I., and de Silva, F. (1988). Atmospheric trace-gas variations as revealed by air trapped in an ice core from Law Dome, Antarctica. *Ann. Glaciol.* 10: 28–33.

79 Etheridge, D.M., Pearman, G.I., and Fraser, P.J. (1992). Changes in tropospheric methane between 1841 and 1978 from a high accumulation-rate Antarctic ice core. *Tellus B* 44 (4): 282–294.

80 Morgan, V.I., Wookey, C.W., Li, J. et al. (1997). Site information and initial results from deep ice drilling on Law Dome, Antarctica. *J. Glaciol.* 43 (143): 3–10.

81 Keeling, C.D., Bacastow, R.B., Bainbridge, A.E. et al. (1976). Atmospheric carbon dioxide variations at Mauna Loa Observatory, Hawaii. *Tellus* 28 (6): 538–551.

82 Keeling, R.F. and Shertz, S.R. (1992). Seasonal and interannual variations in atmospheric oxygen and implications for the global carbon cycle. *Nature* 358 (6389): 723–727.

83 Manning, A. and Keeling, R.F. (2006). Global oceanic and land biotic carbon sinks from the Scripps atmospheric oxygen flask sampling network. *Tellus B* 58 (2): 95–116.

84 Thoning, K.W., Tans, P.P., and Komhyr, W.D. (1989). Atmospheric carbon dioxide at Mauna Loa Observatory, 2. Analysis of the NOAA/GMCC data, 1974–1985. *J. Geophys. Res.* 94 (D6): 8549–8565.

85 Möller, D. (2014). *Chemistry of the Climate System*, 2e. Berlin, New York: De Gruyter.

86 Taylor, J.A. and Orr, J.C. (2000). The natural latitudinal distribution of atmospheric CO_2. *Global Planet. Change* 26 (4): 375–386.

87 Potter, C.S., Randerson, J.T., Field, C.B. et al. (1993). Terrestrial ecosystem production: a process model based on global satellite and surface data. *Global Biogeochem. Cycles* 7 (4): 811–841.

88 Taguchi, S., Murayama, S., and Higuchi, K. (2003). Sensitivity of inter-annual variation of CO_2 seasonal cycle at Mauna Loa to atmospheric transport. *Tellus B* 55 (2): 547–554.

89 Cerveny, R.S. and Coakley, K.J. (2002). A weekly cycle in atmospheric carbon dioxide. *Geophys. Res. Lett.* 29 (2): 4. https://doi.org/10.1029/2001GL013952.

90 Rubner, M. (1907). *Lehrbuch der Hygiene. Systematische Darstellung der Hygiene und ihrer wichtigsten Untersuchungs-Methoden (Textbook of hygiene. Systematic presentation of hygiene and its most important methods of investigation)*, 8e. Leipzig, Wien: Deuticke (in German).

91 Gratani, L. and Varone, L. (2005). Daily and seasonal variation of CO_2 in the city of Rome in relationship with the traffic volume. *Atmos. Environ.* 39 (14): 2619–2624.

92 Miyaoka, Y., Inoue, H.Y., Sawa, Y. et al. (2007). Diurnal and seasonal variations in atmospheric CO_2 in Sapporo, Japan: anthropogenic sources and biogenic sinks. *Geochem. J.* 41 (6): 429–436.

93 Henninger, S. and Kuttler, W. (2004). Mobile measurements of carbon dioxide in the urban boundary layer of Essen, Germany. In: *5th Urban Environment Symposium*, 12.3. (5 pp.). Vancouver, Canada: American Meteorological Society.

94 Idso, S.B., Idso, C.D., and Balling, R.C. Jr. (2002). Seasonal and diurnal variations of near-surface atmospheric CO_2 concentrations within a residential sector of the urban CO_2 dome of Phoenix, AZ, USA. *Atmos. Environ.* 36 (10): 1655–1660.

95 Balling, R.C. Jr., Cerveny, R.S., and Idso, C.D. (2002). Does the urban CO_2 dome of Phoenix, Arizona contribute to its heat island? *Geophys. Res. Lett.* 28 (24): 4599–4601.

96 Moriwaki, R., Kanda, M., and Nitta, H. (2006). Carbon dioxide build-up within a suburban canopy layer in winter night. *Atmos. Environ.* 40 (8): 1394–1407.

97 Widory, D. and Javoy, M. (2003). The carbon isotope composition of atmospheric CO_2 in Paris. *Earth Planet. Sci. Lett.* 215 (1–2): 289–298.

98 Segalstad, T.V. (2009). Correct timing is everything – also for CO_2 in the air. *CO2 Science* 12 (31). http://www.co2science.org/articles/V12/N31/EDIT.php (retrieved 25 February 2017).

99 Prather, M.J. (2007). Lifetimes and time scales in atmospheric chemistry. *Philos. Trans. R. Soc. A* 365 (1856): 1705–1726.

100 Berner, R.A. (2005). A different look at biogeochemistry. *Am. J. Sci.* 305 (6–8): 872–873.

101 Beerling, D.J. and Berner, R.A. (2005). Feedbacks and the coevolution of plants and atmospheric CO_2. *Proc. Natl. Acad. Sci. U.S.A.* 102 (5): 1302–1305.

102 Chen, S. and Hoffman, M.Z. (1973). Rate constants for the reaction of the carbonate radical with compounds of biochemical interest in neutral aqueous solution. *Radiat. Res.* 56 (1): 40–47.

103 Umschlag, T. and Herrmann, H. (1999). The carbonate radical ($HCO_3^{\bullet}/CO_3^{-\bullet}$) as a reactive intermediate in water chemistry: kinetics and modelling. *Acta Hydroch. Hydrob.* 27 (4): 214–222.

104 Czapski, G., Lymar, S.V., and Schwarz, H.A. (1999). Acidity of the carbonate radical. *J. Phys. Chem. A* 103 (18): 3447–3450.

105 Armstrong, D.A., Waltz, W.L., and Rauk, A. (2006). Carbonate radical anion – thermochemistry. *Can. J. Chem.* 84 (12): 1614–1619.

106 Medinas, D.B., Cerchiaro, G., Trindade, D.F., and Augusto, O. (2007). The carbonate radical and related oxidants derived from bicarbonate buffer. *IUBMB Life* 59 (4–5): 255–262.

107 Wu, G., Katsumura, Y., Muroya, Y. et al. (2002). Temperature dependence of carbonate radical in $NaHCO_3$ and Na_2CO_3 solutions: is the radical a single anion? *J. Phys. Chem. A* 106 (11): 2430–2437.

108 Herrmann, H., Ervens, B., Jacobi, H.-W. et al. (2000). CAPRAM 2.3: a chemical aqueous phase radical mechanism for tropospheric chemistry. *J. Atmos. Chem.* 36 (3): 231–284.

109 Draganic, Z.D., Negron-Mendoza, A., Sehested, K. et al. (1991). Radiolysis of aqueous solutions of ammonium bicarbonate overe a large dose range. *Radiat. Phys. Chem.* 38 (3): 317–321.

110 Lilie, J., Hanrahan, R.J., and Henglein, A. (1978). O^- transfer reactions of the carbonate radical anion. *Radiat. Phys. Chem.* 11 (5): 225–227.

111 Sehested, K., Holcman, J., and Hart, E.J. (1983). Rate constants and products of the reactions e_{aq}^-, O_2^- and H with ozone in aqueous solution. *J. Phys. Chem.* 87 (11): 1951–1954.

112 Hart, E.J. and Anbar, M. (1970). *The Hydrated Electron*. New York: Wiley.

113 Todres, Z.V. (2008). *Ion-Radical Organic Chemistry. Principles and Applications*, 2e. Boca Raton, FL: CRC Press.

114 Morkovnik, A.F. and Okhlobystin, O.Y. (1979). Inorganic radical-ions and their organic reactions. *Russ. Chem. Rev.* 48 (11): 1055–1075.

115 Fujita, E. (1999). Photochemical carbon dioxide reduction with metal complexes. *Coord. Chem. Rev.* 185–186: 373–384.

116 Wu, J.C.S. (2009). Photocatalytic reduction of greenhouse gas CO_2 to fuel. *Catal. Surv. Asia* 13 (1): 30–40.

117 Zeebe, R. and Gattuso, J.-P. (2006). Marine carbonate chemistry. In: *Encyclopedia of EarthEnvironmental Information Coalition, National Council for Science and the Environment* (ed. C.J. Cleveland). Washington, DC, Published in the Encyclopedia of Earth 21 September 2006. http://editors.eol.org/eoearth/wiki/Marine_carbonate_chemistry (retrieved 25 February 2017).

118 Gmelin (1943). *Gmelins Handbuch der anorganischen Chemie (Gmelins handbook of inorganic chemistry – oxygen)*, 8. Auflage, System-Nr. 3: Sauerstoff, Lieferung 1–2, 298–300. Weinheim: Verlag Chemie (in German).

119 Carroll, J.J. and Mather, A.E. (1992). The system carbon dioxide-water and the Krichevsky-Kasarnovsky equation. *J. Solution Chem.* 21 (7): 607–621.

120 Gmelin (1973). *Gmelins Handbuch der anorganischen Chemie (Gmelins handbook of inorganic chemistry – carbon)*, 8. Auflage, System-Nr. 14: Kohlenstoff, Teil C 3, 160 pp. Weinheim: Verlag Chemie (in German).

121 Sigg, L. and Stumm, W. (1996). *Aquatische Chemie. Eine Einführung in die Chemie wässriger Lösungen und natürlicher Gewässer (aquatic chemistry, an introduction in the chemistry of aqueous solutions and natural waters)*, 498 pp. Teubner, Stuttgart: Vdf Hochschulverlag an der ETH Zürich (in German).

122 Stumm, W., Morgan, J.J., and Schnoor, J.L. (1983). Saurer Regen, eine Folge der Störung hydrogeochemischer Kreisläufe (acid rain – a result of disturbance of hydrogeochemical cycles). *Naturwissenschaften* 70 (5): 216–223 (in German).

123 Lide, D.R. and Frederikse, H.P.R. (eds.) (1995). *CRC Handbook of Chemistry and Physics*, 76e. Boca Raton, FL: CRC Press.

124 Lundegårdh, H. (1924). *Der Kreislauf der Kohlensäure in der Natur. Ein Beitrag zur Pflanzenökologie und zur landwirtschaftlichen Düngungslehre (circulation of carbonic acid in the nature. A contribution for plant ecology and for agriculture fertilization).* Jena: Verlag von Gustav Fischer (in German).

125 Wagener, K. (1979). The carbonate system of the ocean. In: *The Global Carbon Cycle. Scope 13* (ed. B. Bolin, E.T. Degens, S. Kempe and P. Ketner), 251–258. New York: Wiley.

126 Pearson, P.N. and Palmer, M.R. (1999). Middle Eocene seawater pH and atmospheric carbon dioxide concentrations. *Science* 284 (5471): 1824–1826.

127 Caldeira, K. and Berner, R. (1999). Seawater pH and atmospheric carbon dioxide (Technical comments). *Science* 286 (5547): 2043a.

128 Dore, J.E. (2009). *Hawaii Ocean Time-Series Surface CO_2 System Data Product, 1988–2008, SOEST*. Honolulu, HI: University of Hawaii. http://hahana.soest.hawaii.edu/hot/products/products.html (retrieved 25 February 2017).

129 Dore, J.E., Lukas, R., Sadler, D.W., and Karl, D.M. (2003). Climate-driven changes to the atmospheric CO_2 sink in the subtropical North Pacific Ocean. *Nature* 424 (6950): 754–757.

130 Dore, J.E., Lukas, R., Sadler, D.W. et al. (2009). Physical and biogeochemical modulation of ocean acidification in the central North Pacific. *Proc. Natl. Acad. Sci. U.S.A.* 106 (30): 12235–12240. https://doi.org/10.1073/pnas.0906044106.

131 Marsh, G.E. (2010). Seawater pH and anthropogenic carbon. In: *Climate Change* (ed. S.P. Saikia). Dehradun: International Book Distributors. see also: arXiv:0810.3596v1 [physics.ao-ph].

132 Caldeira, K. and Rau, G.H. (2000). Accelerating carbonate dissolution to sequester carbon dioxide in the ocean: geochemical implications. *Geophys. Res. Lett.* 27 (2): 225–228.

133 Feely, R.A., Sabine, C.L., Hernandez-Ayon, J.M. et al. (2008). Evidence for upwelling of corrosive "acidified" water onto the continental shelf. *Science* 320 (5882): 1490–1492.

134 Le Quéré, C., Rödenbeck, C., Buitenhuis, E.T. et al. (2007). Saturation of the Southern Ocean CO_2 sink due to recent climate change. *Science* 316 (5832): 1735–1738.

135 Schulz, K.G., Riebesell, U., Rost, B. et al. (2006). Determination of the rate constants for the carbon dioxide to bicarbonate inter-conversion in pH-buffered seawater systems. *Mar. Chem.* 100 (1–2): 53–65.

136 Zeebe, R.E. and Wolf-Gladrow, D.A. (2009). Carbon dioxide, dissolved (Ocean). In: *Encyclopedia of Paleoclimatology and Ancient Environments* (ed. V. Gornitz), 123–127. Dordrecht: Springer.

137 http://www.amt-gmbh.com (retrieved 06 October 2016).

138 http://www.analyticon.com (retrieved 06 October 2016).
139 http://www.battelle.org (retrieved 06 October 2016).
140 http://www.contros.eu (retrieved 06 October 2016).
141 http://www.generaloceanics.com (retrieved 06 October 2016).
142 http://www.kimoto-electric.co.jp (retrieved 06 October 2016).
143 http://www.pro-oceanus.com (retrieved 06 October 2016).
144 http://www.subctech.com (retrieved 06 October 2016).
145 http://www.sunburstsensors.com (retrieved 06 October 2016).
146 DOE (1994). Handbook of methods for the analysis of the various parameters of the carbon dioxide system in sea water, version 2 (ed. A.G. Dickson, and C. Goyet). ORNL/CDIAC-74.
147 http://www.kohsieh.com.tw/PDF_Files/LICOR/pCO2ApplicationNote.pdf (retrieved 06 October 2016).
148 Johnson, B. and McNeil, C. (2004). System for the transfer and sensing of gas dissolved in liquid under pressure. US Patent 7,434,446 B2, filed 01 October 2004 and issued 14 October 2008.
149 Lefevre, N., Ciabrini, J.P., Michard, G. et al. (1993). A new optical sensor for pCO_2 measurements in seawater. *Mar. Chem.* 42 (3–4): 189–198.
150 DeGrandpre, M.D. (1993). Measurement of seawater pCO_2 using a renewable-reagent fiber optic sensor with colorimetric detection. *Anal. Chem.* 65 (4): 331–337.
151 DeGrandpre, M.D., Hammar, T.R., Smith, S.P., and Sayles, F.L. (1995). In situ measurements of seawater pCO_2. *Limnol. Oceanogr.* 40 (5): 969–975.
152 Battelle Memorial Institute (2009). System Manual for Seaology pCO_2 Monitor. 635108H1010.
153 Wendy Schmidt Ocean Health XPRIZE (2015). http://oceanhealth.xprize.org/teams (retrieved 06 October 2016).
154 Schindler, W., Bäurich, A., and Mitschke, F. (1990). Elektrodenhalterung zur Druckkompensation an elektrochemischen Messsystemen (Pressure compensation in electrochemical measurement equipment). DE Patent 4,035,447 A1, filed 29 October 1990 and issued 30 April 1992 (in German).
155 Halbach, P., Kuhn, T., Sommer-von Jarmersted, C. et al. (1997). Wissenschaftlich-technische Tiefsee-Erprobung der Hydro-Bottom-Station (HBS) mit FS Le Suroit im Mittelmeer (Scientific technical deep sea test of the Hydro-Bottom Station (HBS) with FS Le Suroit in the Mediterranean Sea). Technischer Report, 50 pp. Freie Universität Berlin, FR Rohstoff- und Umweltgeologie (in German).
156 Dean, J.A. (1992). *Lange's Handbook of Chemistry*, 14e, 5.3–5.4. New York: McGraw-Hill Inc.
157 Cole, J.J., Caraco, N.F., Kling, G.W., and Kratz, T.K. (1994). Carbon dioxide supersaturation in the surface waters of lakes. *Science* 265 (5178): 1568–1570.
158 Schmid, J., Oelßner, W., and Schukraft, G. (2002). Kohlendioxid-Messungen in einem Durchfluß-Meßsystem mit einem neu entwickelten CO_2-Sensor am Beispiel eines Hartwassersees (Willersinnweiher, Ludwigshafen am Rhein) (Carbon dioxide measurements in lake water with a newly developed CO_2

sensor (Lake Willersinnweiher Ludwigshafen/Rhine, Germany)). *Limnologica* 32 (4): 338–349 (in German).

159 Oelßner, W., Schmid, J., and Guth, U. (2003). Determination of carbon dioxide dynamics in lakes. In: *Proceedings of ICGG7*, 50–52. Copernicus GmbH.

160 Parkhurst, D.L. and Appelo, C.A.J. (1999). User's Guide to Phreeqc (Version 2) – A Computer Program for Speciation, Batch-Reaction, One-Dimensional Transport, and Inverse Geochemical Calculations. Water-Resources Investigations Report 99-4259, 312 pp. Denver, CO: U.S. Geological Survey: Earth Science Information Center.

161 Portielje, R. and Lijklema, L. (1995). Carbon dioxide fluxes across the air-water interface and its impact on carbon availability in aquatic systems. *Limnol. Oceanogr.* 40 (4): 690–699.

162 Sellers, P., Hesslein, R.H., and Kelly, C.A. (1995). Continuous measurement of CO_2 for estimation of air-water fluxes in lakes: an in situ technique. *Limnol. Oceanogr.* 40 (3): 575–581.

163 Schreiber, U.C. (2006). *Die Flucht der Ameisen (The run of the ants)*. Berlin: Shayol Verlag (in German).

164 Irwin, W.P. and Barnes, I. (1980). Tectonic relations of carbon dioxide discharges and earthquakes. *J. Geophys. Res.* 85 (B6): 3115–3121.

165 Heinicke, J., Koch, U., and Martinelli, G. (1995). CO_2 and Radon measurements in the Vogtland area (Germany) – a contribution to earthquake prediction. *Geophys. Res. Lett.* 22 (7): 771–774.

166 Heinicke, J. and Koch, U. (2000). Slug flow – a possible explanation for hydrogeochemical earthquake precursors at Bad Brambach, Germany. *Pure Appl. Geophys.* 157 (10): 1621–1641.

167 Heinicke, J., Koch, U., Kaden, H., and Oelßner, W. (2002). Seismizität im sächsischen Vogtland – Einsatz von CO_2-Sensoren für geowissenschaftliche Untersuchungen (Seismicity in the Saxon Vogtland – application of CO_2 sensors for geo-scientific studies). *Akademie-Journal* 1/2002: 52–56 (in German).

13

CO_2 Safety Control

Wolfram Oelßner

Kurt-Schwabe-Institut für Mess- und Sensortechnik e.V. Meinsberg, Kurt-Schwabe-Straße 4, 04736 Waldheim, Germany

13.1 Limit Values for CO_2 Concentrations at Workplaces

The atmospheric CO_2 concentration of currently about 0.04 vol% on average or even somewhat higher as described in Chapters 2 and 12 is harmless for humans and animals and indispensable for many biological processes. However, higher levels of this odourless, colourless, and non-flammable gas can affect the human respiratory function followed by depression of the central nervous system and are dangerous to health and even life-threatening [1].

Carbon dioxide is regulated for diverse purposes but not as a toxic substance. The most important and internationally applied standards for CO_2 concentrations at workplaces are defined by the following US institutions:

- Occupational Safety and Health Administration (OSHA),
- National Institute for Occupational Safety and Health (NIOSH), and
- American Conference of Governmental Industrial Hygienists (ACGIH).

NIOSH is responsible for conducting research and making recommendations for the prevention of work-related illnesses and injuries. OSHA and NIOSH also review and publish specific methods for testing and analysing chemical contaminants in the air, including carbon dioxide. ACGIH is a member-based organization that advances occupational and environmental health. This scientific-based organization reviews available research and studies on airborne contaminants as well as other workplace stressors and establishes exposure levels that are designed to protect most normal, healthy adults in the workplace. Standards promulgated by OSHA have the force of law; the other standards are recommended standards, not laws.

OSHA has set permissible exposure limits (PELs) for carbon dioxide in workplace atmospheres at 10 000 ppm of CO_2 measured as a time-weighted average (TWA) level of exposure and has set 30 000 ppm of CO_2 as a short-term exposure limit (STEL). OSHA has also set a transitional limit of 5000 ppm CO_2 exposure TWA [2].

Carbon Dioxide Sensing: Fundamentals, Principles, and Applications,
First Edition. Edited by Gerald Gerlach, Ulrich Guth, and Wolfram Oelßner.

Table 13.1 Limit values for CO_2 concentrations at workplaces.

Threshold limit value time-weighted average (TLV-TWA) (or German Maximale Arbeitsplatz-Konzentration (MAK) value)	The time-weighted average concentration to which it is believed a worker can be repeatedly exposed over a normal 8 h workday and a 40 h workweek day after day without adverse health effects	0.5 vol% CO_2
Short-term exposure limit TLV-STEL	Allows exposure for duration of 15 min that cannot be repeated more than four times per 8 h workday with more than 60 min between those exposure periods (TWA must still be met)	1.5 vol% CO_2
Ceiling exposure limit TLV-C or maximum exposure concentration	The highest threshold limit value, which should not to be exceeded under any circumstances	3.0 vol% CO_2

NIOSH recommends for the United States a maximum concentration of carbon dioxide of 10 000 ppm or 1% (for the workplace, for a 10 hour work shift with a ceiling of 3.0% or 30 000 ppm for any 10 minute period). These are the highest threshold limit value (TLV) and PEL. Other nations have more protective STEL limits:

- *Germany*: 10 000 ppm for 60 minutes.
- *Sweden*: 10 000 ppm for 15 minutes.
- *United Kingdom*: 15 000 ppm for 10 minutes.

A CO_2 level of 4% is designated by NIOSH as immediately dangerous to life or health; acute toxicity data show that the lethal concentration for CO_2 is 90 000 ppm (9 vol%) over five minutes [3].

The ACGIH has defined STEL as the concentration to which workers can be exposed continuously for a short period of time without suffering from irritation, chronic or irreversible tissue damage, or narcosis of sufficient degree to increase the likelihood of accidental injury, impair self-rescue, or materially reduce work efficiency.

Table 13.1 summarizes the three characteristic limit values for CO_2 concentrations at workplaces. All of these three exposure limit conditions must be satisfied, always and together. A CO_2 concentration of 4% (40 000 ppm) is considered as the maximum instantaneous limit immediately dangerous to life and health.

13.2 CO_2 in Buildings and Workplaces

13.2.1 Air Quality with Respect to CO_2

Medical insights about the effect of CO_2 in the breathing air have been gained already for a long time. As early as 1858 the hygienist Max von Pettenkofer attached great importance to CO_2 as an assessment criterion for the quality of ambient air in buildings [4] and defined a limit value of 0.1 vol% (1000 vol ppm) that even now is essentially accepted. According to standard EN 15251 and other similar international regulations, indoor air quality can be described in three categories as follows [5]:

- *≤750 ppm CO_2*: named "individual" and described as *very good.*
- *(750 to ≤900) ppm CO_2*: named "comfortable" and described as *good.*
- *(900 to ≤1200) ppm CO_2*: named and described as *satisfactory.*

Furthermore, concentrations in the range of 1200–1500 ppm CO_2 can already be described as *unsatisfactory* and higher concentrations of more than 1500 ppm CO_2 as *unacceptable.* Table 13.2 shows how rising levels of carbon dioxide affect the human body.

In office buildings and most workplaces, CO_2 concentrations typically range from the outdoor value up to 1200 vol ppm. Higher carbon dioxide levels are likely to be reached only in unusual circumstances, for instance, when people are enclosed in an airtight small room for a long time or in some industrial workplace settings. An example of usually increased CO_2 concentrations in a laboratory during the working days compared with the outside air is shown in Figure 13.1. The measurements were performed every 10 seconds by a single-channel NDIR device in the flow-through mode with 6 l h^{-1}.

Table 13.2 Critical values, effects, and symptoms of CO_2 exposure.

CO_2 concentration			Critical values and effects of CO_2 exposure
vol%	ppm	mg m^{-3}	
0.038	380	684	Actual CO_2 concentration in air
0.1	1 000	1 800	Pettenkofer value for maximum tolerable indoor air concentration of CO_2
0.5	5 000	9 000	Threshold limit value (TLV-TWA): long-term exposure limit (in Germany, Maximale Arbeitsplatz-Konzentration, MAK-Wert)
1.5	15 000	27 000	Threshold limit value (TLV-STEL): short-term exposure limit, can cause drowsiness with prolonged exposure
3.0	30 000	54 000	Threshold limit value (TLV-C): ceiling exposure limit. Breathing rate doubles, dizzy feeling, heart rate and blood pressure increase, and headaches are more frequent; hearing can be impaired
>4.0	40 000	72 000	Rapid breathing, headaches, tinnitus, and impaired vision, confused in a few minutes, followed by unconsciousness
>8.0	80 000	144 000	Very laboured breathing, headaches, convulsions, fatigue, visual and hearing dysfunctions, and rapid loss of consciousness with risk of death from respiratory failure within minutes of exposure; burning candle goes out
>20.0	200 000	360 000	Fatal within few seconds

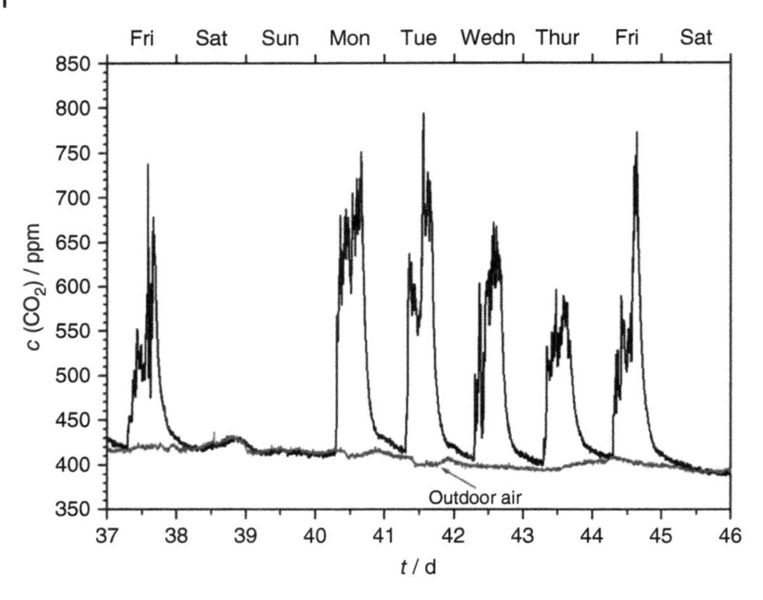

Figure 13.1 CO_2 concentrations in a laboratory and in the outdoor air measured by NDIR single-beam sensor. Source: With courtesy of Gatron GmbH, Greifswald, Germany.

13.2.2 Sick-Building Syndrome

The well-being of humans in office buildings is not only influenced by pollutions and water vapour concentration in the breathing air but also in a high extent by the CO_2 concentration. Even CO_2 concentrations of less than 1000 vol ppm, which are widely below the long-term exposure limit defined by law (5000 ppm), can be felt as bothering, and workers in office buildings often suffer from the so-called sick-building syndrome (SBS). This becomes apparent with various unspecific symptoms such as headache, fatigue, eye problems, nasal and respiratory tract symptoms such as cough or wheeze, and other non-specific symptoms like difficulty in concentration as well as in breathing and tight chest [6–9]. It was demonstrated that even concentrations as low as 1000 ppm can impair thinking, concentration, and logical thought processes. At a concentration of only 2500 ppm for 2.5 hours, most measured cognitive functions were impaired to the extent that the individuals were rendered cognitively marginal or dysfunctional. The functions measured included task orientation, initiative, information usage, and use of basic strategy [10]. In [11] the relation between cardiac function and cerebral blood flow in stroke patients before, during, and after 5% CO_2 inhalation was studied.

The term SBS is used primarily when the symptoms do not indicate a specifically known disease. Although these symptoms can be caused by various gases coming from human respiration and perspiration and not only by CO_2 itself, results of extensive statistical studies indicate a correlation between elevated indoor CO_2 levels and increases in certain SBS symptoms. Most of the carbon dioxide studies suggest that the risk of sick-building syndrome symptoms decreases significantly if the carbon dioxide concentration is below 800 ppm.

Primary source of the increased CO_2 in an enclosed space such as a tight home, office building, school, or workspace is respiration of the building occupants. Especially effects of CO_2 on the indoor air quality in school buildings were investigated in numerous extensive studies [12–16], and from the results requirements for the ventilation rates were derived [17–20]. The ventilation studies reported relative risks for sick-building syndrome symptoms for low and high ventilation rates, respectively. The necessity of such ventilation measures should be demonstrated by the following example, which shows that the bothersome concentration of 0.1% CO_2 in the classroom can be reached already after less than 0.5 hours, provided that no fresh air pours into the room.

Volume of the classroom	225 m^3
Number of occupants	30
Oxygen consumption per person [21]	0.5 l min^{-1}
Respiration quotient (RQ) [21]	0.8
Increase of CO_2 concentration to	0.1% CO_2
Time to reach 0.1% CO_2 in the room	About 19 min

In this context it is worth mentioning that at modestly elevated levels of CO_2 in air in most cases, the corresponding relatively low percentage reduction in available oxygen in that same breathing air is not the cause of SBS or other dangers to health. The above-mentioned properties of CO_2 are fatal, not the drop in oxygen. Harming effects of elevated levels of CO_2 are serious at levels when the oxygen reduction effects are only minor. For example, introduction of 5% CO_2 to the room's air volume will reduce each of the components of the normal air to 95% of their original proportion. According to Table 13.3, the resulting oxygen level is at 19.85%. While a 5% level of CO_2 is directly toxic, the oxygen concentration of 19.85% is above the minimum permissible oxygen level of 19.5% that is defined, e.g. in [22–24] and in other international standards.

Sometimes the effect of a reduced oxygen concentration at sea level is compared with that of the equivalently reduced oxygen content in the air at the top of a high mountain [25]. Such a comparison can only serve as a rough illustration. Indeed, the total air pressure on the mountain is less than at sea level, according to the barometric altitude formula. But the percentage composition of the

Table 13.3 Composition of the room air after introduction of 5% CO_2 to the room's air volume.

Gas component	Standard air percentage (%)	After addition of 5% CO_2 (%)
Nitrogen (N_2)	78.1	74.20
Oxygen (O_2)	20.9	19.85
Others	1.0	0.95
Carbon dioxide (CO_2)	0.04	5.04

atmosphere on the mountain is equal to that at sea level. For this reason, the effect on the human body of 16% oxygen in the breathing air at the top of a mountain cannot be compared with 16% oxygen in the breathing air at sea level.

13.2.3 Dangerous Areas

Dangerous CO_2 concentrations can occur in many confined spaces, e.g. underground mines, caves, cellars, tunnels, sewers, fruit and vegetable storage depots, or wine cellars. The common thread in these locations is the failure to recognize that carbon dioxide gas can accumulate there, resulting in unsuspected exposures to hazardous levels of carbon dioxide where it is dangerous to breathe [26].

Some sources of hazardous concentrations of carbon dioxide in confined spaces are [27] as follows:

- Leaking carbonators, beer keg connections, and any equipment using carbon dioxide in manifold beverage-dispensing machines where CO_2 is used to make the "fizz" in soft drinks, which are widely available in bars, restaurants, etc.
- Leaking fittings, connections, piping/tubing/hoses, or CO_2 storage container plumbing. The CO_2 released there is one of the significant sources of fugitive CO_2 emission.
- Carbon dioxide pressure relief devices (PRDs) or pressure release valves (PRVs), which can sometimes stick open or be open for other reasons and often allow carbon dioxide to be vented directly out of its respective process vessel.
- Fire extinguishers containing CO_2 for use on electrical and flammable-liquid fires. They have sometimes already led to fatal accidents.

In coal mines occasionally high carbon dioxide concentrations can be present. For quite a long time, diggers have been mindful of the risk of "dark moist" [28]. Formerly miners tried to alert themselves to dangerous levels of carbon dioxide by bringing a candle, a mouse, or a caged canary with them when they were working in a mine shaft. The canary would inevitably die before the CO_2 concentration reached levels toxic to humans. Such simple protective measures, however, often indicate a dangerous low oxygen level rather than a critically increased CO_2 concentration. Of course, using CO_2 warning devices is the better method. In the potash salt mining, the monitoring of the CO_2 concentration is indispensable. Sudden CO_2 eruptions have caused several times even deadly accidents [29].

Since the density of CO_2 is higher than that of air, its concentration is particularly high at the bottom of the object concerned. This applies, for example, to basements, sewage systems, and caves. In the famous so-called Grotta del Cane (Dog's grotto), a small cave near Naples, Italy, carbon dioxide of volcanic origin accumulates on the bottom of the cave, because it is denser than air. There the CO_2 concentration amounts to approx. 70 vol%. Small animals such as dogs become unconscious after short time in the cave and must be immediately resuscitated, while humans inhaling air from a higher level are not affected.

Great dangers also lurk in lakes that are saturated with carbon dioxide. Tragic examples are Lake Monoun and Lake Nyos, both in Cameroon, and Lake Kivu in the Democratic Republic of Congo. On 15 August 1984, the Lake Monoun exploded in a limnic eruption, which resulted in the release of a large amount of carbon dioxide that killed 37 people [30]. Only two years later, on 21 August 1986, the sudden, catastrophic release of CO_2 gas from Lake Nyos caused the deaths of at least 1700 people [31]. The reason for the eruptions can be explained as follows: carbon dioxide gas of volcanic origin is dissolved in great amounts in the denser cold water at the lake's bottom. Over long times the lakes were stable. However, the water becomes supersaturated, and if an event such as an earthquake or landslide occurs, the intensively bubbling CO_2 from deep lake waters may lead to a gas burst at the surface, and large amounts of CO_2 may suddenly break out from the lake.

Since CO_2 is a natural by-product of many biological processes, people can routinely be exposed to carbon dioxide not only in facilities in which yeast or grain is processed, for example, in the beer manufacturing as well as in the ethanol production industries, but also in other facilities associated with bioenergy and other biogenic processes. Brewers entering encased regions, for example to clean tanks subsequent to fermentation, could be overcome by elevated amounts of CO_2. Workers in grain elevators, storehouses, and rail cars, where stored grain can produce an atmosphere of 37% CO_2 during oxidation of carbohydrates, are at extreme risk for high CO_2 levels of exposure [3, 8].

In beverage-dispensing systems, CO_2 comes typically via a pressure-reducing valve from a compressed gas bottle. If, due to a leak in the equipment, the CO_2 content in the room exceeds a predetermined value, an alarm must be triggered by a CO_2 sensor. The warning device proposed in [32] uses at least one carbon dioxide sensor, coupled to a control unit, for activating an alarm when a given carbon dioxide level is exceeded. There is simultaneous closure of a blocking valve for cutting off the carbon dioxide feed. The blocking valve may be incorporated in a pressure reduction valve for a carbon dioxide flask or may be provided by a separate valve incorporated in the carbon dioxide line.

Most of the CO_2 produced in a submarine or a manned spacecraft results from respiration and must be removed to maintain a safe CO_2 level. Typically, the limits for spacecraft are lower than limits suggested for submarines for a given time of exposure. The difference can be attributed to the reality that astronauts will need to repair the CO_2 removal system (a relatively sophisticated task), whereas submariners could engage additional scrubbers or possibly surface [1, 33, 34]. To remove CO_2 from submarine atmospheres, regenerative or non-regenerative techniques are used, depending on whether the absorbent can be recycled at sea:

- *LiOH absorbers*: These are a non-regenerative means for removing CO_2 from the gas stream passed through canisters holding the LiOH according to

$$2\text{LiOH} + \text{CO}_2 \rightarrow \text{Li}_2\text{CO}_3 + \text{H}_2\text{O}$$

- *CO_2 scrubbers*: These are regenerative systems that utilize monoethanolamine (MEA, $NH_2CH_2CH_2OH$) to absorb the CO_2 from the air. There are two fundamental mechanisms for the reaction of amines (R-NH_2) with CO_2 [35]:

$$2\text{R}-\text{NH}_2 + \text{CO}_2 \leftrightarrow \text{R}-\text{NH}_3{}^+ + \text{R}-\text{NH}-\text{COO}^-$$
$$\text{R}-\text{NH}_2 + \text{CO}_2 + \text{H}_2\text{O} \leftrightarrow \text{R}-\text{NH}_3{}^+ + \text{HCO}_3{}^-$$

  The MEA is then heated to drive out the gas, and the latter is compressed and ejected overboard [36].

In manned spacecraft, e.g. during the NASA's Mercury, Gemini, and Apollo programmes, also lithium hydroxide (LiOH) canisters have been employed in controlling CO_2 levels. Manned spacecraft built for longer mission durations, e.g. Skylab, Russian Mir, and the ISS, use LiOH as a supplement to regenerative CO_2 scrubbing systems [37, 38].

CO_2 in the form of dry ice is widely used to refrigerate dairy products, meat products, poultry, and other perishable foods while in transit, as coolant in chemical reactions, and also on the stage and in live theatre productions to make fog. If transported in closed vehicles or utilized in unventilated rooms, tight workspaces that have little fresh air exchange, and enclosed areas, the dry ice off-gas can increase the indoor CO_2 concentration to critical levels.

13.3 CO_2 Warning Devices

CO_2 warning devices are available as portable handheld monitors or clip-on monitors for personal safety and also for stationary applications where CO_2 levels are continuously measured and monitored with audible and visual alarms, sometimes combined with ventilation equipment that will automatically be started when critical set points are exceeded.

The warning devices are based on the following sensor principles, which are described in detail in Part II of this book:

- CO_2 detector tubes and dosimeter tubes (Section 6.2),
- Electrochemical CO_2 sensors (Chapters 4 and 5),
- Non-dispersive infrared (NDIR) CO_2 sensors (Chapter 7),
- Solid electrolyte CO_2 sensors (Chapter 5), and
- Gas chromatographs with thermal conductivity detector, GC/TCD (Section 3.2).

Nowadays, NDIR CO_2 sensors and CO_2 detector and dosimeter tubes are dominating, while electrochemical and solid electrolyte CO_2 sensors are hardly used for this purpose yet.

13.3.1 CO_2 Detector and Dosimeter Tubes

Colorimetric gas detector tubes, as described in Section 6.2, are preferred for measuring CO_2 exposures in the workplace as a relatively inexpensive way to test the actual level of CO_2. The tubes are simple to use and quite accurate and can be selected and used down to very low concentrations down to 100 ppm of CO_2, provided that a properly chosen and calibrated gas testing pump is used. For very precise measurements, the influences of temperature and pressure must be corrected. Since short-term detector tubes offer only spot checks of the environment to determine long-term CO_2 concentrations, other methods are needed [39].

Dosimeter tubes work like detector tubes, but instead of a pump the principle of air diffusion is applied. Clipped to the worker's collar near to his breathing zone or attached to the wall of the room, they are used to measure personal exposures to CO_2 as an integral value over a specified time period, usually 4–8 hours [39].

13.3.2 Electrochemical CO_2 Sensors

Figure 13.2 shows a portable CO_2 warning device with an easily exchangeable electrochemical CO_2 sensor that had been developed for application in underground potash salt mining. According to a traffic light, the normal state and the threshold limit values TLV-TWA (0.5 vol% CO_2) and TLV-STEL (1.5 vol% CO_2) were indicated by green, yellow, and red LEDs. Exceeding of the limit values is also signalized acoustically.

The stationary CO_2 warning device with electrochemical CO_2 sensor shown in Figure 13.3 was likewise purpose-built for application also in underground

Figure 13.2 Portable CO_2 warning device with electrochemical CO_2 sensor. Source: KSI Kurt-Schwabe Research Institute Meinsberg, Germany.

Figure 13.3 Stationary CO_2 warning device with electrochemical CO_2 sensor for application in underground potash salt mining. Source: KSI Meinsberg.

potash salt mining. Critical CO_2 concentrations are intensively signalized acoustically, optically by the rotating flashing beacon, and also electrically transferred to the control centre [40–42]. In consequence of the extremely low air moisture in underground salt mines, electrochemical sensors tend to dry out rapidly on this particular usage condition. This must be prevented by reducing the water vapour pressure of the sensor electrolyte, e.g. by adding ethylene glycol (see Section 4.1.2).

Nowadays carbon dioxide warning devices are scarcely equipped with Severinghaus-type electrochemical CO_2 sensors.

13.3.3 NDIR CO_2 Sensors

13.3.3.1 Properties

Currently, the most broadly used technique for real-time measurement of carbon dioxide in facilities that have the potential for carbon dioxide exposures is by means of NDIR sensors because of their long-term stability and widely maintenance-free operation. As described in detail in Chapter 7, this type of sensors measures CO_2 as a function of the absorbance of infrared light at a specified wavelength. Most commercial NDIR products are single-channel sensors due to lower costs because they are smaller and need less energy, which is important for handheld devices in battery operation. On the other hand, two-channel sensors exhibit better long-term stability.

Typical properties of commercial NDIR CO_2 sensors are as follows.

CO_2 measuring range	0–2000 ppm (often even higher up to 50 000 ppm)
Accuracy	±30 ppm ±3%
Temperature range	0–50 °C
Sensor life expectancy	>6 years (even higher, e.g. 15 years)

Now there is a wide variety of NDIR CO_2 sensor devices available in the market. The spectrum comprises the following categories, which will be described below by means of some selected typical examples:

- Portable single-gas CO_2 monitors that can be worn or handheld,
- Portable multi-gas monitors that are excellent for confined space entry and other uses, and
- Stationary single-gas CO_2 monitors that give audible and visible alarms when an area of the facility exceeds set threshold levels of CO_2.

Figure 13.4 shows as an example a single CO_2 gas detector with NDIR sensor (Pac 7000, Dräger, Lübeck, Germany) [43]. The small, robust, personal monitor has user-friendly features including a single-handed, simple two-button operation, a large LCD display with bright backlight, and adjustable intervals for tests and calibration, reminders, and “out of order” message if overdue or failed. Main technical specifications of the single CO_2 gas detector are as follows:

- Sensor range: 0–5 vol%.
- Resolution: 0.1 vol% (1000 ppm).
- Operating temperature range: 0–50 °C.

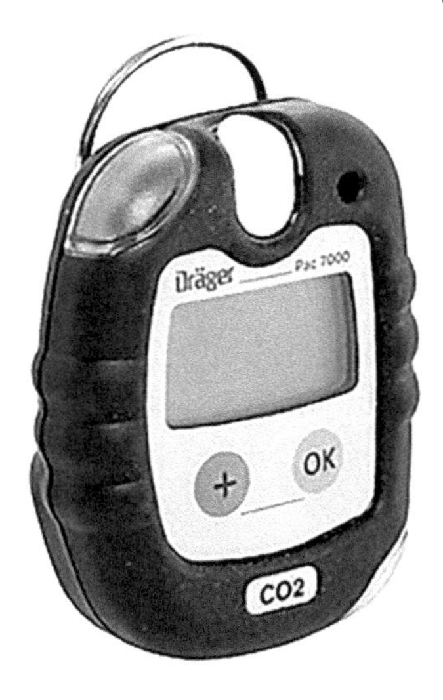

Figure 13.4 Single CO_2 gas detector Pac 7000. Source: © Drägerwerk AG & Co. KGaA, Lübeck [43], all rights reserved.

- Response time t_{50}: 30 seconds.
- Alarm setting adjustable TWA and STEL alarms: 0.5 and 1 vol%, respectively.
- Alarm type: loud audible, bright visual, and robust vibrating alarms.
- 120 hour data logger with date and time stamp.
- Battery type, alkaline lithium; battery life, 13 months.
- Display: LCD display with bright backlight.
- Housing: IP65, dust and water ingress protection.

An example of a multi-gas detector is shown in Figure 13.5. Among other gases (e.g. O_2, H_2S, CO), the Polytector III G999 detects carbon dioxide and combustible hydrocarbons in ambient air by means of an NDIR sensor with only one IR measurement chamber. To that, the sensor uses up to four detector elements. Absorption paths and wavelengths of the radiation are adapted to the gases and the measuring ranges. Optionally, the infrared sensor can be pressure compensated, if in the use case big pressure fluctuations are reasonably likely to occur (e.g. in the mining sector, underground). The typical lifetime of the sensor is more than five years. Four or ten devices can be linked by radio with a GfG-Link, a safe link between devices. At this, the username for the rapid identification of the device as well as the current measurement data will be displayed on the screen of the GfG-Link. The device has been equipped with a life-saving man down alarm. It will be enhanced by an unmistakable acoustical sequence of signals and a red blinking alarm display. The time interval when the alarm shall be triggered can be individually set password-protected for the corresponding requirement situation [44].

Stationary CO_2 detection systems with NDIR sensor are installed anywhere where CO_2 is stored or produced in high volume indoors or in enclosed outdoor locations to notify and protect humans in case of a CO_2 gas release. In general, such systems must fulfil the following requirements and induce appropriate measures:

- Activation of an audible alarm within the room or area in which the carbon dioxide source is located and also at the entrances to the room or area to notify anyone who might try to enter the area of a potential problem.

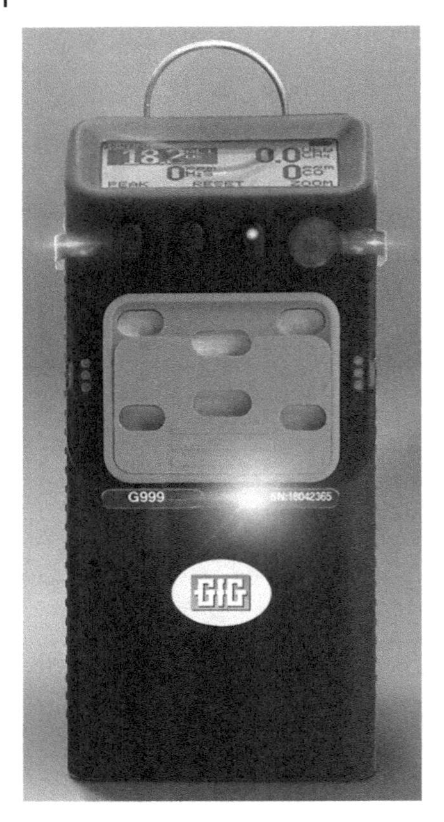

Figure 13.5 Gasmessgerät Polytector III G999 (GfG Gesellschaft für Gerätebau, Dortmund, Germany).

- Carbon dioxide monitoring. These monitors are set to alarm at 15 000 ppm (1.5% concentration) for the low-level alarm and at 30 000 ppm (3% concentration) for high-level alarms.
- At low-level alarm (15 000 ppm), appropriate cross-ventilation to the area should be provided. Personnel may enter area for short periods of time (not to exceed 15 minutes at a time) in order to identify and repair potential leaks.
- At high-level alarm (30 000 ppm), personnel should evacuate the area, and nobody should enter the affected area without proper self-contained breathing apparatus until the area is adequately ventilated and the concentration of CO_2 is reduced below the high alarm limit.

Among numerous others, as an example, the RAD-0102 Remote CO_2 Storage Safety Dual Alarm Meter [45] meets these requirements for dual-level alarms with the following features and specifications:

- Large digital LCD display, clearly indicating both the current CO_2 level and the temperature.
- Two relays to automatically control fans or blowers to ventilate confined spaces.
- 80 dB audible alarm indications.
- 100 cd strobes visual alarm when the first alarm is reached at 1.5% CO_2.

- IP65 enclosure.
- Measurement range: 0–5% (50 000 ppm).
- CO_2 alarms at 1.5 and 3% CO_2 (configurable).
- Alarm response time: <60 seconds.
- Operating temperature: 0–50 °C.
- Sensor life expectancy: 15 years.

13.3.3.2 Calibration of NDIR CO_2 Measuring Devices

Although NDIR sensors are long-term stable and largely maintenance-free, like nearly all chemical sensors, they have to be calibrated occasionally. Depending on the application and the required accuracy of the CO_2 level, reading the calibration intervals can range from weekly for personal safety purposes to annually for manufacturing facility monitoring or indoor air quality measurements. For calibration the following methods are applied [45, 46]:

- Calibration with a known CO_2-free dry gas, typically 100% nitrogen: This method is very accurate but elaborate and expensive. For this reason, it is recommended only for demanding medical or scientific purposes or if CO_2 levels below 400 ppm will be measured.
- Calibration with fresh air: If the maximum accuracy is less important, the CO_2 sensor can be simply calibrated in fresh air. In most cases for CO_2 concentrations in outdoor air, a value of 400 ppm is used.
- Automatic baseline calibration (ABC): It is assumed that each day, if the room is unoccupied, the CO_2 level should return to the outdoor level of about 400 ppm. The lowest CO_2 readings taken over time (typically several days) are stored and used for a simple self-calibration mode of the CO_2 sensor. This method is very comfortable, but it can display inaccurate values if the above assumption is not valid. In [47] a method is proposed to evaluate and eliminate the drift component (Figure 13.6). It states: "Once each cycle, for a number of cycles, the sensor measures the variable at a time when its value should equal the extrinsically-known value. The differences are plotted versus time, and a best-fitting straight line is determined, which indicates the drift. Throughout the next cycle as the variable is continuously sensed, the drift determined from the best-fitting straight line is continuously applied to correct the sensed value".

13.3.3.3 Pressure Dependence

The NDIR type of sensors measures the IR radiation absorbed by CO_2 molecules. When pressure p increases, the number of molecules in a given volume also increases linearly. Therefore, all NDIR CO_2 sensors depend on pressure. In sensor data sheets the accuracy is typically given as percentage of the reading per kPa deviation from normal pressure (100 kPa). This is an approximation that is valid around mean sea level pressure (101.3 kPa), at which the sensors are factory-calibrated. A better approximation of the pressure sensitivity of the SenseAir NDIR CO_2 sensors (SenseAir, Delsbo, Sweden) is given by [48]

$$\text{True reading} = \text{reading}/(4.026 \times 10^{-3} p/\text{kPa} + 5.780 \times 10^{-5} (p/\text{kPa})^2) \quad (13.1)$$

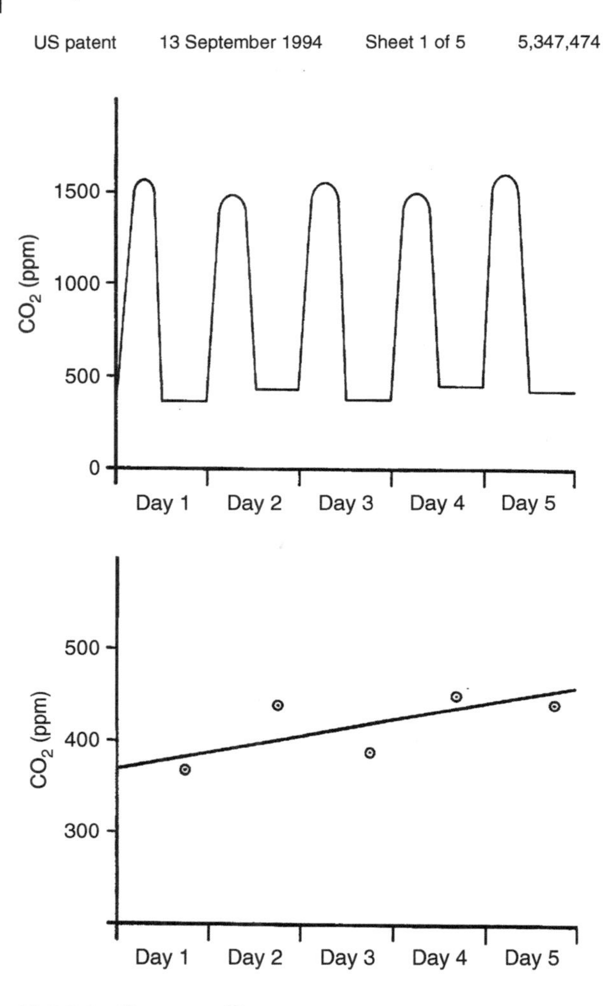

Figure 13.6 Determination of the NDIR sensor drift according to [47].

13.3.3.4 Response Time

The response time of NDIR CO_2 sensor devices is specified by the manufacturers in the range between 30 seconds and 2 minutes. This time is caused by the diffusion of the sample gas into the cuvette. In [49] a CO_2 sensor (GSS C20) is described that can be used also for biological applications, requiring shorter response time. The sensor was used to measure the CO_2 concentration of human respiration. For this purpose, the sensor was held closely to a person who breathed across the sensor cap at a normal rate. Data are captured at a time resolution of 0.5 Hz.

13.3.4 Solid Electrolyte CO_2 Sensors

Solid electrolyte CO_2 sensors have been manufactured (e.g. by Zirox GmbH, Greifswald, Germany) and successfully applied in measurement and control of the indoor and outdoor atmosphere. Their functional principle and set-up is described in detail in Chapter 5. As an example, the concentration in a

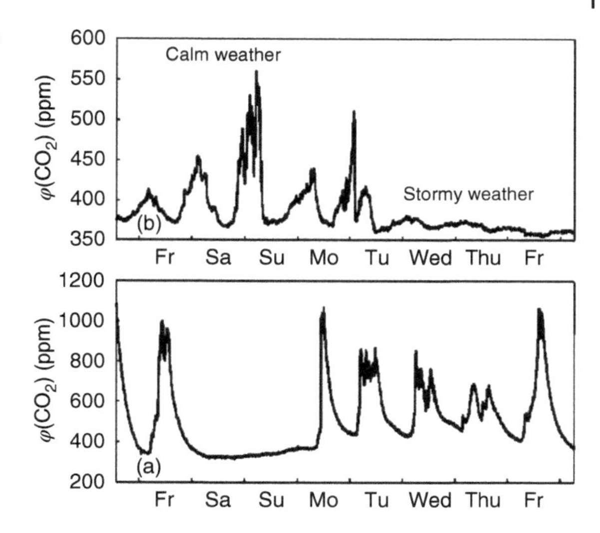

Figure 13.7 CO_2 concentrations (a) in laboratory air and (b) in atmosphere of urban areas near the bottom [50].

chemical laboratory, in which several people are working, was measured over one week (Figure 13.7a). The concentration of CO_2 outside (Figure 13.7b) is strongly influenced by the day–night cycle of the plant activity and also by the distribution of CO_2 due to the weather conditions (calm or windy) [50].

To measure dissolved CO_2 in liquids by means of solid electrolyte gas sensors, the dissolved gas must be transferred into a defined gas atmosphere. For this purpose, a gas-permeable membrane is inserted into the solution to be measured. As shown in Figure 7.22, a carrier gas streams along the reverse side of the membrane [51]. Carbon dioxide permeates through the membrane into the carrier gas in which the CO_2 concentration is measured with the solid electrolyte gas sensor. Instead of using a carrier gas, the CO_2 can also permeate through the membrane into an evacuated vessel in which the gaseous CO_2 is measured.

13.3.5 Gas Chromatograph with Thermal Conductivity Detector

The gas chromatographic method offers an accurate and precise possibility for TWA or STEL CO_2 determinations in the workplace. To assess sample concentrations of CO_2 by means of a pump, samples are collected in gas sampling bags and analysed using a gas chromatograph (GC) with a gas sampling loop and a thermal conductivity detector (TCD, see Table 3.7 in Chapter 3).

The gas chromatographic determination of CO_2 is relatively specific. However, any compound having a similar column retention time and response as CO_2 is a potential interference [39]. Furthermore, the method does not show *in situ* values and is too complex and expensive to be generally widely applicable.

References

1 Canadian Center for Occupational and Health and Safety (CCOHS) (2005). Health effects of carbon dioxide gas. https://www.ccohs.ca/ (retrieved 2 February 2017).

2 OSHA (1989). Carbon dioxide. In: *Industrial Exposure and Control Technologies for OSHA Regulated Hazardous Substances*, Volume I, Substance A – I. Washington, DC, USA: Occupational Safety and Health Administration, U.S. Department of Labor.
3 National Institute for Occupational Safety and Health (NIOSH) (1976). *Criteria for a Recommended Standard: Occupational Exposure to Carbon Dioxide* (ed. U.S. Department of Health, Education, and Welfare) https://www.cdc.gov/niosh/pdfs/76-194a.pdf (retrieved 2 February 2017).
4 von Pettenkofer, M. (1858). *Über den Luftwechsel in Wohngebäuden (About the air change in residential buildings)*. München: Cotta (in German).
5 Comité Européen de Normalisation (CEN) (2007). Indoor environmental input parameters for design and assessment of energy performance of buildings addressing indoor air quality, thermal environment, lighting and acoustics. European Standard. EN 15251-2007.
6 Apte, M.G., Fisk, W.J., and Daisey, J.M. (2000). Associations between indoor CO_2 concentrations and sick building syndrome symptoms in U.S. office buildings: An analysis of the 1994–1996 BASE study data. *Indoor Air* 10 (4): 246–257.
7 Erdmann, C.A., Steiner, K.C., and Apte, M.G. (2002). Indoor CO_2 concentrations and sick building syndrome symptoms in the base study revised: analysis of the 100 building data set. In: *Proceedings 9th International Conference of Indoor Air Quality and Climate*, vol. III (ed. H. Levine), 443–448. CA, USA: Santa Cruz.
8 Nelson, L. (2000). Carbon dioxide poisoning. Summary of physiological effects and toxicology of CO_2 on humans. *Emerg. Med.* 32 (5): 36–38.
9 Priestley, M.A. and Litman, R.S. (2003). *Respiratory Acidosis. Pediatrics: An On-line Medical Reference*, 2e (ed. G.P. Cantwell, M.L. Windle, B. Evans, et al.). Emedicine Online Textbooks Emedicine.com, Inc.
10 Satish, U., Mendell, M., Shekhar, K. et al. (2012). Is CO_2 an indoor pollutant? Direct effects of low-to-moderate CO_2 concentrations on human decision-making performance. *Environ. Health Perspect.* 120 (12): 1671–1677.
11 Cooper, E.S., West, J.W., Jaffe, M.E. et al. (1970). The relation between cardiac function and cerebral blood flow in stroke patients. 1. Effect of CO_2 inhalation. *Stroke* 1 (5): 330–347.
12 Raatikainen, M., Skön, J.-P., Turunen, M. et al. (2013). Evaluating effects of indoor air quality in school buildings and students' health: a study in ten schools of Kuopio, Finland. In: *2013 2nd International Conference on Environment, Energy and Biotechnology IPCBEE*, vol. 51, 80–86. https://doi.org/10.7763/IPCBEE. 2013. V51. 16.
13 Poupard, O., Blondeau, P., Iordache, V., and Allard, F. (2005). Statistical analysis of parameters influencing the relationship between outdoor and indoor air quality in schools. *Atmos. Environ.* 39 (11): 2071–2080.
14 Lee, S.C. and Chang, M. (2000). Indoor and outdoor air quality investigation at schools in Hong Kong. *Chemosphere* 41 (1-2): 109–113.
15 Wargocki, P., Wyon, D.P., Baik, Y.K. et al. (1999). Perceived air quality, sick building syndrome (SBS) symptoms and productivity in an office with two different pollution loads. *Indoor Air* 9 (3): 165–179.

16 Menzies, R., Tamblyn, R., Farant, J.-P. et al. (1993). The effect of varying levels of outdoor-air supply on the symptoms of sick building syndrome. *N. Engl. J. Med.* 328 (12): 821–827.

17 Daisey, J.M., Angell, W.J., and Apte, M.G. (2003). Indoor air quality, ventilation and health symptoms in schools: an analysis of existing information. *Indoor Air* 13 (1): 53–64.

18 Seppänen, O.A., Fisk, W.J., and Mendell, M.J. (1999). Association of ventilation rates and CO_2-concentrations with health and other responses in commercial and institutional buildings. *Indoor Air* 9 (4): 226–252.

19 Technische Regeln für Arbeitsstätten - Lüftung ASR A3.6 (technical rules for work places – ventilation) ASR A3.6 (2012). BAUA Bundesanstalt für Arbeitsschutz und Arbeitsmedizin (in German).

20 Berglund, B., Clausen, G., de Ceaurriz, J. et al. (1997). *Report 19: Total Volatile Organic Compounds (TVOC) in Indoor Air Quality Investigations*, 56. Luxembourg: Office for Official Publications of the European Communities.

21 Müller, R. (2001). Atmung, Stoffwechsel und Blutkreislauf (respiration, metabolism, and blood circuit). *Prax. Naturwiss., Phys. Sch.* 50 (8): 23–26. (in German).

22 DGUV (2013). *Information Arbeiten in sauerstoffreduzierter Atmosphäre* (information about working in oxygen-reduced atmosphere), BGI/GUV-I 5162. Berlin: Deutsche Gesetzliche Unfallversicherung e.V. (in German).

23 DGUV (1994). BGR 160 Sicherheitsregeln für Bauarbeiten unter Tage (safety rules for construction works below ground) (in German). http://www.publikationen.dguv.de/dguv/pdf/10002/bgr-160.pdf. (retrieved 2 February 2017).

24 http://www.raesystems.com/sites/default/files/content/resources/Application-Note-206_Guide-To-Atmospheric-Testing-In-Confined-Spaces_04-06.pdf (retrieved 2 February 2017).

25 Toxicity of Carbon Dioxide Gas Exposure, CO_2 Poisoning Symptoms, Carbon Dioxide Exposure Limits. http://www.inspectapedia.com/hazmat/Carbon_Dioxide_Hazards.php (retrieved 2 February 2017).

26 DeVany, M.C. (2015). *Carbon Dioxide (CO_2) Safety Program*. DeVany Industrial Consultants. https://www.linkedin.com/company/devany-industrial-consultant (retrieved February 2, 2017).

27 Cable, J. (2004). NIOSH report details dangers of carbon dioxide in confined spaces. http://www.ehstoday.com/news/ehs_imp_37358 (retrieved March 12, 2017).

28 Abdullah, F.B. (2015). Effects of CO_2 emission on health & environment: evidence from fuel sources in Pakistani industry. *Pak. J. Eng. Technol. Sci.* 5 (2): 112–126.

29 Duchrow, G., Thoma, K., Marggraf, P., and Salzer, K. (1988). Forschungen zum Phänomen der Salz-Gas-Ausbrüche im Werra-Kaligebiet der DDR (research on the phenomenon of salt-gas eruptions in the Werra potash region of the GDR). *Neue Bergbautechnik* 18 (7): 241–250. (in German).

30 Sigurdsson, H., Devine, J.D., Tchua, F.M. et al. (1987). Origin of the lethal gas burst from Lake Monoun, Cameroun. *J. Volcanol. Geotherm. Res.* 31 (1–2): 1–16.

31 Kling, G.W., Clark, M.A., Wagner, G.N. et al. (1987). The 1986 Lake Nyos Gas Disaster in Cameroon, West Africa. *Science* 236 (4798): 169–175.

32 von Khreninger-Guggenberger, K.-J. (1996). CO_2-Warnanlage (carbon dioxide warning device for drinks dispenser). DE Patent 19,651,192 A1, filed 10 December 1996 and issued 18 June 1998.

33 James, J.T. and Macatangay, A. (2009). *Carbon Dioxide – Our Common "Enemy"*. Houston, USA: NASA/Johnson Space Center https://www.ntrs.nasa.gov/archive/nasa/casi.ntrs.nasa.gov/20090029352.pdf (retrieved March 12, 2017).

34 James, J.T. (2008). Carbon dioxide. In: *Spacecraft Maximum Allowable Concentrations for Selected Airborne Contaminants*, vol. 5, 112–124. Washington, DC: National Academy Press.

35 Huertas, J.I., Gomez, M.D., Giraldo, N., and Garzón, J. (2015). CO_2 absorbing capacity of MEA. *J. Chem.* 2015: 7, 965015. http://dx.doi.org/10.1155/2015/965015.

36 https://en.wikipedia.org/wiki/Carbon_dioxide_scrubber (retrieved 12 March 2017).

37 Matty, C.M. (2008). Overview of long-term lithium hydroxide storage aboard the International Space Station. *38th International Conference on Environmental Systems*, San Francisco, SAE technical paper, Warrendale, PA, USA, 6 pp.

38 Reysa, R., Davis, M., El Sherif, D., and Lewis, J. (2004). International Space Station (ISS) Carbon Dioxide Removal Assembly (CDRA) on-orbit performance. *34th International Conference on Environmental Systems*, Colorado Springs, USA, SAE technical paper, Warrendale, PA, 13 pp.

39 United States Department of Labor, Occupational Safety and Health Administration (OSHA) (revised 1990). Carbon Dioxide in Workplace Atmospheres (OSHA-SLTC Method No. ID-172). https://www.osha.gov/dts/sltc/methods/inorganic/id172/id172.html (retrieved 12 March 2017).

40 Kaden, H., Oelßner, W., and Schindler, W. (1990). Über eine elektrochemische Meßmethode und Geräte zur Bestimmung des Kohlendioxidgehaltes in Grubenwettern des Kalibergbaus (about an electrochemical measuring method and devices for the determination of carbon dioxide in air in potash mining). *Neue Bergbautechnik* 20: 34–37. (in German).

41 Schindler, W., Oelßner, W., and Kaden, H. (1990). Elektrochemische Sensoren und Geräte zur Messung und Kontrolle des Kohlendioxidgehaltes von Luft (electrochemical sensors and devices for the control of the carbon dioxide concentration in air). *Atemschutz Information* 29: 9–12.

42 Oelßner, W., Kaden, H., Lehmann, P. et al. (1988). Handgerät zur Überwachung des Kohlendioxidgehaltes der Luft (Hand-held device for monitoring the carbon dioxide content of the air). DE Patent 3,828,648 A1, filed 24 August 1988 and issued 23 March 1989 (in German).

43 https://www.draeger.com/Products/Content/pac-7000-pi-9072229-de-de.pdf (retrieved 2 February 2017).

44 http://www.gasmessung.de/produkte/tragbare-gasmessgeraete.html (retrieved 24 July 2018).

45 http://www.co2meters.com/Documentation/Datasheets/DS-RAD-0102SX2.pdf (retrieved 2 February 2017).

46 http://www.co2meters.com/Documentation/AppNotes/AN131-Calibration.pdf (retrieved 2 February 2017).

47 Wong, J.Y. (1991) Self-calibration of an NDIR gas sensor. US Patent 5,347,474, filed 19 September 1991 and issued 13 September 1994.

48 http://www.senseair.com/wp-content/.../12/Pressure_d-20120327.pdf (retrieved 2 February 2017).

49 https://www.co2meter.com/collections/0-1-co2 (retrieved 2 February 2017).

50 Brüser V., Klingner, W., Möbius, H.-H., and Guth, U. (1997). Galvanic solid state sensors for potentiometric determination of CO_2, 8th Int Conf Sensors Transducers & Systems Nuremberg, Germany, Proceedings 3 (209–214).

51 Kempe, E. (1989). Probe means for sampling volatile components from liquids or gases. US Patent 4,821,585, filed 12 December 1984 and issued 18 April 1989.

14

CO_2 Measurement in Biotechnology and Industrial Processes

Wolfram Oelßner and Jens Zosel

Kurt-Schwabe-Institut für Mess- und Sensortechnik e.V. Meinsberg, Kurt-Schwabe-Straße 4, 04736 Waldheim, Germany

14.1 Beverage and Food Industry

The use of advanced instrumentation and sensors in the food industry has led to continuing improvement in food quality control, safety, and process optimization [1]. Chemical sensors as well as biosensors are increasingly used for process monitoring and for quality and freshness control of meat, dairy products, beverages, and a variety of other foodstuffs. Apart from pH value, oxygen concentration, and conductivity, the CO_2 concentration is an interesting parameter that, among others, is measured or controlled in:

- The beverage and brewery industry,
- Fruit and vegetable ripening and storage,
- Greenhouses,
- The production of yeast,
- CO_2 monitoring in egg setters,
- Food packaging, and
- Quality control and characterization of food quality and freshness.

In the beverage and brewery industry, carbon dioxide is used to carbonate the beverages in the production of soft drinks. Carbonated mineral water, beer, or cola contains more than 2000 mg/l CO_2. The CO_2 concentration of the beverages is an important quality parameter.

To increase the yield and quality of the products, numerous manufacturers offer special CO_2 measuring devices. In general, when used in the food industry, the sensors must meet very high requirements, concerning in particular:

- Mechanical robustness,
- Reliability,
- Exclusive use of "food safe" and "food contact safe" FDA-compliant materials (FDA: Food and Drug Administration),
- Heat sterilizability up to +180 °C, and
- CIP (cleaning-in-place) applicability.

Carbon Dioxide Sensing: Fundamentals, Principles, and Applications,
First Edition. Edited by Gerald Gerlach, Ulrich Guth, and Wolfram Oelßner.

Apart from product and process control, there is still another need for carbon dioxide measurement in the brewery industry. The fermentation process produces large amounts of carbon dioxide. During the bottling process, CO_2 can be emitted from the fillers into the ambient air of the production hall. As high concentration of CO_2 is hazardous, safety monitoring is essential in beverage filling facilities. For this purpose, infrared CO_2 transmitters, e.g. Vaisala CARBOCAP® carbon dioxide meters, are preferably used.

Furthermore, handling, preparation, processing, and packaging of food must be done in an extremely hygienic manner, with hygienic sensors and devices. The European Hygienic Engineering and Design Group (EHEDG) [2] provides practical guidance on hygienic engineering aspects to help manufacturers of equipment used in the food industry comply to these requirements with directives for hygiene, machines, and food contact materials. The CO_2 measuring devices for application in the beverage and brewery industry are based on different sensor principles [3], most of which are described in detail in Part II of this book.

14.1.1 Sensor Principles

14.1.1.1 Electrochemical Sensors

For example, METTLER TOLEDO's dissolved carbon dioxide measuring system InPro 5000i combines the proven Severinghaus principle for the electrochemical CO_2 sensor (described in detail in Section 4.1 of this book) with an Intelligent Sensor Management (ISM) [4]. It is specially intended for use in the brewery and pharmaceutical industry for sterile/hygienic processes (fermentation). Made of FDA-compliant materials, with highly polished surface finish to facilitate cleaning, the sensor fully meets international regulations such as EHEDG. The main specifications are:

- Accuracy: ±(10% of the reading + 10 mbar)
- Operating temperature: 0–60 °C (32–140 °F)
- Operating pressure: 0.2–2 bar (3–30 psia)
- Mechanical temperature resistance: –20 to 130 °C (–4 to 266 °F)
- Steam sterilizable and autoclavable: Yes
- 90% response time at 25 °C (77°F): <120 s

14.1.1.2 *p/T* (Pressure/Temperature) Sensors

For measuring carbon dioxide, the carbonated liquid flows through the sensor head (measurement chamber). Several times per minute the chamber is closed and its volume rapidly increases. This expansion generates a gas phase in the chamber. The large partial pressure difference of CO_2 forces the carbon dioxide out of the liquid in the chamber into the gas phase. This fundamental scientific principle is described by Henry's law. Within seconds pressure equilibrium in the measurement chamber is reached. The equilibrium pressure corresponds to the content of CO_2 in the sample. The CO_2 content is determined by pressure and temperature measurement in the chamber. After each measurement the sample is completely returned to the product without any loss. p/T-based sensors provide accurate measurements, but they contain moving parts for sample withdrawal/return and chamber expansion. As an example, the CARBOTEC TR-PT (Centec Gesellschaft für Labor- und Prozessmesstechnik mbH, Maintal,

Germany) is offered as a very precise carbon dioxide sensor for measurement of dissolved CO_2 in in fermentation reactors [5]. The main specifications are:

- Measuring range: 0–10 g/l
- Accuracy: ±0.05 g/l
- Repeatability: ±0.01 g/l
- Response time: ≤20 seconds

A similar CO_2 measurement system and its application are described in [6]. It is also based on Henry's law, which defines the relationship between the concentration of a dissolved gas and its saturation pressure. A sample of the product is introduced into a measuring chamber, where a mechanical shock releases a small portion of the dissolved CO_2 until the chamber pressure reaches the corresponding saturation pressure. The instrument electronically detects the sample's pressure and temperature and calculates the volumes CO_2 present in the beverage. The measurement cycle can be repeated up to four times per minute.

14.1.1.3 NIR-Based In-Line CO_2 Measurement

This type of CO_2 sensors is based on attenuated total reflection (ATR) technology. While transmitting a sapphire glass crystal, near-infrared (NIR) light is reflected several times at the surface. The surface of the crystal is in contact to the carbonated liquid. Since the CO_2 molecules in the liquid absorb the specific wavelength of the transmitting light, each reflection decreases the intensity according to the CO_2 content. The measurement result is not influenced by any other gases such as nitrogen in beer. Due to the lack of moving parts, the instrument is virtually maintenance-free. As an example, the NIR-based in-line CO_2 monitor Centec CARBOTEC NIR is offered by Centec to food and beverage manufacturers, especially for breweries and soft drink plants [7]. The main specifications are:

- Measuring range: 0–10 g/l CO_2
- Repeatability: ±0.02 g/l CO_2
- Accuracy: ±0.1 g/l
- Response time: ≤3 seconds
- Temp. range: −10 to 90 °C

14.1.1.4 Thermal Conductivity Sensors

With this type of sensor, the CO_2 concentration in a gas mixture is determined by measuring the thermal conductivity of the sample gas and comparing it with the thermal conductivity of a selected reference gas. Two ultra-stable, precision glass-coated thermistors are used, one in contact with the sample gas and the other in contact with the reference gas. The temperature difference between the two thermistors is detected in a Wheatstone bridge. Hach Ultra has developed [8] and patented [9] a thermal conductivity sensor with no moving parts (TMO2-TC) [10], which was applied for determination of CO_2 in package beer. This principle has also been reported in [11].

14.1.1.5 Other Sensor Principles

Moreover, CO_2 sensors based on the sensor principles described in Chapters 6 and 7 of this book are also used in equipment for the food and beverage industry:

- Opto-chemical YSI 8505 autoclavable CO_2 probe with YSI 8500 CO_2 monitor (Yellow Springs Instruments [YSI], Yellow Springs, OH) or
- Infrared CO_2 transmitters, e.g. Vaisala CARBOCAP carbon dioxide meter.

14.1.2 Application Examples

Fruits and vegetables are stored and ripened in specially controlled atmosphere (CA) rooms that have systems for controlling humidity, temperature, and CO_2 concentration. CO_2 must be monitored to determine the level of ventilation, since high CO_2 levels can preserve the fruits in storage but also retard the ripening. Depending upon the product being stored, CA conditions typically include decreased oxygen and increased carbon dioxide concentrations plus high relative humidity (RH) levels that have to be precisely maintained throughout the storage period. Research has shown that specific fruits require optimized storage conditions. As an example, the objective of a study [12] was to determine what storage conditions and pretreatments would permit long-term storage of New Zealand limes with minimal loss of quality. Limes are an attractive fruit crop but generally suffer a loss in value as their colour changes from green to yellow. Various approaches were taken to slow degreening including low temperature storage, use of CA environments, and treatment of fruit with physiologically active agents. However, the cold storage life of lime fruit can also be restricted by a number of factors including chilling injury. CA storage (10% O_2 with 0 or 3% CO_2) was compared with regular air storage and treatments with varying durations across a range of temperatures. Although some CA storage regimes could assist in delaying degreening, none of the treatments provided protection against frost damage. CA storage at 3% CO_2 delayed yellowing and gave better fruit quality than the low CO_2 treatment. High CO_2 CA treatments at 5 or 7 °C decreased the rate of colour change compared with other constant temperature treatments but did not protect against chilling injury.

In greenhouses carbon dioxide supplementation and controlling the CO_2 concentration can enhance plant growth and ripening and thus increase productivity. At the normal CO_2 level in the ambient outside air of about 340 vol ppm, all plants grow well, but if the CO_2 level is increased from 340 to 1000 vol ppm for the majority of greenhouse crops, net photosynthesis increases by about 50% over ambient CO_2 levels, resulting in more sugars and carbohydrates available for plant growth. Increased CO_2 levels will shorten the growing period, improve crop quality and yield, and increase leaf size and leaf thickness. But if the carbon dioxide level is too high, the growth of plants can be stunted or they can be even damaged. Plants during photosynthesis use carbon dioxide. The level to which the CO_2 concentration should be raised depends on the crop, light intensity, mineral fertilizer, temperature, stage of the crop growth, and the economics of the crop [13]. Since photosynthesis normally occurs only during daylight hours, CO_2 addition is not required at night. The supplementary carbon dioxide can be obtained by burning carbon-based fuels such as natural gas, propane, and kerosene or directly from tanks of pure CO_2. It must be distributed in the greenhouse by means of an adequate distribution system. Air movement around the plants will also improve the CO_2 uptake by bringing the CO_2 molecules closer to the leaf [14]. For the CO_2 control in greenhouses, special devices are commercially available. As an example, the Day/Night CO_2 Monitor & Controller for Greenhouses RAD-0501 (CO_2 Meter, Inc. Ormond Beach, FL, USA) uses for this purpose non-dispersive infrared (NDIR) CO_2 sensing technology [15]. When CO_2 levels get too low, power is supplied to a CO_2 generator or regulator, and when the target level is reached, the power is turned off. In addition, the CO_2 Monitor & Controller has a built-in photosensor that overrides the CO_2 control and shuts off the CO_2 when it senses darkness. Thus, CO_2 is only supplied during the light cycle when it is needed. The main specifications of this greenhouse CO_2 controller are:

- Measurement range: 0–10 000 ppm
- NDIR CO_2 sensing technology
- Accuracy: ±70 ppm or ±5% CO_2
- Sensor response time: <2 minutes
- Operating temperature: 32–122 °F
- Calibration: *in situ* with fresh air
- Photosensor that turns off CO_2 at night
- Sensor life expectancy: 15 years

Controlling the carbon dioxide concentration is also very important for growth and quality in mushroom farms. This is especially necessary during the pinning phase, the most elusive part of the mushroom's growth cycle. According to the respiration reaction

$$6O_2 + C_6H_{12}O_6 \rightarrow 6H_2O + 6CO_2 \quad (14.1)$$

CO_2 is produced during the mushroom's growth at night or in the dark, which negatively affects mushroom growth during pinning. Therefore proper ventilation is required to lower the CO_2 concentration in this phase up to 400 ppm. When the mushroom growth becomes visible, CO_2 concentrations of 800–1500 ppm is an optimum level for growth and development; however there are some differences among the fungi types and the different stages of growth. Generally, at CO_2 concentrations of less than 800 ppm, fungi are too small and numerous. Above 2000 ppm, the quality of the mushrooms is poor (the stem is too long and the cap is too small). At CO_2 concentrations of 4000–5000 ppm, mushroom development is inhibited. New mycelium begins to grow if the CO_2 concentration is higher than 5000 ppm [16]. Also for greenhouses, especially for application in mushroom farms, designed measuring instruments are offered commercially. For instance, the stationary CO_2 Monitor & Controller for Mushroom Farms and Growers RAD-0501A (CO_2 Meter, Inc. Ormond Beach, FL, USA) automatically exhausts grow room air when carbon dioxide levels get too high. The main specifications of this mushroom CO_2 controller that is based on NDIR CO_2 sensing technology are the same as those of the greenhouse controller given above [17]. Of course, generally applicable portable CO_2 meters are also used for this purpose [18].

Yeast is an important microorganism delivering an essential contribution for our nutrition. It is used, e.g. for bread dough processing and making alcoholic drinks. During the fermentation processes, CO_2 is formed, especially in the aerobic phase. High as well as very low carbon dioxide concentrations influence the fermentation of yeast. The effect of carbon dioxide on yeast growth and fermentation has been examined in numerous studies already for a long time [19–21]. For instance, in [19] portions of the nitrogen in the airstream were replaced by carbon dioxide while maintaining the oxygen content at the normal 20% level. Inhibition of yeast growth was negligible below 20% CO_2 in the aeration mixture. Slight inhibition was noted at the 40% CO_2 level, and significant inhibition was noted above the 50% CO_2 level, corresponding to 1.6×10^{-2} M of dissolved CO_2 in the fermenter broth. High carbon dioxide content in the gas phase also inhibited the fermentation activity of baker's yeast. In [21] the effects of pCO_2 were investigated in chemostat cultivation. An increase in pCO_2 from 44 to 195 kPa resulted in 25% decrease in cell concentration, 8% increase in ethanol concentration, and 50% decrease in glycerol concentration. While most of the studies have been concerned with the inhibitory effects of CO_2 on yeast growth, in a study cited in [22], the effect of CO_2 concentrations in comparison with 100% air in batch cultures of *Saccharomyces cerevisiae* was investigated. The data presented show that even 2.5% CO_2 causes a significant increase of yeast concentration. Optimum CO_2 content proved to be 5%. The aim of the research published in [23] was to study the effects of CO_2 pressure on *S. cerevisiae* at different pressures and temperatures for a designated time in potato dextrose broth. As a result, carbon dioxide was suitable for inactivation of *S. cerevisiae.*

Proper control of humidity, temperature, and carbon dioxide levels in egg setters and hatcheries during egg incubation can ensure an improved hatching rate and increased profitability. While slightly elevated levels of CO_2 during the first few days of incubation stimulate the growth of embryos, too high CO_2 concentration will inhibit growth. Numerous studies have indicated a clear correlation between the amount of CO_2 in the air and embryonic development [24]. Just after laying, the albumen of an egg contains a considerable amount of CO_2, mostly present as hydrogen carbonate. Since eggshell is porous and the concentration of CO_2 in the environment is much lower, CO_2 diffuses out of the egg, resulting in an increase in the pH value of the albumen. As this process of passive CO_2 diffusion from the egg continues during storage and in the first days of incubation, the concentration of CO_2 inside the fully sealed setter will gradually increase. While slightly elevated levels of CO_2 during the first few days of incubation stimulate the growth of embryos, too high CO_2 concentration will inhibit growth. Carbon dioxide levels above 0.5% in the setter reduce hatchability, with significant reductions at 1.0%. Total embryo mortality occurs at 5.0% CO_2. After days 10–12, embryo metabolism increases exponentially, and accordingly the production of CO_2 increases. In this period, the ventilation rate of the setter should gradually increase to maintain a maximum CO_2 level of 0.4 vol%. CO_2 NDIR sensor devices, such as they are also used in greenhouses and mushroom farms, are available and often standard equipment for controlling the CO_2 content of the air in egg setters and hatcheries.

The study [25] investigated the effect of non-ventilation of the chicken incubator during the first 10 days of incubation on carbon dioxide concentrations in the incubator and its effects on the embryonic and post-hatch development of the chicken (*Gallus gallus*). Two different incubation conditions were created – one incubator was kept at standard conditions, with adequate ventilation (V), and a second incubator was non-ventilated (NV) during the first 10 days of incubation, allowing the CO_2 to rise. From these results, it is clear that higher levels of CO_2 during the first 10 days of incubation have persistent (epigenetic) effects during the incubation and early post-hatch period. The levels of CO_2, O_2, and RH were continually monitored in both incubators using a computerized monitoring system with a CO_2 sensor (Vaisala GMM221, Waarloos, Belgium). The GMM221 utilizes the Vaisala CARBOCAP sensor, a silicon-based NDIR sensor. It is designed for high CO_2 concentration measurements and can be calibrated to operate within one of five concentration ranges from 0–2 vol% CO_2 to 0–20 vol% CO_2.

While most of the numerous published papers on egg setters refer to broiler eggs, the study [26] focuses on the relationship between CO_2 concentration and the biological activity in a red-legged partridge incubator. For this purpose a total amount of 43 316 eggs of red-legged partridge (*Alectoris rufa*) were supervised during five actual incubations. An infrared CO_2 sensor (model EE82-5C2, E+E ELEKTRONIK GmbH, Engerwitzdorf, Austria) was used to supervise the CO_2 concentration inside the incubator. It was calibrated in the range of 0–5000 vol ppm with an accuracy of 50 vol ppm.

In the food industry carbon dioxide is the most important protective gas in the packaging of food under modified atmospheres (MAP). Adding CO_2 to food packaging extends considerably the storage and shelf life of meat, cheese, and fruits and vegetables. In meat packaging, for example, a high concentration of CO_2 in the packaging inhibits bacterial growth and retains the natural colouring of the meat. Leak detection system for packages based on CO_2 like LEAK-MASTER® [27] feature non-destructive detection of leaks in the packages directly after the packing process. If the test sample is leaking, the pressure difference will result in a gas flow from the package into a test chamber, and the CO_2 concentration within the chamber rises. By means of a highly sensitive infrared CO_2 sensor (measuring range 0–5000 ppm, resolution 1 ppm), even smallest leaks can easily be detected.

In the review paper [28], progress on the development of different types of CO_2 sensors such as optical sensors, polymer opal films, polymer hydrogels, etc., which can be readily applicable to food packaging applications, is discussed. Aim of such developments is the smart or intelligent food packaging technology that also helps to trace a product's history through the critical points in the food supply chain. For this the development of CO_2 sensors specifically for food packaging applications, which can intelligently monitor the gas concentration changes inside a food package, is essential.

The freshness of milk and dairy products can be tested by various electrochemical and enzymatic methods and sensors [29]. In comparative studies, the pH value, the chloride, oxygen, and lactate concentrations, the temperature, conductivity, and carbon dioxide were measured simultaneously during the souring process of the milk [30]. Figure 14.1 shows results of measurement of the souring process of fresh, untreated cow's milk. The souring process begins gradually after approximately 15 hours. The biochemical reactions taking place in the milk during the souring process result in a considerable increase of the lactate and decrease in the oxygen concentration. As expected, the pH value decreases with increasing milk souring, while the CO_2 concentration rises.

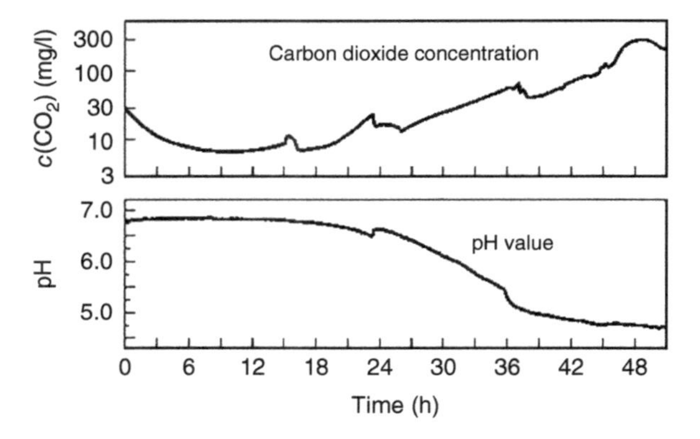

Figure 14.1 Results of measurement (performed at KSI Meinsberg Kurt Schwabe Research Institute, Germany) of the souring process of fresh, untreated cow's milk.

Table 14.1 Changes in CO_2 concentration and pH value of some liquid and paste-like perishable foods in the course of the spoiling process.

Food	CO_2 concentration (mg/L)	pH value
Milk fresh, untreated	30	6.8
After 2 d	277	4.4
Buttermilk, fresh	260	4.4
After 7 d	16	4.4
Apple puree, fresh	17	3.3
After 2 d	1700	3.3
Tomato puree, fresh	16	4.5
After 2 d	1600	4.4

As generally in electrochemical measuring technique, in food industry, the pH value is the most measured variable. But in some cases, as shown in Table 14.1, the CO_2 concentration indicates the state of spoilage much more significantly than the pH value, which remains nearly constant in some cases.

14.2 Bioreactors

It has been known for many years that the growth and metabolism of microorganisms is accompanied by the uptake and/or evolution of CO_2. To study and indeed to exploit the effects of CO_2 on microbial metabolism, it is necessary to control the level of (dissolved) CO_2 within the culture medium. CO_2 measurement and control can keep fermentation cultures in their optimal growth range. This consequently increases process yield and efficiency of the fermentation process. The review article [1] gives an overview of the role, measurement, and control of the magnitude of $p\mathrm{CO}_2$ during laboratory and industrial fermentations and describes in detail the various CO_2 equilibria and the question of CO_2 absorption rates.

Carbon dioxide is a product of cellular metabolism of microorganisms used in biotechnology. During fermentation the carbon dioxide content is the result of carbon dioxide formation by microorganisms and its transport by aeration. Apart from pH and dissolved oxygen measurements, reliable monitoring and control of the carbon dioxide partial pressure is important for successful fermentation and attracts more and more attention for the large-scale production of monoclonal antibodies and pharmaceutical products. Precise, real-time data on CO_2 concentration increase an understanding of critical fermentation and cell culture processes and can help in gaining insight into cell metabolism, cell culture productivity, and other changes within bioreactors. By means of online CO_2 measurements in the cell suspension culture with a CO_2 probe, it is possible to gain additional data for process characterization and to detect such carbon dioxide concentrations that inhibit metabolism and growth of microorganisms. The extension of observed process states to dissolved CO_2 concentration, HCO_3^- concentration, carbon dioxide production rate (CPR), and respiratory quotient offers the possibility of closed mass balances for the bioreactor and its liquid and gas phase. So the user is able to start compensating steps, e.g. aeration, to get information about the growth and activity of microorganisms during fermentation and to realize the control of anaerobic fermentations [31–35].

Fed-batch cultures have been intensively studied and optimized with the aid of kinetic models [32]. Dissolved CO_2 concentration, hydrogen carbonate (HCO_3^-) concentration, and CPR of mammalian cell suspension culture have great impact on growth conditions. But the carbon dioxide transfer rate (CTR), which can be directly calculated from off-gas measurement, is not necessarily equal to the interesting CPR. Hence, various mathematical methods have been developed for the estimation of the CPR and related parameters, such as the concentrations of dissolved CO_2 and HCO_3^- and total dissolved carbonate, directly from the off-gas data [33–35]. In the papers [34, 35] a mathematical model for the determination of dissolved CO_2 concentration, HCO_3^- concentration, and respiratory activity of mammalian cell suspension culture based on off-gas measurement is presented. Additionally, the calculated dissolved CO_2 concentrations are compared with results of online CO_2 measurements by using an electrochemical CO_2 probe.

In order to test the developed simulation, batch experiments were carried out in a 2 l stirred bioreactor as shown in Figure 14.2.

Off-gas CO_2 measurement was provided by a dual-beam IR CO_2 sensor (0–5%). With the accuracy 0.05 vol%, this CO_2 sensor was especially suitable for the high demands of measurements of mammalian cell culture. Online dissolved CO_2 concentration measurements were carried out by an electrochemical CO_2 probe (Meinsberg Kurt Schwabe Research Institute) as is shown in Figure 4.5. Although this sensor cannot be steam sterilized or autoclaved, it could be used successfully for determination of dissolved CO_2 concentration in mammalian cell suspension in the bioreactor. For this purpose, the sensor was sterilized chemically with ethyl alcohol and connected to the reactor by a sterile adapter. In the diagram in Figure 14.3, results of measured and calculated dissolved CO_2 concentrations during cultivation time of a reactor batch experiment are compared.

Figure 14.3 shows generally satisfactory conformity between simulated and measured curves. The online dissolved CO_2 concentration measurements approved the simulations, which provide valuable information for process

Figure 14.2 Two litres stirred bioreactor (VSF 2000, Bioengineering AG, CH) with electrochemical carbon dioxide sensor (KSI Meinsberg Kurt Schwabe Research Institute, Germany).

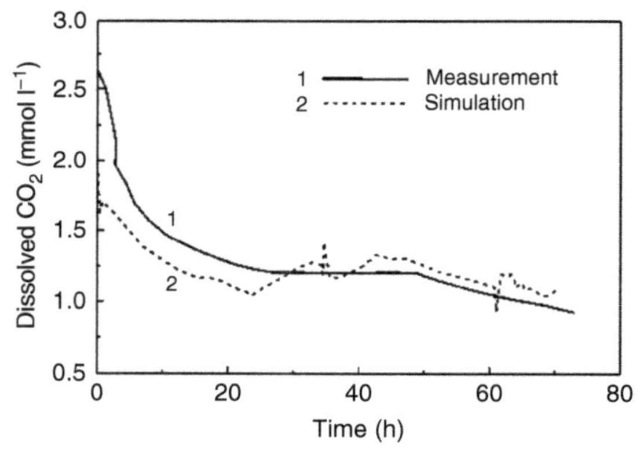

Figure 14.3 Comparison of simulated and measured dissolved CO_2 concentrations in a reactor batch experiment. Source: Frahm et al. 2002 [34]. Reproduced with permission of Elsevier.

modelling and model improvement of cell culture growth. Only during the first 25 hours of the experiment the measured CO_2 concentration was higher. Not yet formed CO_2 mass transfer and CO_2 oversaturation exceeding simulated conditions can cause the slightly underestimated simulations of dissolved CO_2. Furthermore, a spatial CO_2 profile in the liquid phase can evolve with preceding cultivation time.

In the study [36] the effect of elevated CO_2 concentrations on insect cell growth and metabolism and the roles of oxidative stress and intracellular pH (pHi) in CO_2 inhibition were investigated. A critical problem in the mass production of these products is CO_2 accumulation to inhibitory levels within the bioreactor. *Spodoptera frugiperda* Sf9 insect cells were cultured in a 3 l bioreactor controlled at 20% air saturation, 27 °C, and a pH of 6.2. The cells were exposed to a constant

CO_2 concentration by purging the medium with CO_2 and the headspace with air. The experiments were repeated for different CO_2 concentrations, and samples were taken every 24 hours to determine cell density, viability, metabolism, and oxidative stress. It was demonstrated that CO_2 accumulation inhibits the growth of uninfected Sf9 cells, but does not affect cell viability. The CO_2 concentration in the cell culture medium was monitored online using YSI 8505 autoclavable CO_2 probe with YSI 8500 CO_2 monitor (Yellow Springs Instruments, Yellow Springs, OH), which is described in detail in Section 6.1.

14.3 Biogas Plants

Biogas production processes transform organic matter (agricultural products, residues from landscape management, farm animal and human excrements, organic wastes from food production, and others) into a gas mixture of methane and carbon dioxide as the main components and a digestate with depleted carbon content [37]. The overall conversion of this multistep microbial-driven process can be described by the Boyle equation (14.2) [38]:

$$\begin{aligned} &C_aH_bO_cN_dS_e + 1/4(4a - b - 2c + 3d + 2e)H_2O \\ &\quad \rightleftharpoons 1/8(4a + b - 2c - 3d - 2e)CH_4 \\ &\qquad + 1/8(4a - b + 2c + 3d + 2e)CO_2 + dNH_3 + eH_2S \end{aligned} \tag{14.2}$$

That anaerobic fermentation process can be subdivided into four different steps: hydrolysis, acetogenesis, acidogenesis, and methanogenesis. Several hundred kinds of different microbes can be involved in the steps sharing the side conditions of the process like temperature, pressure, hydrodynamics, pH value, chemical composition, etc. Since the microbes have different optimum conditions and limitations, only a relatively small window of conditions enables a long-term stable and efficient process. Therefore, the sensory control of biogas production plants is essential for effectiveness and safety of this important source of renewable energy.

Especially online monitoring by means of a variety of different sensors enables the early detection of critical feeding situations and helps to prevent cost-intensive biogas production breakdowns. Today a variety of methods are used to characterize and control the biogas process in liquid-filled reactors by measuring a number of critical parameters in the gas phase as well as in the liquid phase [39, 40]. Figure 14.4 shows the schematic design of a temperature-controlled biogas laboratory plant for the investigation of biogas production kinetics with sensors for the liquid phase (pH, temperature, dissolved hydrogen) and the gas phase (CH_4, CO_2, H_2, flow rate) [40, 41]. For measuring the CO_2 concentration in the biogas, an IR sensor (AGM 32, Sensors Europe GmbH) was used.

Apart from measuring CH_4 and H_2 as the most interesting components, it is very helpful to determine additionally the concentration of CO_2 in the biogas too. On the one hand, by summarization of these components, it can be determined whether the biogas contains further components. On the other hand, the correct

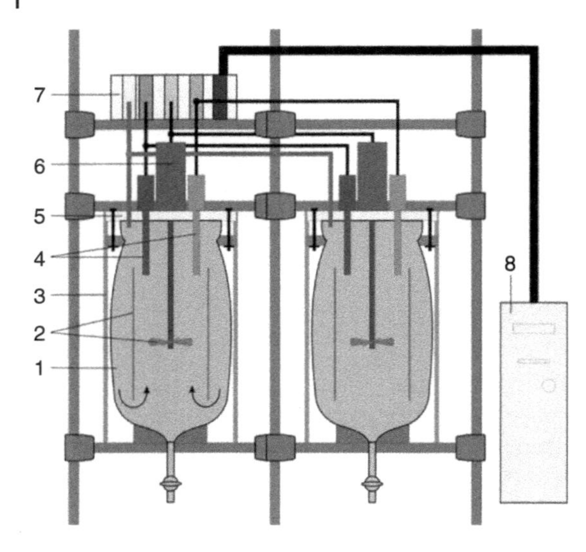

Figure 14.4 Schematic construction of the biogas laboratory plant: (1) two glass reactors with (2) stirrers, (3) thermal insulation, (4) sensors for the liquid phase (pH, temperature, dissolved hydrogen), (5) reactor cover, (6) stirring motor, (7) multi-parameter measuring system with sensors for the gas phase (CH_4, CO_2, H_2, and flow rate) and transducers, and (8) PC with measuring cards.

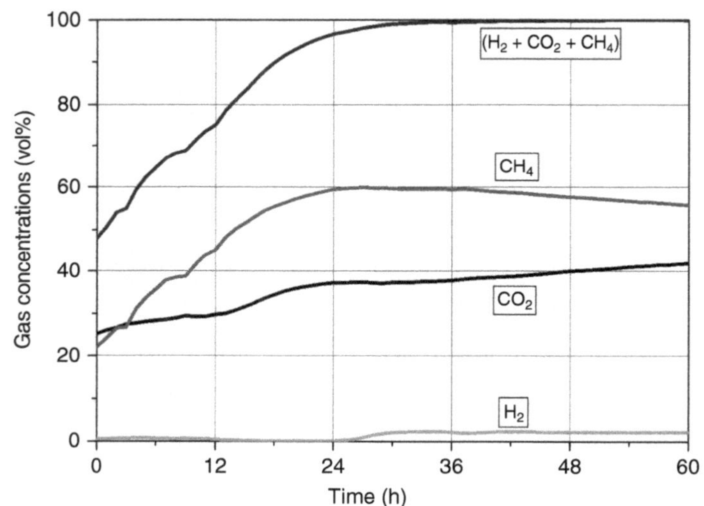

Figure 14.5 Course of the development of the gases H_2, CO_2, and CH_4 during the start-up period of a mesophilic biogas process in the biogas laboratory plant.

function of the other sensors can be controlled. Figure 14.5 shows the course of the biogas production during the start-up period of a mesophilic biogas process in the biogas laboratory plant. After about 36 hours, the air has been completely expelled from the reactor vessels by the developed biogas consisting of about 2% H_2, 42% CO_2, and 56% CH_4. The sum of almost exactly 100% indicates that the biogas process runs as expected and that the hydrogen and methane sensors work perfectly.

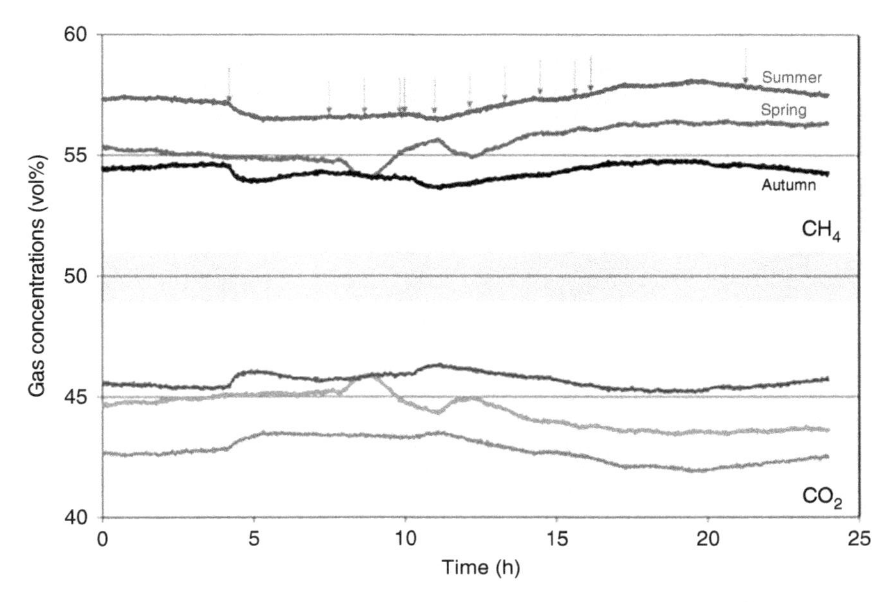

Figure 14.6 Courses of CO_2 and CH_4 concentrations during feeding events of cowshed manure (dark gray arrows) and corn silage (light gray arrows) into a 1.5 m^3 fermenter during three different Sundays in spring, summer, and autumn.

According to the Boyle equation (14.2), the ratio between methane and carbon dioxide gives information on the composition of the organic input material and therefore can be used as an additional stability-related information from the process. Process measurements at pilot plants with enhanced temporal resolution indicate that the ratio between CH_4 and CO_2 contains also information on the transfer kinetics within the different steps and the oxygen intake during feeding. As shown in Figure 14.6, two temporal courses of the CH_4 and CO_2 concentration during feeding of the process with different oxygen intake demonstrate that ability.

Larger commercial biogas production plants with more than 100 kW power are nearly invariably equipped with online monitoring systems for biogas composition. One example [42] of a system for the continuous measurement of CH_4, CO_2, H_2, O_2, and H_2S concentrations in biogas is given in Figure 14.7. The sensors used for CO_2 measurement in that application are based solely on the principle of the absorption of IR radiation as described in Section 7.3.

If the biogas is produced to be lead into the natural gas grid, all non-methane components have to be diminished below fixed limits. For CO_2, that limit amounts to 6 vol% in Germany [43]. Most technical devices for biogas filtration are designed to enable CO_2 concentrations below 1 vol% at the point of transfer into the natural gas grid [44]. These devices are controlled by IR absorption CO_2 sensors.

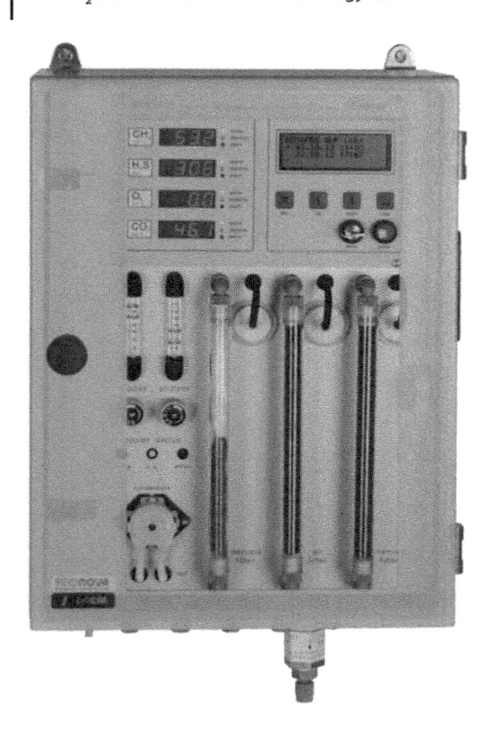

Figure 14.7 Multi-gas monitor SSM 6000 from Pronova Analysentechnik GmbH & Co. KG, Berlin, Germany, for measurement of CH_4, CO_2, H_2, O_2, and H_2S concentrations in gases from biogas facilities, sewage gas, or landfill gas.

References

1 Kress-Rogers, E. (1993). *Instrumentation and Sensors for the Food Industry.* Oxford: Butterworth-Heinemann Ltd.
2 www.ehedg.org/ (retrieved 25 February 2017).
3 http://www.mt.com/de/en/home/supportive_content/news/po/pro/CO2.html (retrieved 25 February 2017).
4 http://www.mt.com/dam/MTPRO/PDF/TD/TD_CO2_Sensor_InPro5000i_EN_52002457_July2012.pdf (retrieved 9 March 2017).
5 http://www.centec.de/sensors/pharma-biotech/carbotec-tr-pt/ (retrieved 25 February 2017).
6 Bloder, S. (1999). CO_2-brix-diet. *Getränkeindustrie* 10 (99): 619.
7 http://www.centec.de/fileadmin/user_upload/Downloads/CO2_Measurement/Datasheet_CARBOTEC_NIR.pdf (retrieved 9 March 2017).
8 Hach Ultra, Application Not Beverage No. 4: Dissolved carbon dioxide measurement in the brewing and beverage industry. https://www.hach.com/asset-get.download.jsa?id=7639984651 (retrieved 9 March 2017).
9 Stehle, G., Studemann, E., Michaud, E., and Fraternale, O. (2007). Method and device for measuring the amount of gas in a sealed liquid container. WO Patent 2,009,050,530 A1, filed 16 October 2007 and issued 23 April 2009.
10 http://www.veronics.com/products/Gas_Analyzers-Thermal_Conductivity/tmo2-tc.pdf (retrieved 9 March 2017).

11 Robert, E. and Klein, C. (2007). Improved Method for CO_2 measurements. *BBI International* 2007 (5): 32–35.

12 Pranamornkith, T. (2009). Effects of postharvest treatments on storage quality of lime (Citrus latifolia Tanaka) fruit. PhD thesis. Massey University, New Zealand.

13 Frantz, J.M. (2011). Elevating carbon dioxide in a commercial greenhouse reduced overall fuel carbon consumption and production cost when used in combination with cool temperatures for lettuce production. *HortTechnology* 21 (5): 647–651.

14 OMAFRA Factsheet Carbon Dioxide in Greenhouses. http://www.omafra.gov.on.ca/english/crops/facts/00-077.htm (retrieved 15 February 2017).

15 http://www.co2meter.com/products/day-night-co2-monitor-controller (retrieved February 15, 2017).

16 Rotronic Application Note N F037 Mushroom Farming (2013) http://www.venta-oprema.com/uploads/6/5/0/0/6500536/application_note_f037_-_co2_in_mushroom_farming.pdf (retrieved 9 March 2017).

17 http://www.co2meter.com/blogs/news/36681665-monitoring-co2-levels-critical-for-mushroom-farm-success (retrieved 25 February 2017).

18 pSense Portable CO_2 Meter. http://www.co2meters.com/Documentation/Datasheets/DS-pSense.pdf (retrieved 7 July 2015).

19 Chen, S.L. and Gutmanis, F. (1976). Carbon dioxide inhibition of yeast growth in biomass production. *Biotechnol. Bioeng.* 18 (10): 1455–1462.

20 Jones, R.P. and Greenfield, P.F. (1982). Effect of carbon dioxide on yeast growth and fermentation. *Enzyme Microb. Technol.* 4 (4): 210–223.

21 Kuriyama, H., Mahakarnchanakul, W., Matsui, S., and Kobayashi, H. (1993). The effects of pCO_2 on yeast growth and metabolism under continuous fermentation. *Biotechnol. Lett.* 15 (2): 189–194.

22 Halasz, A. and Lasztity, R. (1991). *Use of Yeast Biomass in Food Production*, 61. Boca Raton, USA: CRC Press.

23 Erkmen, O. (2002). Effects of carbon dioxide pressure on *Saccharomyces cerevisiae. Food Sci. Tech. Int.* 8 (6): 361–364.

24 De Lange, G. (2016). Managing carbon dioxide in the setter. https://www.pasreform.com/en/knowledge/13/managing-carbon-dioxide-in-the-setter (retrieved 16 February 2017).

25 De Smit, L., Bruggeman, V., Tona, J.K. et al. (2006). Embryonic developmental plasticity of the chick: increased CO_2 during early stages of incubation changes the developmental trajectories during prenatal and postnatal growth. *Comp. Biochem. Physiol. A* 145 (2): 166–175.

26 Garcia-Hierro, J., Barreiro, P., Moya-Gonzalez, A., and Robla, J.I. (2016). Prospectives of monitoring biological activity in a red-legged partridge incubator with a carbon dioxide probe. *AgricEngInt: CIGR Journal* 18 (1): 353–362.

27 Leak detection system LEAK-MASTER® *Technical data sheet,* WITT-Gastechnik GmbH & Co KG, Witten, Germany. http://www.wittgas.com/products/package-leak-detectors.html (retrieved 25 February 2017).

28 Puligundla, P., Jung, J., and Ko, S. (2012). Carbon dioxide sensors for intelligent food packaging applications. *Food Control* 25 (1): 328–333.

29 Belitz, H.-D. and Grosch, W. (1999). *Food Chemistry*, 2e, 471–526. Berlin: Springer.
30 Herrmann, S., Oelßner, W., Schulz, B., and Guth, U. (2001). Comparative studies on the souring process of milk by means of enzymatic and electrochemical sensors. 2nd BioSensor Symposium Tübingen 2001. https://publikationen.uni-tuebingen.de/xmlui/handle/10900/48304 (retrieved 9 March 2017).
31 Dixon, N.M. and Kell, D.B. (1989). The control and measurement of 'CO_2' during fermentations. *J. Microbiol. Meth.* 10 (3): 155–176.
32 Pörtner, R., Schilling, A., Lüdemann, I., and Märkl, H. (1996). High density fed-batch cultures for hybridoma cells performed with the aid of a kinetic model. *Bioprocess Eng.* 15 (3): 117–124.
33 Wu, L., Lange, H.C., van Gulik, W.M., and Heijnen, J.J. (2003). Determination of in vivo oxygen uptake and carbon dioxide evolution rates from off-gas measurements under highly dynamic conditions. *Biotechnol. Bioeng.* 81 (4): 448–458.
34 Frahm, B., Blank, H.-C., Cornand, P. et al. (2002). Determination of dissolved CO_2 concentration and CO_2 production rate of mammalian cell suspension culture based on off-gas measurement. *J. Biotechnol.* 99 (2): 133–148.
35 Frahm, B. and Pörtner, R. (2002). Messung und Simulation des Kohlendioxidausstoßes von Zellkulturprozessen (measurement and simulation of the carbon dioxide output of cell culture processes). In: *Biosystemtechnik. Beiträge des Innovations forums 2001* (ed. H. Kaden), 39–47. Waldheim (in German).
36 Vajrala, S.G. (2010) Mechanism of CO_2 inhibition in insect cell culture. Masters thesis, University of Iowa, USA.
37 United Nations Asian and Pacific Centre for Agricultural Engineering and Machinery (2007). *Recent Developments in Biogas Technology for Poverty Reduction and Sustainable Development.* Beijing: APCAEM.
38 Boyle, W.C. (1977). Energy recovery from sanitary landfills. In: *Microbial Energy Conversion* (eds. A.G. Schlegel and J. Barnea), 119–138. Unitar.
39 Hannsson, M., Nordberg, A., Sundh, I., and Mathisen, B. (2002). Early warning of disturbances in a laboratory-scale MSW biogas process. *Water Sci. Technol.* 45 (10): 255–260.
40 Zosel, J., Oelßner, W., Zimmermann, P. et al. (2008). *New concepts and sensors for online control of biogas plants.* 12th International Meeting Chemical Sensors, 13–16 July 2008, Columbus, OH, USA. Ext. Abstracts, pp. 350–351.
41 Zosel, J., Oelßner, W., Guth, U. et al. (2007). Biogas-Laboranlage zur Verfahrensoptimierung der Biogasgewinnung (Biogas laboratory plant for optimizing biogas production). *Chem.-Ing.-Tech.* 79 (9): 1339. (in German).
42 PRONOVA Analysentechnik GmbH & Co. KG, brochure on the SSM 6000 series. https://pronova.de/media/pdf/dc/d7/37/SSM6000-device-series_EN.pdf (retrieved 6 June 2017).

43 Deutscher Verein des Gas- und Wasserfaches e.V. Bonn (2013). Technische Regel - Arbeitsblatt Gasbeschaffenheit (Technical rule – work sheet gas quality) DVGW G260 (A)(2013-03) (in German).

44 Krayl, P. (2009). Einspeisung von Biogas in das Erdgasnetz - Erfahrungsberichte der ersten Biomethananlagen. (Feeding biogas into the natural gas network – experiences of the first biomethane plants). *Energie|Wasser-Praxis* 4: 20. (in German).

15

CO_2 Measurements in Biology

Wolfram Oelßner

Kurt-Schwabe-Institut für Mess- und Sensortechnik e.V. Meinsberg, Kurt-Schwabe-Straße 4, 04736 Waldheim, Germany

15.1 Aquatic Animals

15.1.1 Fish

For the assessment of the physiological condition of aquatic animals, the measurement of respiration is essential, which is influenced considerably by the water quality. For this reason aquaculture operations demand precise control of water quality, and the monitoring of water parameters becomes increasingly important especially for effective and high-quality rearing and production of freshwater nutritious fish.

15.1.1.1 Influence of CO_2 Concentration on Fish

Apart from pH value, oxygen concentration, biochemical oxygen demand, salinity, etc., carbon dioxide concentration is an important parameter in fish farming [1–5]. In fish, the exchange of respiratory gases between the blood and the water takes place through the gills. Not only sufficient supply of dissolved oxygen O_2 but also the elimination of the breathing product CO_2 over the gills of the fish must be assured, which is only possible if an adequate CO_2 concentration gradient exists between the blood of fish and the surrounding water. If, due to increased CO_2 concentrations in water, CO_2 from the fish's own metabolism cannot be sufficiently released from the blood through the gills, the CO_2 in the blood increases (known as hypercapnia), resulting in a drop in the blood pH, an acidosis. At the same time the oxygen-carrying ability of the haemoglobin in the blood is reduced. High CO_2 concentrations in the feed water, occurring especially in lime-rich environments or in systems with large numbers of fish and relatively slow water turnover, reduce the capacity of the blood to transport oxygen in the brain and in other organs of fish. Consequences are increased liability to infectious diseases and reduced effectiveness of food intake and utilization despite the presence of sufficient oxygen [6–8]. However, a lack of free carbon dioxide in water, which occurs, e.g. when too much CO_2 is utilized for photosynthetic activity by the phytoplankton, can be also harmful for fish. Free carbon dioxide

Carbon Dioxide Sensing: Fundamentals, Principles, and Applications,
First Edition. Edited by Gerald Gerlach, Ulrich Guth, and Wolfram Oelßner.

concentrations below 1 mg l^{-1} affect the acid–base balance in the fish blood and tissues and cause alkalosis.

The maximum admissible CO_2 concentration in trout breeding ponds depends among other parameters to a great extent on the acid binding capacity or the water hardness. Guideline CO_2 concentrations values for rainbow trout are [2]:

<10–30 mg l^{-1} CO_2 for fish spawn,
<15–35 mg l^{-1} CO_2 for adult fish (> 5 cm), and
>50–150 mg l^{-1} CO_2 lethal concentration.

The lower values are valid for water with low water hardness (acid binding capacity <0.5 mval l^{-1}), and the indicated upper limits are admissible at high water hardness (acid binding capacity >3.5 mval l^{-1}).

15.1.1.2 Methods to Determine CO_2 Concentrations

Electrochemical carbon dioxide sensors are often unfavourable for practical use in the fishing industry since they require permanent maintenance and calibration as well as experimental skills. For this reason, to determine the CO_2 concentration in fish waters, various other methods are applied besides the well-known principles:

- Commercially available electrochemical CO_2 sensors can be used for *in situ* measurement and control of CO_2 in fish farming provided that the prerequisites for the correct calibration and maintenance of the sensor are available in the plant. For long-term carbon dioxide measurements in lake water, for example, an electrochemical CO_2 sensor proved to be very suitable [9]. In the evaluation and comparison of CO_2 measurement results obtained with different methods, it must be taken into consideration that the electrochemical CO_2 sensor measures the partial pressure, whereas with alternative methods often the CO_2 concentration is determined. Both values can be quite different.
- According to the German Industrial Standards DIN 38405 (D8) [10] and DIN 38409 (H7) [11], carbon dioxide dissolved in water and the anions of the carbonic acid can be calculated and quantified indirectly by measuring pH value, temperature, and conductivity and by the analytical determination of the acid and base capacity.
- In practice, even today the CO_2 concentration is often taken from tables. In this case it is generally necessary to determine the total alkalinity analytically by titration. In [12] a set of tables is presented for the easy calculation of total carbon dioxide, the partial pressure of free CO_2, and other components of the carbon dioxide system in freshwater. The tables are based on measuring values of pH, temperature, total alkalinity, and electrolytic conductivity.
- The SRAC Publication No. 464 offers also a table with factors for calculating carbon dioxide concentrations in water with known pH, temperature, and alkalinity [3].
- Alternatively, the SRAC Publication No. 464 presents a simple graphical technique for estimating carbon dioxide concentration from pH value and alkalinity [4]. It also gives recommendations for the removal of carbon dioxide from pond waters by chemical treatment with liming agents such as quicklime, hydrated lime, or sodium carbonate.

- For studies on the carbon dioxide supersaturation in the surface waters of lakes, the value of pCO_2 was calculated from pH and dissolved inorganic carbon (DIC) or acid-neutralizing capacity (ANC) with corrections for other physical and chemical variables [13]. Water was collected and equilibrated with ambient air. Gas chromatography was used to measure CO_2 on the equilibrated head space.
- The Model 503 pH/CO_2 Analyzer (Royce Instruments, New Orleans, USA) utilizes the fact that the concentration of dissolved CO_2 in water is a function of pH value, salinity, temperature, and alkalinity. Assuming that salinity and alkalinity are known and do not change in the short term, the CO_2 content can be calculated by measuring only the pH value and the temperature. However, the Octopus Marine Consultancy in Trondheim, Norway, reported that with such an instrument in a fish tank, a too high CO_2 concentration was measured but that this could be significantly reduced with a simple aerator [1].

15.1.1.3 Behaviour of Fish in Regions with Increased CO_2 Concentration

To test experimentally whether fish instinctively escape from regions with increased CO_2 concentration, a special aquarium was developed at the Meinsberg Kurt-Schwabe Research Institute (KSI Meinsberg) (Figure 15.1). It was installed at the Trout Farm Trostadt (Germany), a certified disease-free farm and hatchery with an annual production of about four million rainbow trout fingerlings. This farm is located in the southern part of Thuringia, a limestone

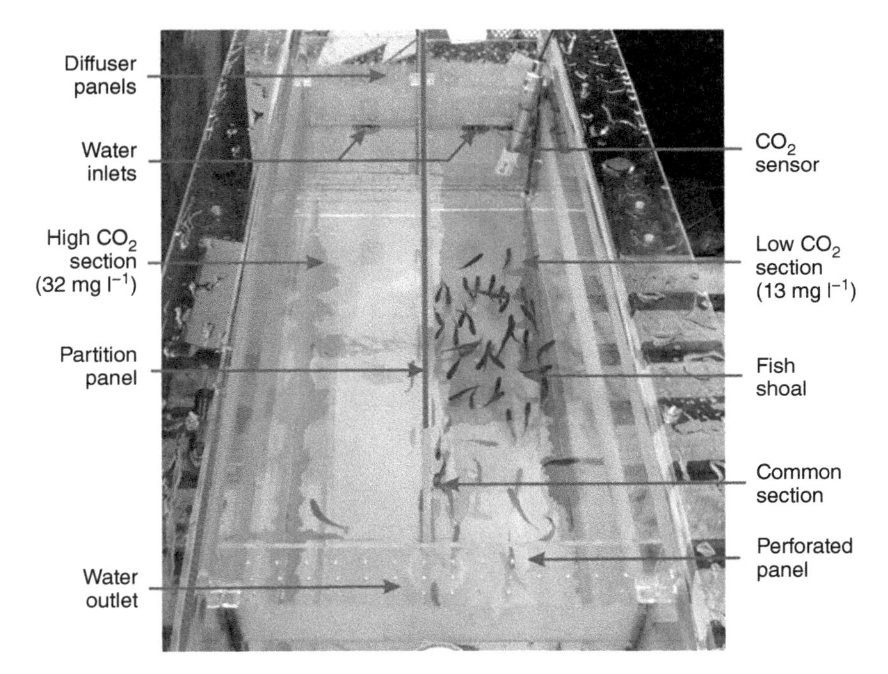

Figure 15.1 Photograph of the aquarium with sections of different CO_2 concentration and the distribution of fish in it about one minute after the fish were inserted into the common section of the aquarium [5].

area, which delivers water with high alkalinity and CO_2 content. Carbon dioxide can be removed by chemical treatment of pond water with liming agents or by ventilation. Since these measures are complicated and expensive, methods to control the carbon dioxide level were studied [5].

The aquarium itself is divided by a panel into three sections. Through the left section water flows with high CO_2 concentration of 32 mg l^{-1} (pH = 7.44), and through the right one degassed water from the same spring with only 13 mg l^{-1} CO_2 (pH = 8.11). In the common section the two different waters admix and then flow off. The feed water for the aquarium was pumped from two reservoirs with different CO_2 concentrations and fed through hoses into the aquarium. In order to exclude the possibility that the behaviour of the fish might be influenced by such factors as different lighting, the water inlets and hence the CO_2 concentrations in the two sections could be reversed during the tests. The feed water to the aquarium was spring water from a well. Low CO_2 levels were achieved by degassing the water by means of trickling and the use of a degassing cascade for very low CO_2 concentrations. The higher and the lower concentration was set at 32 mg l^{-1} (pH = 7.44) and 13 mg l^{-1} (pH = 8.11), respectively.

The CO_2 concentrations in the waters were measured by means of a handheld CO_2 metre with an electrochemical carbon dioxide sensor, developed by KSI Meinsberg. Figure 15.2 shows the measuring device. The sensor system with integrated temperature probe for automatic temperature compensation is arranged within a robust shaft consisting of stainless steel. At the upper end of the electrode shaft, the connecting cable is waterproof-mounted, allowing the sensor to be submerged up to a depth of 1 m. For purposes of comparison, a commercial measuring instrument (Model 503 CO_2 Analyzer, Royce Instruments, New Orleans, USA) was used. In addition, the pH value and the temperature (10 °C) of the water were determined.

The fish used in the experiment were rainbow trout (*Oncorhynchus mykiss*), aged 60 days and with an average weight of 2.5 g geared from eggs from Troutlodge Inc., Sumner, WA, USA. A shoal of 58 fishes was initially exposed in the common section of the aquarium, where the fish had free choice to remain in this region or to move to one of the regions with either higher or lower CO_2

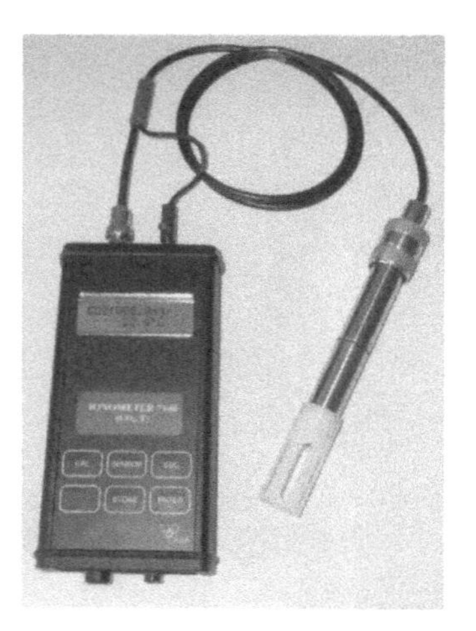

Figure 15.2 Electrochemical carbon dioxide sensor and handheld CO_2 metre [5].

concentrations. As can be seen in Figure 15.1, already after less than one minute, the shoal had left the common section to gather in the section with the lower CO_2 concentration. Individual fish were moving occasionally into the transition region to the section with the higher CO_2 concentration, where their breathing frequency visibly increased, along with the speed of movement of their gills. It was observed that usually these "curious" fish only made such an excursion once and returned quickly into the section of the aquarium with the lower level of CO_2.

From this observation, the conclusion could be drawn that, contrary to human beings, fish are able to perceive carbon dioxide gradients. The experiment demonstrated that fish escape from regions with higher levels of CO_2 even at concentrations that are not yet life-threatening. It concerns the question of the welfare of the fish, defined as its state as regards its attempts to cope with its environment. The scientific assessment of animal welfare, including animal health, has developed rapidly in recent years, and the sophistication of this will improve further. Public concern about the welfare of animals has increased and is increasing rapidly. Methods of assessment of welfare on farm and in other animal keeping places are also developing and will provide better tools for animal keepers and inspectors in future [14]. In this context, the question is also discussed whether fish are capable of experiencing pain and suffering. Although there are basic neurobehavioural differences between fish and humans and it is implausible that fish can experience pain or emotions like humans, they display robust, nonconscious, neuroendocrine, and physiological stress responses to noxious stimuli. Thus, avoidance of potentially injurious stress responses is an important issue with respect to the welfare of fish [15]. Unlike for fish, invertebrates, such as crustaceans, did not show such a behaviour with respect to noxious stimuli (see Section 15.1.2) [16].

Fish rid themselves of carbon dioxide through the gills in response to a difference in carbon dioxide concentration between fish blood and the surrounding water. There are chemoreceptors on the gills, and in ambient conditions of low pO_2 and high pCO_2, gill ventilation can increase more than 10-fold. Some fish species open the mouth during swimming at high speeds to force the water over the gills by the forward movement of the fish [17]. An important receptor site is the pseudobranch, a rudimentary gill that is perfused by oxygenated blood and probably monitors arterial pO_2 but is also sensitive to pCO_2 and a number of other chemical and physical parameters [18].

To investigate effects of carbon dioxide exposure on intensively cultured rainbow trout, fish were exposed to three CO_2 concentrations: control ($2.8\,mg\,l^{-1}$), medium ($3.8\,mg\,l^{-1}$), or high ($4.4\,mg\,l^{-1}$). Growth and physiological results showed that increasing elevated CO_2 concentrations result in corresponding decreased growth rates and CO_2-specific physiological parameters [19]. In recent extensive studies the acute effects of elevated CO_2 on the escape responses of juvenile fish were considered as well as the question whether such effects were altered by exposure of parents to increased CO_2 (transgenerational acclimation). It was found that elevated CO_2 negatively affected the reactivity and locomotor performance of juvenile fish, but parental exposure to high CO_2 reduced the effects in some traits, indicating the potential for acclimation of behavioural impairment across generations. However, transgenerational acclimation does not completely compensate the effects of high CO_2 on escape responses [20].

15.1.2 Mussels

Zebra mussels (*Dreissena polymorpha*) are small clam-like freshwater molluscs named for the striped pattern of their shell. They can grow up to 5 cm long, but most often they are less than 2.5 cm. Their lifespan ranges from three to nine years. Using their glue-like fibres called byssal threads, the mussels attach to almost any hard underwater surface forming dense colonies, called druses, of up to several 100 000 individuals per m^2 [21].

On the one hand, zebra mussels can be a great nuisance and threat to aquiculture and several aquatic equipment. Mussel colonies clog pipes and valves; damage pumps, generators, and motors; encrust boats and navigational buoys; and cause corrosion and other costly problems. In fish farming they can reduce the food for newly hatched fry or other fish or may enhance the efficiency of fish-eating birds by increasing the water transparency.

On the other hand, zebra mussels are cultivated [22, 23] and used to renaturize waters and to improve water quality, because they are very efficient in removing nutrients and improving water clarity [24–28]. This process involves the flowing of wastewater across a mass of zebra mussels that are attached to a fixed media such as a steel plate. The filtering action of the zebra mussels will remove contaminants, which will be used as food and incorporated into the mussel's body tissue or excreted as faeces. Filtered material that is not used as food is encapsulated and excreted as pseudofaeces. Periodically, the accumulated faeces and pseudofaeces, along with dead mussels, will have to be removed from the system.

In a simulation study on water quality objectives and cost-effectiveness of zebra mussel farming in the Szczecin (Oder) lagoon in the southern Baltic Sea, it was concluded that mussel farming could potentially remove about 2 % of the present N and P loads, and it would have the additional benefit of improving water transparency [29, 30].

15.1.2.1 Respiratory Quotient

For the assessment of the effectiveness of zebra mussels, the respiratory quotient RQ is used. This dimensionless number is calculated from the ratio of moles of carbon dioxide produced $[CO_2]_{prod}$ per mole of oxygen consumed $[O_2]_{cons}$:

$$RQ = [CO_2]_{prod}/[O_2]_{cons} \tag{15.1}$$

Due to its strong dependence on the composition of the metabolized food, the RQ is used to characterize the animals' nutritional condition in the habitat and the quantity and quality of food. If carbohydrates are burnt, then the RQ is around 1.0, whereas an RQ of 0.7 indicates that fats (lipids) are metabolized. For proteins, the RQ ranges between 0.8 and 0.9. The calculation of these values will be explained by a few examples.

Carbohydrates The complete oxidation of compounds containing only the elements carbon C, hydrogen H, and oxygen O is described by

$$C_xH_yO_z + (x + y/4 - z/2)O_2 \rightarrow x\,CO_2 + (y/2)\,H_2O \tag{15.2}$$

The metabolism of such compounds results in

$$RQ = x/(x + y/4 - z/2) \tag{15.3}$$

Considering glucose with $C_6H_{12}O_6 + 6\ O_2 \rightarrow 6\ CO_2 + 6\ H_2O$, this leads to

$$RQ = 6\,(CO_2)/6\,(O_2) = 1.00 \tag{15.4}$$

Fats The RQ of fats is lower than that of hydrocarbons as fats contain fewer oxygen atoms in proportion to atoms of carbon and hydrogen. It amounts for triolein ($C_{57}H_{104}O_6 + 80\ O_2 \rightarrow 57\ CO_2 + 52\ H_2O$) to

$$RQ = 57\,(CO_2)/80\,(O_2) = 0.71 \tag{15.5}$$

and for palmitic acid ($C_{16}H_{32}O_2 + 23\ O_2 \rightarrow 16\ CO_2 + 16\ H_2O$) to

$$RQ = 16\,(CO_2)/23\,(O_2) = 0.70 \tag{15.6}$$

Proteins Besides carbon, hydrogen, and oxygen, proteins contain nitrogen and other elements in a variety of different compounds. For albumin with its reaction equation

$$C_{72}H_{112}N_{18}O_{22}S + 77\ O_2 \rightarrow 63\ CO_2 + 38\ H_2O + SO_3 + 9\ CO(NH_2)_2$$

the RQ yields

$$RQ = 63\,(CO_2)/77\,(O_2) = 0.82 \tag{15.7}$$

15.1.2.2 Respiratory Exchange

The zebra mussel can adapt quite easily to rapidly changing biotic and abiotic conditions and has certain qualities such as hardiness under artificial conditions, ease of measurement and manipulation, and marked regularity of physiological rhythm, which render it suitable for experimental work. Therefore, it is one of the most studied organisms in literature.

Results of detailed studies on the respiratory exchange of mussels were published already in 1926 [31]. By the use of large numbers of mussels in each experiment and by statistical treatment of the results, individual variations have been reduced to a minimum. In order to ensure a reasonable uniformity of size and condition in the experimental material, a special stock of about 500 mussels was selected from a hundredweight taken from the beds at Ramsey, Isle of Man. The respiratory chamber consisted of a glass vessel of about 6.5 l capacity. For convenience of handling, the mussels were not placed directly into the glass vessel, but into a special cage, made of perforated sheet celluloid, which closely fits the respiratory chamber. The determinations were made at intervals approximately monthly throughout a complete reproductive cycle of one year, under conditions of temperature and oxygen pressure similar to those obtaining at the time.

For measuring the oxygen, Winkler's original method [32] was applied with very slight modification. Unfortunately, at that time, the determination of CO_2 was much less satisfactory than that of oxygen. For this purpose, a method depending upon the determination of pH by indicators was adopted with limited absolute accuracy, but that was capable of giving suitable results under standard conditions, especially in serial experiments. The indicator employed was m-cresol purple in 0.04 % solution in 20 % alcohol, half-neutralized with NaOH. The carbon dioxide results have been expressed in cubic centimetres of the gas evolved by a given weight of tissue per hour.

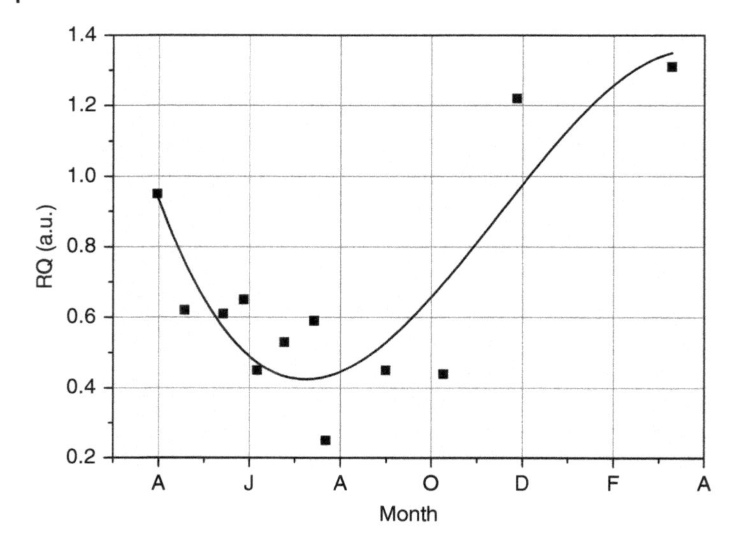

Figure 15.3 Variations in the respiratory quotient of the mussel between April 1925 and March 1926. Source: Bruce 1926 [31]. Reproduced with permission of Portland Press.

Figure 15.3 shows the calculated RQ values, supplemented by an additionally inserted roughly approximated annual metabolic curve.

In [31] it was stated that the method of CO_2 determination was admittedly not of a high order of accuracy, but the results obtained, and the corresponding RQs, were relatively valid, though the numerical scale must be regarded as arbitrary. The trend of the values is unquestionably to be associated with the changing chemical composition of the tissues. The rising quotient, from July to March, coincides with the period when carbohydrate (glycogen) reserves are being laid down, preparatory to the rapid metabolic changes leading to fat formation and a correspondingly low RQ, from March onwards to the actual spawning period in May. The wide disparity between the spring and autumn values of RQ points to a profound difference of metabolism at these periods. The seasonal changes in absolute oxygen requirement and RQ are intimately associated with concurrent changes in the chemical composition of the tissues.

In a similar study performed in 2002 at the Technische Universität Dresden in cooperation with the KSI Meinsberg, a miniaturized electrochemical Severinghaus-type carbon dioxide sensor with a diameter of 10.5 mm, developed at KSI Meinsberg, was used. The oxygen content was recorded by means of a WTW 196 probe. Furthermore, the pH value was determined during the measurements. According to

$$CO_2(aq) + H_2O \rightarrow H_2CO_3 \rightarrow H^+ + HCO_3^- \qquad (15.8)$$

the carbon dioxide excretion of the animals results in an increase of H^+ ions and thus lowers the pH value of the sample. As described in Section 3.3, the relationship between the changes in the CO_2 concentration and the pH value can be determined, taking into account other parameters such as temperature, conductivity, and total carbonate concentration in the water. In the test experiments, the agreement between the computed and the measured CO_2 changes was satisfactory, which was a confirmation for the correct function of the CO_2 sensor.

Oxygen consumption and carbon dioxide excretion of the zebra mussels were tested in Winkler bottles (c. 120 ml) as used for the determination of dissolved oxygen in water [33, 34]. Figure 15.4 shows schematically the measurement set-up for the *in vitro* determination of RQs of zebra mussels. Due to the fact that the measurements of oxygen and carbon dioxide concentrations were carried out in the same vessel on the same animals, it was possible to compare the original data of oxygen consumption and carbon dioxide excretion.

The test species originated from various sewage polishing ponds, which were supposed to be rich in organic suspended solids and, therefore, should be good habitats for *Dreissena polymorpha*, and from a harbour of the river Elbe (Neustädter Hafen) near Dresden, Germany, where they had been exposed for periods of at least one month prior to the experiments. The animals were transported in cooling boxes to the laboratory where they were kept in standard water [35] at 20 °C for a 48 hour adaptation period before the RQ has been determined. Table 15.1 shows results of RQ measurements at different seasons on diverse sampling sites.

In accordance with [31] there were significant differences between RQ values measured in summer and in winter, indicating that the mussels preferably

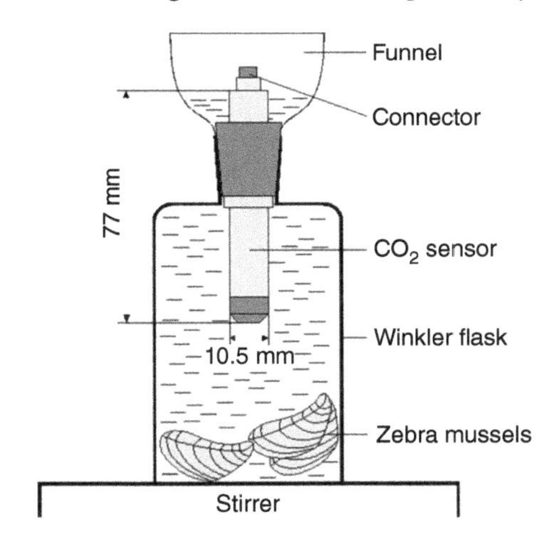

Figure 15.4 Schematic drawing of the measurement set-up for mussels with inserted carbon dioxide sensor from KSI Meinsberg.

Table 15.1 Results of RQ measurements in summer and winter on different sampling sites.

Month		**June**	**November**
Sampling site	River Elbe	0.96	0.75
	Sewage pond 1	0.75	0.63
	Sewage pond 2	0.75	0.71

Source: From Kusserow 2004. Determination of the respiratory quotient in aquatic animals using electrochemical sensors. Short note in the context of [33], not published.

metabolize carbohydrates in summer and proteins in winter. Furthermore, the RQ values of the mussels from the river Elbe are higher than those of the mussels from the two sewage ponds, which possibly may be attributed to better food conditions in the harbour of the river Elbe. The RQ values depend also on the size of the mussels. Statistical analysis of the determined RQ values showed a rather wide range of variation. Surprisingly sometimes even not only RQ values above 1.00 but also values as low as 0.32 were measured. The RQ value can rise above 1.00 if the denominator of the fraction in Eq. (15.1) is below the estimated value due to an intermediate transformation of carbohydrates, which are comparably rich in oxygen, into fats. Vice versa, the RQ value will decrease, if fats and proteins are transformed into carbohydrates or if carbon dioxide is stored in tissue or in the shell of the mussels [17]. The binding of CO_2 as $CaCO_3$ in the standard water during the experiments may have caused also a lower RQ in this test group. Turbidity, temperature, and the chemical composition of the water can influence the oxygen consumption in the zebra mussel, too. [36, 37].

These examples show that care must be taken in assessing the determined RQ values. Nevertheless, in many cases, the knowledge of the RQ value allows a rough estimation of the contribution of the different nutrients to the total metabolism of zebra mussels [38, 39].

15.2 Insects

Insects, like most other animals, inhale oxygen to live, but they have a unique respiratory system. CO_2 measurements on insects have been performed with quite different methods, instrumentation, and goals. As two characteristic examples, results of scientifically fundamental studies on butterfly pupae and more phenomenological investigations on honeybees are presented in this section.

Many insects are capable to detect elevated levels of carbon dioxide [40, 41] and react reasonably on this sensory perception. Some use elevated CO_2 concentrations or CO_2 gradients to locate their vertebrate hosts [41–43] or to evaluate floral quality [44, 45]. Others, like honeybees and ants, regulate potentially lethal CO_2 concentrations in their social colonies [46, 47]. Recently, it has been found that some insects possess and utilize special carbon dioxide chemoreceptors [48–50], but honeybees do not so. It is supposed that they must have evolved other senses to detect carbon dioxide.

15.2.1 CO_2 Measurements on Butterfly Pupae

Butterfly pupae breathe through small openings in the cuticle, called spiracles. These are connected to the inner organs by a system of highly branched tubes, called tracheae. Oxygen enters through the spiracles and diffuses into the blood. The cells release carbon dioxide, which is carried back to the spiracles. Insect spiracles can open and close. They behave like valves; opening and closing allow or restrict the insect's gas exchange [51].

A flow-through measurement set-up was constructed for simultaneous and continuous measurement of oxygen partial pressure, carbon dioxide output, and

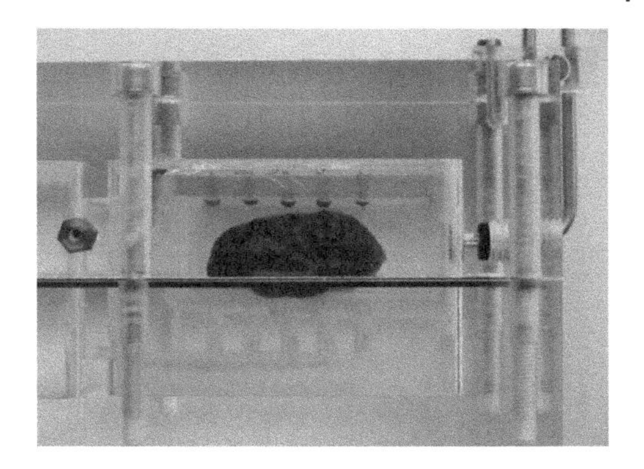

Figure 15.5 Measuring chamber with Atlas moth pupa, length 4.5 cm. Source: Photograph by KSI Meinsberg.

intratracheal hydrostatic pressure on lepidopterous pupae [51, 52]. Figure 15.5 shows the transparent measuring chamber with an Atlas moth (*Attacus atlas*) pupa in it. The chamber was permanently rinsed with CO_2-free moisturized air at a constant flow rate. For intratracheal measurement of oxygen and carbon dioxide, miniaturized electrochemical sensors from the Meinsberg Kurt-Schwabe Research Institute, Germany (KSI Meinsberg), were used. Eight pupae of the giant silk moth *Attacus atlas* (Lepidoptera, Saturniidae) were used in the experiments. The insects were raised from eggs and reared in the laboratory. Two spiracles of each pupa were intubated for the experiments and connected to the sensors via a short piece of polyethylene tubing. For oxygen measurement, a miniaturized electrochemical sensor according to the Clark principle with an outer diameter of 1.2 mm and very low oxygen consumption was inserted into the spiracle of the butterfly pupae. The amount of carbon dioxide released by the pupae was measured using a flow-through IR CO_2 gas analyser URAS 3G (Hartmann & Braun, Frankfurt, Germany). High flow rate allowed good time resolution of the carbon dioxide output. Compared with the previous studies cited below, this method of simultaneous data acquisition of several respiratory parameters over a long period provided results with improved accuracy and time resolution from individual pupae.

Figure 15.6 shows typical results of *in vivo* measured respiratory cycles of an Atlas moth pupa. During the experiments six of the eight pupae showed discontinuous carbon dioxide discharge with clear constriction, fluttering, and opening phases. Two pupae showed continuous carbon dioxide release. This unusual respiratory behaviour, which has been observed in many adult insects as well as in resting butterfly and moth pupae, is referred to as discontinuous gas exchange cycle (DGC). In insects exhibiting DGC, the spiracles close for long periods up to several hours or even days and open occasionally for only a few minutes. Initiated by a critically high amount of CO_2 in blood, a burst of CO_2 release was observed. During the longer closed phase, the CO_2 release is very low. Initiated by critically low levels of oxygen, the closed phase is followed by a flutter phase during which CO_2 is released in short intervals. A very similar picture is also shown in [51]

Already long before, Schneiderman, Levy, and Sláma have done extensive pioneering work on discontinuous respiration cycles in insects [53–57]. Sláma has

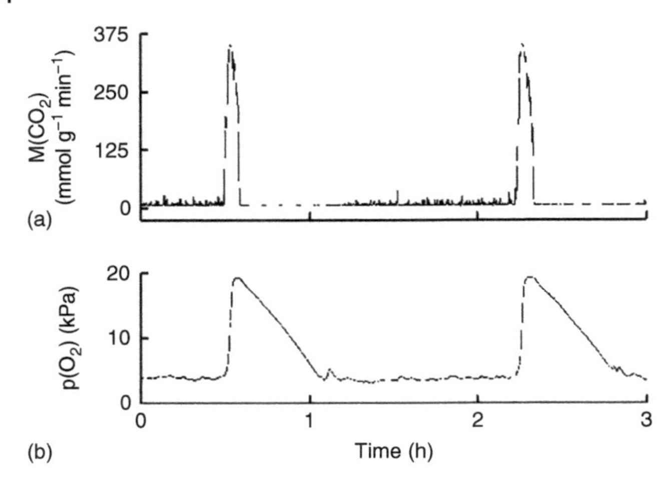

Figure 15.6 *In vivo* measured respiratory cycles of an Atlas moth pupa, (a) carbon dioxide output and (b) intratracheal oxygen partial pressure. Source: Hetz et al. 1993 [51]. Reproduced with permission of IOP Publishing.

summarized this work and acknowledged the merits of Schneiderman in this field in [58]. In their experiments, the authors repeatedly removed small volumes of gas from the tracheal system. The composition was analysed using a micro-respirographic method that revealed both the O_2 consumption and the CO_2 output are in the sub-nanolitres per minute range [59]. The released CO_2 was absorbed in solution of KOH present within the respiratory compartment. Despite this method of taking probes being punctual and non-simultaneous, the data presented are in excellent agreement with the results of the more recent experiments described above.

For a long time scientists had suggested that this opening and closing of the spiracles helps to prevent water loss. But recently, based on experimental facts, it is assumed that insects do it to protect their tissues from getting too much oxygen [60]. The cyclical pattern of open and closed spiracles observed in resting insects is supposed to be a necessary consequence of the need to rid the respiratory system of accumulated CO_2, followed by the need to reduce oxygen toxicity.

Since it is very difficult to investigate flying insects, most respiration measurements have been performed on resting insects and developing instars. A highly sophisticated method and experimental set-up for CO_2 measurements on insects also during flight activities has been published by Wasserthal [61]. For measuring tracheal pressure and CO_2 emission from specified spiracles of hawkmoths, he used a special split-specimen chamber with a controlled constant speed airflow and adjustable pressure and temperature. In the chamber the healthy moths were suspended at the descaled mesoscutellum. The tubes leading from the anterior spiracles were connected to a small anterior compartment of the split-specimen chamber, whereas the posterior thoracic and abdominal spiracles opened into a larger posterior specimen compartment. CO_2 emission from the anterior spiracles and the posterior thoracic and abdominal spiracles was measured separately with the split-specimen chamber. For CO_2 measurement, the airflow from either the anterior or the posterior compartment was conveyed directly to an infrared gas analyser, while the air from the other compartment was conveyed to a CO_2-absorbing vessel containing NaOH. Flight was initiated some hours after the moths had been mounted in the chamber. All moths were used at first for tracheal pressure measurements and then for CO_2 emission measurements. Furthermore, the wingbeat was recorded by projecting the shadow of the moving wings onto photocells installed on the bottom of the Perspex specimen chamber. As an interesting result of the studies, it was stated that during steady flight, air is inspired through the anterior spiracles and expired through the posterior thoracic spiracles. CO_2 is emitted only at the posterior spiracles. No CO_2 emission was recorded from the anterior spiracles, which opened into the anterior chamber.

15.2.2 CO_2 Measurements on Honeybees

As already mentioned above, insects utilize different carbon dioxide chemoreceptors or other senses to detect carbon dioxide. Studies on the gene lineage encoding the carbon dioxide receptor of insects showed that the three-gene lineage of other insects is entirely absent from the genome of the honeybees that have receptor neurons in their antennae that can detect carbon dioxide [50]. The antennal carbon dioxide receptors are stimulated by high carbon dioxide concentrations (but not low oxygen concentrations) to ventilate the nest cavity [62].

CO_2 measurements on honeybees showed that the carbon dioxide levels in a beehive can rise to excessive levels. Figure 15.7 presents results of CO_2 measurements in two beehive chambers, which had been carried out by KSI Meinsberg in cooperation with Freie Universität Berlin, Faculty of Biology, Department of Animal Physiology [52]. For the measurements an electrochemical carbon dioxide sensor (EMCO2, KSI Meinsberg) as can be seen in Figure 15.8 was used. The

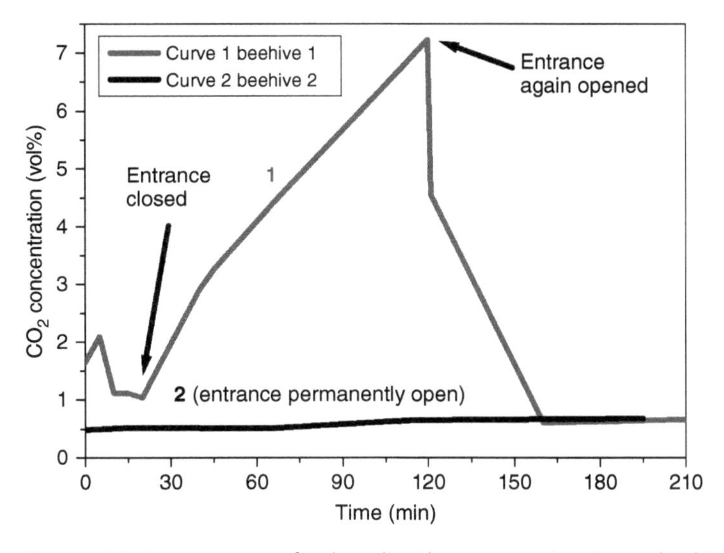

Figure 15.7 Measurement of carbon dioxide concentrations in two beehives. Source: Zosel et al. 2011 [52]. Reproduced with permission of IOP Publishing.

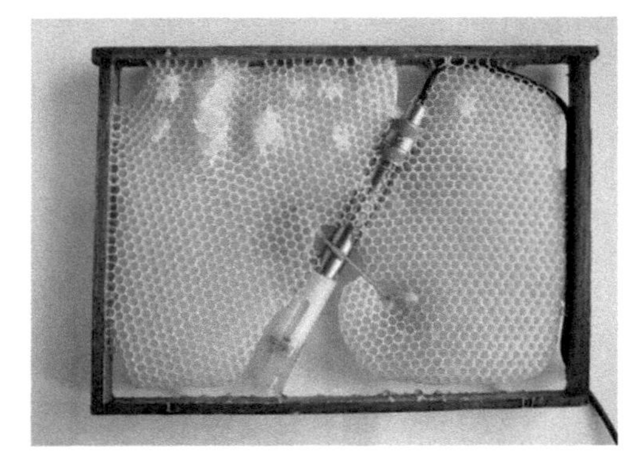

Figure 15.8 Electrochemical carbon dioxide sensor (EMCO2, KSI Meinsberg) in a honeycomb. Source: Zosel et al. 2011 [52]. Reproduced with permission of IOP Publishing.

measured carbon dioxide concentration in the open beehive was permanently 0.5 vol% or even higher. This is just the threshold limit value/time-weighted average (TLV-TWA) value, the maximum allowable concentration for human beings. After the entrance of the beehive had been closed, the CO_2 concentration in the beehive rose within 90 minutes to more than 7 vol%. While such a high CO_2 concentration would have been lethal for human beings within a short time, the bees got over this concentration without recognizable injury.

Southwick and Moritz used a similar measuring chamber for respiratory measurements to investigate the social control of air ventilation of small colonies of honeybees. Oxygen was measured using an applied electrochemistry oxygen analyser, whereas carbon dioxide was measured in a Beckman infrared analyzer (Model 864) [62].

The very high CO_2 concentration and heat generated by honeybees in the hive is also a very effective weapon with which bees can defend against hornets and kill them. In related studies it was found that giant hornets (*Vespa mandarinia japonica*) are killed in less than 10 minutes when they are trapped in a bee ball created by the Japanese honeybees *Apis cerana japonica*. The effect is that the lethal temperature of the hornet (45–46 °C) is somewhat lower than that of the honeybee (50–51 °C) under the condition of high CO_2 concentration (3.7–0.44%) in the experiments produced using human expiratory air. A portable gas detector (COSMOS XP-3140) was used to measure the CO_2 concentration in open-nest bee balls. It was concluded that CO_2 produced inside the bee ball by honeybees is a major factor together with the temperature involved in defence against giant hornets [63].

Bees determine whether ventilation in a hive is adequate from CO_2 concentration and regulate potentially lethal CO_2 concentrations in their colonies [47]. If concentration in a beehive rises above 0.5%, young worker bees orient in a single direction and beat their wings to ventilate (fan) the hive [64]. Nevertheless, it is assumed that high carbon dioxide concentrations in the hive could be as detrimental for the bees as high temperature or humidity and could cause reductions in honey production and pollination. To reduce these stress factors, it is recommended to install a beehive ventilator at the hive entrance to exhaust the stale air [65, 66]. Furthermore, the exhausted air should help bees to orient to their domestic hive. For this reason, exhausted warm air is preferably directed towards

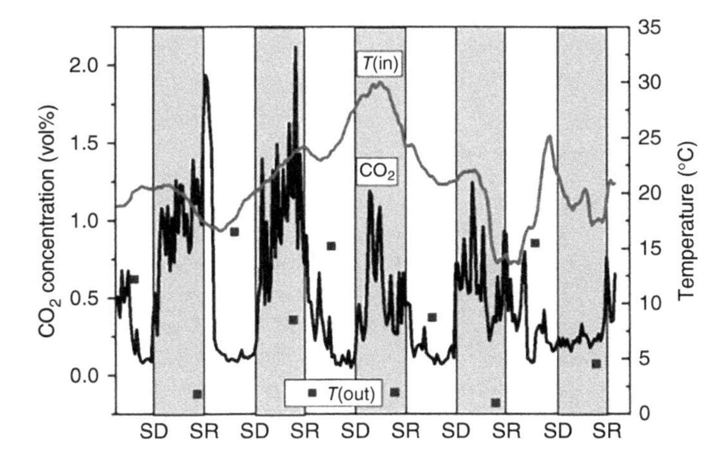

Figure 15.9 CO_2 concentration in a beehive over a period of five days. SR, sunrise; SD, sundown; *T*(in), temperature inside the hive; *T*(out), outside temperature.

the direction of the hive entrance. Pheromones of the particular beehive are carried on the exhausted air towards the front exterior of the hive, where they may be detected by the bees of that beehive.

For more detailed measurements of carbon dioxide concentration in beehives, an EMCO2 (KSI Meinsberg) was embedded in a customary honeycomb according to Figure 15.8 [52]. Special precautions had to be taken to prevent bees from wrapping up the sensor in honey. Apart from CO_2, the temperature inside and outside the hive, the humidity, and the time of sunrise and sundown were recorded, and a bee counter was installed at the entrance of the hive.

Figure 15.9 shows typical results of CO_2 concentration measurements in the beehive over a period of five days. From the diagram no direct relationship between CO_2 concentration and temperature may be derived. Obviously, the bees attend the hive exactly in accordance with sunrise and sundown.

In the literature it has been reported extensively on CO_2 measurements relating to bees. For example, temporal changes in temperature, humidity, and CO_2 concentration have been measured simultaneously within and out of a honeybee hive using an NDIR radiation analyzer (GMT220, GMD20, Vaisala, Finland) for CO_2 measurement. The hive CO_2 concentration fluctuated corresponding to the hive temperature even when atmospheric CO_2 concentration was stable [67].

CO_2 is a commonly used anaesthetic in instrumental insemination of honeybees. In numerous studies the effect of different concentrations of carbon dioxide upon the survival of the honeybee and changes in the behaviour of honeybees following their recovery from anaesthesia has been examined [68, 69].

15.3 Plants

Photosynthesis and plant respiration are of utmost importance for environment and living organisms and play a central role for living systems. Using solar energy, plants transform CO_2 absorbed from the atmosphere via photosynthesis into chemical energy stored in organic carbon compounds like glucose. Furthermore, the photosynthetic process balances the consumption of oxygen during respiration by continually supplying the atmosphere with O_2. According to the equation

$$6\,CO_2 + 6\,H_2O + \text{photons} \rightarrow C_6H_{12}O_6 + 6\,O_2\uparrow \tag{15.9}$$

during the process of photosynthesis, leaves convert carbon dioxide and water with the help of light energy to glucose and free oxygen.

The study of photosynthesis can be carried out almost anywhere there are plants. In a simple demonstration experiment as shown in Figure 15.10, the fundamental significance of the light for this process has been illustrated [52]. A miniaturized electrochemical CO_2 sensor from the Meinsberg Kurt-Schwabe Research Institute (KSI Meinsberg) with a diameter of only 10 mm was used for carbon dioxide measurement on indoor plants. The room fern was arranged near a window in a transparent thermostated vessel that provided for constant temperature during the measurements.

Figure 15.11 shows a typical result of such carbon dioxide measurements on indoor plants. As expected, due to the photosynthetic assimilation of CO_2 under the influence of day and room light, it was observed that the brighter the daylight was, the lower was the carbon dioxide concentration in the test vessel.

To permit CO_2 measurements behind stomatal pores of plant leaves, a miniaturized potentiometric CO_2 biosensor was built with a tip diameter of 2 μm [70]. It consists of a H^+ carrier-based pH microelectrode concentrically arranged within a sheathing micropipette, the tip of which is filled with

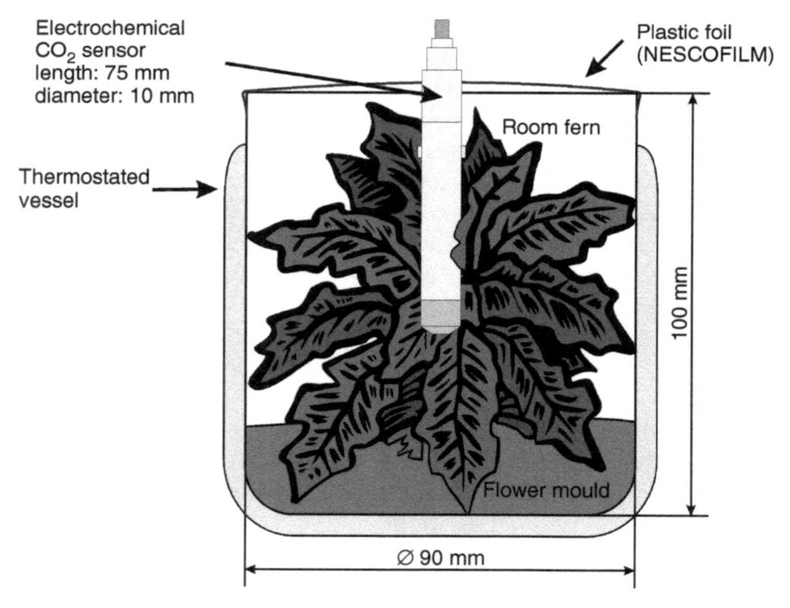

Figure 15.10 Carbon dioxide measurement on plants using a miniaturized electrochemical CO_2 sensor (KSI Meinsberg). Source: Zosel et al. 2011 [52]. Reproduced with permission of IOP Publishing.

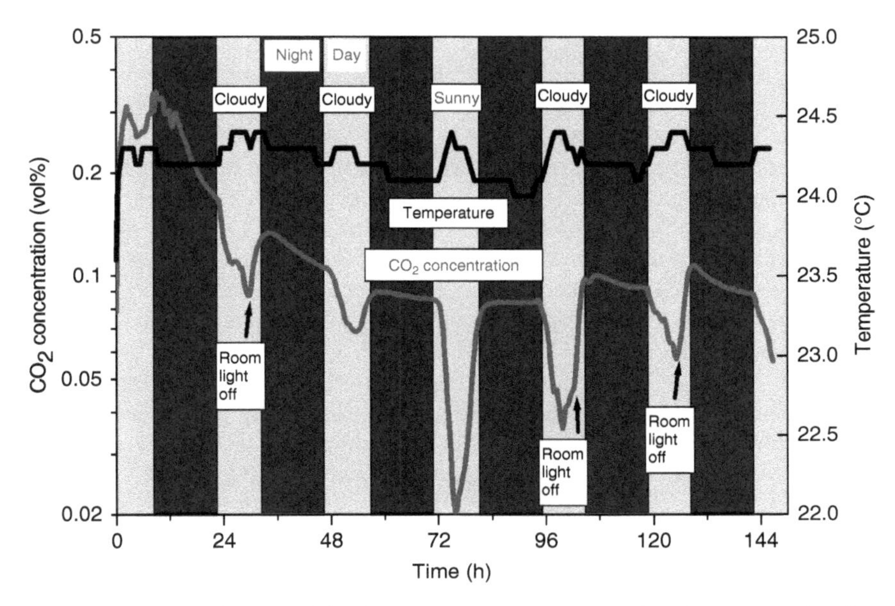

Figure 15.11 Typical course of the CO_2 concentration on a room fern. Source: Zosel et al. 2011 [52]. Reproduced with permission of IOP Publishing.

carbonic anhydrase solution. Due to incorporated carbonic anhydrase, the pH of the solution responds quickly to CO_2 concentration changes. The CO_2 microsensor shows a logarithmic response to CO_2 in the physiologically relevant concentration range of 50–800 ppm. Its 90% response time varied between 18 and 63 seconds. Sensor calibration and leaf experiments were performed in an open-flow tube-like minicuvette, allowing tangential airflow along the leaf surface with controlled gas mixtures and flow rates of choice. At an external CO_2 concentration of 800 ppm, CO_2 concentration within the leaf was close to this value when stomatal pores were wide open. During stomatal closure the concentration dropped to 350 ppm due to CO_2 consumption by photosynthesis, thus demonstrating distinct sensing of internal leaf CO_2. Following "light off", internal CO_2 rapidly rose close to 700 ppm, which was completely reversed by "light on". It was concluded that this sensor is a suitable tool for CO_2 monitoring in places too small to be accessible by conventional tools.

While aquatic plants like algae absorb CO_2 dissolved in the water, the terrestrial green plants take up CO_2 from the air through tiny openings on their leaves. Through these stomata, which are able to rapidly change their aperture, carbon dioxide penetrates leaves, and – in the presence of water – it is dissolved and enters into cells. Once inside the leaf, CO_2 has to diffuse from the intercellular air spaces to the sites of carboxylation, where photosynthesis occurs. The CO_2 gradient within the leaf and the internal resistance to carbon dioxide diffusion affect the efficiency of the leaf, which depends on the partial pressure of CO_2 at the sites of carboxylation. For this reason, the internal diffusion paths of CO_2 in the leaf are not only of scientific but also of practical interest and the topic of numerous publications [71].

Apparently, it is not possible to measure directly the gradient of CO_2 partial pressure to the sites of carboxylation in leaves with the electrochemical CO_2 sensors and methods described in Chapter 4. In [72] two other possibilities are discussed and applied. The first method is based on carbon isotope discrimination and measures the change in carbon isotopic composition of CO_2 passing over the leaf, while the second one is a non-intrusive optical method and measures fluorescence during photosynthesis. In [73] the effect of CO_2 upon photosynthetic electron transport was examined. It was concluded that there is a complex interaction between CO_2 and O_2 in the regulation of this process.

CO_2 measurement provides an alternative with important advantages over measurements of dry weight change by which growth of crops is usually determined. The principles, equipment, and procedures for measuring carbon dioxide assimilation and the gas exchange in the leaves by plants in the field and the laboratory have been described in numerous publications, e.g. [74–76]. The measurement of the maximum quantum yield of CO_2 uptake and the construction and use of field systems for measuring photosynthesis were also described. The majority of CO_2 exchange studies have involved enclosure methods. The rate of CO_2 assimilation by the material enclosed is determined by measuring the change in the CO_2 concentration of the air flowing across the chamber. Alternatively, CO_2 exchange of large areas of vegetation may be measured without enclosure, using micrometeorological techniques.

In portable photosynthesis measurement systems for CO_2 measurement, infrared gas analysis, as described in Chapter 7, is the means of choice. For this purpose, a series of commercial, portable, open gas exchange systems with options for controlling CO_2, humidity, temperature, and light are on the market. As an example, the sensor head of the LI-6400XT (Li-Cor Inc., Lincoln, Nebraska, USA) comprises two absolute CO_2 and two absolute H_2O NDIR analyzers. CO_2 and H_2O concentrations are continuously controlled at the leaf surface, and due to the fast response changes in the leaf, dynamics are measured in real time [77].

Cavity-enhanced Raman multi-gas spectrometry (CERS) is a versatile technique for monitoring of $^{13}CO_2$ isotope-labelling experiments [78]. It allows the online determination of very small gas exchange fluxes on plants during respiration. In [79] it was applied for the *in situ* determination of the RQ (CO_2 released per O_2 consumed during respiration; see Eq. (15.1)) in drought-tolerant pine (*Pinus sylvestris*) and drought-intolerant spruce (*Picea abies*). RQ is an excellent indicator of the nature of the respiration substrate. Two different treatments, drought and shading, were applied to reduce photosynthesis and force dependency on stored substrates. Changes in respiration rates and RQ values were continuously monitored over periods of several days. First results suggest a physiological explanation for greater drought tolerance in pine. Cavity-enhanced Raman gas spectrometry was also applied for online multi-gas analysis of the gas exchange rates of O_2 and CO_2 and the labelling of *Fagus sylvatica L.* (European beech) seedlings with $^{13}CO_2$ [80]. The rapid monitoring of all these gases simultaneously allowed for the separation of photosynthetic uptake of CO_2 by the beech seedlings and a constant $^{12}CO_2$ efflux via respiration and thus for a correction of the measured $^{12}CO_2$ concentrations in course of the labelling experiment. The effects of aphid infestation with the woolly beech aphid (*Phyllaphis fagi L.*) as well as the effect of a drought period on the respirational gas exchange were investigated. A slightly decreased respirational activity of drought-stressed seedlings in comparison with normally watered seedlings was found already for a low drought intensity. Cavity-enhanced Raman gas monitoring of O_2, $^{12}CO_2$, and $^{13}CO_2$ was proven to be a powerful new tool for studying the effect of drought stress and aphid infestation on the respirational activity of European beech seedlings as an example of important forest species in the Central Europe.

The measurement of CO_2 as a product of metabolism of plants and animals is also possible with high temperature sensors as described in Chapter 5, because these sensors are very small and the transfer of heat into the analysing system is very low (c. 3 W).

It should be mentioned that a plant leaf itself can be used as CO_2 gas sensing probe [81]. In a corresponding experiment a leaf of spiderwort (*Commelina communis*) was attached on a glass slide under a microscope. The stem of the leaf was dipped in a bathing solution. Then the intracellular potential (ICP) was measured with a glass capillary electrode referred to another electrode dipped in the bathing solution. When CO_2 gas was supplied to the leaf under illumination, a considerable shift of the ICP was observed. This process was reversible; when the CO_2 supply was stopped, the ICP returned to the initial level immediately.

References

1 Steen, J.E. (1997). Why CO_2 must be controlled. *Fish Farm. Int.* 24: 16.
2 Bauer, K. (1981). Zur Bedeutung der freien Kohlensäure in Forellenzuchtbetrieben (To the relevance of free carbonic acid for the fish farming of trout). *Zeitschrift für Binnenfischerei* 31: 1–5 (in German).
3 Wurts, W.A. and Durborow, R.M. (1992). *Interactions of pH, Carbon Dioxide, Alkalinity and Hardness in Fish Ponds*, 4 p. Southern Regional Aquaculture Center (SRAC Publication No. 464).
4 Hargreaves, J. and Brunson, M. (1996). *Carbon Dioxide in Fish Ponds*, 6p. Southern Regional Aquaculture Center (SRAC Publication No. 468).
5 Oelßner, W., Kaden, H., Sauer, R., and Tautenhahn, A. (2002). Effects of CO_2 sensitivity on trout. *Fish Farmer* 25: 12–13.
6 Heisler, N. (1986). *Acid-Base Regulation in Animals*, 328. Amsterdam: Elsevier.
7 Wilson, W.K. (1995). *Advances in Comparative and Environmental Physiology*, 21e, 70. Berlin: Springer-Verlag.
8 Crocker, C.E. and Cech, J.J. Jr., (1996). The effects of hypercapnia on growth of juvenile white sturgeon, *Acipenser transmontanus*. *Aquaculture* 47 (3–4): 293–299.
9 Schmid, J., Oelßner, W., and Schukraft, G. (2002). Kohlendioxid-Messungen in einem Durchfluß-Meßsystem mit einem neu entwickelten CO_2-Sensor am Beispiel eines Hartwassersees (Willersinnweiher, Ludwigshafen am Rhein) (Carbon dioxide measurements in a flow-through measuring system with

a newly developed CO_2 sensor in lake water (Lake Willersinnweiher Ludwigshafen/Rhine, Germany)). *Limnologica* 32 (4): 338–349 (in German).

10 Deutsches Institut für Normung DIN 38409 D8 (1971). Deutsche Einheitsverfahren zur Wasser-, Abwasser- und Schlammuntersuchung. Die Berechnung des gelösten Kohlendioxids (der freien Kohlensäure), des Carbonat- und Hydrogencarbonat-Ions (German standards for water, waste water and sludge investigations. The calculation of dissolved CO_2 [free carbonic acid], of carbonate and hydrocarbonate), 1–11. Berlin: Beuth-Verlag (in German).

11 Deutsches Institut für Normung Determination of the acid-base capacity) 38409 H7 (1979). *Deutsche Einheitsverfahren zur Wasser- Abwasser- und Schlammuntersuchung. Bestimmung der Säure- und Basekapazität* (German standards for Water, Waste water and sludge investigations, 1–9. Berlin: Beuth-Verlag (in German).

12 Rebsdorf, A. (1972). *The Carbon Dioxide System in Freshwater*. A set of tables for easy computation of total carbon dioxide and other components of the carbon dioxide system, 66 pp. Hillerod, DK: Freshwater Biological Laboratory.

13 Cole, J.J., Caraco, N.F., Kling, G.W., and Kratz, T.K. (1994). Carbon dioxide supersaturation in the surface waters of lakes. *Science* 265 (5178): 1568–1570.

14 Broom, D.M. (1988). The scientific assessment of animal welfare. *Appl. Anim. Behav. Sci.* 20 (1–2): 5–19.

15 Rose, J.D. (2002). The neurobehavioral nature of fishes and the question of awareness and pain. *Rev. Fish. Sci.* 10 (1): 1–38.

16 Puri, S. and Faulkes, Z. (2010). Do decapod crustaceans have nociceptors for extreme pH? *PLoS ONE* 5 (4): e10244. https://doi.org/10.1371/journal.pone.0010244.

17 Penzlin, H. (1991). *Lehrbuch der Tierphysiologie* (Textbook of Animal Physiology), 5e, 265. Jena: G. Fischer Verlag (in German).

18 Roberts, R.J. (2001). *Fish Pathology*, 3e. London: Harcourt Publishers Ltd.

19 Danley, M.L., Kenney, P.B., Mazik, P.M. et al. (2005). Effects of carbon dioxide exposure on intensively cultured rainbow trout *Oncorhynchus mykiss*: physiological responses and fillet attributes. *J. World Aquacult. Soc.* 36 (3): 249–261.

20 Allan, B.J.M., Miller, G.M., McCormick, M.I. et al. (2014). Parental effects improve escape performance of juvenile reef fish in a high-CO_2 world. *Proc. R. Soc. B: Biol. Sci.* 281 (1777): 20132179. https://doi.org/10.1098/rspb.2013.2179.

21 Benson, A.J., Raikow, D., Larson, J., and Fusaro, A. (2014). *Dreissena Polymorpha*. Gainesville, FL: USGS Nonindigenous Aquatic Species Database. http://nas.er.usgs.gov/queries/factsheet.aspx?speciesid=5 (accessed 6 July 2012).

22 Bean, R.A. (1997). Process for treatment of wastewater utilizing zebra mussels (*Dreissena polymorpha*). US Patent 5,628,904 A, filed 18 December 1995 and issued 13 May 1997.

23 Mählmann, J., Kusserow, R., Arnold, R. et al. (2001). Biologisches filter. DE Patent 101 61 239 A1, filed 13 December 2001, issued 26 June 2003.

24 Fenske, C. (2003). Die Wandermuschel (*Dreissena polymorpha*) im Oderhaff und ihre Bedeutung für das Küstenzonenmanagement. (The zebra mussel

(*Dreissena polymorpha*) in the Szczecin lagoon and its importance for the coastal zone) Thesis. Germany: Ernst-Moritz-Arndt-Universität Greifswald (in German).

25 Haamer, J. (1996). Improving water quality in a eutrophied fjord system with mussel farming. *Ambio* 25 (5): 356–362.

26 Lindahl, O., Hart, R., Hernroth, B. et al. (2005). Improving marine water quality by mussel farming: a profitable solution for Swedish society. *Ambio* 34 (2): 131–138.

27 Schröter-Bobsin, U. (2005). Untersuchungen zur Einsatzmöglichkeit der Dreikantmuschel Dreissena polymorpha als biologischer Filter und Wasserhygienemonitor. (Investigations about the applicability of the zebra mussel Dreissena polymorpha as biological filter and monitoring system for water hygiene) Thesis. Germany: Technische Universität Dresden (in German).

28 Reeders, H.H. and Bij de Vaate, A. (1990). Zebra mussels (*Dreissena polymorpha*): a new perspective for water quality management. *Hydrobiologia* 200–201: 437–450.

29 Schernewski, G., Stybel, N., and Neumann, T. (2012). Zebra mussel farming in the Szczecin (Oder) Lagoon: water-quality objectives and cost-effectiveness. *Ecol. Soc.* 17 (2): 4, 13 pp. https://doi.org/10.5751/ES-04644-170204.

30 Stybel, N., Fenske, C., and Schernewski, G. (2009). Mussel cultivation to improve water quality in the Szczecin Lagoon. *J. Coastal Res.* SI56 (ICS 2009): 1459–1463.

31 Bruce, J.R. (1926). The respiratory exchange of the mussel (*Mytilus edulis*, L.). *Biochem J.* 20 (4): 829–846.

32 *DIN EN 25813 (*1993–01). Water quality; determination of dissolved oxygen; iodometric method (ISO 5813:1983).

33 Kusserow, R. (2004). Möglichkeiten und Grenzen des Einsatzes bodenbesiedelnder filtrierender Organismen in der Gewässer- und Abwasserreinigung am Beispiel der Dreikantmuschel *(Dreissena polymorpha).* (Chances and limits of the utilization of ground settling and filtering organisms in the water and wastewater treatment using the example of zebra mussels (*Dreissena* polymorpha)). PhD thesis. Germany: Technische Universität Dresden (in German).

34 Kusserow, R., Mählmann, J., Bobsin, U. et al. (2003). Einsatzmöglichkeiten der Dreikantmuschel (*Dreissena polymorpha*) als biologisches Filter und Wasserhygiene-Monitor. (Application possibilities of zebra mussels (*Dreissena polymorpha*) as biological filter and water hygiene monitor). Deutsche Gesellschaft für Limnologie (DGL) Tagungsbericht 2002 (Braunschweig), Tutzingen 2003; Bd. II, 688–692 (in German).

35 CEN European Committee for Standardization (1997). Water quality – Determination of the acute lethal toxicity of substances to a freshwater fish [*Brachydanio rerio* Hamilton-Bachanan (Teleostei, Cyprinidae)] – Part 1: Static method (ISO 7346-1:1996). Berlin: Beuth-Verlag.

36 Vinogradov, G.A., Smirnova, N.F., Sokolov, V.A., and Bruznitsky, A.A. (1993). Influence of chemical composition of the water on the mollusk *Dreissena polymorpha*. In: *Zebra Mussels - Biology, Impacts, and Control* (ed. T.F. Nalepa and D.W. Schloesser), 283–294. Boca Raton, FL: Lewis Publishers.

37 Alexander, J.E. Jr.,, Thorp, J.H., and Fell, R.D. (1994). Turbidity and temperature effects on oxygen consumption in the zebra mussel (*Dreissena polymorpha*). *Can. J. Fish. Aquat. Sci.* 51 (1): 179–184.

38 Sprung, M. (1995). Physiological energetics of the zebra mussel *Dreissena polymorpha* in lakes III. Metabolism and net growth efficiency. *Hydrobiologia* 304 (2): 147–158.

39 Wacker, A. and von Elert, E. (2003). Food quality controls reproduction of the zebra mussel (*Dreissena polymorpha*). *Oecologia* 135 (3): 332–338.

40 Stange, G. (1996). Sensory and behavioural responses of terrestrial invertebrates to biogenic carbon dioxide gradients. In: *Advances in Bioclimatology*, vol. 4 (ed. G. Stanhill), 223–253. Berlin: Springer-Verlag.

41 Stange, G. and Stowe, S. (1999). Carbon-dioxide sensing structures in terrestrial arthropods. *Microsc. Res. Tech.* 47 (6): 416–427.

42 Gillies, M.T. (1980). The role of carbon dioxide in host-finding by mosquitoes (Diptera: Culicidae): a review. *Bull. Entomol. Res.* 70 (4): 525–532.

43 Vale, G.A. and Hall, D.R. (1985). The role of 1-octen-3-ol, acetone and carbon dioxide in the attraction of tsetse flies, *Glossina spp.* (Diptera: Glossinidae), to ox odour. *Bull. Entomol. Res.* 75 (2): 209–218.

44 Stange, G., Monro, J., Stowe, S., and Osmond, C.B. (1995). The CO_2 sense of the moth *Cactoblastis cactorum* and its probable role in the biological control of the CAM plant *Opuntia stricta*. *Oecologia* 102 (3): 341–352.

45 Thom, C., Guerenstein, P.G., Mechaber, W.L., and Hildebrand, J.G. (2004). Floral CO_2 reveals flower profitability to moths. *J. Chem. Ecol.* 30 (6): 1285–1288.

46 Kleineidam, C. and Tautz, J. (1996). Perception of carbon dioxide and other "air-condition" parameters in the leaf cutting ant *Atta cephalotes*. *Naturwissenschaften* 83 (12): 566–568.

47 Seeley, T.D. (1974). Atmospheric carbon dioxide regulation in honey-bee (*Apis mellifera*) colonies. *J. Insect Physiol.* 20 (11): 2301–2305.

48 Kwon, J.Y., Dahanukar, A., Weiss, L.A., and Carlson, J.R. (2007). The molecular basis of CO_2 reception in *Drosophila*. *Proc. Natl. Acad. Sci. U.S.A.* 104 (9): 3574–3578.

49 Jones, W.D., Cayirlioglu, P., Grunwald Kadow, I., and Vosshall, L.B. (2007). Two chemosensory receptors together mediate carbon dioxide detection in *Drosophila*. *Nature* 445 (7123): 86–90.

50 Robertson, H.M. and Kent, L.B. (2009). Evolution of the gene lineage encoding the carbon dioxide receptor in insects. *J. Insect Sci.* 9 (1): 19, 14p. https://doi.org/10.1673/031.009.1901.

51 Hetz, S.K., Wasserthal, L.T., Herrmann, S. et al. (1993). Direct oxygen measurements in the tracheal system of lepidopterous pupae using miniaturized amperometric sensors. *Bioelectrochem. Bioenerg.* 33 (2): 165–170.

52 Zosel, J., Oelßner, W., Decker, M. et al. (2011). The measurement of dissolved and gaseous carbon dioxide concentration. *Meas. Sci. Technol.* 22: 072001, 45pp. https://doi.org/10.1088/0957-0233/22/7/072001.

53 Schneiderman, H.A. and Williams, C.M. (1955). An experimental analysis of the discontinuous respiration of the cecropia silkworm. *Biol. Bull.* 109 (1): 123–143.

54 Schneiderman, H.A. (1956). Spiracular control of discontinuous respiration in insects. *Nature* 177 (4521): 1169–1171.

55 Levy, R.I. and Schneiderman, H.A. (1958). An experimental solution to the paradox of discontinuous respiration in insects. *Nature* 182 (4634): 491–493.

56 Levy, R.I. and Schneiderman, H.A. (1966). Discontinuous respiration in insects II-IV. *J. Insect Physiol.* 12 (1): 83–121.

57 Levy, R.I. and Schneiderman, H.A. (1966). Discontinuous respiration in insects II-IV. *J. Insect Physiol.* 12 (4): 465–492.

58 Sláma, K. (2010). A new look at discontinuous respiration in pupae of *Hyalophora cecropia* (Lepidoptera: Saturniidae): haemocoelic pressure, extracardiac pulsations and O_2 consumption. *Eur. J. Entomol.* 107 (4): 487–507.

59 Sláma, K. (1984). Microrespirometry in small tissues and organs. In: *Measurement of Ion Transport and Metabolic Rate in Insects* (ed. T.J. Bradley and T.A. Miller), 101–129. New York, Berlin: Springer-Verlag.

60 Hetz, S.K. and Bradley, T.J. (2005). Insects breathe discontinuously to avoid oxygen toxicity. *Nature* 433 (7025): 516–519.

61 Wasserthal, L.T. (2001). Flight-motor-driven respiratory air flow in the hawkmoth *Manduca sexta*. *J. Exp. Biol.* 204 (13): 2209–2220.

62 Southwick, E.E. and Moritz, R.F.A. (1987). Social control of air ventilation in colonies of honey bees, *Apis Mellifera*. *J. Insect Physiol.* 33 (9): 623–626.

63 Sugahara, M. and Sakamoto, F. (2009). Heat and carbon dioxide generated by honeybees jointly act to kill hornets. *Naturwissenschaften* 96 (9): 1133–1136.

64 Bloom, A. (2012). Carbon Dioxide Sensing: Life's Sixth Sense for Carbon Levels. http://editors.eol.org/eoearth/wiki/Carbon_Dioxide_Sensing:_Life's_Sixth_Sense_for_Carbon_Levels (retrieved March, 28, 2017).

65 Jamison, M. and Sharp, C.C. (1984). Beehive ventilator. US Patent 4,512,050 A, filed 23 January 1984 and issued 23 April 1985.

66 Stearns, G.D. (1995). Solar-powered beehive cooler and ventilator. US Patent 5,575,703 A, filed 15 March 1995 and issued 19 November 1996.

67 Ohashi, M., Ikeno, H., Kimura, T. et al. (2008). Control of hive environment by honeybee (*Apis mellifera*) in Japan. In: *Proceedings of Measuring Behavior 2008, 6th International Conference on Methods and Techniques of Behavioral Research*, Maastricht, NL (26–29 August 2008), 243.

68 Niño, E.L., Tarpy, D.R., and Grozinger, C. (2011). Genome-wide analysis of brain transcriptional changes in honey bee (*Apis mellifera* L.) queens exposed to carbon dioxide and physical manipulation. *Insect Mol. Biol.* 20 (3): 387–398.

69 Madras-Majewska, B., Kamiński, Z., and Zajdel, B. (2011). The survival and the awaking time of the worker bees after carbon dioxide anesthesia and gas treatment with different oxygen and nitrogen concentration. *Ann. Warsaw Univ. Life Sci. - SGGW, Anim. Sci.* 49: 109–114.

70 Hanstein, S., de Beer, D., and Felle, H.H. (2001). Miniaturised carbon dioxide sensor designed for measurements within plant leaves. *Sens. Actuators, B* 81 (1): 107–114.

71 Parkhurst, D.F. (1994). Diffusion of CO_2, and other gases in leaves. *New Phytol.* 126 (3): 449–479.

72 Evans, J.R. and von Caemmerer, S. (1996). Carbon dioxide diffusion inside leaves. *Plant Physiol.* 110 (2): 339–346.

73 Ireland, C.R., Baker, N.R., and Long, S.P. (1987). Evidence for a physiological role of CO_2 in the regulation of photosynthetic electron transport in intact leaves. *Biochim. Biophys. Acta, Bioenerg.* 893 (3): 434–443.

74 Long, S.P. and Hällgren, J.-E. (1993). Measurement of CO_2 assimilation by plants in the field and the laboratory. In: *Photosynthesis and Productivity in a Changing Environment: A Field and Laboratory Manual* (ed. D.O. Hall, J.M.O. Scurlock, H.R. Bolhar-Nordenkampf, et al.), 129–167. London, New York: Chapman and Hall.

75 Long, S.P., Farage, P.K., and Garcia, R.L. (1996). Measurement of leaf and canopy photosynthetic CO_2 exchange in the field. *J. Exp. Bot.* 47 (11): 1629–1642.

76 Long, S.P. and Bernacci, C.J. (2003). Gas exchange measurements, what can they tell us about the underlying limitations to photosynthesis? Procedures and sources of error. *J. Exp. Bot.* 54 (392): 2393–2401.

77 LI-COR. LI-6400XT system.www.licor.com/6400XT (Rev. 8 03/16) (retrieved 14 February 2017).

78 Keiner, R., Frosch, T., Massad, T. et al. (2014). Enhanced Raman multigas sensing – a novel tool for control and analysis of $^{13}CO_2$ labeling experiments in environmental research. *Analyst* 139 (16): 3879–3884.

79 Hanf, S., Fischer, S., Hartmann, H. et al. (2015). Online investigation of respiratory quotients in *Pinus sylvestris* and *Picea abies* during drought and shading by means of cavity-enhanced Raman multi-gas spectrometry. *Analyst* 140 (13): 4473–4481.

80 Keiner, R., Gruselle, M.-C., Michalzik, B. et al. (2015). Raman spectroscopic investigation of $^{13}CO_2$ labeling and leaf dark respiration of *Fagus sylvatica L.* (European beech). *Anal. Bioanal. Chem.* 407 (7): 1813–1817.

81 Matsuoka, H., Homma, T., Takekawa, Y., and Ai, N. (1986). Use of plant leaf as CO_2 gas sensing probe. *Biosensors* 2 (4): 197–210.

16

CO_2 Sensing in Medicine

Gerald Urban[1], Josef Guttmann[2], Jochen Kieninger[1], Andreas Weltin[1], Jürgen Wöllenstein[3], and Jens Zosel[4]

[1] *University of Freiburg, IMTEK – Department of Microsystems Engineering, Georges-Köhler-Allee 103, 79110 Freiburg im Breisgau, Germany*
[2] *University of Freiburg, Department of Anesthesiology and Critical Care, Medical Center, Hugstetter Street 55, 79106 Freiburg, Germany*
[3] *Universität Freiburg IMTEK, Fraunhofer-Institut für Physikalische Messtechnik IPM, Heidenhofstraße 8, 79110 Freiburg im Breisgau, Germany*
[4] *Kurt-Schwabe-Institut für Mess- und Sensortechnik e.V. Meinsberg, Kurt-Schwabe-Straße 4, 04736 Waldheim, Germany*

16.1 Introduction

Besides the measurement of non-electrical physical parameters such as temperature or blood pressure, the analysis of chemical and biochemical parameters of patients has gained an increasing importance in the past. The exact knowledge of metabolic clinical parameters, especially small molecules in blood and tissue, is essential to assess the patient's health. Additionally, for emergency patients, the doctor needs a fast, secure, and reliable decision support, not only for diagnosis but also for rapid therapy access.

The quantitative and reliable assessment of biochemical parameters is a technical and logistical challenge and mainly based on discrete blood sampling and subsequent analysis. The next desired step would be a continuous monitoring of such metabolites, which is a much more difficult task and only possible for selected analytes.

For the metabolic parameter CO_2, a standard blood analysis as well as a monitoring method is available, capable to perform monitoring in blood and in the expired human breath. The latter is called capnometry and is routinely used during anaesthesia. The measurement of biochemical parameters is nowadays primarily done in centralized laboratories (laboratory medicine), in intensive care units (ICUs), and in the operating theatre with near-patient point-of-care testing (POCT). The measurement of CO_2 in blood as a part of the so-called blood gas panel, including the parameters pO_2, pH, and pCO_2, is a typical domain for POCT or bedside analytics. POCT is a patient-centred diagnostic test that is not performed in a centralized laboratory, but in the hospital immediately near the operating room or intensive care or in the ambulance. Such immediate near-patient analytics is necessary because imbalance of the so-called acute

Carbon Dioxide Sensing: Fundamentals, Principles, and Applications,
First Edition. Edited by Gerald Gerlach, Ulrich Guth, and Wolfram Oelßner.

parameters such as O_2, pH, and CO_2 is a life-threatening event that must be immediately balanced. Therefore, such applications are located in functional units, such as ICUs, outpatient clinics, operating rooms, delivery rooms and neonatal units, emergency departments, and ambulances [1, 2]. Consequently, the POCT analysis is a particularly interesting and fast-growing market [3]. During surgical procedures CO_2 monitoring devices are used in blood as well as for exhaled air in mechanically ventilated patients.

Besides the analysis of blood gases, oxygen saturation, electrolytes, glucose, lactate, creatinine, and coagulation parameters should be measured in realtime and are prominent additional parameters for POCT analysis. In a standard clinical chemical analysis setting in a central laboratory, the analysis process is divided into the pre-analysis, the actual analysis, and the post-analytical procedures using miniaturization of measuring systems with simultaneous use of digital control technology to achieve a high degree of automation.

The POCT in the hospital centres, ICUs, and operation rooms start their analysis immediately after sampling without any pre-analytics. The patient-centred implementation requires neither sample preparation nor pipetting. For such applications ready-to-use reagents are available in special measurement settings. The handling of POCT devices can be performed by unskilled personnel, because their operation is simple and requires minor training only. The analysis time is only a few minutes, so that the turnaround time (TAT), which indicates the time between sample receipt and result, is minimal. The post-analytical processes in POCT systems are linked in any case to the hospital network and data processing unit for proper quality control.

The units have an automatic calibration in general and are maintenance-free. In all clinical settings, quality control and data transfer via networks for advanced system management is mandatory [4]. The reliability of their results must always be verified according to general standards. POCT systems have a small sample volume and, therefore, also less pre-analytical error in unstable analyte in the sample. An important point in clinical analysis is to ensure the quality standards, which should be guaranteed by regulative measures. Internationally, the EN ISO 15189 applies, and in Germany additionally also the RiliBÄK (Richtlinie der Bundesärztekammer zur Qualitätssicherung) [5]. RiliBÄK are guidelines of the German Medical Council to assure the quality of medical laboratory tests. All devices, including POCT, must satisfy the *in vitro* diagnostic (IVD) guidelines for authorization and provide accurate and reliable measurement results by untrained personnel. Good cooperation of doctors, nurses, laboratory staff, and biomedical engineers is therefore crucial for practical success of POCT systems [6]. The EC Directive 98/79/EC on IVD medical devices is internationally called *in vitro* diagnostic medical devices directive (IVDD) [7].

16.2 Physiological Background of CO_2 Sensing

Carbon dioxide (CO_2) is a gaseous waste product of metabolism and physiologically very tightly controlled. The CO_2 balance is maintained by a carbonate buffer system in blood:

$$CO_2 + 2H_2O \rightleftarrows H_2CO_3 + H_2O \rightleftarrows HCO_3^- + H_3O^+ \quad (16.1)$$

CO_2 is dissolved in aqueous solutions (blood) as carbonic acid (H_2CO_3), which is in equilibrium with the hydrogen carbonate ion (HCO_3^-) depending on and controlling the pH.

In the body, the control of CO_2 is maintained by the pulmonary and the renal system. The blood carries carbon dioxide to the lungs, where it is exhaled. More than 90% of carbon dioxide in the blood exists in the form of hydrogen carbonate ion (HCO_3^-). The rest of the carbon dioxide is either dissolved carbon dioxide gas (CO_2) or carbonic acid (H_2CO_3). The kidneys and lungs balance the levels of carbon dioxide, hydrogen carbonate, and carbonic acid in the blood [8]. The hydrogen carbonate buffer system keeps the pH of blood extremely stable within the % range (Eq. (16.1)) [9].

The respiratory system consists of two parts, each of which can be independently impaired: (i) the lungs, which facilitate gas exchange, and (ii) the respiratory muscle pump, which drives ventilation.

Type I respiratory failure or pulmonary failure is associated with insufficient blood oxygenation, with pCO_2 either remaining normal or decreasing because of compensatory hyperventilation. In contrast, type II respiratory failure or ventilatory failure leads to alveolar hypoventilation, thus resulting in increased pCO_2 levels. In the case of acute ventilatory failure, the accumulation of CO_2 induces a decrease in pH levels by the formation of carbonic acid, which is called respiratory acidosis. In chronic ventilatory failure, hydrogen carbonate retention occurs to attempt normalization of blood pH [10–12]. Metabolic acidosis is a condition that occurs when the body produces excessive quantities of acid or when the kidneys do not remove enough acid from the body. If untreated, metabolic acidosis leads to a blood pH less than 7.35 due to increased production of hydrogen ions by the body or the inability of the body to form hydrogen carbonate (HCO_3^-) in the kidney. Together with respiratory acidosis, it is one of the general causes of severe acidosis, which is a life-threatening event including coma and death [13]. Therefore, the rapid and near-patient measurement and control of CO_2 is of utmost importance.

16.3 Measuring Principles

The main measurement principles of CO_2 in blood are the electrochemical principle according to the Severinghaus method, which is presented in more detail in Section 4.1, and the optical methods like fluorescence quenching or absorbance, which are presented in more detail in Section 6.3. For capnography, the reliable method of dispersive IR spectroscopy is applied, which is presented in more detail in Chapter 7.

16.3.1 Electrochemical Principle: Severinghaus Method

The Severinghaus electrode was developed to measure carbon dioxide (CO_2) in blood (see Section 4.1.1) [14]. For medical applications it is important that only CO_2 and water vapour are transported through the membrane of the electrode exhibiting an excellent selectivity, and therefore the Severinghaus electrode must be kept at elevated temperature. However, the critical issue is the membrane that

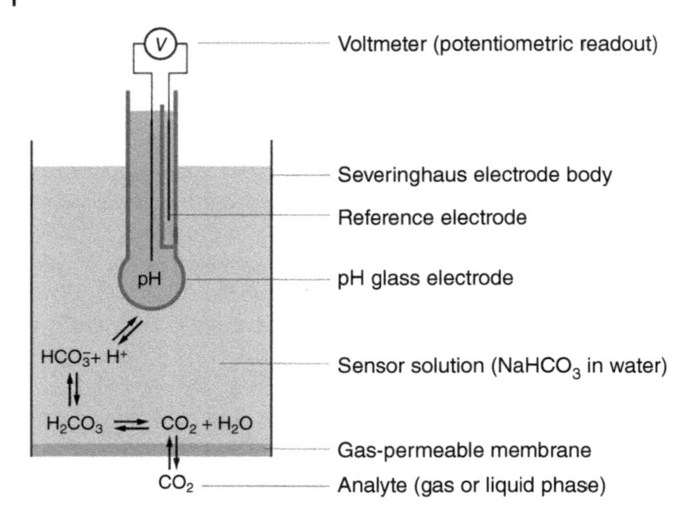

Figure 16.1 Set-up of a Severinghaus electrode showing the important chemical processes that are all temperature dependent.

should not be affected by blood or tissue components (Figure 16.1), which can be maintained by using a proper Teflon membrane.

The pH electrode is generally a miniaturized glass pH electrode [15] or an ion-selective field effect transistor (ISFET; see Section 10.2) in an integrated microdevice [16].

Since diffusion of the CO_2 into the electrolyte solution is required, the response time of the sensor is slow with time constants in the range of 2–3 minutes. The diffusion is strongly temperature dependent, and, therefore, the temperature should be stabilized to get an accurate reading. In general, the temperature dependency of the sensor sensitivity is quite complex. Apart from the temperature dependency of the diffusion, the CO_2 solubility in the sensor solution, the pH of the hydrogen carbonate buffer, and the pH and reference electrode contribute to the overall temperature dependency. Ideally, the temperature dependency of the pH and the reference electrode compensate in first approximation. In case of an equilibrium (constant CO_2 concentration in the analyte), the permeation of CO_2 in the gas-permeable membrane does not contribute to the temperature dependency of the sensitivity in equilibrium.

16.3.2 Optical Principles

Instead of using a potentiometric pH electrode in a Severinghaus set-up, the pH in a hydrogen carbonate solution can also be measured optically by fluorescence quenching [17, 18] (for details see Chapter 6). In medical devices such optical methods have seen a widespread use in commercial analysers and are competing with electrochemical technologies. The detector has the form of either an optical-fibre device or a planar device and employs different sensing techniques including fluorescence intensity detection [19], fluorescence resonance energy transfer [20], or dual-luminophore referencing (DLR) [21]. As measuring signal, the fluorescence lifetime measurement is the best to get a high accuracy and a low drift [22]. A commonly used pH-sensitive fluorescent dye is 1-hydroxy-3,6,8-pyrene trisulphonic acid trisodium salt (pyranine or HPTS), which has distinct absorption/emission bands in the visible light region [23, 24]. Such optical methods are used in extracorporeal CO_2 monitoring instruments (see Section 16.4.2) as well as in point-of-care testing (POCT) analysers [25–27]. Optical principles for measuring gaseous CO_2 using spectroscopic methods are explained in more detail in Section 16.5.

16.3.3 New and Unconventional CO_2 Measuring Principles

There are a lot of new and unconventional measuring principles that are in an exploratory state and are not used in clinical settings (see also Chapters 5 and 10). Solid electrolyte CO_2 sensors use sodium superionic conductor membranes (NASICON) [28] or Li phosphates as material [29] (see also Chapter 5). An interesting detection strategy uses impedance methods [30] or surface plasmon principles as transducing elements [31]. A sophisticated electrochemical approach utilizes the adsorption of CO_2 at copper electrodes and the desorption of CO_2 reaction products by means of anodic voltammetric stripping methods [32]. Other approaches are optrodes integrated into CMOS devices [33] or optical set-ups based on up-converting fluorescent nanoparticles [34]. Such optrode-based devices were also tested in clinical settings [35]. A further interesting approach uses a pressure sensor as transducing element, which has the advantage to be a very robust and reliable sensor element suitable for clinical applications in principle [36].

16.4 Clinical Applications

Clinical blood gas analysis systems can be grouped into devices for single analysis of a discrete blood sample and continuous monitoring devices. The continuous monitoring devices can be distinguished further by their invasiveness. Invasive monitoring of blood gas can be performed by intracorporeal (*in vivo*) or extracorporeal (*ex vivo*) measurement in the bloodstream. Non-invasive monitoring is possible by measurement through the skin (transcutaneous) or by direct CO_2 measurement in exhaled air (capnometry).

16.4.1 Blood Gas Analysing Devices

Blood gas analysis for discrete samples can be distinguished by the origin of the used blood sample. Samples from arteries are used in arterial blood gas (ABG) analysis, from veins in venous blood gas (VBG) analysis, and from arterialized capillaries in capillary blood gas (CBG) analysis. Most common parameters are the partial pressure of dissolved arterial oxygen (paO_2), carbon dioxide ($paCO_2$), and acidity (pH). Such information is vital when caring for patients with critical illness or respiratory disease. Therefore blood gas analysis is one of the most common point-of-care testing (POCT) analysis performed on patients in ICUs. The

Severinghaus electrode described in Section 4.1 is used to measure the $paCO_2$ electrochemically. The system consists of a sensor cartridge, which has to be replaced frequently, an electronic data acquisition unit, and as most important components the calibration solutions. To maintain the clinical quality standards, an automatic one-point calibration is done after every blood gas sample. This is generally accompanied by a two-point calibration every day. The data is nowadays transmitted to a clinical information system for quality assurance and is part of the clinical standard equipment maintaining POCT in all medical settings [7, 37].

16.4.2 Monitoring Devices

As stated above, the discrete measurement of CO_2 is mandatory for POCT application. However, the knowledge of current CO_2 concentration and the trend in concentration of blood gases can be of outstanding importance to get information about the patient's outcome in ICU [38]. Quality standards are maintained by a stringent calibration procedure that can be done for each measurement. To get a continuous CO_2 reading, called monitoring, such quality standards are not easily guaranteed. More details of the importance of extracorporeal monitoring, regulations, and directives are given in [39].

There are two possibilities to access CO_2 in blood or tissue in a continuous way: (i) transcutaneously and (ii) directly by accessing arterial or venous blood.

16.4.2.1 Transcutaneous pCO_2 Measurement (tcpCO_2)

Besides the use of CO_2 sensors in POCT devices for emergency applications, the continuous monitoring of CO_2 can give deep insights into normal and pathological metabolic events [38]. To get direct and continuous access to blood is one of the major problems of monitoring at patients. One method to overcome such challenges is the use of transcutaneous devices measuring the CO_2 and O_2 that penetrates the skin [40].

The principle of transcutaneous measurement is based on the fact that carbon dioxide gas is able to diffuse through the body tissue and skin and can be detected by a sensor at the skin surface. By warming up the sensor, a local hyperaemia is induced, which increases the supply of arterial blood to the dermal capillary bed below the sensor. The first devices were dedicated to measurements of tcpCO_2 on embryos, newborns, or babies [41, 42].

The transcutaneous CO_2 value (tcpCO_2) has to be interpreted primarily as the partial pressure pCO_2 prevailing at the level of the arterialized skin tissue. Such sensors consist of a Severinghaus-type electrode as described above. The electrolyte is separated within a thin hydrophilic spacer, which is placed over the sensor surface and is coupled to the skin via a highly gas-permeable hydrophobic membrane (Figure 16.2). The membrane can be protected by a thin metal plate to eliminate any mechanical damage to the measuring site.

The sensor is calibrated in a gas mixture of known CO_2 concentrations. To overcome the issue of the temperature-dependent sensitivity and to maintain a proper transport of arterial blood CO_2, the whole transcutaneous measurement is heated to at least 37 °C. This increased temperature also offers the advantage of improved response time because the system reacts much quicker to a sudden

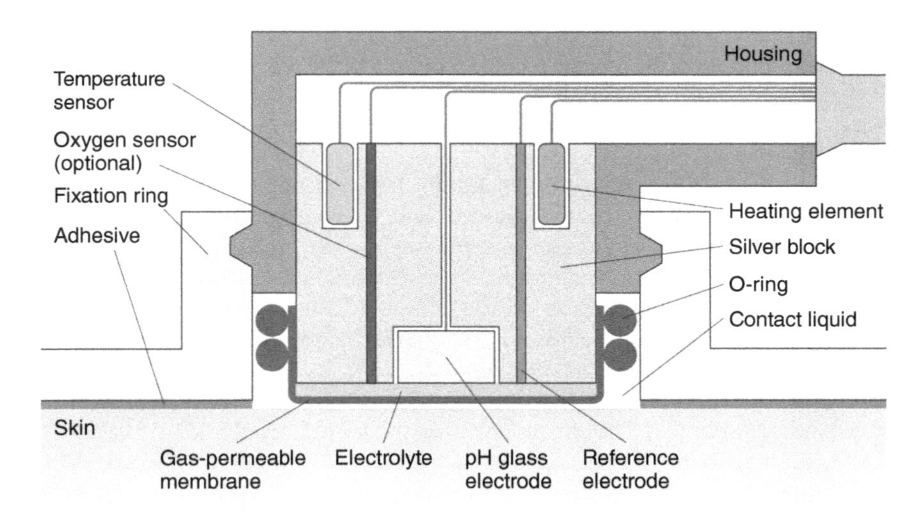

Figure 16.2 Sensor set-up for the transcutaneous measurement of pCO_2 and pO_2. The device is fixed to the patient's skin. CO_2 diffuses through the membrane into the electrolyte. The corresponding pH change is measured by the glass electrode via the Severinghaus method. An amperometric oxygen sensor is often included.

change in the pCO_2. However, the production of CO_2 is greatly increased at this higher temperature due to increased tissue metabolism. The electrical power needed to heat the sensor to a constant over-temperature depends also to some extent on the local tissue perfusion due to heat transport. At constant ambient temperature, deviations of the heating power from a stored reference value may indicate changes in perfusion, which can be used to measure tissue perfusion in parallel.

A high correlation between transcutaneous pCO_2 (tcpCO_2) and arterial pCO_2 (paCO_2) was found, especially in young patients or newborns. However, the recent generation of tcCO_2 sensors shows a good correlation also in adult patients. Such systems are used in clinical settings, with the risk of skin burning due to overheating [43–47]. Currently there are several devices on the market using such principles [48, 49].

In general, electrodes based on the Severinghaus principle (see Section 4.1) suffer from a high dependency on temperature as described in Section 16.3.1. Due to the additional heating of the sensor to an elevated temperature, the transcutaneous pCO_2 (tcpCO_2) is higher than the arterial value (paCO_2). Therefore, a correction of the transcutaneous value is necessary to provide a monitor read-out that corresponds as closely as possible to arterial paCO_2. The shift of tcpCO_2 towards higher values is attributed to two main factors [50]: (i) the elevated temperature raises local blood and tissue pCO_2 by approx. 4.5% K^{-1} (called anaerobic factor), and (ii) the living epidermal cells also produce carbon dioxide, which contributes to the capillary CO_2 level by a constant background (metabolic constant). This metabolic contribution may change with age, gender, or skin thickness. A generally accepted estimation is that skin metabolism raises the transcutaneous pCO_2 by approx. 667 Pa (5 mm Hg) [51].

16.4.2.2 Blood Monitoring Devices: Direct Venous or Arterial Monitoring of Blood Gases

Transcutaneous sensors are established for special applications with good correlation of skin pCO_2 to the arterial one. Early attempts at continuous CO_2 monitoring in the bloodstream were made with intravascular optical in vivo sensors [52, 53]. A few commercial devices following this approach were developed. An early system was the CDI 1000 (CDI-3M Healthcare). Other devices followed, e.g. the Gas-STAT GSM-100 (American Bentley) and the Paratrend 7 (Diametrix Medical) [54–60].

The early type of monitoring device relied on fibre-optic-based fluorescence measurements (optodes). Disadvantages were the complicated and brittle set-up and very high costs for a disposable sensor element. Along with unclear clinical relevance and improving alternative methods, this led to the replacement of intracorporeal devices by POCT methods and emerging extracorporeal devices. An extracorporeal arterial or venous line is connected to a flow-through system with integrated optical sensors (Figure 16.3).

Blood is guided through an array of optical microsensor patches. The optical sensor signal is determined by an attached LED/photodetector array, which reads the sensor patches through a transparent window. The CO_2 sensor is based on the diffusion of CO_2 through a gas-permeable membrane into a matrix-containing fluorescent pH-sensitive dye and the measurement of the corresponding pH change (Severinghaus principle).

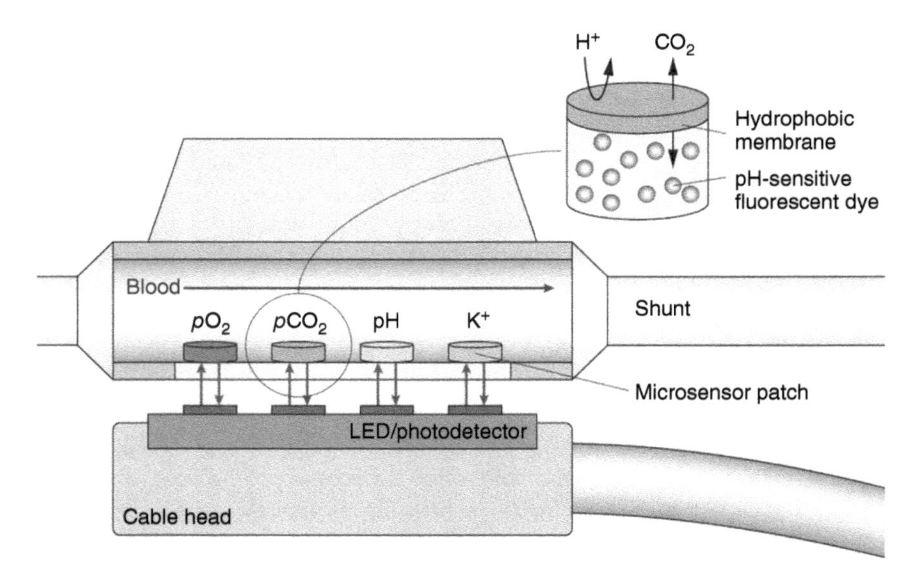

Figure 16.3 Set-up of an extracorporeal sensor system measuring pCO_2, pO_2, pH, and potassium ions in patient blood (Terumo CDI Blood Parameter Monitoring System 500).

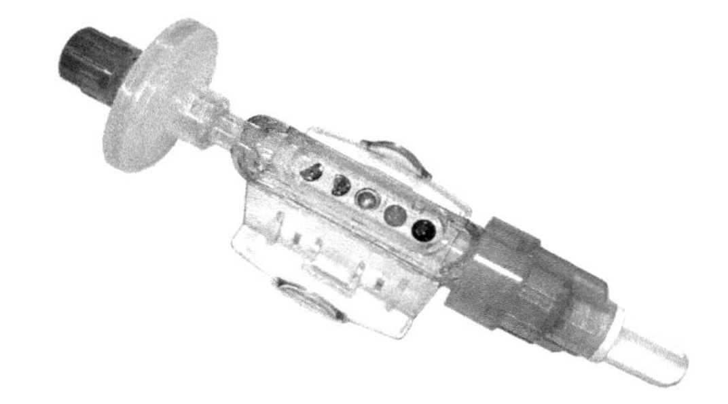

Figure 16.4 Photograph of an *ex vivo* optical CO_2 sensor (Terumo CDI Blood Parameter Monitoring System 500). Source: IMTEK-Sensors, University of Freiburg.

Clinically employed systems use extracorporeal flow systems (*ex vivo* systems), in which blood gases can be measured over a period of several hours during operations, CDI Blood Parameter Monitoring System 500 (Terumo) [26, 61], as shown in Figure 16.4, or the BMU 40 device (Maquet) [27, 62]. Electrochemically operating sensor systems are no longer in use for such monitoring applications. However, the use of continuously operating sensor systems on patients is generally associated with difficulties in terms of safety and reliability, which can lead to recalls [63], and systems are used only for special applications in standard clinical settings.

16.5 Comparison of Methods and Conclusions

Severinghaus-based CO_2 sensor systems using electrochemical or optical approaches are well established in clinical settings for both discrete blood gas analysis and continuous monitoring. For POCT applications, discrete blood gas measurements have become the gold-standard procedure. They are reliable and cost-effective and are integrated into the laboratory and data handling system in a hospital.

Continuous monitoring devices are on the market and can be separated into transcutaneous and invasive monitoring. The electrochemical transcutaneous approach is expensive and more challenging, but less invasive, and is still relevant for special applications. The more invasive *ex vivo* measurements by optical methods are used in intensive care and have largely replaced intracorporeal approaches. However, continuous measurements can be done in the exhaled breath of ventilated patients, which is nowadays a standard procedure and is presented in the next chapter.

16.6 CO_2 Analysis in Human Breath

The measurement of CO_2 in human breath – called capnometry – is nowadays common practice in respiratory gas monitoring in intensive care medicine, emergency medicine, and anaesthesiology. The medical procedure of measuring the concentration of CO_2 in the inspiratory and expiratory gas of a patient is also an indirect monitor of the CO_2 partial pressure in the arterial blood. Such a method overcomes the problems related to invasive blood access.

Capnography means the graphical representation of the concentration of CO_2 in the respiratory gas over time. The corresponding temporal course is called capnogram. The expiratory CO_2 level is an important indicator for proper lung ventilation and the patient's circulation and metabolism. Thus, it can aid in the diagnosis of low cardiac output states and pulmonary embolism. Further capnography helps in detecting accidental oesophageal intubation and other airway management problems in mechanically ventilated patients [64].

The end-expiratory (also end-tidal) CO_2 (Etp CO_2) measured outside of the body is determined primarily by the current pulmonary CO_2 clearance (VCO_2[pulmonary]), which is the product of minute volume V and CO_2 concentration.

Furthermore, important is the endogenous CO_2 clearance (VCO_2 [endogenous]), i.e. the product of cardiac output Q and arteriovenous CO_2 difference ($CvCO_2$-$CaCO_2$) [65]. Therefore, the end-expiratory CO_2 concentration is an indicator not only for lung ventilation (represented by minute volume and $CaCO_2$) but also for the current state of the elementary body functions such as circulation (represented by the cardiac output) and metabolism (represented by $CvCO_2$). When two of these three systems are in equilibrium, a change in $EtpCO_2$ indicates a dysfunction in the third regulatory circuit.

16.6.1 Methods for CO_2 Detection in Breath

16.6.1.1 Qualitative and Semi-Quantitative Detection

The simplest method for the detection of CO_2 in the respiratory gas is based on a colorimetric sensor indicating reversibly changes in colour with a breath-to-breath response (EasyCAP, Nellcor, Medtronic, USA). The sensor is based on the colour change of metacresol purple from purple to yellow when CO_2 is present (Figure 16.5). The easy-to-use CO_2 detector attaches directly to the endotracheal tube and responds quickly to exhaled CO_2 for up to two hours in a CO_2 concentration range from 0.03% to 5%.

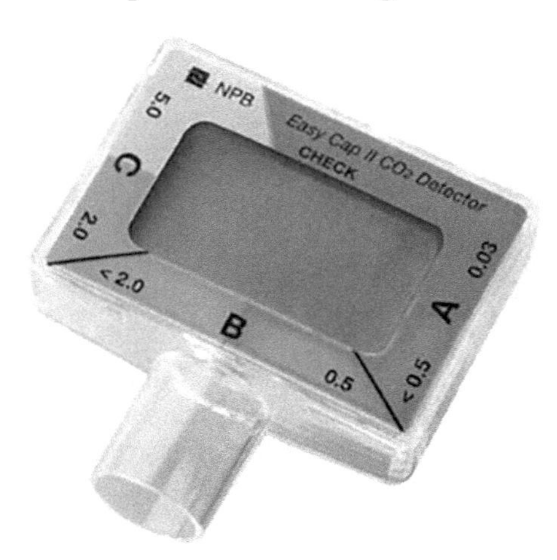

Figure 16.5 Colorimetric sensor for CO_2 analysis in breath (Nellcor™ adult/pediatric colorimetric CO_2 detector). This single-use sensor provides direct visual feedback of the exhaled CO_2 in a semi-quantitative way through a colour change from purple (in the black/white figure dark gray) to yellow (in the black/white figure light gray in the upper left corner).

16.6.1.2 Quantitative Detection by Non-Dispersive Infrared Absorption

If the CO_2 content in the respiratory gas must be quantified, then sensor systems based on non-dispersive infrared absorption (NDIR) are used (see Section 7.1). The breathing air is passed through a measuring chamber that is irradiated by an infrared source. This method was first described in [66] and is today considered as safe and practical. It is available in the form of small handheld instruments. IR radiation is absorbed by CO_2 at defined wavebands, the most prominent around 4.3 μm. If the breathing gas in the measuring chamber contains CO_2, energy is absorbed by the IR radiation, and the signal intensity at the IR detector is directly correlated with the concentration of CO_2. In terms of their set-up, a distinction can be made between single- and two-channel systems: the two-channel system includes two IR detectors equipped with wavelength-selective optical filters, one at the absorption range of CO_2 and a reference. This allows, for example, compensating temperature changes or ageing effects in the optical components. The single-channel set-up has no reference and only measures the CO_2 absorption, which is accompanied by a lower accuracy.

Capnograph analysers can be configured as either mainstream or side-stream systems. In mainstream analysers, the CO_2 sensor is integrated directly into the respiratory flow of the patient, and the CO_2 concentration is analysed over the entire flow cross section, i.e. the mainstream analyser is more or less attached to the patient and is connected to a monitor (Figure 16.6). This method is characterized by its high dynamics. Side-stream analysers are located away from the patient, and the respiratory gases are suctioned at a constant flow rate through a narrow tube to the CO_2 sensor. Therefore, side-stream analysers show a delay in their response time, which can be minimized by using short and narrow tubes, or a higher suction flow rate.

The typical measurement range amounts to 0–100 mm Hg (13.33 kPa) CO_2 partial pressure or 0–15 vol% CO_2. The typical accuracy is ±2 mm Hg (266.64 Pa), and the relative measurement uncertainty is specified depending on the measuring range as 4–8% of the displayed value. Typical sample gas flows in side-stream devices range from 50 ml min^{-1} (pediatric) to 180 ml min^{-1} and in devices for use in animal studies (small animals) from 10 ml min^{-1} to 20 ml min^{-1}. Typical response times are around 60 ms.

The water vapour from the patient's exhaled breath is problematic since H_2O condenses in the sampling tubes. To prevent condensed water from entering the measurement chamber of the CO_2 analyser, the respiratory gas sample is directed through a water trap (approx. 10 ml). In mainstream devices the sensor head is usually heated in order to prevent the condensation of water vapour. CO_2 analysers usually have a moisture compensation, in which the measured value corresponding to the expiratory vapour saturation pressure pH_2O (47 mm Hg or 6.27 kPa at 37 °C body temperature) is automatically converted to the correct CO_2 partial pressure (pH_2O correction). In addition, the measured CO_2 partial pressure is automatically referenced to the current barometric pressure.

An example for a clinical side-stream capnography analyser is shown in Figure 16.7.

An enhancement of capnography, the volumetric capnography, is increasingly applied in the differentiated respiratory monitoring of mechanically ventilated

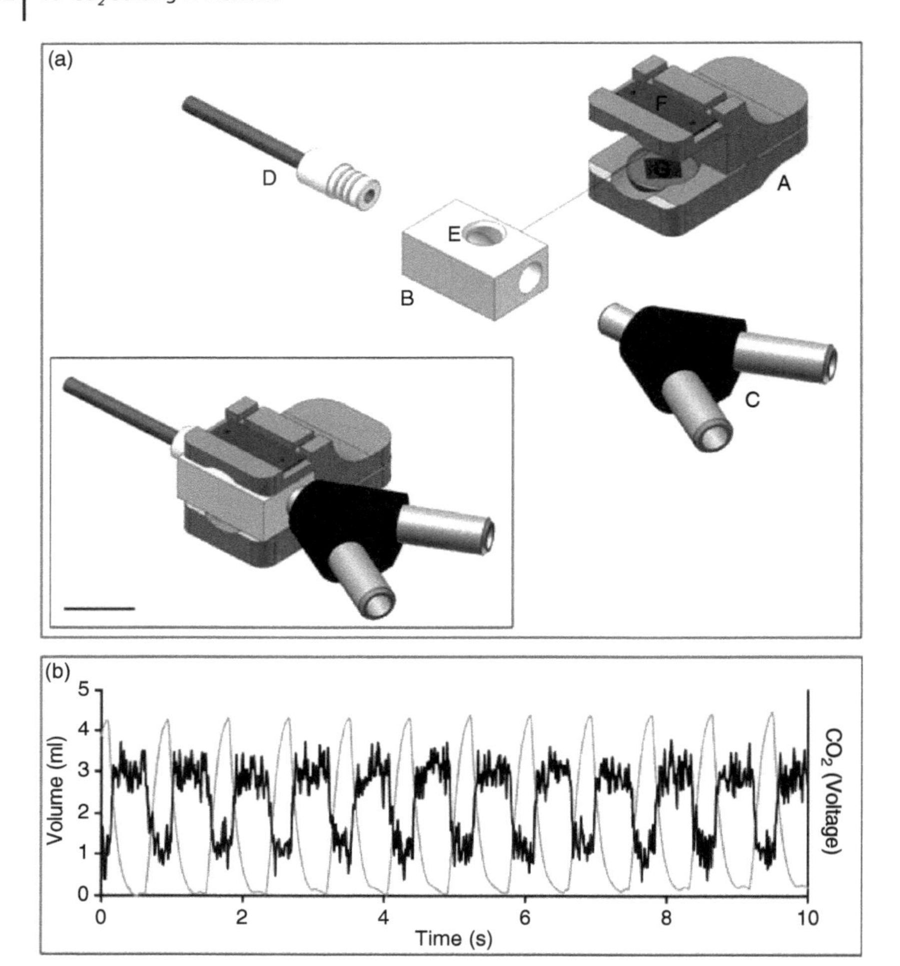

Figure 16.6 Schematic of a miniaturized CO_2 mainstream sensor based on infrared absorption. (a) CO_2 sensor with cell (B) and clamped sensor receptacle (A). A diode (F) emits infrared radiation through the cell (E) and an optical filter on a photoresistor (G). The cuvette is placed between the tracheal cannula (D) and the Y-piece (C) of the fan. (b) Typical capnogram [67].

intensive care patients. Here, a CO_2 sensor is combined with a breathing gas flow sensor. From the simultaneously recorded history of the expiratory CO_2 concentration and the exhaled volume determined from the flow of breathing gas, the pulmonary dead space volume is determined. The knowledge of the dead volume, i.e. the volume of the respiratory system not participating in the physiological gas exchange, is very helpful for adequate adjustment of mechanical ventilation.

Generally, the exhaled CO_2, as shown by the capnogram, has passed the following phases in the organism:

- CO_2 production in the somatic cells, dependent on the metabolism.
- CO_2 transport from the cells to the lungs, dependent on the circulation.
- CO_2 elimination, dependent on the state of the complete respiratory system.

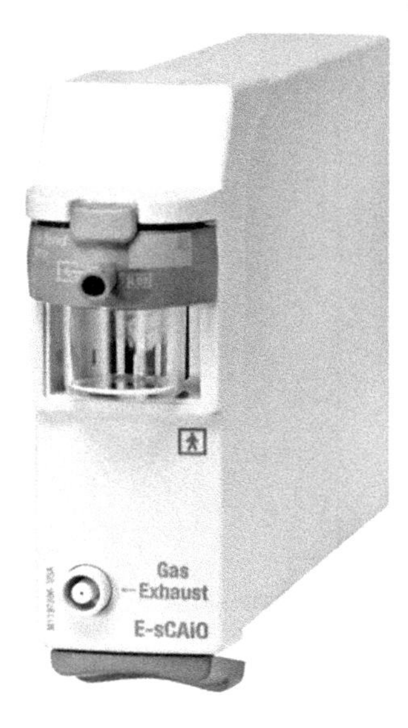

Figure 16.7 Capnograph with one compact module for respiratory monitoring of O_2, CO_2, and N_2O, anaesthetic agents sucking the breathing gas sample with 120 ml min^{-1} by the side-stream method (GE Healthcare). Courtesy Reference.

Accordingly, the level of capnogram falls at elevated ventilation and increases with reduced ventilation. A wealth of clinically serious events – such as air embolism, disorders of the respiratory regulation, and airway obstruction of any kind – can be identified in the capnogram (Figure 16.6b). In the field of emergency medicine – such as in the acute life-threatening situation of cardiac arrest, capnography is extremely useful: the capnogram clearly and unequivocally shows the manifest cardiac arrest, revitalization measures can be initiated immediately, and the effect of these measures on the circulation can be monitored continuously. In addition, the capnogram provides information on possible technical malfunctions of a ventilator (intensive care) or an anaesthesia ventilator (anaesthesiology).

Within the capnogram, the quantitative determination of one particular point is of great practical importance: the maximum concentration of CO_2 at the end of exhalation, which is called end-tidal CO_2 concentration ($EtpCO_2$). The particular clinical importance of the end-tidal CO_2 value lies in its relation to the arterial partial pressure of CO_2, because the alveolar space is in equilibrium with the capillary blood in the lungs. However, this equilibrium would be destabilized by a manifest circulatory disorder. As long as circulation problems – asserted by clinical examination methods – can be excluded, the end-tidal CO_2 concentration can be used for trend analysis of the arterial partial pressure of CO_2 and, thus, provides valuable service in the operating room and in the ICU. The distinction of a capnogram into inspiratory and expiratory phases is rather arbitrary and also depends on the response time of the sensor technology, yet facilitates a prompt detection of rebreathing and allows the use of standardized and physiologically appropriate nomenclature for better understanding and interpretation of capnograms [68]. Figure 16.8 shows the different phases of a typical capnogram with the important $EtpCO_2$ level at the end of the expiration phase.

In Figure 16.8, after inspiration and at the beginning of expiration (phase I), only the CO_2-free gas from the upper airways that has not undergone gas exchange

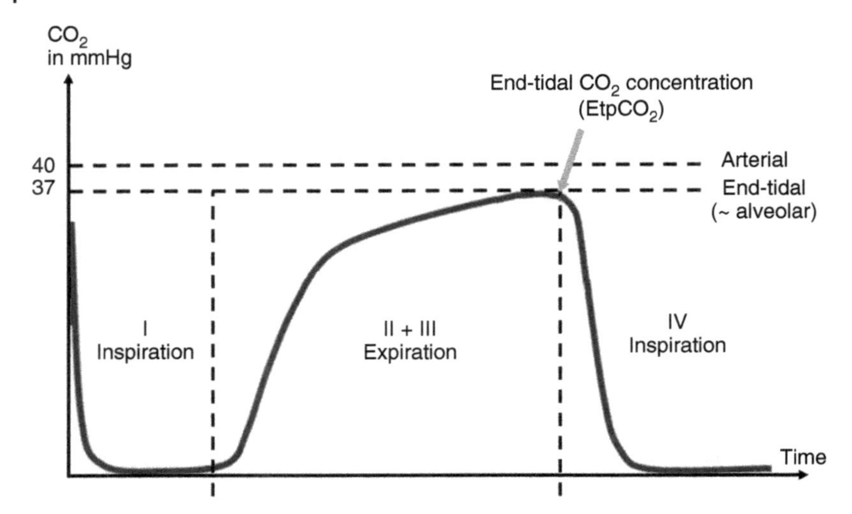

Figure 16.8 Phases of a capnogram.

(dead space volume) is present. Then, the CO_2 signal rapidly increases with the increasing emptying of the alveoli (phase II). Phase III represents the alveolar plateau, in which the gas from the alveoli is expired. The highest signal at the end of the expiration phase corresponds to the end-tidal CO_2 ($EtpCO_2$). Subsequently, the next inspiration phase (phase IV) starts with a rapid decrease in the CO_2 signal since the inspiratory gas is CO_2-free.

Capnometry also gives a good monitoring of the condition of the respiratory and circulatory system during anaesthesia, which is why it is an indispensable part of the modern operating room. The clinical objectives pursued primarily with intraoperative capnometry relate to ensuring adequate ventilation and to monitoring the metabolic and cardiovascular state of the patient. In addition, the effect of anaesthetics on the capnogram allows the correct dosage of inhalative anaesthetics [69].

Ergospirometry – as a combination of ergometry (measuring physical performance) and spirometry (lung function measurements) – is a method of performance diagnostics for analysing the responses of the heart, circulation, respiration, and metabolism during muscular work. It allows a precise analysis of cardiopulmonary capacity. In ergospirometry two methods of physical stress are distinguished: treadmill and cycle ergometry. Both steady-state stress and stress tests with every minute increase (ramp protocols) are used. During the ergospirometry, the inhaled air and exhaled air of the patient are analysed through a mouthpiece or a mouth–nose mask. Four measurement signals form the basis of each ergospirometry: (i) respiratory gas flow, (ii) oxygen consumption, (iii) CO_2 output, and (iv) heart rate (exercise ECG). The data is collected by most devices with each respiratory cycle.

The detection of respiratory gas exchange, oxygen uptake, and CO_2 output also plays a central role in the clinical determination of energy turnover (metabolic intensity) in the context of indirect calorimetry. The indirect calorimetry assumes that all energy-generating processes in the organism are based on oxidative processes, so the oxygen uptake is a valid measure of energy turnover. The total energy turnover is calculated as the product of oxygen consumption and caloric equivalent of food and amounts on average to 20.2 kJ per litre of oxygen [70] (see also Section 9.4).

16.7 CO_2 Measurements on Baby Mattresses

Rebreathing of exhaled air is one of numerous possible reasons for the sudden infant death syndrome (SIDS), which is marked by the sudden death of an infant that is unexpected by history and remains unexplained after a detailed death scene investigation. Typically, the infant is found dead after having been put to bed and exhibits no signs of having suffered. Some observations and theories indicate that the rebreathing of carbon dioxide plays a role in the occurrence of SIDS among prone sleeping infants. Under unfavourable conditions, while face down, babies again and again inhale parts of their exhaled breath from a pool forming around them in which the oxygen content drops and the carbon dioxide concentration continuously increases with each breath. In numerous scientific investigations, the permeation of air and CO_2 through different baby mattresses and bedding materials has been studied. Research has shown that the mattress influences SIDS outcomes. As a result, different kinds of specially structured baby mattresses came on the market, which are equipped with "climatic channels" to improve the CO_2 permeability and thus to reduce the risk of SIDS [71–73]. Generally, a firm mattress lowers SIDS risk. Kemp and Thach showed an increased risk of child death on polystyrene-filled pillows [74], on sheepskins [75], and on waterbeds. Natural fibre mattresses have also been associated with an increased risk of childhood mortality [76–78]. This was noted in particular for sheepskins. Since 2003 sheep and lamb skins in Germany can only be sold with the remark "not suitable as a shelter for sleeping babies".

In [79] it was evaluated how CO_2 dispersal was affected by a conventional crib mattress and some commercial products marketed to prevent prone rebreathing. In the tests an infant dummy with its nares connected via tubing to a reservoir filled with 5% CO_2 was positioned prone face down or near face down on the different sleep surfaces. The fall in percentage end-tidal CO_2 was measured as the reservoir was ventilated with a piston pump. The half-time for CO_2 dispersal ($t_{1/2}$) was regarded as an index of the ability to cause or prevent rebreathing. Not only the firm mattress but also nearly all of the surfaces designed to prevent rebreathing consistently showed $t_{1/2}$ values above thresholds for the onset of CO_2 retention. Thus, for infants placed prone or rolling to the prone position, significant rebreathing of exhaled air would be likely on all surfaces studied, except one.

The US Consumer Product Safety Commission, Washington, recommended already in 1998 a standardized test procedure for the examination of the CO_2 permeability of baby mattresses by using a mechanical baby model to simulate infant breathing. Using this standardized model, in the paper [80], extensive inspection results of different baby mattresses and materials and effects of bedding on exhaled air retention are published. Under simulated rebreathing

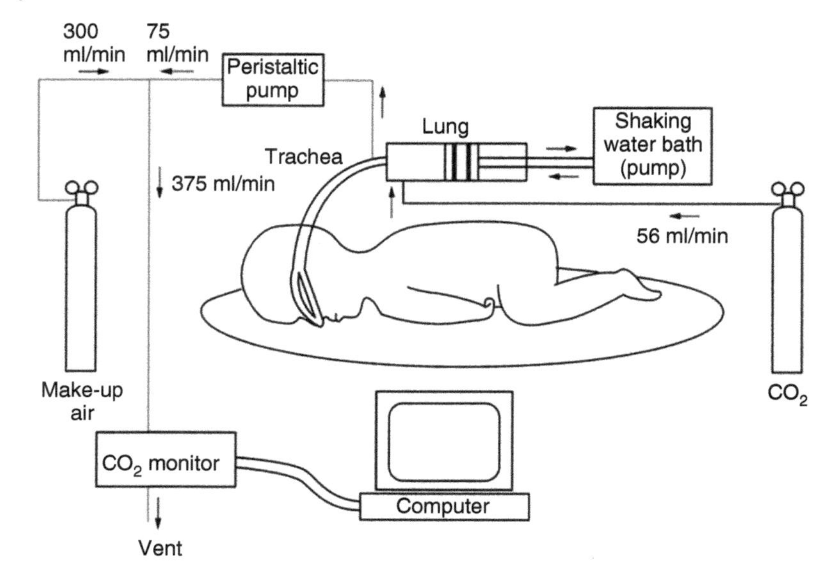

Figure 16.9 Schematic representation of the mechanical model. Source: Carleton et al. 1998 [80]. Reproduced with permission of BMJ.

conditions, the model allows the monitoring of raised carbon dioxide inside an artificial lung–trachea system. Resulting levels of CO_2 suggest that common bedding materials vary widely in inherent rebreathing potential. In face-down tests, maximum airway CO_2 ranged from less than 5% on sheets and waterproof mattresses to over 25% on sheepskins and some pillows and comforters. The magnitude of this CO_2 increase was found to vary as a function of bedding type and to a lesser extent between different items of the same general type. Concentrations of CO_2 decreased with increasing head angle of the doll, away from the face-down position. Figure 16.9 provides a schematic representation of the model. The model employed a brass syringe mounted on a laboratory shaking water bath as a mechanical "lung", with the bath functioning as a pumping mechanism. The head was capable of rotating sideways relative to the body so that head angle could be precisely set. The bath motion was set such that 35 ml "breaths" were delivered at a constant rate of 45 min^{-1}. CO_2 was fed from a tank into the lung at 56 ml min^{-1} to simulate a mean lung concentration of 5%, a normal alveolar value. Constancy of the gas flow was assured with an in-line flowmeter. For carbon dioxide measurements, the gas was routed through an infrared monitor (VacuMed Co 17600, Ventura, CA, USA) that measured the CO_2 content of the sampled gas. The monitor was interfaced to a personal computer, allowing CO_2 concentrations inside the trachea to be recorded as a function of time.

Figure 16.10 shows an alternative measuring equipment that had been developed for this purpose at the Meinsberg Kurt-Schwabe Research Institute [81]. A cylindrical measuring chamber (volume 1 litre), containing a solid electrolyte CO_2 sensor (ZIROX GmbH, Greifswald, Germany) on the top, is placed with defined mechanical pressure closely on the centre of the mattress. At the

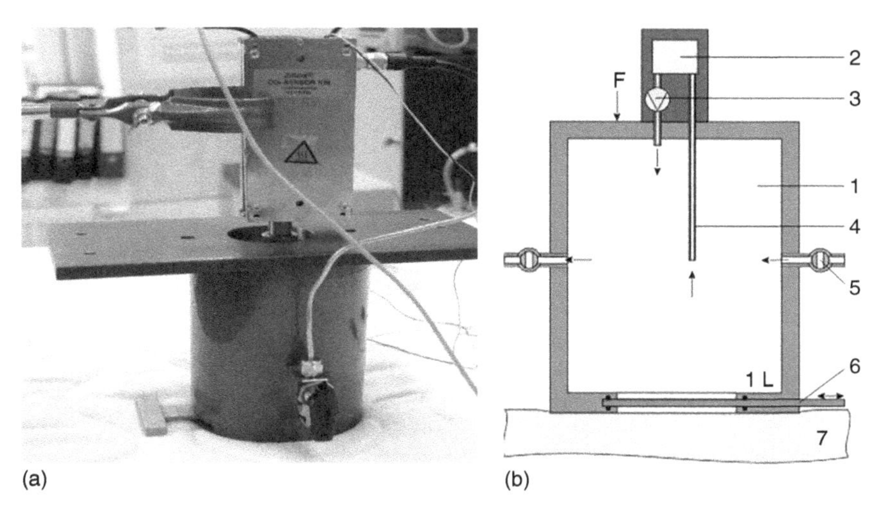

Figure 16.10 Experimental set-up for investigation of CO_2 diffusion through mattresses and other textile materials. (a) Photograph of the measuring equipment. (b) Schematic drawing: (1) cylindrical measuring chamber, (2) CO_2 solid electrolyte sensor (ZIROX GmbH, Greifswald, Germany), (3) gas micropump, (4) intake and outlet pipes to the sensor, (5) gas inlet and outlet valves, (6) slider, and (7) test mattress; F, defined pressing force [81].

beginning of the measurement, the slider at the bottom of the chamber is closed, and the chamber is filled with a gas containing a defined concentration of CO_2. After adjustment of a constant concentration, the rinsing was terminated, the chamber was completely closed, and, after adjustment of the equilibrium, a slide located in the chamber wall facing the mattress was opened. When the chamber is opened by removing the slider, the CO_2 diffuses through the mattress into the surrounding air. The rate of the concentration decrease in the measuring chamber was recorded as a measure of the diffusion volume flow through the mattress and compared for the individual mattresses.

Consequently, the CO_2 concentration in the measuring chamber decreases depending on the permeability of the mattress material as shown in Figure 16.11. For comparison the unhindered diffusion of CO_2 from the measuring chamber through the opening into the surrounding air without any mattress has been measured (curve 4 in Figure 16.11). *T* is measuring method, and set-up can be used to characterize and to compare the CO_2 permeability of baby mattresses. Of course, it can be utilized also for measurements with other gases as well as for the investigation of other textile materials.

At the Dresden University of Technology, Institute of Occupational Medicine, a photoacoustic multi-gas monitor (Innova AirTech Instruments A/S) was used for investigations on the CO_2 permeability of three differently structured baby mattresses [82]. As proposed in [80], baby dummies with body mass 4.0, 6.0, and 8.0, respectively, were positioned on the mattresses in different positions, and the CO_2 concentrations were measured above the mattress near the nose of the baby and immediately below the mattress for 30 minutes in each case. Surprisingly it was found that the CO_2 permeabilities of the three baby mattresses were 92%, 46%, and 20% and thus differed considerably.

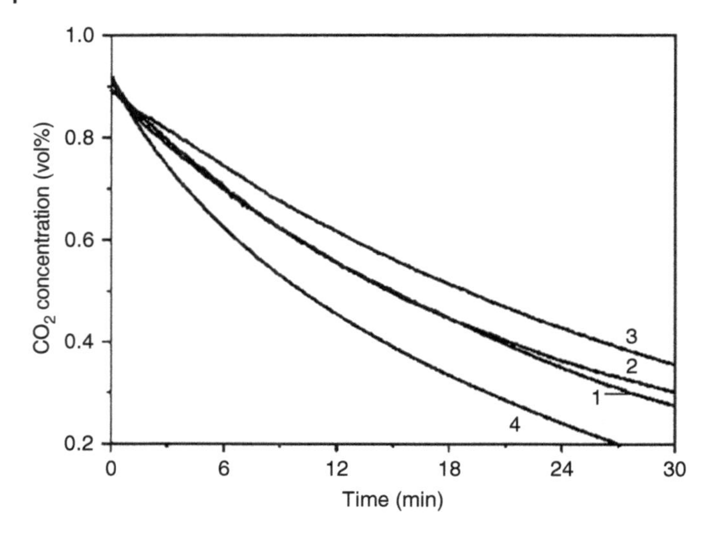

Figure 16.11 Course of the CO_2 concentration in the opened measuring chamber with different mattresses (curves 1 and 2), mattress 2 with cover (curve 3), and without mattress (curve 4).

References

1 Kost, G.J. (ed.) (2002). Goals, guidelines and principles for point-of-care testing. In: *Principles & Practice of Point-of-Care Testing*, 3–12. Philadelphia, PA: Lippincott Williams & Wilkins.

2 Roche in Deutschland. Accu-Chek® Inform II System. https://www.roche.de/diagnostics/systeme/point-of-care-diagnostik/accu-chekinform2.html (retrieved 31 March 2017).

3 Siemens Healthcare. RAPIDComm Data Management System. http:/www.healthcare.siemens.com/point-of-care/information-technology/rapidcomm-data-management (retrieved 31 March 2017).

4 Abbott Point of Care *i-Stat* Handheld. https://www.pointofcare.abbott/us/en/offerings/istat/istat-handheld (retrieved 31 March 2017).

5 DIN-Normenausschuss Medizin (NAMed) (2014). Medizinische Laboratorien – Anforderungen an die Qualität und Kompetenz (ISO 15189:2012, korrigierte Fassung 2014-08-15); Deutsche Fassung EN ISO 15189:2012 DIN EN ISO 15189:2014-11, (Medical Laboratories – Requirements for Quality and Competence (ISO 15189:2012, Corrected version 2014-08-15); German version EN ISO 15189:2012), 1–66. Berlin: Beuth-Verlag (in German).

6 von Eiff, W. and Henke, V. (2011). Erfolgsfaktoren und Hindernisse bei der Einführung vernetzter Point-of-Care Glucosemessgeräte: Erkenntnisse und Empfehlungen aus der GLUMO-Studie (factors of success and obstacles for the introduction of networked point-of-care glucose analysers: findings and recommendations of the GLUMO-study). *LaboratoriumsMedizin* 35 (2): 93–105 (in German).

7 The European Parliament and the Council of The European Union (1998). Directive 98/79/EC of the European Parliament and of the Council of 27 October 1998 on in vitro diagnostic medical devices. *The Official Journal of the European Communities.* L331/31–37.
8 Bear, R.A. and Dyck, R.F. (1979). Clinical approach to the diagnosis of acid-base disorders. *Can. Med. Assoc. J.* 120 (2): 173–182.
9 Heinemann, H.O. and Goldring, R.M. (1974). Bicarbonate and the regulation of ventilation. *Am. J. Med.* 57 (3): 361–370.
10 Roussos, C. (1982). The failing ventilatory pump. *Lung* 160 (1): 59–84.
11 Windisch, W. (2013). Home mechanical ventilation. In: *Principles and Practice of Mechanical Ventilation*, 3e (ed. M.J. Tobin), 683–698. New York: McGraw-Hill.
12 Huttmann, S.E., Windisch, W., and Storre, J.H. (2014). Techniques for the measurement and monitoring of carbon dioxide in the blood. *Ann. Am. Thorac. Soc.* 11 (4): 645–652.
13 Kraut, J.A. and Madias, N.E. (2010). Metabolic acidosis: pathophysiology, diagnosis and management. *Nat. Rev. Nephrol.* 6 (5): 274–285.
14 Severinghaus, J.W. and Bradley, A.F. (1958). Electrodes for blood pO_2 and pCO_2 determination. *J. Appl. Physiol.* 13 (3): 515–520.
15 Buck, R.P., Rondinini, S., Covington, A.K. et al. (2002). Measurement of pH. Definition, standards, and procedures (IUPAC Recommendations 2002). *Pure Appl. Chem.* 74 (11): 2169–2200.
16 Gumbrecht, W., Peters, D., Schelter, W. et al. (1994). Integrated pO_2, pCO_2, pH sensor system for online blood monitoring. *Sens. Actuator B Chem.* 19 (1–3): 704–708.
17 Opitz, N. and Lübbers, D.W. (1987). Theory and development of fluorescence based optoochemical oxygen sensors: oxygen optodes. *Int. Anesthesiol. Clin.* 25 (3): 177–197.
18 Borisov, S.M. and Wolfbeis, O.S. (2008). Optical biosensors. *Chem. Rev.* 108 (2): 423–461.
19 Nivens, D.A., Schiza, M.V., and Angel, S.M. (2003). Multilayer sol-gel membranes for optical sensing applications: single layer pH and dual layer CO_2 and NH_3 sensors. *Talanta* 58 (3): 543–550.
20 von Bültzingslöwen, C., McEvoy, A.K., McDonagh, C., and MacCraith, B.D. (2003). Lifetime-based optical sensor for high-level pCO_2 detection employing fluorescence resonance energy transfer. *Anal. Chim. Acta* 480 (2): 275–283.
21 von Bültzingslöwen, C., McEvoy, A.K., McDonagh, C. et al. (2002). Sol-gel based optical carbon dioxide sensor employing dual luminophore referencing for application in food packaging technology. *Analyst* 127 (11): 1478–1483.
22 Holst, G.A., Köster, T., Voges, E., and Lübbers, D.W. (1995). FLOX—an oxygen-flux-measuring system using a phase-modulation method to evaluate the oxygen-dependent fluorescence lifetime. *Sens. Actuator B Chem.* 29 (1–3): 231–239.
23 He, X. and Rechnitz, G.A. (1995). Linear response function for fluorescence-based fiber-optic CO_2 sensors. *Anal. Chem.* 67 (13): 2264–2268.

24 Malins, C. and MacCraith, B.D. (1998). Dye-doped organically modified silica glass for fluorescence based carbon dioxide gas detection. *Analyst* 123 (11): 2373–2376.

25 OPTI Medical Systems, OPTI™ CCA-TS Analyzer. http://www.optimedical.com/pdf/product/opti-cca-ts-analyzer-ops-manual-english.pdf (retrieved 31 March 2017).

26 Terumo Cardiovascular Systems. CDI® Blood Parameter Monitoring System 500. http://www.terumo-cvs.com/products/ProductDetail.aspx?groupId=3&familyID=47&country=1 (retrieved 31 March 2017).

27 Maquet. Blutüberwachungsgerät BMU 40. http://www.maquet.com/de/produkte/blood-monitoring-unit (retrieved 31 March 2017).

28 Sahner, K., Schulz, A., Kita, J. et al. (2008). CO_2 Selective potentiometric sensor in thick-film technology. *Sensors* 8 (8): 4774–4785.

29 Lee, H.K., Choi, N.J., Moon, S.E. et al. (2012). A solid electrolyte potentiometric CO_2 gas sensor composed of lithium phosphate as both the reference and the solid electrolyte materials. *J. Korean Phys. Soc.* 61 (6): 938–941.

30 Ishizu, K., Takei, Y., Honda, M. et al. (2013). Ionic gel based carbon dioxide gas sensor. In: *Transducers & Eurosensors XXVII: The 17th International Conference on Solid-State Sensors, Actuators and Microsystems 2013*, Barcelona, Spain, 1633–1636.

31 Herminjard, S., Sirigu, L., Herzig, H.P. et al. (2009). Surface plasmon resonance sensor showing enhanced sensitivity for CO_2 detection in the mid-infrared range. *Opt. Express* 17 (1): 293–303.

32 Fasching, R., Kohl, F., and Urban, G. (2003). A miniaturized amperometric CO_2 sensor based on dissociation of copper complexes. *Sens. Actuator B Chem.* 93 (1–3): 197–204.

33 Ratterman, M., Shen, L., Klotzkin, D. et al. (2011). CMOS-based luminescent CO_2 sensor. In: *15th International Conference on Miniaturized Systems for Chemistry and Life Sciences*, Seattle, WA, USA, 1260–1262.

34 Ali, R., Saleh, S.M., Meier, R.J. et al. (2010). Upconverting nanoparticle based optical sensor for carbon dioxide. *Sens. Actuator B Chem.* 150 (1): 126–131.

35 Cajlakovic, M., Bizzarri, A., and Ribitsch, V. (2006). Luminescence lifetime-based carbon dioxide optical sensor for clinical applications. *Anal. Chim. Acta* 573–574: 57–64.

36 Herber, S., Bomer, J., Olthuis, W. et al. (2005). A miniaturized carbon dioxide gas sensor based on sensing of pH-sensitive hydrogel swelling with a pressure sensor. *Biomed. Microdevices* 7 (3): 197–204.

37 Radiometer Medical. Radiometer ABL800 FLEX blood gas analyzer, http://www.radiometer.com/en/products/blood-gas-testing/abl800-flex-blood-gas-analyzer (retrieved 31 March 2017).

38 Frost, M.C. and Meyerhoff, M.E. (2015). Real-time monitoring of critical care analytes in the bloodstream with chemical sensors: progress and challenges. *Annu. Rev. Anal. Chem.* 8: 171–192.

39 Baker, R.A., Bronson, S.L., Dickinson, T.A. et al. (2013). Report from AmSECT's International Consortium for evidence-based perfusion: American

Society of ExtraCorporeal Technology standards and guidelines for perfusion practice. *J. Extra Corpor. Technol.* 45 (3): 156–166.

40 Huch, A., Lübbers, D.W., and Huch, R. (1973). Patientenüberwachung durch transkutane PCO_2 Messung bei gleichzeitiger Kontrolle der relativen lokalen Perfusion (monitoring of transcutaneous PCO_2 measurement of patients and simultaneous control of the relative local perfusion). *Anaesthesist* 22: 379–380 (in German).

41 Bromley, I. (2008). Transcutaneous monitoring – understanding the principles. *Infant* 4 (3): 96–98.

42 Stieglitz, S., Matthes, S., Priegnitz, C. et al. (2016). Comparison of transcutaneous and capillary measurement of PCO_2 in hypercapnic subjects. *Respir. Care* 61 (1): 98–105.

43 Bobbia, X., Claret, P.-G., Palmier, L. et al. (2015). Concordance and limits between transcutaneous and arterial carbon dioxide pressure in emergency department patients with acute respiratory failure: a single-center prospective observational study. *Scand. J. Trauma Resusc. Emerg. Med.* 23: 40, 7p. https://doi.org/10.1186/s13049-015-0120-4.

44 Domingo, C., Canturri, E., Moreno, A. et al. (2010). Optimal clinical time for reliable measurement of transcutaneous CO_2 with ear probes: counterbalancing overshoot and the vasodilatation effect. *Sensors* 10 (1): 491–500.

45 Domingo, C., Roig, J., Coll, R. et al. (1996). Evaluation of the use of three different devices for nocturnal oxygen therapy in COPD patients. *Respiration* 63 (4): 230–235.

46 Bendjelid, K., Schütz, N., Stotz, M. et al. (2005). Transcutaneous PCO_2 monitoring in critically ill adults: clinical evaluation of a new sensor. *Crit. Care Med.* 33 (10): 2203–2206.

47 Choi, S.-H., Kim, J.-Y., Yoon, Y.-H. et al. (2014). The use of transcutaneous CO_2 monitoring in cardiac arrest patients: a feasibility study. *Scand. J. Trauma Resusc. Emerg. Med.* 22: 70, 5p. https://doi.org/10.1186/s13049-014-0070-2.

48 Radiometer. TCM TOSCA monitor. https://www.radiometer.com/en/products/transcutaneous-monitoring/tcm-tosca (retrieved 31 March 2017).

49 Radiometer. TCM5 FLEX monitor. http://www.radiometer.de/de-de/produkte-und-lösungen/transkutanes-monitoring (retrieved 31 March 2017).

50 Severinghaus, J.W. (1982). Transcutaneous blood gas analysis. *Respir. Care* 27 (2): 152–159.

51 Drysdale, D. (2014). Transcutaneous carbon dioxide monitoring: literature review. *Oral Health Dent. Manag.* 13 (2): 453–457.

52 Gehrich, J.L., Lübbers, D.W., Opitz, N. et al. (1986). Optical fluorescence and its application to an intravascular blood gas monitoring system. *IEEE Trans. Biomed. Eng.* 33 (2): 117–132.

53 Opitz, N. and Lübbers, D.W. (1987). Theory and development of fluorescence based optochemical oxygen sensors: oxygen optodes. *Int. Anesthesiol. Clin.* 25 (3): 177–197.

54 Tusa, J.K. and He, H. (2005). Critical care analyzer with fluorescent optical chemosensors for blood analytes. *J. Mater. Chem.* 15 (27–28): 2640–2647.

55 Gefke, K., Waaben, J., Andersen, L.I. et al. (1988). An investigation of the Bentley® Gas-STAT™ monitoring system GSM-100. *J. Extra Corpor. Technol.* 20 (2): 59–62.
56 Weiss, I.K., Fink, S., Edmunds, S. et al. (1996). Continuous arterial gas monitoring: initial experience with the Paratrend 7 in children. *Intensive Care Med.* 22 (12): 1414–1417.
57 Pappert, D., Rossaint, R., Lewandowski, K. et al. (1995). Preliminary evaluation of a new continuous intra-arterial blood gas monitoring device. *Acta Anaesthesiol. Scand. Suppl.* 107: 67–70.
58 Gilbert, H.C. and Vender, J.S. (1996). Pro: is continuous intra-arterial blood gas and pH monitoring justifiable? *J. Clin. Monit.* 12 (2): 179–181.
59 Hoffer, J.E. and Norfleet, E.A. (1996). Con: is continuous intra-arterial blood gas and pH monitoring justifiable? *J. Clin. Monit.* 12 (2): 183–189.
60 Venkatesh, B., Clutton-Brock, T.H., and Hendry, S.P. (1995). Continuous intra-arterial blood gas monitoring during cardiopulmonary resuscitation. *Resuscitation* 29 (2): 135–138.
61 Ottens, J., Tuble, S.C., Sanderson, A.J. et al. (2010). Improving cardiopulmonary bypass: does continuous blood gas monitoring have a role to play? *J. Extra Corpor. Technol.* 42 (3): 191–198.
62 Schaarschmidt, J., Große, F.O., and Müller, T. (2010). Funktionsprinzip des kontinuierlichen In-Line-Monitorings am Beispiel des Blutanalysemonitors BMU 40 (theory of operation of continuous in-line monitoring using the example of the blood analysis monitor BMU 40). *Kardiotechnik* 4: 107–110 (in German).
63 U.S. Food and Drug Administration. (2016). Class 2 Device Recall CDI Blood Parameter Monitoring System 500, No. Z-1250-2016. www.accessdata.fda.gov/scripts/cdrh/cfdocs/cfRes/res.cfm?id=139412 (retrieved 29 July 2017).
64 Whitaker, D.K. (2011). Time for capnography – everywhere. *Anaesthesia* 66 (7): 544–549.
65 Brambrink, A.M. (1997). Die CO_2-Messung im Atemgas: Ein wichtiger Globalmonitor in der Notfallmedizin: theoretischer Hintergrund, Indikationen und Übersicht über verfügbare, transportable Meßsysteme (CO_2 measurement in exhaled breath: an important global monitor in emergency medicine: theoretical background, indications and overview about available transportable analysis systems). *Anaesthesist* 46 (7): 604–612 (in German).
66 Luft, K.F. (1943). Über eine neue Methode der registrierenden Gasanalyse mit Hilfe der Absorption ultraroter Strahlen ohne spektrale Zerlegung (about a new method of recording gas analysis by means of absorption of ultrared radiation without spectral diffraction). *Z. Tech. Phys.* 24: 97–104 (in German).
67 Dassow, C., Schwenninger, D., Runck, H., and Guttmann, J. (2013). Time and volume dependence of dead space in healthy and surfactant-depleted rat lungs during spontaneous breathing and mechanical ventilation. *J. Appl. Physiol.* 115 (9): 1268–1274.
68 Bhavani-Shankar, K. and Phillip, J.H. (2000). Defining segments and phases of a time capnogram. *Anesth. Analg.* 91 (4): 973–977.

69 Kay, B., Healy, T.E.J., and Bolder, P.M. (1985). Blocking the circulatory responses to tracheal intubation: a comparison of fentanyl and nalbuphine. *Anaesthesia* 40 (10): 960–963.
70 Ferrannini, E. (1988). The theoretical bases of indirect calorimetry: a review. *Metabolism* 37 (3): 287–301.
71 Colditz, P.B., Joy, G.J., and Dunster, K.R. (2002). Rebreathing potential of infant mattresses and bedcovers. *J. Paediatr. Child Health* 38 (2): 192–195.
72 Djupesland, P.G. and Børresen, B.A. (2000). Computational simulation of accumulation of expired air in the infant cot. *Acta Otolaryngol. Suppl.* (543): 183–185.
73 Funayama, M., Mimasaka, S., Iwashiro, K., and Nozawa, R. (1998). Inhaled air trapping effect of Japanese bedding as a risk of sudden unexpected death in infancy. *Tohoku J. Exp. Med.* 185 (1): 55–65.
74 Kemp, J.S. and Thach, B.T. (1991). Sudden death in infants sleeping on polystyrene-filled cushions. *N. Engl. J. Med.* 324 (26): 1858–1864.
75 Kemp, J.S. and Thach, B.T. (1993). A sleep position-dependent mechanism for infant death on sheepskins. *Am. J. Dis. Child.* 147: 642–646.
76 Ponsonby, A.L., Dwyer, T., Gibbons, L.E. et al. (1993). Factors potentiating the risk of sudden infant death syndrome associated with the prone position. *N. Engl. J. Med.* 329 (6): 377–382.
77 Mitchell, E.A., Thompson, J.M.D., Becroft, D.M.O. et al. (2008). Head covering and the risk for SIDS: findings from the New Zealand and German SIDS case-control studies. *Pediatrics* 121 (6): 1478–1483.
78 Mitchell, E.A., Thompson, J.M.D., Ford, R.P.K., Taylor, B.J., and other members of the New Zealand Cot Death Study Group (1998). Sheepskin bedding and the sudden infant death syndrome. *J. Pediatr.* 133 (5): 701–704.
79 Carolan, P.L., Wheeler, W.B., Ross, J.D., and Kemp, J.S. (2000). Potential to prevent carbon dioxide rebreathing of commercial products marketed to reduce sudden infant death syndrome risk. *Pediatrics* 105 (4): 774–779.
80 Carleton, J.N., Donoghue, A.M., and Porter, W.K. (1998). Mechanical model testing of rebreathing potential in infant bedding materials. *Arch. Dis. Child.* 78 (4): 323–328.
81 Zosel, J., Oelßner, W., Decker, M. et al. (2011). The measurement of dissolved and gaseous carbon dioxide concentration. *Meas. Sci. Technol.* 22: 072011, 45p. https://doi.org/10.1088/0957-0233/22/7/072001.
82 Dietze, P., Schütze, P., Scheuch, K., and Paditz, E. (2006). Einfluss der Struktur von drei Babymatratzen auf den CO_2-Durchlass und auf die Temperaturaufnahme (Influence of the structure of three baby mattresses on the CO_2 permeation and the temperature intake), 3. Expertentagung zur Prävention des plötzlichen Kindstodes. *Dresden* (in German). http://docplayer.org/57810047-Einfluss-der-struktur-von-drei-babymatratzen-auf-den-co-2-durchlass-und-auf-die-temperaturaufnahme.html.

Index

Carbon Dioxide Sensing: Fundamentals, Principles, and Applications,
First Edition. Edited by Gerald Gerlach, Ulrich Guth, and Wolfram Oelßner.